AF361638

Membrane–Peptide Interactions

Membrane–Peptide Interactions

From Basics to Current Applications

Editors

Nuno C. Santos
Sónia Gonçalves

MDPI • Basel • Beijing • Wuhan • Barcelona • Belgrade • Manchester • Tokyo • Cluj • Tianjin

Editors
Nuno C. Santos
Universidade de Lisboa
Portugal

Sónia Gonçalves
Universidade de Lisboa
Portugal

Editorial Office
MDPI
St. Alban-Anlage 66
4052 Basel, Switzerland

This is a reprint of articles from the Special Issue published online in the open access journal *International Journal of Molecular Sciences* (ISSN 1422-0067) (available at: https://www.mdpi.com/journal/ijms/special_issues/Membrane_Peptide).

For citation purposes, cite each article independently as indicated on the article page online and as indicated below:

LastName, A.A.; LastName, B.B.; LastName, C.C. Article Title. *Journal Name* **Year**, *Article Number*, Page Range.

ISBN 978-3-03943-022-2 (Hbk)
ISBN 978-3-03943-023-9 (PDF)

Contents

About the Editors . . . . **vii**

Preface to "Membrane–Peptide Interactions" . . . . **ix**

Marlon H. Cardoso, Beatriz T. Meneguetti, Bruna O. Costa, Danieli F. Buccini, Karen G. N. Oshiro, Sergio L. E. Preza, Cristiano M. E. Carvalho, Ludovico Migliolo and Octávio L. Franco
Non-Lytic Antibacterial Peptides That Translocate Through Bacterial Membranes to Act on Intracellular Targets
Reprinted from: *Int. J. Mol. Sci.* **2019**, *20*, 4877, doi:10.3390/ijms20194877 . . . . **1**

Marie-Lise Jobin, Lydie Vamparys, Romain Deniau, Axelle Grélard, Cameron D. Mackereth, Patrick F.J. Fuchs and Isabel D. Alves
Biophysical Insight on the Membrane Insertion of an Arginine-Rich Cell-Penetrating Peptide
Reprinted from: *Int. J. Mol. Sci.* **2019**, *20*, 4441, doi:10.3390/ijms20184441 . . . . **25**

Edit Pári, Kata Horváti, Szilvia Bősze, Beáta Biri-Kovács, Bálint Szeder, Ferenc Zsila and Éva Kiss
Drug Conjugation Induced Modulation of Structural and Membrane Interaction Features of Cationic Cell-Permeable Peptides
Reprinted from: *Int. J. Mol. Sci.* **2020**, *21*, 2197, doi:10.3390/ijms21062197 . . . . **47**

Thiru Sabapathy, Evelyne Deplazes and Ricardo L. Mancera
Revisiting the Interaction of Melittin with Phospholipid Bilayers: The Effects of Concentration and Ionic Strength
Reprinted from: *Int. J. Mol. Sci.* **2020**, *21*, 746, doi:10.3390/ijms21030746 . . . . **65**

Elżbieta Kamysz, Emilia Sikorska, Maciej Jaśkiewicz, Marta Bauer, Damian Neubauer, Sylwia Bartoszewska, Wioletta Barańska-Rybak and Wojciech Kamysz
Lipidated Analogs of the LL-37-Derived Peptide Fragment KR12—Structural Analysis, Surface-Active Properties and Antimicrobial Activity
Reprinted from: *Int. J. Mol. Sci.* **2020**, *21*, 887, doi:10.3390/ijms21030887 . . . . **85**

Yeon-Jee Yoo, Hiran Perinpanayagam, Jue-Yeon Lee, Soram Oh, Yu Gu, A-Reum Kim, Seok-Woo Chang, Seung-Ho Baek and Kee-Yeon Kum
Synthetic Human β Defensin-3-C15 Peptide in Endodontics: Potential Therapeutic Agent in *Streptococcus gordonii* Lipoprotein-Stimulated Human Dental Pulp-Derived Cells
Reprinted from: *Int. J. Mol. Sci.* **2020**, *21*, 71, doi:10.3390/ijms21010071 . . . . **107**

Felicitas Vernen, Peta J. Harvey, Susana A. Dias, Ana Salomé Veiga, Yen-Hua Huang, David J. Craik, Nicole Lawrence and Sónia Troeira Henriques
Characterization of Tachyplesin Peptides and Their Cyclized Analogues to Improve Antimicrobial and Anticancer Properties
Reprinted from: *Int. J. Mol. Sci.* **2019**, *20*, 4184, doi:10.3390/ijms20174184 . . . . **119**

Virginia Sara Grancieri do Amaral, Stephanie Alexia Cristina Silva Santos, Paula Cavalcante de Andrade, Jenifer Nowatzki, Nilton Silva Júnior, Luciano Neves de Medeiros, Lycia Brito Gitirana, Pedro Geraldo Pascutti, Vitor H. Almeida, Robson Q. Monteiro and Eleonora Kurtenbach
Pisum sativum Defensin 1 Eradicates Mouse Metastatic Lung Nodules from B16F10 Melanoma Cells
Reprinted from: *Int. J. Mol. Sci.* **2020**, *21*, 2662, doi:10.3390/ijms21082662 . . . . **145**

**Da Hyeon Choi, Dongwoo Lee, Beom Soo Jo, Kwang-Sook Park, Kyeong Eun Lee,
Ju Kwang Choi, Yoon Jeong Park, Jue-Yeon Lee and Yoon Shin Park**
A Synthetic Cell-Penetrating Heparin-Binding Peptide Derived from BMP4 with
Anti-Inflammatory and Chondrogenic Functions for the Treatment of Arthritis
Reprinted from: *Int. J. Mol. Sci.* **2020**, *21*, 4251, doi:10.3390/ijms21124251 . . . . **169**

Lili Dai, Zhe Sun and Ping Zhou
Modification of Luffa Sponge for Enrichment of Phosphopeptides
Reprinted from: *Int. J. Mol. Sci.* **2020**, *21*, 101, doi:10.3390/ijms21010101 . . . . **187**

**Chia-Ru Chung, Jhih-Hua Jhong, Zhuo Wang, Siyu Chen, Yu Wan, Jorng-Tzong Horng and
Tzong-Yi Lee**
Characterization and Identification of Natural Antimicrobial Peptides on Different Organisms
Reprinted from: *Int. J. Mol. Sci.* **2020**, *21*, 986, doi:10.3390/ijms21030986 . . . . **201**

Ruidan Wang, Xin Lu, Qiang Sun, Jinhong Gao, Lin Ma and Jinian Huang
Novel ACE Inhibitory Peptides Derived from Simulated Gastrointestinal Digestion in Vitro of
Sesame (*Sesamum indicum* L.) Protein and Molecular Docking Study
Reprinted from: *Int. J. Mol. Sci.* **2020**, *21*, 1059, doi:10.3390/ijms21031059 - **227**

**Stefano Borocci, Giulia Della Pelle, Francesca Ceccacci, Cristina Olivieri,
Francesco Buonocore and Fernando Porcelli**
Structural Analysis and Design of Chionodracine-Derived Peptides Using Circular Dichroism
and Molecular Dynamics Simulations
Reprinted from: *Int. J. Mol. Sci.* **2020**, *21*, 1401, doi:10.3390/ijms21041401 . . . . **247**

Chetna Tyagi, Tamás Marik, Csaba Vágvölgyi, László Kredics and Ferenc Ötvös
Accelerated Molecular Dynamics Applied to the Peptaibol Folding Problem
Reprinted from: *Int. J. Mol. Sci.* **2019**, *20*, 4268, doi:10.3390/ijms20174268 . . . . **267**

About the Editors

Nuno C. Santos (Ph.D.) graduated with a degree in Biochemistry from the Faculty of Science, University of Lisbon, Portugal, in 1995 and received his Ph.D. in Theoretical and Experimental Biochemistry in 1999 from the same university, although all the experimental work was conducted at Instituto Superior Técnico (Technical University of Lisbon) and University of California (Santa Barbara). Currently, he is Associate Professor with Habilitation of the Faculty of Medicine, University of Lisbon, and Head of the Biomembranes & Nanomedicine Unit at the Institute of Molecular Medicine (iMM). His research work includes the characterization and development of antimicrobial and antiviral peptides, as well as nanomedicine-based strategies for clinical prognosis in cardiovascular diseases.

Sónia Gonçalves (Ph.D.) graduated with a degree in Chemistry at the Central University of Venezuela in 1998 and obtained her Ph.D. in Physical Chemistry from the same university in 2002. In 2005, she joined the Faculty of Medicine of the University of Lisbon as Assistant Researcher. She is now also Staff Scientist at the Institute of Molecular Medicine (Lisbon, Portugal). Her research is focused on the study of naturally isolated or synthetically designed biomolecules as potential sources of antimicrobial and anticancer agents with therapeutic properties; a key part of this research is the characterization of their molecular interaction with lipid membranes through the use of fluorescence spectroscopy, circular dichroism, light scattering, and atomic force microscopy.

Preface to "Membrane–Peptide Interactions"

The interactions between peptides and membranes are of fundamental importance in the mechanisms of numerous membrane-mediated cellular processes, including antimicrobial peptide action, hormone–receptor interactions, drug bioavailability across the blood–brain barrier, and viral fusion processes. Membrane-interacting peptides comprise a large family of diverse peptides exhibiting a broad range of biological activities; therefore, they continue to attract growing interest for their biomedical applications. Moreover, a major goal of modern biotechnology is to obtain new potent pharmaceutical agents whose biological activity is dependent on the interaction of peptides with lipid bilayers. Several issues need to be addressed, such as eventual changes in peptide secondary structure, orientation, oligomerization, and localization inside the membrane. At the same time, the structural effects that the peptides induce on the lipid bilayer are important for the interactions and need to be elucidated. The structural characterization of peptides in membranes is challenging from an experimental point of view. It is well known that no single experimental technique can give a complete structural picture of the interaction; rather, a combination of different techniques is necessary.

In this Special Issue, peptides obtained from different sources (plants and animals, as well as in silico designed) are considered as potential therapeutic molecules for the improvement of human health. Antimicrobial, anticancer, antirheumatoid, anti-inflammatory, and immunomodulatory peptide applications, as well as applications in the food industry, are all addressed in this Special Issue. Peptides for human health improvement show great potential, but the mechanisms underlying their mode of action are far from fully described. It is important to combine different experimental and computational tools to better understand the interaction between peptides and membranes. Here, authors bring into play biological approaches together with biophysical methodologies to understand peptide–membrane interactions.

During peptide–membrane interactions, both the peptide and the membrane may experience a series of changes. Hence, experimental and theoretical studies of peptide–membrane interactions encounter challenges in attempting to completely understand the relationship between the structure of the peptide and the mechanism of interaction with membranes, and the molecular details of this process sometimes remain unclear. However, it is important to reveal the biological functions of membrane-active peptides to improve the design of peptides with optimized and customized functionalities that may be exploited for different applications, among other reasons.

This Special Issue book, Membrane–Peptide Interactions: From Basics to Current Applications, includes a selection of 14 articles, namely 13 original research articles and 1 review, exploring the determinants for peptide–membrane interactions. In the first article, Cardoso et al. [1] review the non-lytic antibacterial peptides that translocate bacterial membranes. The authors focused on the description of in vivo and in vitro assays of non-lytic peptides, as well as antibiofilm activity, focusing the action on intracellular targets. The related cell-penetrating peptides (CPPs) are capable of translocating across the cell membrane, as carriers or alone, to deliver drugs to their target. Jobin et al. [2] investigated the insertion of RW16, a CPP with antibacterial and antitumor activities, into zwitterionic membranes. Using complementary approaches, such as NMR, fluorescence, and circular dichroism spectroscopies, together with molecular dynamics simulations, the authors give important insights into these actions. Pári et al. [3] improved the mode of action of isoniazid (INH), an antibacterial agent used against tuberculosis, by testing its conjugation with a set of CCPs as drug carriers. Sabapathy et al. [4] explored the effects of peptide concentration and ionic strength in the interaction of melittin with phospholipid bilayers. Kamysz et al. [5] improved the activity of a small synthetic peptide, KR12, by its conjugation with different n-alkyl and aromatic acids. Yoo et al. [6] demonstrated the therapeutic potential of the synthetic human β defensin-3-C15 as an inhibitor of the inflammatory response induced by Streptococcus gordonii.

Some antimicrobial peptides (AMPs) from natural sources or synthetically designed are able to recognize and selectively kill many pathogens. Due to the similarities in their modes of action, some AMPs can also act as anticancer peptides (ACPs), therefore attracting further interest in their biomedical applications. Vernen et al. [7] characterized the cyclized analogs of tachyplesin to improve its antimicrobial and anticancer properties. Amaral et al. [8] used a mouse model of metastatic lung cancer to assess the eradication properties of Psd1, a defensin with demonstrated antifungal properties. Choi et al. [9] evaluated a synthetic cell-penetrating heparin-binding peptide derived from BMP4, showing anti-inflammatory and chondrogenic actions for the treatment of arthritis. Recently explored properties of the luffa sponge demonstrated a great potential for use as a solid-phase extraction material due to its physical and chemical properties. To this end, Dai et al. [10] investigated a strategy for the quaternization of the luffa sponge for selective enrichment of phosphopeptides.

So far, peptides have been demonstrated to be important molecules with diverse biological functions and biomedical uses. Several databases have been created in order to organize peptide sequences, combining them with their associated biological data. This information can be used to improve existing peptides for a given purpose. To achieve this, Chung et al. [11] developed an algorithm to identify AMPs in different organisms, including bacteria, plants, insects, fish, amphibians, humans, and other mammals. Their proposed method yielded more than 92% accuracy in predicting AMPs in each category, complementing the existing tools in the characterization and identification of AMPs in different organisms.

As previously mentioned, spectroscopic techniques have been developed and have contributed to the determination of many structural details of peptide–membrane interactions. These advances allow computational strategies to permeate all aspects of drug discovery today. In this sense, molecular docking and molecular dynamics simulations have been demonstrated to be as important as experimental approaches in the study of peptide–membrane interactions, becoming a useful predictor of such interactions. An example application is the improvement of functional food for hypertension treatment. For this, Wang et al. [12] isolated and purified a set of angiotensin-I-converting enzyme (ACE) inhibitor peptides from sesame protein, simulated the gastrointestinal digestion in vitro, and explored the underlying mechanisms by molecular docking. From the tested peptides, GHIITVAR, derived from 11S globulin, exhibited superior ACE inhibitory activity. An example of how computational approaches together with experimental methodologies may be complementary is the study developed by Borocci et al. [13]. These authors put together circular dichroism and molecular dynamics to design chionodracine-derived peptides and analyzed their structural properties upon interacting with lipid membranes. In the last selected article, Tyagi et al. [14] developed molecular dynamics work, studying fungal peptaibol structures to understand their folding dynamics in Escherichia coli mimicking membrane models.

We hope these articles will be used by professionals and research students wishing to characterize peptide–membrane systems by using the practical approaches contained in this book.

Reference

1. Cardoso, M.; Meneguetti, B.; Costa, B.; Buccini, D.; Oshiro, K.; Preza, S.; Carvalho, C.; Migliolo, L.; Franco, O. Non-Lytic Antibacterial Peptides That Translocate Through Bacterial Membranes to Act on Intracellular Targets. *Int. J. Mol. Sci.* **2019**, *20*, 4877, doi:10.3390/ijms20194877.

2. Jobin, M.; Vamparys, L.; Deniau, R.; Grélard, A.; Mackereth, C.; Fuchs, P.; Alves, I. Biophysical Insight on the Membrane Insertion of an Arginine-Rich Cell-Penetrating Peptide. *Int. J. Mol. Sci.* **2019**, *20*, 4441, doi:10.3390/ijms20184441.

3. Pári, E.; Horváti, K.; Bősze, S.; Biri-Kovács, B.; Szeder, B.; Zsila, F.; Kiss, É. Drug Conjugation Induced Modulation of Structural and Membrane Interaction Features of Cationic Cell-Permeable Peptides. *Int. J. Mol. Sci.* **2020**, *21*, 2197, doi:10.3390/ijms21062197.

4. Sabapathy, T.; Deplazes, E.; Mancera, R. Revisiting the Interaction of Melittin with Phospholipid Bilayers: The Effects of Concentration and Ionic Strength. *Int. J. Mol. Sci.* **2020**, *21*, 746, doi:10.3390/ijms21030746.

5. Kamysz, E.; Sikorska, E.; Jaśkiewicz, M.; Bauer, M.; Neubauer, D.; Bartoszewska, S.; Barańska-Rybak, W.; Kamysz, W. Lipidated Analogs of the LL-37-Derived Peptide Fragment KR12—Structural Analysis, Surface-Active Properties and Antimicrobial Activity. *Int. J. Mol. Sci.* **2020**, *21*, 887, doi:10.3390/ijms21030887.

6. Yoo, Y.; Perinpanayagam, H.; Lee, J.; Oh, S.; Gu, Y.; Kim, A.; Chang, S.; Baek, S.; Kum, K. Synthetic Human *β* Defensin-3-C15 Peptide in Endodontics: Potential Therapeutic Agent in Streptococcus gordonii Lipoprotein-Stimulated Human Dental Pulp-Derived Cells. *Int. J. Mol. Sci.* **2020**, *21*, 71, doi:10.3390/ijms21010071.

7. Vernen, F.; Harvey, P.; Dias, S.; Veiga, A.; Huang, Y.; Craik, D.; Lawrence, N.; Troeira Henriques, S. Characterization of Tachyplesin Peptides and Their Cyclized Analogues to Improve Antimicrobial and Anticancer Properties. *Int. J. Mol. Sci.* **2019**, *20*, 4184, doi:10.3390/ijms20174184.

8. Amaral, V.; Santos, S.; de Andrade, P.; Nowatzki, J.; Júnior, N.; de Medeiros, L.; Gitirana, L.; Pascutti, P.; Almeida, V.; Monteiro, R.; Kurtenbach, E. Pisum sativum Defensin 1 Eradicates Mouse Metastatic Lung Nodules from B16F10 Melanoma Cells. *Int. J. Mol. Sci.* **2020**, *21*, 2662, doi:10.3390/ijms21082662.

9. Choi, D.; Lee, D.; Jo, B.; Park, K.; Lee, K.; Choi, J.; Park, Y.; Lee, J.; Park, Y. A Synthetic Cell-Penetrating Heparin-Binding Peptide Derived from BMP4 with Anti-Inflammatory and Chondrogenic Functions for the Treatment of Arthritis. *Int. J. Mol. Sci.* **2020**, *21*(12), 4251, doi:10.3390/ijms21124251.

10. Dai, L.; Sun, Z.; Zhou, P. Modification of Luffa Sponge for Enrichment of Phosphopeptides. *Int. J. Mol. Sci.* **2020**, *21*, 101, doi:10.3390/ijms21010101.

11. Chung, C.; Jhong, J.; Wang, Z.; Chen, S.; Wan, Y.; Horng, J.; Lee, T. Characterization and Identification of Natural Antimicrobial Peptides on Different Organisms. *Int. J. Mol. Sci.* **2020**, *21*, 986, doi:10.3390/ijms21030986.

12. Wang, R.; Lu, X.; Sun, Q.; Gao, J.; Ma, L.; Huang, J. Novel ACE Inhibitory Peptides Derived from Simulated Gastrointestinal Digestion in Vitro of Sesame (Sesamum indicum L.) Protein and Molecular Docking Study. *Int. J. Mol. Sci.* **2020**, *21*, 1059, doi:10.3390/ijms21031059.

13. Borocci, S.; Della Pelle, G.; Ceccacci, F.; Olivieri, C.; Buonocore, F.; Porcelli, F. Structural Analysis and Design of Chionodracine-Derived Peptides Using Circular Dichroism and Molecular Dynamics Simulations. *Int. J. Mol. Sci.* **2020**, *21*, 1401, doi:10.3390/ijms21041401.

14. Tyagi, C.; Marik, T.; Vágvölgyi, C.; Kredics, L.; Ötvös, F. Accelerated Molecular Dynamics Applied to the Peptaibol Folding Problem. *Int. J. Mol. Sci.* **2019**, *20*, 4268, doi:10.3390/ijms20174268.

Nuno C. Santos, Sónia Gonçalves
Editors

International Journal of
Molecular Sciences

Review

Non-Lytic Antibacterial Peptides That Translocate Through Bacterial Membranes to Act on Intracellular Targets

Marlon H. Cardoso [1,2], Beatriz T. Meneguetti [1], Bruna O. Costa [1], Danieli F. Buccini [1], Karen G. N. Oshiro [1,3], Sergio L. E. Preza [1], Cristiano M. E. Carvalho [1], Ludovico Migliolo [1,4] and Octávio L. Franco [1,2,3,*]

[1] S-inova Biotech, Programa de Pós-Graduação em Biotecnologia, Universidade Católica Dom Bosco, Campo Grande 79117-900, Brazil; marlonhenrique6@gmail.com (M.H.C.); biatmeneguetti@gmail.com (B.T.M.); ocostab@gmail.com (B.O.C.); dfbuccini@gmail.com (D.F.B.); oshiro.kgn@gmail.com (K.G.N.O.); dyrosha@gmail.com (S.L.E.P.); rf7085@ucdb.br (C.M.E.C.); ludovico.migliolo@gmail.com (L.M.)
[2] Centro de Análises Proteômicas e Bioquímicas, Pós-Graduação em Ciências Genômicas e Biotecnologia, Universidade Católica de Brasília, Brasília 70790-160, Brazil
[3] Programa de Pós-Graduação em Patologia Molecular, Faculdade de Medicina, Universidade de Brasília, Brasília 70910-900, Brazil
[4] Programa de Pós-Graduação em Bioquímica, Universidade Federal do Rio Grande do Norte, Natal 59078-900, Brazil
* Correspondence: ocfranco@gmail.com; Tel.: +55-61-34487167 or +55-61-34487220

Received: 30 July 2019; Accepted: 14 September 2019; Published: 1 October 2019

Abstract: The advent of multidrug resistance among pathogenic bacteria has attracted great attention worldwide. As a response to this growing challenge, diverse studies have focused on the development of novel anti-infective therapies, including antimicrobial peptides (AMPs). The biological properties of this class of antimicrobials have been thoroughly investigated, and membranolytic activities are the most reported mechanisms by which AMPs kill bacteria. Nevertheless, an increasing number of works have pointed to a different direction, in which AMPs are seen to be capable of displaying non-lytic modes of action by internalizing bacterial cells. In this context, this review focused on the description of the in vitro and in vivo antibacterial and antibiofilm activities of non-lytic AMPs, including indolicidin, buforin II PR-39, bactenecins, apidaecin, and drosocin, also shedding light on how AMPs interact with and further translocate through bacterial membranes to act on intracellular targets, including DNA, RNA, cell wall and protein synthesis.

Keywords: antimicrobial peptides; non-lytic peptides; bacterial membranes

1. Introduction

The World Health Organization (WHO) has identified antimicrobial resistance as one of the three major threats to human health [1]. Bacteria can be efficient in the synthesis and sharing of genes involved in the development of antibiotic resistance mechanisms, leading to negative outcomes in the clinic [2]. This inefficiency may be related to the intrinsic resistance of a bacterium to a specific antibiotic, which can be explained by its ability to resist the action of this drug as a result of inherent structural or functional characteristics [3]. Therefore, the dissemination of antibiotic resistance factors, along with the misuse of these drugs, has made drug design a broad field of research [4]. In this scenario, the antimicrobial peptides (AMPs) have been considered as an alternative to conventional antibacterial treatments [5].

AMPs can be produced as part of the host's defense system (innate immune system) during an infection process [6]. These peptides belong to a broad group of molecules produced by many tissues and cell types in a variety of organisms, including plants, invertebrates, vertebrates, fungi, and bacteria [7]. The majority of AMPs are composed of relatively small (<10 kDa), cationic and amphipathic molecules, mostly consisting of 6 to 50 amino acid residues [8]. Moreover, AMPs have often been reported for their diverse biological activities, more specifically, antibacterial activities [9]. The different amino acid compositions lead to structural properties in terms of amphipathicity, net positive charge, shape and size, which favor interaction with microbial surfaces, insertion into lipid bilayers and induction of membrane damage [10]. It is proposed that AMPs firstly bind to biological membranes and then, due to their amphipathic arrangement, insert into the bilayer by breaking the lipid chain interactions [11]. The mechanisms of action associated with destabilization and disruption of bacterial membranes have been widely described, triggering mechanisms known as the carpet model, toroidal pore, barrel type, detergent, and several other variations [12–14]. In addition to AMPs' membrane disruptive properties, studies have suggested that these peptides may also affect bacterial viability by acting via non-lytic pathways [15].

Diverse works assume that AMPs may present intracellular targets [15]. However, the mechanisms by which some AMPs are capable of penetrating bacterial cells are still under investigation [16]. It has been suggested that some peptides (e.g., proline-rich AMPs) can bind to the bacterial surface followed by their translocation into the cell through the formation of transient pores and, finally, acting on intracellular targets [17,18]. Additionally, works have proposed that AMPs can translocate through the membrane without forming pores, which may include receptor-mediated processes [19]. Once these molecules cross the bacterial membranes, they may target intracellular macromolecules and bioprocesses, including DNA replication and transcription inhibition [20,21]. Additionally, AMPs have been proved to inactivate bacterial chaperones involved in protein folding, also leading to bactericidal effects by inhibiting the synthesis of proteins [18,22].

In this context, some advantages have been attributed to non-lytic AMPs in terms of clinical applications. From the therapeutic point of view, AMPs may present great specificity with their intracellular target, which may hinder the development of resistance mechanisms. Moreover, this specificity for intracellular bacterial targets could also lead to reduced toxicity toward healthy human cells [15]. Therefore, this review will focus on the main non-lytic AMPs described to date, including indolicidin, buforin II, PR-39, bac7, apidaecin, and drosocin. Thus, although previous review articles have extensively described AMP intracellular mechanisms of action, here we provide an all-in-one overview of how non-lytic AMPs first interact with and further translocate across bacterial membranes to act on intracellular bacterial components, finally leading to cell death. We also provide a detailed description of the antibacterial, antibiofilm and anti-infective properties of these peptides in vitro and in vivo. Taken together, the data here summarized may provide useful information on the most promising non-lytic AMPs, and how these peptides could be used as model molecules for drug design strategies aiming at antibacterial therapies.

2. Indolicidin—A Tryptophan/Proline-Rich Peptide

The first indolicidin was isolated from cytoplasmic granules of bovine neutrophils and, at that time, it was considered the shortest peptide discovered [23]. Indolicidin is a tryptophan/proline-rich AMP belonging to the cathelicidin family and constituted of 13 amino acid residues [23] that has shown antibacterial properties against Gram-positive and -negative bacteria [24]. In terms of structural profile, indolicidin is dynamic, as in an aqueous solution it is unstructured [25], but it adopts a poly-L-proline type II helix or extended structures (Table 1) in membrane-like conditions [26,27].

Table 1. Summary of the non-lytic AMPs here described in terms of antibacterial potential and applicability, the design strategies for the generation of improved analogues, structural profiles, modes of translocation across bacterial membrane and known intracellular targets.

Non-Lytic AMPs	Antibacterial Potential	Treatment Strategies	Design Strategies	Structural Profile	Membrane Translocation	Intracellular Target	References
Indolicidin	Bacteriostatic; bactericide; anti-bacteremia	Monotherapy; synergism between two indolicidin analogues	Amino acid substitution; Amide bond modification; Hybrid peptides	poly-L-proline type II helix; extended structures; β-turn	Transmembrane orientation followed by cell internalization	DNA binding; DNA biosynthesis inhibition	[27–32]
Buforin II	Bacteriostatic; bactericide; anti-sepsis	Monotherapy; synergism with rifampicin; additive effects when combined with ranalexin, amoxicillin-clavulanate, ceftriaxone, meropenem, doxycycline, and clarithromycin; conjugation with PNA	Amino acid substitution; truncated analogues	Helical-helix-propeller structure	Formation of transient toroidal pores	DNA and RNA binding	[33–36]
PR-39	Bacteriostatic; bactericide; anti-sepsis; toxin neutralization; wound healing	Monotherapy	Truncated analogues; amino acid substitution	Extended	Receptor-mediated (SbmA)	Protein and DNA synthesis inhibition	[37–39]
Bac7	Bacteriostatic; bactericide; anti-sepsis; immunomodulatory	Monotherapy; synergism; association with PEG	Truncated analogues; amino acid substitution	Extended	Receptor-mediated (SbmA)	Protein and DNA synthesis inhibition; Ribosome; Binding to lipid II precursor; cell wall synthesis	[37,40–44]
Apidaecin	Bacteriostatic; bactericide	Monotherapy	Chemical modifications; amino acid substitution; peptide–peptoid hybrids	Extended	Oligomers formation (OM); Interaction with IM permeases and transporters	DnaK and GroEL, leading to bacterial protein misfolding; Protein synthesis; Ribosome	[45–47]
Drosocin	Bacteriostatic; bactericide	Monotherapy	Chemical modifications	Extended	Receptor-mediated (unknown)	DnaK and GroEL, leading to bacterial protein misfolding	[48–50]

PEG: polyethyleneglycol; OM: outer membrane; IM: inner membrane; PNA: peptide nucleic acid; SbmA: peptide antibiotic transporter.

2.1. Indolicidin Interacts with and Translocates through Bacterial Membranes

Structural studies of indolicidin in contact with lipid bilayers started in the 1990s. At first, it was proposed that indolicidin adopted a poly-L-proline type II helix upon interaction with 1-pamitoyl -2-oleoyl-sn-glycero-3-phosphocholine(POPC)/1-palmitoyl-2-oleoyl-sn-glycero-3-phosphoglycerol (POPG) liposomes (7:3 lipid-to-lipid ratio), which was further correlated to indolicidin's ability to bind lipopolysaccharides and cross the *Escherichia coli* outer membrane by self-promoted uptake [25]. Years later, this poly-L-proline type II helix structural profile was reviewed, opening a new perspective on indolicidin's structure by the formation of extended and β-turn structures [27]. Nuclear magnetic resonance (NMR) studies with indolicidin have enabled researchers to clarify this controversy regarding indolicidin's structure in membrane-like conditions, including zwitterionic dodecylphosphocholine (DPC) and anionic sodium dodecyl sulfate (SDS) micelles. NMR spectra of indolicidin in these conditions have shown that this peptide adopts an extended conformation from residues 3 to 11, with two half-turns at residues 5 and 8 when in contact with DPC [28]. A similar extended conformation was observed in SDS from residues 5 to 11. However, it lacks the bend in the C-terminal region. Further investigations revealed that, in contact with DPC, indolicidin's conformation seems to be ideal for its intercalation between the DPC molecules. Moreover, based on hydrogen bond pattern, peptide–lipid charge distribution and membrane location, two main membrane insertion modes have been proposed, including the direct penetration into one of the bilayer leaflets via a "boat" structural orientation, and a transmembrane orientation (Table 1 and Figure 1) [28]. In addition to indolicidin–micelle interactions, evidence of multiple structural profiles involved in membrane binding has also been reported in aqueous solution and 50% 2,2,2-trifluoroethanol (TFE) in water [29]. Therefore, it has been suggested that such structural plasticity seems to be related to different combinations of contact between the two WPW motifs in indolicidin's sequence [29].

The trajectory of indolicidin has also been investigated through molecular dynamics (MD) in membrane environments. Hsu and Yip [30] developed a study with indolicidin, in which the simulations were performed on single lipid bilayers of dioleoylphosphatidylcholine (DOPC), distearoylphosphatidylcholine (DSPC), dioleoylphosphatidylglycerol (DOPG), and distearoylphosphatidylglycerol (DSPG) for 50 ns. The results indicated that indolicidin was partitioned between water and bilayer for all systems. The results suggest that electrostatic interactions are important in the initial attraction of the peptide/membrane. This approach was faster with the anionic lipids (DOPG and DSPG) and there was hydrogen bonding between the peptide side chains and the phospholipid head groups in all simulations. Intermolecular hydrogen bonds were formed between the tryptophan residues from indolicidin, indicating that it is not only by electrostatic interactions that the association with anionic membranes occurs. The authors also observed a decrease in membrane thickness caused by this peptide, along with interdigitation of lipid tails. However, intermolecular hydrogen bonds were not observed when simulating indolicidin in contact with zwitterionic DOPC and DSPC membranes.

More recently, Tsai et al. [31] performed a work with a synthetic analogue of indolicidin, called SAP10, which preserved the non-lytic behavior of the parent peptide but reduced its cytotoxicity against mouse fibroblasts (NIH/3T3). MD simulations of these two peptides (parent and analogue) were performed in the presence of POPC lipid bilayers and the results compared with small-angle X-ray scattering (SAXS). Carbon-carbon order parameters of the lipid acyl chains were used to measure the perturbation in the membrane. For indolicidin, there was a decrease in the values of lipid acyl chains when compared to the isolated membrane, whereas for SAP10, the values did not change significantly [31]. This indicates that both molecules interact with the membrane. However, the indolicidin disturbance is more evident than the SAP10 peptide. The authors associated this lower perturbation with the fact that SAP10 has fewer tryptophan residues, an amino acid that is usually associated with peptide stability in membranes [31].

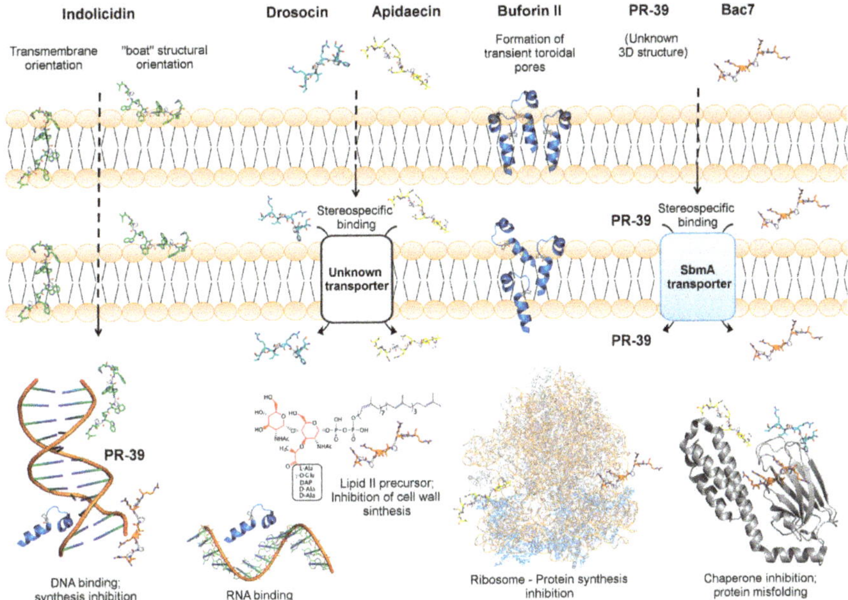

Figure 1. Representation of the membrane translocation mechanisms and intracellular targets for indolicidin (green sticks—PDB 1g8c), PR-39 (name only), bac7 (orange sticks—PDB: 5f8k), apidaecin (yellow sticks—PDB: 5o2r), drosocin (cyan sticks—PDB: 4ezr) and buforin II (blue—PDB: 4kha). Indolicidin adopts a "boat-type" arrangement or transmembrane orientations to cross both the outer membrane (OM) and inner membrane (IM) to bind DNA, whereas buforin II forms transient toroidal pores, thus internalizing the bacterial cell to target DNA and RNA. Apidaecin and drosocin require an IM transporter (e.g., membrane permease) to reach the bacterial cytosol and target chaperones and ribosomes. Similarly, bac7 and PR-39 require an SbmA transporter to cross the IM and then interact with DNA, chaperones and lipid II precursors (the later is exclusive to bac7). Proline residues are highlighted as white sticks in all peptides. The tridimensional structure of buforin II is not deposited in the Protein Data Bank. Therefore, buforin II structure was extracted from the N-terminus region of the histone H2A (from which this peptide is derived), for representation purposes.

2.2. Indolicidin Antibacterial Properties

Diverse studies have focused on the biological characterization of indolicidin. In the first studies conducted with this AMP, indolicidin showed antibacterial activity against *E. coli*, *Pseudomonas aeruginosa*, *Staphylococcus aureus*, *Staphylococcus epidermidis*, and *Salmonella typhimurium* (Table 2) [25,51]. These activities have been directly correlated to the large number of tryptophan and proline residues in indolicidin's sequence [51]. Nonetheless, the presence of these residues has also been related to the hemolytic activity of this peptide, thus representing an obstacle for its application in clinical trials [52].

Considering that hemolytic and cytotoxic effects represent a bottleneck in taking AMPs into the clinic, indolicidin analogues have been developed. Over the years, different strategies have been carried out to enhance the therapeutic potential of this peptide (Table 1) and, during these investigations, important findings were reported. In a study by Ryge et al. [53], indolicidin analogues were developed based on the sequence ILPXKXPXXPXRR-NH$_2$, where tryptophan (labeled with X) residues were altered by an N-substituted class of non-proteogenic residues or by glycine. A total of 28 indolicidin analogues were developed, out of which 22 presented improved antibacterial properties against *S. aureus* and *E. coli* (Table 2). In that same work, non-natural constructs were further submitted to modifications to boost the antibacterial activity of the analogue peptides. For this, phenylalanine residues were added at positions 4, 6, 8, 9, and 11 [53]. As a result, the authors observed that tryptophan might not be essential to maintain the antibacterial activity of the parent indolicidin, as the phenylalanine-containing analogues presented higher minimal inhibitory concentration (MIC) values against *E. coli* and *S. aureus,* as well as lower hemolytic activities.

Amide bond modifications have also been performed aiming at generating analogues with greater stability and antibacterial activity [54,55]. Kim et al. [56], for instance, altered the amide bonds of indolicidin by reduced amide bonds $\psi[CH_2 NH]$ in the pseudopeptide analogues, called ID, ID-I, ID-W and ID-IW (Table 2). Among them, the pseudopeptide (ID-IW) containing two reduced amide bonds not only presented reduced hemolytic effects, but also improved resistance to enzymatic degradation [56]. Moreover, the antibacterial potential of the parent peptide (indolicidin) was preserved in ID-IW, which was capable of inhibiting *Bacillus subtilis, Micrococcus luteus, S. aureus* and *E. coli* strains.

More recently, indolicidin has also been used for the generation of hybrid AMPs. In a study by Jindal et al. [57], 13 hybrid peptides were developed based on two AMPs, indolicidin and ranalexin, which is derived from bullfrog (*Rana catesbeiana*) skin [58]. Among them, five analogues (RN7-IN10, RN7-IN9, RN7-IN8, RN7-IN7 and RN7-IN6) presented antibacterial activity against 30 clinical isolates from the genus *Pneumococcus* (Table 2). These authors also used the analogues RN7-IN10 and RN7-IN8 (lower cytotoxicity) to treat mice infected with *Streptococcus pneumoniae*. It was observed that, at the concentration of 5 mg·kg^{-1}, the peptides showed no activity. On the other hand, 10% of the animals survived after treatment with RN7-IN10, at 10 mg·kg^{-1}, whereas a 30% survival rate was observed for those animals treated with RN7-IN8 at the same concentration. Finally, the highest survival rates of 30% and 50% were reported for the groups treated with 20 mg·kg^{-1} of RN7-IN10 and RN7-IN8 [57]. Interestingly, it was also shown that RN7-IN10 and RN7-IN8 synergize (Table 1), as animals treated with 10 mg·kg^{-1} of each peptide in combination presented a survival rate of 60%. Among all the tests performed, RN7-IN8 presented the most promising activities, besides being highly effective in the treatment of bacteremia [57].

Apart from the antibacterial activity of indolicidin against bacteria in their planktonic mode of growth, studies have also evaluated this AMP on bacterial biofilms. However, in contrast to the promising antibacterial effects of indolicidin and its analogues, antibiofilm studies have shown that the mechanisms by which this AMP acts are not effective on biofilms. Pompilio et al. [59] analyzed the antibiofilm activity of indolicidin against clinical isolates of *P. aeruginosa, Stenotrophomonas maltophilia,* and *S. aureus,* but no activity was observed at the maximal concentration tested. In a study by Dosler et al. [60], indolicidin was tested against *S. aureus* and methicillin-resistant *Staphylococcus aureus* (MRSA) biofilms. Despite presenting low MICs, antibiofilm properties were reported only at 40-fold higher concentrations. Overall, these data reinforce the theory that antibacterial and antibiofilm properties in AMPs are most likely to be independent.

Table 2. Non-lytic peptides, their source organisms, class, analogue peptides, and antibacterial activity spectrum.

Peptide	Organism	Source	Class	Analogues	Antibacterial Activity Spectrum	MIC Range (μM)	References
Indolicidin	*Bos taurus*	Neutrophils cytoplasmic granules	Tryptophan-rich	N-substituted class of non-proteogenic residues or by glycine; ID, ID-I, ID-W and ID-IW; RN7-IN6 to RN7-IN10	*E. coli, P. aeruginosa, S. aureus, S. epidermidis, S. pneumoniae, M. luteus* and *S. typhimurium*	0.2 to 50	[22,51,53,56–58]
Buforin II	*Bufo bufo gargarizans*	Stomach tissue	Helical-helix-propeller peptide	BF2-A; BF2-C; BUF(5–21); BUF(5–13)-[RLLR]₃; Buf-IIIa to Buf-IIId	*A. baumannii,B. subtilis, C. neoformans, E. coli, L. monocytogene, P. putida, S. aureus, S. dysenteriae, S. hemolyticus, S. marcescens, S. mutans, S. pneumoniae, S. typhimurium* and *Serratia sp.*	0.2 to 3.2	[34–36,61–67]
PR-39	*Sus scrofa*	Porcine neutrophils	Proline/arginine-rich	PR-39 (1–26); PR-39 (1–22); PR-39 (1–18); PR-39 (1–15); PR35	*B. globigii, E. coli, S.typhimurium* and *S.choleraesuis*	1.25 to 20	[68–70]
Bac7	*Bos taurus*	Bovine neutrophils	Proline/arginine-rich	Bac7 (1–35); Bac7 (5.35); Bac7 (1–23); Bac7 (5–23); Bac7 (1–16); Bac7 (1–18)	*A. baumannii, E. coli, K. pneumoniae, P. aeruginosa, S. aureus, S. enterica* and *S. maltophilia*	0.06 to 64	[71–73]
Apidaecin	*Apis mellifera*	Lymph fluid	Proline/arginine-rich	api6; api7; api39; api88; api137; apidaecin Ib;	*E. coli, K. pneumoniae, P. aeruginosa, S. enteritidis, S. typhimurium*	0.27 to 64	[50,74–77]
Drosocin	*Drosophila melanogaster*	Abdomen and thoraxes	Proline/arginine-rich	Thr⁶-glycosylated drosocin; β-Ala drosocin; M-drosocin; Di-drosocin	*E. coli, Erwinia herbicola, S. enteritidis, S. infantis, S. montevideo, S. panama, S. typhimurium,* and *M. luteus*	0.25 to 100	[78,79]

2.3. Indolicidin Targets Bacterial DNA

Some AMPs are capable of directly interacting with DNA and/or RNA, thus interfering with their synthesis, replication and translation processes [80,81]. Indolicidin, at high concentrations, increases the permeability of the bacterial cell and, consequently, reaches the cytosol to inhibit, exclusively, DNA biosynthesis (Table 1 and Figure 1) [82]. Hsu et al. [29] performed gel retardation and fluorescence studies to confirm that indolicidin binds to DNA. Besides, different single- and duplex-strand DNAs were immobilized on a biosensor surface and the association/dissociation of indolicidin was monitored. It was demonstrated that indolicidin bound strongly to ds-[AT], ds-[CG] and ds-[AG], but only weakly to ds-[GT]. The authors further suggested that indolicidin's amphipathicity plays a crucial role in its ability to bind to nucleic acid and, thereby, kill bacteria. Moreover, the data reported by those authors suggest that indolicidin's mechanism of action involves an initial stage of electrostatic binding to the DNA duplex phosphate groups, followed by its insertion into the DNA groove [29]. More recently, the structural and mechanistic features that favor indolicidin's DNA-binding property were investigated through the combination of spectroscopy and microscopy methods [32]. It has been shown that the central PWWP motif plays a key role in the indolicidin/duplex DNA stabilization, as mutations in the central WW pair significantly impaired indolicidin's DNA-binding activity [32].

3. Buforin II—A Frog-Derived Peptide that Internalizes Bacterial Cells

Buforin has been described as an effective non-lytic AMP family. The buforin family comprises AMPs that have complete sequence identity with the N-terminal region of the histone H2A, which interacts directly with nucleic acids [83]. Among buforins, buforin II has attracted particular interest due to its broad-spectrum activities against microorganisms when compared to other α-helical AMPs [61]. This peptide was obtained by treating buforin I, which is derived from the stomach tissue of the Asian toad *Bufo bufo gargarizans*, with an endoproteinase Lys-C, thus resulting in the generation of a 21 amino acid residue peptide (TRSSRAGLQFPVGRVHRLLRK), named buforin II [61]. Buforin II has a helical-helix-propeller structure (Table 1), which is amphipathic in hydrophobic environments. In addition, the suggested mechanisms of action of this peptide against bacteria include DNA- and RNA-binding properties after translocation across the lipid bilayer, without causing cell lysis (Figure 1) [84,85].

3.1. Buforin II Translocates Membranes by the Formation Of Transient Toroidal Pores

The first NMR structural study performed with buforin II revealed a coil-to-helix transition when this peptide is transferred from hydrophilic (water) to hydrophobic (TFE/water mixtures) conditions [33]. Although buforin II is a non-proline-rich AMP, it presents a proline residue at position 11 in its sequence that acts as a helix breaker. Therefore, the amphipathic structure of buforin II consists of a random coil region from Thr^1 to Ser^4, followed by a distorted helical structure from Arg^5 to Phe^{10} and a well-defined α-helix from Val^{12} to Lys^{21} after a hinge at Pro^{11} [33,34]. The presence of a proline hinge in buforin II has been reported as a crucial factor for its cell-penetrating ability. Interestingly, although the proline acts as a translocation promoter in buforin II, its cis-trans isomerization does not affect the translocation mechanism [35]. Confocal microscopy studies have shown that, by performing a single amino acid substitution for proline in buforin II sequence, this peptide's mechanism of action on bacteria changes from intracellular to membrane active [34]. Similar findings were observed through the investigation of buforin II in contact with membrane bilayers [36]. Compared to magainin II, buforin II translocates more efficiently across lipid bilayers, without inducing lipid flip-flop, suggesting non-membranolytic mechanisms [36].

Additional studies with lipid bilayers have also demonstrated that buforin II causes a positive curvature on membranes [35]. As mentioned above, Pro^{11} distorts the helical segment in buforin II at the N-terminal region, leading to the concentration of basic residues in a limited amphipathic region, which destabilizes pore formation due to peptide–peptide electrostatic repulsions [35]. Therefore, it is proposed that buforin II translocates membranes by the formation of transient toroidal pores with extremely short lifetime to act on intracellular targets (Table 1 and Figure 1). These findings have also been observed in computational studies [84].

3.2. Buforin II as a Promising Scaffold for Antibacterial Therapies

The first study to evaluate the antimicrobial activity of buforin II was developed by Park et al. [61], who determined the MICs against diverse Gram-positive and -negative bacteria, and fungi. In addition, that study also revealed that, compared to the AMP magainin II, buforin II was approximately 10-fold more potent against a wide range of microorganisms [61].

Moreover, in a direct comparison with the model AMP magainin II, buforin II has been evaluated regarding its membrane permeabilization, and its hemolytic and antibacterial properties [36]. In this context, studies have shown that buforin II is more efficient at translocating through lipid bilayers than magainin II [36]. Regarding their antibacterial activity against *E. coli*, buforin II exhibited significantly greater activity than magainin II [36]. Interestingly, however, despite their different modes of action on bacteria, both buforin II and magainin II were not hemolytic at concentrations 25-fold higher than their MICs [36].

Over the years, an increasing number of pharmacologic strategies have been applied to AMPs, including their administration in combination with conventional antibiotics (Table 1) [62]. In this context, Cirioni et al. [63] investigated both in vitro and in vivo the antibacterial activity of buforin II (Table 2) and the antibiotic rifampicin (alone and in combination) against *A. baumannii* strains. As a result, in vitro experiments with buforin II showed higher antibacterial activity when compared to rifampicin against susceptible and multidrug-resistant *A. baumannii*. Moreover, the combination of these two antimicrobial agents resulted in a synergistic effect (fractionary inhibitory concentration (FIC) index of 0.312) [63]. In vivo assays were carried out using a model of sepsis in rats, in which the animals were infected with susceptible and multidrug-resistant *A. baumannii*. The groups treated with buforin II had a lower percentage of lethality (40% and 46.6%, respectively) when compared to the control groups (100% and 100%, respectively) and treated with antibiotic rifampicin (93.3% and 93.3%, respectively) [63]. In addition, the treatment with this peptide also reduced bacterial endotoxin and plasma cytokine concentrations when compared to the other groups [63]. As observed in vitro, the combinatory therapy buforin II and rifampicin was more promising (20% lethality rate for susceptible and resistant *A. baumannii*) compared to the results obtained by the treatment with these antimicrobial agents alone. This combination was also reflected in a significant reduction in the concentrations of bacterial endotoxin and plasma cytokines [63]. Therefore, these results demonstrate that buforin II combined with rifampicin has superior efficacy to monotherapy (Table 1).

In another study, Zhou et al. [64] investigated the interaction of buforin II with the conventional antibiotics ranalexin, amoxicillin-clavulanate, ceftriaxone, meropenem, doxycycline, and clarithromycin (Table 1), which are all commonly used in the clinic for the treatment of Gram-positive and -negative bacteria. The combination of buforin II and the above-mentioned antibiotics against 120 clinical isolates was not synergistic, but additive [64]. However, this potent effect of one treatment agent over another still supports the hypothesis that the combination of peptides with antibiotics may represent a promising alternative to antimicrobial monotherapies.

Studies have also been conducted with buforin II analogue sequences (Table 2). For instance, Park et al. [34] have developed buforin II analogues to shed some light on the structural characteristics of buforin II that are crucial for its potent antimicrobial activity. Therefore, a series of N- and C-terminal truncated buforin II fragments or analogues with amino acid substitutions were designed and evaluated for their antimicrobial activity and mechanism of action [34]. As a result, the analogues BUF (5–21—N-terminal truncation), BUF (5–13—N-terminal truncation) with three repeats of the C-terminal regular RLLR motif, named BUF (5–13)-[RLLR]$_3$), were more potent against bacteria than their parent peptide, buforin II (Table 2). In contrast, additional N-terminal truncation, or removal of four amino acids from the C-terminal of buforin II, resulted in analogues with progressive decrease or null antimicrobial activity [34]. These results demonstrate the importance of the C-terminal helical region (residues 18 to 21) in buforin II antimicrobial activity, whereas the N-terminal random coil region seems not to play a key role [34]. Therefore, the systematic study of the structure-activity relationship of buforin II and its analogues has shown that the effectiveness of cell penetration in terms of antimicrobial potency depends on the α-helical content of this AMP [34].

Based on the findings cited above, Hao et al. [65] designed and synthesized a novel, 21-amino acid residue buforin II analogue, called BF2-C. This analogue is constituted by the N-terminal residues 5–13 from buforin II, in addition to three repeats of the C-terminal α-helical motif (RLLR) from this same peptide. Moreover, BF2-C also presents a single substitution, in which a valine residue was replaced by a leucine residue at position 12 in the parent peptide buforin II [65]. These modifications resulted in increased hydrophobicity of the amphipathic α-helix at the C-terminal region of BF2-C. It was observed that BF2-C showed remarkable antimicrobial activities against Gram-positive and -negative bacteria (Table 2), compared to its parent peptide [65]. These results suggest that the α-helical content in buforin-like peptides may be directly correlated with their increased antibacterial potential. Furthermore, structure-activity ratio analyses revealed that cell penetration efficacy and DNA affinity were critical factors in determining the antimicrobial potency of BF2-C. Therefore, these results provide important information on the development of novel potent peptide-based drugs that act intracellularly [65].

Strategies of amino acid replacement have also been applied for the generation of buforin II analogues. Jang et al. [66] designed four analogues, named Buf-IIIa to Buf-IIId, based on the buforin IIb (BUF2-B) respecting the following criteria: (i) the non-alteration of the structural characteristics important for the antimicrobial activity of buforin IIb, and (ii) the conservation of global hydrophobicity, which provides the effective antimicrobial activity of AMPs (Table 2). In that study, all Buf-III analogues had similar structures and mechanisms of action to buforin IIb. Regarding their antimicrobial activity against the tested pathogens (bacteria and fungi), Buf-IIIb and Buf-IIIc presented ≥2-fold higher antibacterial and antifungal activities compared to the parent peptide. Moreover, the hemolytic activity against human erythrocytes was decreased in those analogues, resulting in a 7-fold improvement in their therapeutic index (62.5 for buforin IIb and 444 for Buf-IIIb and IIIc). Therefore, these results suggest that Buf-III analogues may be promising candidates to complement conventional antimicrobial therapy [66].

A buforin II analogue (BF2-A) has also been evaluated in an alternative drug design approach (Table 1), involving its conjugation with a peptide nucleic acid (PNA) to inactivate *E. coli* strains [67]. Due to BF2-A's intracellular mechanism of action, this peptide would be an efficient vehicle for the release of PNA within the bacterial cells, which in turn targets the *acpP* gene. This gene is essential in fatty acid biosynthesis and, therefore, its regulation interferes with the cell wall organization. Thus, the antimicrobial activity observed against *E. coli* treated with BF2-A and PNA were successfully achieved by the silencing of the target gene promoted by the conjugate [67].

3.3. Buforin II Targets DNA and RNA

Apart from indolicidin, the AMP buforin II binds to DNA after its translocation through *E. coli* membranes [34]. The proposed model for buforin II is the formation of a transient toroidal pore (Table 1 and Figure 1), similar to magainin II. The lifetime of the pore is shorter and, as a consequence, the translocation rate is increased due to the disintegration of the pores [35,36]. Once in the cytosol, buforin II binds to DNA and RNA (Figure 1), as shown by Park et al. [85]. The strong affinity of this peptide for nucleic acids has been shown to depend highly on the complementarity between the sequences of buforin II and the N-terminal region of the H2A histone [83].

4. PR-39 and bac7—Two Proline/Arginine-Rich Peptides

Proline/arginine-rich peptides have been described and characterized by the presence of a repeating PRPR motif [17]. Arginine and proline residues can facilitate access to the intracellular region of the target bacteria to effectively inactivate these pathogens [17]. The proline/arginine AMP, named PR-39, was firstly isolated from pigs' intestines [86]. This peptide is constituted of 39 amino acids, with high contents of proline and arginine residues [87]. The large amount of proline residues gives the PR-39 greater stability for degradation by serine proteases, leading to a longer half-life [68,88]. Over the years, it has been shown that PR-39 acts as an antibacterial and wound healing agent (Table 1) [69]. Moreover, when targeting bacteria, PR-39 acts on DNA and/or protein synthesis (Figure 1) [89].

Similarly, the bactenecin-like peptide, bac7, was firstly isolated from bovine neutrophils, and also constitutes a proline/arginine-rich AMP [90,91]. This peptide has shown antibacterial potential toward *E. coli, Klebsiella* sp. [90] and may also be effective against *S. epidermidis* [92]. Moreover, the mechanisms by which bac7 exerts its antibacterial properties have been elucidated, and involve a DnaK-binding mode of action (Figure 1) [93].

4.1. PR-39 and bac7 Membrane Translocation Require an Inner Membrane Transporter

The bacterial inner membrane (IM) transporter, SbmA, is required for bac7 and PR-39 cellular uptake (Table 1, Figure 1). This IM protein is constituted of 406 amino acid residues with seven or eight transmembrane-spanning domains that facilitate the internalization of glycopeptides, AMPs and PNA oligomers into Gram-negative bacterial cells [40]. To investigate and confirm the role of SbmA in bac7(1–35) (a bac7 truncated fragment) and PR-39 internalization in bacteria, studies have shown that *E. coli* carrying a point mutation in the *sbmA* gene, along with other *sbmA*-null mutants, are resistant to the administration of these two AMPs [37]. These findings have been further confirmed by fluorescence analyses, in which fluorescently labeled bac7(1–35) revealed lower cell internalization properties in *sbmA* mutated *E. coli* [37]. More recently, the functional characterization of SbmA in the presence of bac7(1–35) was carried out [40]. In that work, it was proposed that bac7(1–35) uptake is not ATP-dependent, but requires the presence of a transmembrane electrochemical gradient [40]. Moreover, it was found that bac7(1–35) directly binds to SbmA with high affinity, finally leading to conformational changes in this transporter [40].

4.2. PR-39 Antibacterial Properties

One of the first studies conducted with PR-39 demonstrated that this AMP inhibits *E. coli, S. typhimurium* and *Salmonella choleraesuis* growth (Table 2) [69]. In addition, this AMP also causes bacterial death, with the highest activities reported against *E. coli* [69]. Similar findings were observed by Jeon et al. [94], who considered the antibacterial potential of PR-39 similar to those obtained for ampicillin and gentamicin.

As for indolicidin, PR-39 analogues have also been generated aiming at an optimized therapeutic index. Studies have reported the evaluation of PR-39 truncated analogues, including PR-39 (1–26), PR-39 (1–22), PR-39 (1–18), PR-39 (1–15), PR-39 (16–39), PR-39 (20–39) and PR-39 (24–39), against different bacterial strains (Table 2). As a result, the most effective analogues were PR-39 (1–26), PR-39 (1–22), PR-39 (1–18) and PR-39 (1–15), presenting similar minimal bactericidal concentrations of PR-39 against *E. coli* and *Bacillus globigii* [68]. These findings suggest that shorter N-terminal fragments from the parent PR-39 could be developed aiming at conserved/improved antibacterial properties, allied to a lower cost of synthesis.

In terms of in vivo antibacterial properties, PR-39 has been used for the treatment of sepsis in mice through endotoxin neutralization. Studies have shown that PR-39, when administered with lipopolysaccharides (LPS), leads to a decreased release of nitric oxide (NO) by mice cells, thus reducing the cellular stress and, consequently, improving the survival rates of the treated animals in a sepsis model (Table 1) [70].

4.3. Bac7 Antibacterial Properties

Bac7 and its truncated analogues have been tested against numerous Gram-negative bacteria, including *E. coli*, *A. baumannii*, *K. pneumoniae* and *Salmonella enterica*, revealing the highest inhibitory potential for bac7(1–35) (Table 2) [95]. Moreover, the antibiofilm activity of bac7 has also been investigated against clinical isolates of *S. maltophilia* and *S. aureus* and *P. aeruginosa*. However, as for indolicidin, promising results were not obtained at the maximal concentration tested [59]. On the other hand, an in vivo study demonstrated that treatment with bac7 protects rats against *E. coli* endotoxins, thus avoiding septic shock (Table 1) [71].

The antibacterial activity of bac7(1–35) has also been evaluated in vivo using a murine model of *Salmonella* infection, resembling systemic infections in humans [96]. Therefore, it has been observed that untreated animals (control) survived for 10 days post-infection, whereas those animals treated with bac7 (75 mg·kg^{-1}) survived for 24.5 days post-infection. In addition, 35% of the animals treated with bac7 recovered completely from the infection, thus significantly reducing the mortality rates [96]. Years later, Benincasa et al. [96] used a 20 kDa polyethyleneglycol (PEG) to improve the effectiveness of bac7 against *Salmonella* infection in mice models. After intraperitoneal administration, the animals were observed for 24 h. Greater activity of bac7 and PEG were observed, although it was found in organs (e.g., kidneys and liver) for longer periods. Therefore, the association of PEG with bac7 proved to be a promising modification for the therapeutic applicability of this AMP (Table 1) [97].

4.4. PR-39 and bac7 Target Bacterial Protein Synthesis

One of the mechanisms by which non-lytic AMPs lead bacteria to death is the inhibition of protein synthesis. The proline/arginine-rich AMP PR-39 is known to rapidly cross bacterial cell membranes, without causing significant damage. Once in the intracellular compartment, this AMP inhibits proteins involved in DNA replication (Table 1 and Figure 1). The mechanism of action is attributed to PR-39's proteolytic activity, which causes the degradation of proteins associated with DNA replication, leading to the secondary inhibition of DNA synthesis [72]. In an attempt to find out the exact mechanism by which PR-39 exerts its antibacterial properties, Ho et al. [73] carried out a proteome microarray study with *E. coli* to systematically identify the intracellular protein targeted by this AMP. The inhibitory effects of PR-39 on diverse metabolic pathways have been confirmed, including those for translation, transport and metabolism of nucleotides, transport, and metabolism of coenzymes and others [73].

Protein and RNA synthesis have also been targeted by the non-lytic AMP bac7 (Table 1 and Figure 1) [38,90]. Mardirossian et al. [39] showed that bac7 (1–35) blocks protein synthesis by targeting ribosomal proteins. Moreover, the authors also proposed that this mechanism could prevent additional co-linear events, including the interaction of cotranslational chaperones with ribosomes, which is a known mechanism to ensure the translation of any polypeptide chain [39]. More recently, this mechanism was further explored, revealing that bac7(1–35) blocks the peptide exit tunnel in 70S ribosomes from *Thermus thermophiles* [98]. It was also concluded that this mechanism occurs through the interaction of bac7(1–35) with antibiotic-binding sites, thus interfering with the initial step of translation [98]. In addition, it has been proposed that bactenecins also target cell wall synthesis by binding to the lipid II precursor (Table 1 and Figure 1) [41]. These data support the idea that a single AMP may have multiple mechanisms of action simultaneously, which contributes to the lower occurrence of bacterial resistance to AMPs.

Although bac7(1–35) has been widely described as a non-lytic AMP that internalizes bacterial cells through the transporter SbmA, it has been shown that this mechanism varies depending on the characteristics of the target bacterial species. For instance, Runti et al. [42] reported that *P. aeruginosa* (PAO1) cells are inactivated by bac7(1–35) through cellular membrane disruption, which differs from what has been observed against *E. coli*. Interestingly, by expressing the SbmA transporter in *P. aeruginosa* (PAO1) it was found that bac7(1–35) internalization was enhanced, along with higher bacterial resistance to membrane disruption [42]. Therefore, this evidence supports the idea of bac7's (1–35) multiple mechanisms of action, which are highly dependent on the strain tested.

5. Apidaecin and Drosocin—Two Non-Lytic AMPs Derived from Insects

Apidaecin was the first proline-rich AMP isolated from bees in the mid-1980s. Apidaecin comprises an 18–20 amino acid residue peptide with proline and arginine repetitions along its sequence [43]. In contrast, drosocin is a peptide isolated from the fruit fly (*Drosophila melanogaster*), which was first reported by Bulet et al. [44]. Drosocin is composed of 19 amino acid residues in length and shares a high degree of sequence homology with apidaecin [44]. This peptide is characterized by three PRP motif repeats and glycosylation of threonine residues, which is suggested to be intrinsically related to its antibacterial properties [44,99]. Moreover, cytotoxic effects have not been reported for this peptide, reinforcing its therapeutic applicability [99,100].

5.1. Apidaecin and Drosocin Depend on Membrane Receptors to Internalize the Target Cell

Initial studies on the structure and membrane translocation of apidaecin peptides have suggested that the antibacterial activities of these peptides are intrinsically related to the presence of PXP motifs, which contribute to the ordered formation of oligomers that facilitates the entry through the bacterial outer membrane (OM) [101]. Nevertheless, although apidaecin's functional oligomers are capable of translocating across the OM, evidence suggests that its internalization and translocation across the IM are facilitated by specific interaction with membrane permeases and transporters (Table 1 and Figure 1) [78]. Moreover, it seems that such interaction is energy-driven, irreversible and stereospecific (Figure 1), as all-D-apidaecin (apidaecin constituted entirely of D-amino acids) does not bind to periplasmic or IM components [78].

As for apidaecin, drosocin has also been suggested to internalize bacterial cells through interactions with IM receptor/channels [48]. Drosocin is glycosylated at Thr^{11}, which has been characterized as a key factor for its antibacterial activities and, therefore, has been investigated in NMR structural studies. In general, spectra recorded in water indicate a high population of random coil arrangements for both glycosylated and non-glycosylated forms, whereas the presence of 50% TFE/water mixtures induces the formation of turns [49]. Although no significant differences were detected for the random coil arrangements in water, the glycosylated and non-glycosylated forms differ greatly in the folded conformations, especially at residues 10 to 13 (extended turn) and 17 to 19 (tightening of the downstream turn) in the glycosylated form. Additional studies have also shown that not only is the glycosylation at Thr^{11} crucial for drosocin's internalization into bacterial cells, but also its chirality [48]. Similarly to apidaecin, it has been reported that drosocin's action on bacterial cells is stereospecific, as its D-enantiomers are incapable of internalizing bacterial cells. These findings re-emphasize the receptor-driven mechanism by which drosocin acts (Table 1 and Figure 1). However, although this mechanism has been proposed for both apidaecin and drosocin, the specific target of these non-lytic AMPs on the periplasmic space or IM is still under investigation.

5.2. Apidaecin Antibacterial Properties

The first study carried out with apidaecin demonstrated that the activity of this peptide does not depend on cell membrane lysis [45,72]. Years later, when tested against bacteria, apidaecin was proved to cause bacterial cell death without triggering membrane destabilization [78]. It is presumed that the apidaecin C-terminal region (PRPPHPR (L/I)) is responsible for its non-lytic mechanism of action [45,46]. In terms of biological activities, apidaecin has been characterized for its antibacterial effects against numerous Gram-negative bacteria, including *E. coli* [102], *K. pneumoniae* [103] and *P. aeruginosa* (Table 2) [74].

Apidaecin analogues have been developed for improved antibacterial properties (Table 1). Czihal et al. [102], for instance, performed a robust study regarding the comparison between apidaecin and its analogues (api6, api7, api39 and api88) (Table 2). By modifying the C-terminal region through the inversion of an amide in the analogue api6, the authors reported a 32-fold and 4-fold higher antibacterial potential against resistant *E. coli* and *K. pneumoniae* when compared to the parent peptide, apidaecin [102]. In contrast, by acetylating the N-terminal of the analogue api7 greater stability was observed. However, the antibacterial activity of this analogue was abolished. Interestingly, by performing further modification on api7, including the replacement of Gly^1 by ornithine or lysine, the antibacterial potential of this analogue was reestablished [102]. Similar findings were obtained for the api39 analogue when replacing the Glu^{10} by an arginine, leading to improved inhibitory effects toward bacteria [102]. Finally, the api88 analogue, which presented the highest net positive charge among all analogues, underwent modifications in the N-terminal region, where acetyl amide ($CH_3CONH–$) was replaced by N, N, N', N'-tetramethylguanidine ($(((CH_3)_2N)_2\text{-}CNH\text{-})$). As a result, a remarkable improvement was observed in the antibacterial activity of api88, which revealed low MIC values against the three strains tested and, therefore, was pinpointed as a promising AMP for therapeutic purposes [102].

Additional studies have also shown that replacing the N-terminal glycine of apidaecin by tetramethylguanidino-L-ornithine led to the generation of a structurally stable analogue, named api137, with promising activity against *E. coli* (Table 2) [75]. Moreover, further investigations demonstrated that removing the C-terminal Leu^{18} residue resulted in a substantial loss of antibacterial activity, suggesting the crucial role of the api137 C-terminal region for its antibacterial potential [76]. Structural stability and resistance to enzymatic degradation have also been investigated in apidaecin Ib by substitutions of arginine/leucine residues with peptoid residues (Table 1). Gobbo et al. [104] engineered peptide–peptoid hybrids based on apidaecin Ib and observed that, although presenting higher stability to degradation, the position at which the peptoid residues lie in the apidaecin hybrids impairs their antibacterial activities. The authors reported that modifications at the N-terminal region of apidaecin Ib only slightly reduced the antibacterial property of the hybrids, whereas peptoid residues in the C-terminal region drastically reduced this property [104], once again reinforcing the relevance of a conserved C-terminal for apidaecin peptides' antibacterial potential.

5.3. Drosocin Antibacterial Properties

As described above, the glycosylation of drosocin residues seems to directly interfere with its biological activities against bacteria. In a study by Bikker et al. [100] the glycosylation of Tyr^6 and Ser^7 from drosocin was performed. As a result, the antibacterial activities of Tyr^6 glycosylated and N-terminal β-Ala drosocin analogues against *E. coli*, *Erwinia herbicola* (currently classified as *Pantoea agglomerans*) and several *S. enterica* serovars, namely *S. panama*, *S. infantis*, *S. montevideo*, *S. typhimurium* and *S. enteritidis* (Table 2), were improved compared to the parent non-glycosylated drosocin (Table 2) [100]. More recently, it was shown that the addition of a monosaccharide at Thr^{11} (GKPRPYSPRPT (αGalNAc)SHPRPIRV) led to a remarkable improvement of antibacterial potential against numerous Gram-negative strains compared to non-glycosylated drosocin [50]. Similar findings were reported for a drosocin analogue with the addition of a disaccharide at Thr^{11} [GKPRPYSPRPT (βGal (1 → 3) αGalNAc) SHPRPIRV] (Table 2) [50]. Taken together, these reports highlight the advantages of modulating drosocin's structure aiming at screening for optimized activities against pathogenic bacteria (Table 1).

5.4. Apidaecin and Drosocin Interact with Bacterial Chaperones

As described above, apidaecin translocation across membranes is receptor-mediated and, according to Castle et al. [78], probably has a component of the permease-type carrier system. It has been shown that apidaecin peptides are capable of causing bacterial protein misfolding and aggregation by interacting with bacterial chaperones (Table 1 and Figure 1). Dnak and GroEL are chaperones that aid in the correct folding and assembly of proteins and, consequently, affect many cellular processes including DNA replication, RNA transcription and protein transport. Structural studies involving the molecular complex DnaK/apidaecin have revealed two binding modes, indicating that DnaK is quite unspecific in terms of peptide-binding. Cross-linking and free-cell translation assays have demonstrated that Api88 and Api137 (apidaecin analogues) bind to the 70S ribosome, leading to protein synthesis inhibition (Figure 1) [76]. Apidaecin and drosocin share a high degree of sequence homology, as well as similarities in their antibacterial activity spectra [77]. Therefore, as for apidaecin, drosocin interacts with intracellular proteins, including the heat-shock proteins DnaK and GroEL to inhibit the DnaK ATPase activity and chaperone-assisted protein folding, respectively [79]. Apart from its chaperone-binding property, apidaecin has also been shown to inhibit release factors in bacteria. Matsumoto et al. [105] reported this unusual mechanism through the in vivo target exploration of apidaecin based on acquired resistance induced by gene overexpression (ARGO assay). In that work, recombinant *E. coli* strains overexpressing proteins involved in translation were treated with apidaecin, among which only one clone overexpressing a peptide chain release factor 1 (PrfA) was selected as a positive candidate. PrfA is known to bind to ribosomes to terminate translation processes by recognizing stop codons in mRNA. Therefore, it was proposed that apidaecin probably binds to ribosomes, competitively, thus inhibiting the termination step of translation [105].

6. Conclusions and Future Prospects

Here, the antibacterial properties, membrane translocation processes and intracellular mechanisms of action of specific non-lytic AMPs were reviewed. In general, membrane active and non-lytic AMPs present similar physicochemical properties and, therefore, have a high affinity for membrane-like environments. Membrane active AMPs, including magainin, cecropin, and melittin, are known to firstly establish electrostatic interactions with the target bacterial cell, followed by the accumulation of peptides aiming to achieve a critical concentration that favors peptides' self-association and penetration into the membrane core [47]. From this point on, different modes of action have been described, including barrel-stave/toroidal pores, "carpet"-like mechanism, peptide–lipid aggregation and amyloid models [106]. It has also been shown that synthetic AMPs are capable of delocalizing membrane-bound proteins, leading to bacterial cell envelope stress response [107,108]. In addition, membrane-associated mechanisms not necessarily lead to cell lysis, as observed for lactoferricin and daptomycin, which cause non-lytic membrane depolarization [109,110], and the human α-defensin 6 (HD6), which forms nanonets that interact with membrane proteins to entangle bacteria [111]. Taken together, these membrane-associated mechanisms trigger a series of negative effects on bacterial homeostasis, including disturbance of ion gradient, loss of metabolites, phospholipid flip-flop, membrane depolarization and loss of membrane symmetry [112].

Although AMPs can rapidly display their actions on bacterial membranes, an increasing number of reports have highlighted that bacteria can easily evade membrane-associated mechanisms by adapting the constitution and proportion of phospholipids in their OM and IM [113], as reported for *E. coli* strains resistant to magainin [114]. Therefore, non-lytic AMPs have been pinpointed for their ability to inactivate bacteria by interrupting vital cellular process, instead of membrane destabilization and disruption. Considering the alarming scenario imposed by bacterial infections, the intracellular mechanisms displayed by non-lytic AMPs appear as an advantage over membrane-active AMPs, as those peptides are less likely to induce bacterial resistance. Moreover, a primary non-lytic mechanism (e.g., peptide–protein interactions aiming at compromising bacterial viability) may trigger a secondary mechanism, thus imposing an additional obstacle for bacterial adaptation to non-lytic AMP administration. In terms of bacterial internalization, we emphasized the role of proline residues in all peptides here described, as this residue has been proved to be a membrane translocation promoter and, therefore, is considered a key feature that could be used for future drug design strategies.

Here we summarize the main molecular mechanisms by which non-lytic AMPs translocate across membranes. These mechanisms involve different AMP arrangements (e.g., "boat-like" and transmembrane orientations, for indolicidin) [28] and the formation of transient toroidal pores, which facilitates non-lytic AMPs (e.g., buforin II) in crossing both the bacterial OM and IM to act on intracellular targets [84]. In addition, the stereospecific binding of AMPs to IM transporters (e.g., apidaecin, bac7, and PR-39) has also been highlighted as a strategy by which these peptides reach the bacterial cytosol to exert their functions [37,78]. These mechanisms have also been reported for another non-lytic AMP, called pyrrhocoricin, which is derived from the European firebug *Pyrrhocoris apterus*. As for bac7, apidaecin, and drosocin, pyrrhocoricin binds stereospecifically to an IM target protein and further enters the cytosol to inhibit chaperone-assisted protein folding by interacting with the molecular chaperone DnaK [115]. Similar findings have been reported for oncocin, a proline-rich AMP derived from the milkweed bug, *Oncopeltus fasciatus* [116].

Apart from the chaperone activity of DnaK, the non-lytic AMPs here described are also capable of binding to lipid II precursor, as well as interfering with DNA, RNA and protein synthesis. Although this review focused on eukaryotic-derived AMPs, it is worth noting that bacteriocins (bacteria-derived AMPs) also present intracellular mechanisms of action. Nisin, for instance, represents a bacteriocin derived from *Lactococcus lactis* that inhibits cell wall synthesis by targeting lipid II [117]. Nevertheless, this AMP has also shown membrane-associated mechanisms by the formation of pores and, therefore, is not considered a non-lytic AMP. In contrast, studies have reported that nukacin ISK-1, which is produced by *Staphylococcus warneri*, is also capable of inhibiting cell wall synthesis, but with no membrane-associated properties [118]. Additionally, in terms of peptide–DNA interaction, the bacteriocin microcin B17, originally isolated from *E. coli*, has been shown to inhibit bacterial DNA gyrase, thus interfering with DNA replication [119]. Finally, bacteriocins have also been proved to act as DNase and RNase, as is the case of colicin family members [120,121].

In general, the non-lytic AMPs here presented have demonstrated promising antibacterial effects on both susceptible and resistant strains, whereas reports of antibiofilm activities are scarce and somewhat insubstantial. Although none of them have effectively reached the market, some have been used as lead molecules for the engineering of antimicrobial agents that have achieved advanced clinical trials. Indolicidin, for instance, was used as a model molecule for the design of omiganan, a 12-amino acid residue peptide rich in tryptophan and proline residues. Compared to indolicidin, omiganan N-terminal tryptophan and proline residues were removed, along with the addition of a lysine residue at the C-terminal and a K5R substitution [122]. Omiganan has been submitted to a total of 11 clinical trials as an antimicrobial agent to prevent and treat *Acnes vulgaris*, atopic dermatitis, seborrheic dermatitis, sepsis, fungaemia, among others (please check, DrugBank accession code DB0661). In addition, bactenecin and an innate defense regulator peptide, called IDR1, have been used as parent peptides for the development of a synthetic host defense peptide, IMX942/SGX942 (dusquetide) [123]. This drug candidate has been indicated for oral mucositis and, currently, is in phase III trials (please check, DrugBank accession code DB11879).

Allied to that, an increasing number of studies have highlighted the great therapeutic potential of the other non-lytic AMPs here described. PR-39, for instance, has shown promising anti-sepsis effects on mice, which are related to endotoxin neutralization [70], whereas apidaecin [75] and drosocin [50] have been used for proof-of-concept studies, defining which determinants modulate the generation of improved analogues aiming at antibacterial therapies. Finally, buforin II has shown a wide applicability, as both the parent peptide and its analogues have revealed synergistic effects with conventional antibiotics [63] and have also been proposed as carrier molecules aiming at gene regulation via PNA [67]. Conversely, it is worth mentioning that, in some aspects, non-lytic AMPs still require further attention. The identification of specific binding sites on the target proteins, ribosomes, DNA and RNA would allow the guided design of improved, strain-specific analogue peptides. Moreover, although efforts have been made on optimized non-lytic AMP analogues, their failure to reach the market, in some cases, still relies on poor in vivo effectiveness, nonspecific cytotoxicity, and bioavailability.

Overall, the data here summarized indicate the biotechnology and pharmaceutical potential of non-lytic AMPs as promising drug leads. However, it also reveals the need for deeper investigations aiming at generating candidates that could be successfully translated to the clinic.

Funding: This work was supported by grants from Fundação de Apoio à Pesquisa do Distrito Federal (FAPDF), Coordenação de Aperfeiçoamento de Pessoal de Nível Superior (CAPES) (to M.H.C. 88887.351521/2019-00), Conselho Nacional de Desenvolvimento e Tecnológico (CNPq) and Fundação de Apoio ao Desenvolvimento do Ensino, Ciência e Tecnologia do Estado de Mato Grosso do Sul (FUNDECT), Brazil.

Conflicts of Interest: The authors declare no conflict of interest.

References

1. Shrivastava, S.; Shrivastava, P.; Ramasamy, J. World health organization releases global priority list of antibiotic-resistant bacteria to guide research, discovery, and development of new antibiotics. *J. Med. Soc.* **2018**, *32*, 76. [CrossRef]

2. El-Halfawy, O.M.; Valvano, M.A. Antimicrobial heteroresistance: An emerging field in need of clarity. *Clin. Microbiol. Rev.* **2015**, *28*, 191–207. [CrossRef] [PubMed]

3. Nicoloff, H.; Hjort, K.; Levin, B.R.; Andersson, D.I. The high prevalence of antibiotic heteroresistance in pathogenic bacteria is mainly caused by gene amplification. *Nat. Microbiol.* **2019**, *4*, 504–514. [CrossRef] [PubMed]

4. Boucher, H.W.; Talbot, G.H.; Bradley, J.S.; Edwards, J.E.; Gilbert, D.; Rice, L.B.; Scheld, M.; Spellberg, B.; Bartlett, J. Bad bugs, no drugs: no ESKAPE! An update from the Infectious Diseases Society of America. *Clin. Infect. Dis.* **2009**, *48*, 1–12. [CrossRef] [PubMed]

5. Lai, Y.; Gallo, R.L. AMPed up immunity: How antimicrobial peptides have multiple roles in immune defense. *Trends Immunol.* **2009**, *30*, 131–141. [CrossRef] [PubMed]

6. Hancock, R.E.; Diamond, G. The role of cationic antimicrobial peptides in innate host defences. *Trends Microbiol.* **2000**, *8*, 402–410. [CrossRef]

7. Brogden, K.A. Antimicrobial peptides: Pore formers or metabolic inhibitors in bacteria? *Nat. Rev. Microbiol.* **2005**, *3*, 238–250. [CrossRef] [PubMed]

8. Bradshaw, J.P. Cationic antimicrobial peptides. *BioDrugs* **2003**, *17*, 233–240. [CrossRef]

9. Izadpanah, A.; Gallo, R.L. Antimicrobial peptides. *J. Am. Acad. Dermatol.* **2005**, *52*, 381–390. [CrossRef]

10. Shagaghi, N.; Palombo, E.A.; Clayton, A.H.A.; Bhave, M. Antimicrobial peptides: Biochemical determinants of activity and biophysical techniques of elucidating their functionality. *World J. Microbiol. Biotechnol.* **2018**, *34*, 62. [CrossRef]

11. Henriques, S.T.; Melo, M.N.; Castanho, M.A.R.B. Cell-penetrating peptides and antimicrobial peptides: How different are they? *Biochem. J.* **2006**, *399*, 1–7. [CrossRef] [PubMed]

12. Shai, Y. Mode of action of membrane active antimicrobial peptides. *Pep. Sci.* **2002**, *66*, 236–248. [CrossRef] [PubMed]

13. Matsuzaki, K. Magainins as paradigm for the mode of action of pore forming polypeptides. *Biochim. Biophys. Acta Biomembr.* **1998**, *1376*, 391–400. [CrossRef]

14. Huang, H.W. Action of antimicrobial peptides: Two-state model. *Biochemistry* **2000**, *39*, 8347–8352. [CrossRef] [PubMed]

15. Le, C.-F.; Fang, C.-M.; Sekaran, S.D. Intracellular targeting mechanisms by pntimicrobial Peptides. *Antimicrob. Agents Chemother.* **2017**, *61*, e02340-16. [CrossRef] [PubMed]

16. Neundorf, I. Antimicrobial and cell-penetrating peptides: How to understand two distinct functions despite similar physicochemical properties. In *Antimicrobial Peptides. Advances in Experimental Medicine and Biology;* Matsuzaki, K., Ed.; Springer: Singapore, 2019; Volume 1117, pp. 93–109.

17. Scocchi, M.; Tossi, A.; Gennaro, R. Proline-rich antimicrobial peptides: Converging to a non-lytic mechanism of action. *Cell. Mol. Life Sci.* **2011**, *68*, 2317–2330. [CrossRef] [PubMed]

18. Otvos, L. The short proline-rich antibacterial peptide family. *Cell. Mol. Life Sci.* **2002**, *59*, 1138–1150. [CrossRef]

19. Ulmschneider, J.P. Charged antimicrobial peptides can translocate across membranes without forming channel-like pores. *Biophys. J.* **2017**, *113*, 73–81. [CrossRef]

20. Friedrich, C.L.; Rozek, A.; Patrzykat, A.; Hancock, R.E. Structure and mechanism of action of an indolicidin peptide derivative with improved activity against gram-positive bacteria. *J. Biol. Chem.* **2001**, *276*, 24015–24022. [CrossRef]

21. Ulvatne, H.; Samuelsen, Ø.; Haukland, H.H.; Krämer, M.; Vorland, L.H. Lactoferricin B inhibits bacterial macromolecular synthesis in *Escherichia coli* and *Bacillus subtilis*. *FEMS Microbiol. Lett.* **2004**, *237*, 377–384. [CrossRef]

22. Graf, M.; Wilson, D.N. Intracellular antimicrobial peptides targeting the protein synthesis machinery. *Adv. Exp. Med. Biol.* **2019**, *1117*, 73–89. [PubMed]

23. Selsted, M.E.; Novotny, M.J.; Morris, W.L.; Tang, Y.Q.; Smith, W.; Cullor, J.S. Indolicidin, a novel bactericidal tridecapeptide amide from neutrophils. *J. Biol. Chem.* **1992**, *267*, 4292–4295. [PubMed]

24. Mishra, A.K.; Choi, J.; Moon, E.; Baek, K.H. Tryptophan-rich and proline-rich antimicrobial peptides. *Molecules* **2018**, *23*, 815. [CrossRef] [PubMed]

25. Falla, T.J.; Karunaratne, D.N.; Hancock, R.E. Mode of action of the antimicrobial peptide indolicidin. *J. Biol. Chem.* **1996**, *271*, 19298–19303. [CrossRef]

26. Ladokhin, A.S.; Selsted, M.E.; White, S.H. Bilayer interactions of indolicidin, a small antimicrobial peptide rich in tryptophan, proline, and basic amino acids. *Biophys. J.* **1997**, *72*, 794–805. [CrossRef]

27. Ladokhin, A.S.; Selsted, M.E.; White, S.H. CD spectra of indolicidin antimicrobial peptides suggest turns, not polyproline helix. *Biochemistry* **1999**, *38*, 12313–12319. [CrossRef]

28. Rozek, A.; Friedrich, C.L.; Hancock, R.E. Structure of the bovine antimicrobial peptide indolicidin bound to dodecylphosphocholine and sodium dodecyl sulfate micelles. *Biochemistry* **2000**, *39*, 15765–15774. [CrossRef] [PubMed]

29. Hsu, C.H.; Chen, C.; Jou, M.L.; Lee, A.Y.; Lin, Y.C.; Yu, Y.P.; Huang, W.T.; Wu, S.H. Structural and DNA-binding studies on the bovine antimicrobial peptide, indolicidin: Evidence for multiple conformations involved in binding to membranes and DNA. *Nucleic Acids Res.* **2005**, *33*, 4053–4064. [CrossRef]

30. Hsu, J.C.; Yip, C.M. Molecular dynamics simulations of indolicidin association with model lipid bilayers. *Biophys. J.* **2007**, *92*, L100–L102. [CrossRef]

31. Tsai, C.W.; Lin, Z.W.; Chang, W.F.; Chen, Y.F.; Hu, W.W. Development of an indolicidin-derived peptide by reducing membrane perturbation to decrease cytotoxicity and maintain gene delivery ability. *Colloids Surf. B.* **2018**, *165*, 18–27. [CrossRef]

32. Ghosh, A.; Kar, R.K.; Jana, J.; Saha, A.; Jana, B.; Krishnamoorthy, J.; Kumar, D.; Ghosh, S.; Chatterjee, S.; Bhunia, A. Indolicidin targets duplex DNA: Structural and mechanistic insight through a combination of spectroscopy and microscopy. *ChemMedChem* **2014**, *9*, 2052–2058. [CrossRef] [PubMed]

33. Yi, G.S.; Park, C.B.; Kim, S.C.; Cheong, C. Solution structure of an antimicrobial peptide buforin II. *FEBS Lett.* **1996**, *398*, 87–90. [CrossRef]

34. Park, C.B.; Yi, K.S.; Matsuzaki, K.; Kim, M.S.; Kim, S.C. Structure-activity analysis of buforin II, a histone H2A-derived antimicrobial peptide: The proline hinge is responsible for the cell-penetrating ability of buforin II. *Proc. Natl. Acad. Sci. USA* **2000**, *97*, 8245–8250. [CrossRef] [PubMed]

35. Kobayashi, S.; Chikushi, A.; Tougu, S.; Imura, Y.; Nishida, M.; Yano, Y.; Matsuzaki, K. Membrane translocation mechanism of the antimicrobial peptide buforin 2. *Biochemistry* **2004**, *43*, 15610–15616. [CrossRef] [PubMed]

36. Kobayashi, S.; Takeshima, K.; Park, C.B.; Kim, S.C.; Matsuzaki, K. Interactions of the novel antimicrobial peptide buforin 2 with lipid bilayers: Proline as a translocation promoting factor. *Biochemistry* **2000**, *39*, 8648–8654. [CrossRef] [PubMed]

37. Mattiuzzo, M.; Bandiera, A.; Gennaro, R.; Benincasa, M.; Pacor, S.; Antcheva, N.; Scocchi, M. Role of the *Escherichia coli* SbmA in the antimicrobial activity of proline-rich peptides. *Mol. Microbiol.* **2007**, *66*, 151–163. [CrossRef] [PubMed]

38. Skerlavaj, B.; Romeo, D.; Gennaro, R. Rapid membrane permeabilization and inhibition of vital functions of gram-negative bacteria by bactenecins. *Infect. Immun.* **1990**, *58*, 3724–3730.

39. Mardirossian, M.; Grzela, R.; Giglione, C.; Meinnel, T.; Gennaro, R.; Mergaert, P.; Scocchi, M. The host antimicrobial peptide Bac71–35 binds to bacterial ribosomal proteins and inhibits protein synthesis. *Chem. Biol.* **2014**, *21*, 1639–1647. [CrossRef]

40. Runti, G.; Lopez Ruiz Mdel, C.; Stoilova, T.; Hussain, R.; Jennions, M.; Choudhury, H.G.; Benincasa, M.; Gennaro, R.; Beis, K.; Scocchi, M. Functional characterization of SbmA, a bacterial inner membrane transporter required for importing the antimicrobial peptide Bac7(1–35). *J. Bacteriol.* **2013**, *195*, 5343–5351. [CrossRef]

41. de Leeuw, E.; Li, C.; Zeng, P.; Li, C.; Buin, M.D.-d.; Lu, W.-Y.; Breukink, E.; Lu, W. Functional interaction of human neutrophil peptide-1 with the cell wall precursor lipid II. *FEBS Lett.* **2010**, *584*, 1543–1548. [CrossRef]

42. Runti, G.; Benincasa, M.; Giuffrida, G.; Devescovi, G.; Venturi, V.; Gennaro, R.; Scocchi, M. The Mechanism of killing by the proline-rich peptide bac7(1–35) against clinical strains of *Pseudomonas aeruginosa* differs from that against other gram-negative bacteria. *Antimicrob. Agents Chemother.* **2017**, *61*. [CrossRef] [PubMed]

43. Li, W.F.; Ma, G.X.; Zhou, X.X. Apidaecin-type peptides: Biodiversity, structure-function relationships and mode of action. *Peptides* **2006**, *27*, 2350–2359. [CrossRef] [PubMed]

44. Bulet, P.; Dimarcq, J.L.; Hetru, C.; Lagueux, M.; Charlet, M.; Hegy, G.; Van Dorsselaer, A.; Hoffmann, J.A. A novel inducible antibacterial peptide of Drosophila carries an O-glycosylated substitution. *J. Biol. Chem.* **1993**, *268*, 14893–14897. [PubMed]

45. Casteels, P.; Tempst, P. Apidaecin-type peptide antibiotics function through a non-poreforming mechanism involving stereospecificity. *Biochem. Biophys. Res. Commun.* **1994**, *199*, 339–345. [CrossRef] [PubMed]

46. Piantavigna, S.; Czihal, P.; Mechler, A.; Richter, M.; Hoffmann, R.; Martin, L.L. Cell penetrating apidaecin peptide interactions with biomimetic phospholipid membranes. *I. J. P. R. Ther.* **2009**, *15*, 139–146. [CrossRef]

47. Lee, T.H.; Hall, K.N.; Aguilar, M.I. Antimicrobial peptide structure and mechanism of action: A focus on the role of membrane structure. *Curr. Top. Med. Chem.* **2016**, *16*, 25–39. [CrossRef] [PubMed]

48. Lele, D.S.; Talat, S.; Kumari, S.; Srivastava, N.; Kaur, K.J. Understanding the importance of glycosylated threonine and stereospecific action of Drosocin, a proline rich antimicrobial peptide. *Eur. J. Med. Chem.* **2015**, *92*, 637–647. [CrossRef]

49. McManus, A.M.; Otvos, L.; Hoffmann, R.; Craik, D.J. Conformational studies by NMR of the antimicrobial peptide, drosocin, and its non-glycosylated derivative: Effects of glycosylation on solution conformation. *Biochemistry* **1999**, *38*, 705–714. [CrossRef]

50. Lele, D.S.; Kaur, G.; Thiruvikraman, M.; Kaur, K.J. Comparing naturally occurring glycosylated forms of proline rich antibacterial peptide, Drosocin. *Glycoconj. J.* **2017**, *34*, 613–624. [CrossRef]

51. Subbalakshmi, C.; Krishnakumari, V.; Nagaraj, R.; Sitaram, N. Requirements for antibacterial and hemolytic activities in the bovine neutrophil derived 13-residue peptide indolicidin. *FEBS Lett.* **1996**, *395*, 48–52. [CrossRef]

52. Subbalakshmi, C.; Bikshapathy, E.; Sitaram, N.; Nagaraj, R. Antibacterial and hemolytic activities of single tryptophan analogs of indolicidin. *Biochem. Biophys. Res. Commun.* **2000**, *274*, 714–716. [CrossRef] [PubMed]

53. Ryge, T.S.; Doisy, X.; Ifrah, D.; Olsen, J.E.; Hansen, P.R. New indolicidin analogues with potent antibacterial activity. *J. Pep. Res.* **2004**, *64*, 171–185. [CrossRef] [PubMed]

54. Frecer, V. QSAR analysis of antimicrobial and haemolytic effects of cyclic cationic antimicrobial peptides derived from protegrin-1. *Bioorg. Med. Chem.* **2006**, *14*, 6065–6074. [CrossRef] [PubMed]

55. Oren, Z.; Hong, J.; Shai, Y. A repertoire of novel antibacterial diastereomeric peptides with selective cytolytic activity. *J. Biol. Chem.* **1997**, *272*, 14643–14649. [CrossRef] [PubMed]

56. Kim, S.M.; Kim, J.M.; Joshi, B.P.; Cho, H.; Lee, K.H. Indolicidin-derived antimicrobial peptide analogs with greater bacterial selectivity and requirements for antibacterial and hemolytic activities. *Biochim. Biophys. Acta* **2009**, *1794*, 185–192. [CrossRef] [PubMed]

57. Jindal, H.M.; Le, C.F.; Mohd Yusof, M.Y.; Velayuthan, R.D.; Lee, V.S.; Zain, S.M.; Isa, D.M.; Sekaran, S.D. Antimicrobial activity of novel synthetic peptides derived from indolicidin and ranalexin against *Streptococcus pneumoniae*. *PLoS ONE* **2015**, *10*, e0128532. [CrossRef] [PubMed]

58. Clark, D.P.; Durell, S.; Maloy, W.L.; Zasloff, M. Ranalexin. A novel antimicrobial peptide from bullfrog (*Rana catesbeiana*) skin, structurally related to the bacterial antibiotic, polymyxin. *J. Biol. Chem.* **1994**, *269*, 10849–10855.

59. Pompilio, A.; Scocchi, M.; Pomponio, S.; Guida, F.; Di Primio, A.; Fiscarelli, E.; Gennaro, R.; Di Bonaventura, G. Antibacterial and anti-biofilm effects of cathelicidin peptides against pathogens isolated from cystic fibrosis patients. *Peptides* **2011**, *32*, 1807–1814. [CrossRef] [PubMed]

60. Dosler, S.; Mataraci, E. In vitro pharmacokinetics of antimicrobial cationic peptides alone and in combination with antibiotics against methicillin resistant *Staphylococcus aureus* biofilms. *Peptides* **2013**, *49*, 53–58. [CrossRef]

61. Park, C.B.; Kim, M.S.; Kim, S.C. A novel antimicrobial peptide from *Bufo bufo gargarizans*. *Biochem. Biophys. Res. Commun.* **1996**, *218*, 408–413. [CrossRef]

62. Hollmann, A.; Martinez, M.; Maturana, P.; Semorile, L.C.; Maffia, P.C. Antimicrobial peptides: Interaction with model and biological membranes and synergism with chemical antibiotics. *Front. Chem.* **2018**, *6*, 204. [CrossRef] [PubMed]

63. Cirioni, O.; Silvestri, C.; Ghiselli, R.; Orlando, F.; Riva, A.; Gabrielli, E.; Mocchegiani, F.; Cianforlini, N.; Trombettoni, M.M.; Saba, V.; et al. Therapeutic efficacy of buforin II and rifampin in a rat model of *Acinetobacter baumannii* sepsis. *Crit. Care Med.* **2009**, *37*, 1403–1407. [CrossRef] [PubMed]

64. Zhou, Y.; Peng, Y. Synergistic effect of clinically used antibiotics and peptide antibiotics against Gram-positive and Gram-negative bacteria. *Exp. Ther. Med.* **2013**, *6*, 1000–1004. [CrossRef] [PubMed]

65. Hao, G.; Shi, Y.H.; Tang, Y.L.; Le, G.W. The intracellular mechanism of action on *Escherichia coli* of BF2-A/C, two analogues of the antimicrobial peptide Buforin 2. *J. Microbiol.* **2013**, *51*, 200–206. [CrossRef] [PubMed]

66. Jang, S.A.; Kim, H.; Lee, J.Y.; Shin, J.R.; Kim, D.J.; Cho, J.H.; Kim, S.C. Mechanism of action and specificity of antimicrobial peptides designed based on buforin IIb. *Peptides* **2012**, *34*, 283–289. [CrossRef] [PubMed]

67. Hansen, A.M.; Bonke, G.; Larsen, C.J.; Yavari, N.; Nielsen, P.E.; Franzyk, H. Antibacterial peptide nucleic acid-antimicrobial peptide (PNA-AMP) conjugates: Antisense targeting of fatty acid biosynthesis. *Bioconjugate Chem.* **2016**, *27*, 863–867. [CrossRef] [PubMed]

68. Veldhuizen, E.J.; Schneider, V.A.; Agustiandari, H.; van Dijk, A.; Tjeersdsma-van Bokhoven, J.L.; Bikker, F.J.; Haagsman, H.P. Antimicrobial and immunomodulatory activities of PR-39 derived peptides. *PLoS ONE* **2014**, *9*, e95939. [CrossRef]

69. Shi, J.; Ross, C.R.; Leto, T.L.; Blecha, F. PR-39, a proline-rich antibacterial peptide that inhibits phagocyte NADPH oxidase activity by binding to Src homology 3 domains of p47 phox. *Proc. Natl. Acad. Sci. USA* **1996**, *93*, 6014–6018. [CrossRef]

70. James, P.E.; Madhani, M.; Ross, C.; Klei, L.; Barchowsky, A.; Swartz, H.M. Tissue hypoxia during bacterial sepsis is attenuated by PR-39, an antibacterial peptide. *Adv. Exp. Med. Biol.* **2003**, *530*, 645–652.

71. Ghiselli, R.; Giacometti, A.; Cirioni, O.; Circo, R.; Mocchegiani, F.; Skerlavaj, B.; D'Amato, G.; Scalise, G.; Zanetti, M.; Saba, V. Neutralization of endotoxin in vitro and in vivo by Bac7(1–35), a proline-rich antibacterial peptide. *Shock* **2003**, *19*, 577–581. [CrossRef]

72. Boman, H.G.; Agerberth, B.; Boman, A. Mechanisms of action on *Escherichia coli* of cecropin P1 and PR-39, two antibacterial peptides from pig intestine. *Infect. Immun.* **1993**, *61*, 2978–2984. [PubMed]

73. Ho, Y.-H.; Shah, P.; Chen, Y.-W.; Chen, C.-S. Systematic analysis of intracellular-targeting antimicrobial peptides, bactenecin 7, hybrid of pleurocidin and dermaseptin, proline–arginine-rich peptide, and lactoferricin B, by using *Escherichia coli* proteome microarrays. *Mol. Cell. Proteom.* **2016**, *15*, 1837–1847. [CrossRef]

74. Graf, M.; Mardirossian, M.; Nguyen, F.; Seefeldt, A.C.; Guichard, G.; Scocchi, M.; Innis, C.A. Proline-rich antimicrobial peptides targeting protein synthesis. *Nat. Prod. Rep.* **2017**, *34*, 702–711. [CrossRef] [PubMed]

75. Berthold, N.; Czihal, P.; Fritsche, S.; Sauer, U.; Schiffer, G.; Knappe, D.; Alber, G.; Hoffmann, R. Novel apidaecin 1b analogs with superior serum stabilities for treatment of infections by gram-negative pathogens. *Antimicrob. Agents Chemother.* **2013**, *57*, 402–409. [CrossRef]

76. Krizsan, A.; Volke, D.; Weinert, S.; Strater, N.; Knappe, D.; Hoffmann, R. Insect-derived proline-rich antimicrobial peptides kill bacteria by inhibiting bacterial protein translation at the 70S ribosome. *Angew. Chem. Int. Ed. Engl.* **2014**, *53*, 12236–12239. [CrossRef]

77. Scocchi, M.; Mardirossian, M.; Runti, G.; Benincasa, M. Non-membrane permeabilizing modes of action of antimicrobial peptides on bacteria. *Curr. Top. Med. Chem.* **2016**, *16*, 76–88. [CrossRef] [PubMed]

78. Castle, M.; Nazarian, A.; Yi, S.S.; Tempst, P. Lethal effects of apidaecin on *Escherichia coli* involve sequential molecular interactions with diverse targets. *J. Biol. Chem.* **1999**, *274*, 32555–32564. [CrossRef]

79. Otvos, L.; O, I.; Rogers, M.E.; Consolvo, P.J.; Condie, B.A.; Lovas, S.; Bulet, P.; Blaszczyk-Thurin, M. Interaction between heat shock proteins and antimicrobial peptides. *Biochemistry* **2000**, *39*, 14150–14159. [CrossRef]

80. Haney, E.F.; Petersen, A.P.; Lau, C.K.; Jing, W.; Storey, D.G.; Vogel, H.J. Mechanism of action of puroindoline derived tryptophan-rich antimicrobial peptides. *Biochim. Biophys. Acta Biomembr.* **2013**, *1828*, 1802–1813. [CrossRef]

81. Yonezawa, A.; Kuwahara, J.; Fujii, N.; Sugiura, Y. Binding of tachyplesin I to DNA revealed by footprinting analysis: Significant contribution of secondary structure to DNA binding and implication for biological action. *Biochemistry* **1992**, *31*, 2998–3004. [CrossRef]

82. Subbalakshmi, C.; Sitaram, N. Mechanism of antimicrobial action of indolicidin. *FEMS Microbiol. Lett.* **1998**, *160*, 91–96. [CrossRef] [PubMed]

83. Cho, J.H.; Sung, B.H.; Kim, S.C. Buforins: Histone H2A-derived antimicrobial peptides from toad stomach. *Biochim. Biophys. Acta Biomembr.* **2009**, *1788*, 1564–1569. [CrossRef] [PubMed]

84. Elmore, D.E. Insights into buforin II membrane translocation from molecular dynamics simulations. *Peptides* **2012**, *38*, 357–362. [CrossRef] [PubMed]

85. Park, C.B.; Kim, H.S.; Kim, S.C. Mechanism of action of the antimicrobial peptide buforin II: Buforin II kills microorganisms by penetrating the cell membrane and inhibiting cellular functions. *Biochem. Biophys. Res. Commun.* **1998**, *244*, 253–257. [CrossRef] [PubMed]

86. Agerberth, B.; Lee, J.Y.; Bergman, T.; Carlquist, M.; Boman, H.G.; Mutt, V.; Jornvall, H. Amino acid sequence of PR-39. Isolation from pig intestine of a new member of the family of proline-arginine-rich antibacterial peptides. *Eur. J. Biochem.* **1991**, *202*, 849–854. [CrossRef] [PubMed]

87. Gudmundsson, G.H.; Magnusson, K.P.; Chowdhary, B.P.; Johansson, M.; Andersson, L.; Boman, H.G. Structure of the gene for porcine peptide antibiotic PR-39, a cathelin gene family member: Comparative mapping of the locus for the human peptide antibiotic FALL-39. *Proc. Natl. Acad. Sci. USA* **1995**, *92*, 7085–7089. [CrossRef]

88. Sang, Y.; Blecha, F. Porcine host defense peptides: Expanding repertoire and functions. *Dev. Comp. Immunol.* **2009**, *33*, 334–343. [CrossRef]

89. Gallo, R.L.; Ono, M.; Povsic, T.; Page, C.; Eriksson, E.; Klagsbrun, M.; Bernfield, M. Syndecans, cell surface heparan sulfate proteoglycans, are induced by a proline-rich antimicrobial peptide from wounds. *Proc. Natl. Acad. Sci. USA* **1994**, *91*, 11035–11039. [CrossRef]

90. Gennaro, R.; Skerlavaj, B.; Romeo, D. Purification, composition, and activity of two bactenecins, antibacterial peptides of bovine neutrophils. *Infect. Immun.* **1989**, *57*, 3142–3146.

91. Litteri, L.; Romeo, D. Characterization of bovine neutrophil antibacterial polypeptides which bind to *Escherichia coli*. *Infect. Immun.* **1993**, *61*, 966–969.

92. Price, R.; Bugeon, L.; Mostowy, S.; Makendi, C.; Wren, B.; Williams, H.; Willcocks, S. In vitro and in vivo properties of the bovine antimicrobial peptide, Bactenecin 5. *PLoS ONE* **2019**, *14*, e0210508. [CrossRef] [PubMed]

93. Zahn, M.; Kieslich, B.; Berthold, N.; Knappe, D.; Hoffmann, R.; Strater, N. Structural identification of DnaK binding sites within bovine and sheep bactenecin Bac7. *Protein Pept. Lett.* **2014**, *21*, 407–412. [CrossRef] [PubMed]

94. Jeon, H.; Le, M.T.; Ahn, B.; Cho, H.S.; Le, V.C.Q.; Yum, J.; Hong, K.; Kim, J.H.; Song, H.; Park, C. Copy number variation of PR-39 cathelicidin, and identification of PR-35, a natural variant of PR-39 with reduced mammalian cytotoxicity. *Gene* **2019**, *692*, 88–93. [CrossRef] [PubMed]

95. Benincasa, M.; Scocchi, M.; Podda, E.; Skerlavaj, B.; Dolzani, L.; Gennaro, R. Antimicrobial activity of Bac7 fragments against drug-resistant clinical isolates. *Peptides* **2004**, *25*, 2055–2061. [CrossRef]

96. Benincasa, M.; Pelillo, C.; Zorzet, S.; Garrovo, C.; Biffi, S.; Gennaro, R.; Scocchi, M. The proline-rich peptide Bac7(1–35) reduces mortality from *Salmonella typhimurium* in a mouse model of infection. *BMC Microbiol.* **2010**, *10*, 178. [CrossRef] [PubMed]

97. Benincasa, M.; Zahariev, S.; Pelillo, C.; Milan, A.; Gennaro, R.; Scocchi, M. PEGylation of the peptide Bac7(1–35) reduces renal clearance while retaining antibacterial activity and bacterial cell penetration capacity. *Eur. J. Med. Chem.* **2015**, *95*, 210–219. [CrossRef] [PubMed]

98. Gagnon, M.G.; Roy, R.N.; Lomakin, I.B.; Florin, T.; Mankin, A.S.; Steitz, T.A. Structures of proline-rich peptides bound to the ribosome reveal a common mechanism of protein synthesis inhibition. *Nucleic Acids Res.* **2016**, *44*, 2439–2450. [CrossRef]

99. Bulet, P.; Urge, L.; Ohresser, S.; Hetru, C.; Otvos, L. Enlarged scale chemical synthesis and range of activity of drosocin, an O-glycosylated antibacterial peptide of Drosophila. *Eur. J. Biochem.* **1996**, *238*, 64–69. [CrossRef]

100. Bikker, F.J.; Kaman-van Zanten, W.E.; de Vries-van de Ruit, A.M.; Voskamp-Visser, I.; van Hooft, P.A.; Mars-Groenendijk, R.H.; de Visser, P.C.; Noort, D. Evaluation of the antibacterial spectrum of drosocin analogues. *Chem. Biol. Drug Des.* **2006**, *68*, 148–153. [CrossRef]

101. Dutta, R.C.; Nagpal, S.; Salunke, D.M. Functional mapping of apidaecin through secondary structure correlation. *Int. J. Biochem. Cell Biol.* **2008**, *40*, 1005–1015. [CrossRef]

102. Czihal, P.; Knappe, D.; Fritsche, S.; Zahn, M.; Berthold, N.; Piantavigna, S.; Muller, U.; van Dorpe, S.; Herth, N.; Binas, A.; et al. Api88 is a novel antibacterial designer peptide to treat systemic infections with multidrug-resistant Gram-negative pathogens. *ACS Chem. Biol.* **2012**, *7*, 1281–1291. [CrossRef] [PubMed]

103. Ostorhazi, E.; Nemes-Nikodem, E.; Knappe, D.; Hoffmann, R. In vivo activity of optimized apidaecin and oncocin peptides against a multiresistant, KPC-producing *Klebsiella pneumoniae* strain. *Protein Pept. Lett.* **2014**, *21*, 368–373. [CrossRef] [PubMed]

104. Gobbo, M.; Benincasa, M.; Bertoloni, G.; Biondi, B.; Dosselli, R.; Papini, E.; Reddi, E.; Rocchi, R.; Tavano, R.; Gennaro, R. Substitution of the arginine/leucine residues in apidaecin Ib with peptoid residues: Effect on antimicrobial activity, cellular uptake, and proteolytic degradation. *J. Med. Chem.* **2009**, *52*, 5197–5206. [CrossRef] [PubMed]

105. Matsumoto, K.; Yamazaki, K.; Kawakami, S.; Miyoshi, D.; Ooi, T.; Hashimoto, S.; Taguchi, S. In vivo target exploration of apidaecin based on Acquired Resistance induced by Gene Overexpression (ARGO assay). *Sci. Rep.* **2017**, *7*, 12136. [CrossRef] [PubMed]

106. Chairatana, P.; Nolan, E.M. Molecular basis for self-assembly of a human host-defense peptide that entraps bacterial pathogens. *J. Am. Chem. Soc.* **2014**, *136*, 13267–13276. [CrossRef] [PubMed]

107. Omardien, S.; Drijfhout, J.W.; van Veen, H.; Schachtschabel, S.; Riool, M.; Hamoen, L.W.; Brul, S.; Zaat, S.A.J. Synthetic antimicrobial peptides delocalize membrane bound proteins thereby inducing a cell envelope stress response. *Biochim. Biophys. Acta Biomembr.* **2018**, *1860*, 2416–2427. [CrossRef] [PubMed]

108. Xhindoli, D.; Pacor, S.; Guida, F.; Antcheva, N.; Tossi, A. Native oligomerization determines the mode of action and biological activities of human cathelicidin LL-37. *Biochem. J.* **2014**, *457*, 263–275. [CrossRef] [PubMed]

109. Gifford, J.L.; Hunter, H.N.; Vogel, H.J. Lactoferricin: A lactoferrin-derived peptide with antimicrobial, antiviral, antitumor and immunological properties. *Cell. Mol. Life Sci.* **2005**, *62*, 2588–2598. [CrossRef]

110. Jeu, L.; Fung, H.B. Daptomycin: A cyclic lipopeptide antimicrobial agent. *Clin. Ther.* **2004**, *26*, 1728–1757. [CrossRef] [PubMed]

111. Chu, H.; Pazgier, M.; Jung, G.; Nuccio, S.P.; Castillo, P.A.; de Jong, M.F.; Winter, M.G.; Winter, S.E.; Wehkamp, J.; Shen, B.; et al. Human alpha-defensin 6 promotes mucosal innate immunity through self-assembled peptide nanonets. *Science* **2012**, *337*, 477–481. [CrossRef]

112. Melo, M.N.; Ferre, R.; Castanho, M.A. Antimicrobial peptides: Linking partition, activity and high membrane-bound concentrations. *Nat. Rev. Microbiol.* **2009**, *7*, 245–250. [CrossRef] [PubMed]

113. Candido, E.S.; de Barros, E.; Cardoso, M.H.; Franco, O.L. Bacterial cross-resistance to anti-infective compounds. Is it a real problem? *Curr. Opin. Pharmacol.* **2019**, *48*, 76–81. [CrossRef] [PubMed]

114. Cardoso, M.H.; de Almeida, K.C.; Candido, E.S.; Murad, A.M.; Dias, S.C.; Franco, O.L. Comparative NanoUPLC-MS(E) analysis between magainin I-susceptible and -resistant *Escherichia coli* strains. *Sci. Rep.* **2017**, *7*, 4197. [CrossRef] [PubMed]

115. Chesnokova, L.S.; Slepenkov, S.V.; Witt, S.N. The insect antimicrobial peptide, L-pyrrhocoricin, binds to and stimulates the ATPase activity of both wild-type and lidless DnaK. *FEBS Lett.* **2004**, *565*, 65–69. [CrossRef] [PubMed]

116. Roy, R.N.; Lomakin, I.B.; Gagnon, M.G.; Steitz, T.A. The mechanism of inhibition of protein synthesis by the proline-rich peptide oncocin. *Nat. Struct. Mol. Biol.* **2015**, *22*, 466–469. [CrossRef] [PubMed]

117. Scherer, K.M.; Spille, J.H.; Sahl, H.G.; Grein, F.; Kubitscheck, U. The lantibiotic nisin induces lipid II aggregation, causing membrane instability and vesicle budding. *Biophys. J.* **2015**, *108*, 1114–1124. [CrossRef] [PubMed]

118. Islam, M.R.; Nishie, M.; Nagao, J.; Zendo, T.; Keller, S.; Nakayama, J.; Kohda, D.; Sahl, H.G.; Sonomoto, K. Ring A of nukacin ISK-1: A lipid II-binding motif for type-A(II) lantibiotic. *J. Am. Chem. Soc.* **2012**, *134*, 3687–3690. [CrossRef]

119. Collin, F.; Thompson, R.E.; Jolliffe, K.A.; Payne, R.J.; Maxwell, A. Fragments of the bacterial toxin microcin B17 as gyrase poisons. *PLoS ONE* **2013**, *8*, e61459. [CrossRef] [PubMed]

120. Ng, C.L.; Lang, K.; Meenan, N.A.; Sharma, A.; Kelley, A.C.; Kleanthous, C.; Ramakrishnan, V. Structural basis for 16S ribosomal RNA cleavage by the cytotoxic domain of colicin E3. *Nat. Struct. Mol. Biol.* **2010**, *17*, 1241–1246. [CrossRef]

121. Garza-Sanchez, F.; Gin, J.G.; Hayes, C.S. Amino acid starvation and colicin D treatment induce A-site mRNA cleavage in *Escherichia coli*. *J. Mol. Biol.* **2008**, *378*, 505–519. [CrossRef]

122. Melo, M.N.; Dugourd, D.; Castanho, M.A. Omiganan pentahydrochloride in the front line of clinical applications of antimicrobial peptides. *Recent Pat. Antiinfect. Drug Discov.* **2006**, *1*, 201–207. [CrossRef] [PubMed]
123. Yeung, A.T.; Gellatly, S.L.; Hancock, R.E. Multifunctional cationic host defence peptides and their clinical applications. *Cell. Mol. Life Sci.* **2011**, *68*, 2161–2176. [CrossRef] [PubMed]

International Journal of
Molecular Sciences

Article

Biophysical Insight on the Membrane Insertion of an Arginine-Rich Cell-Penetrating Peptide

Marie-Lise Jobin [1,*,†], Lydie Vamparys [2], Romain Deniau [2], Axelle Grélard [1],
Cameron D. Mackereth [3], Patrick F.J. Fuchs [4,5] and Isabel D. Alves [1,*]

[1] Institute of Chemistry & Biology of Membranes & Nanoobjects (CBMN), CNRS UMR5248, University of Bordeaux, Bordeaux INP, allée Geoffroy St-Hilaire, 33600 Pessac, France
[2] University of Paris, Institut Jacques Monod, CNRS, 75013 Paris, France
[3] ARNA Laboratory, INSERM U1212, CNRS UMR5320, University of Bordeaux, IECB, 2 rue Robert Escarpit, 33600 Pessac, France
[4] Laboratoire des biomolécules (LBM), CNRS UMR7203, Sorbonne University, École normale supérieure, PSL University, 75005 Paris, France
[5] University of Paris, UFR Sciences du Vivant, 75013 Paris, France
* Correspondence: marie-lise.jobin@u-bordeaux.fr (M.-L.J.); i.alves@cbmn.u-bordeaux.fr (I.D.A.);
   Tel.: +33-5-3351-4735 (M.-L.J.); +33-5-4000-6849 (I.D.A.)
† Present address: Interdisciplinary Institute for Neuroscience (IINS), CNRS UMR5297, University of Bordeaux, 33000 Bordeaux, France.

Received: 31 July 2019; Accepted: 4 September 2019; Published: 9 September 2019

Abstract: Cell-penetrating peptides (CPPs) are short peptides that can translocate and transport cargoes into the intracellular milieu by crossing biological membranes. The mode of interaction and internalization of cell-penetrating peptides has long been controversial. While their interaction with anionic membranes is quite well understood, the insertion and behavior of CPPs in zwitterionic membranes, a major lipid component of eukaryotic cell membranes, is poorly studied. Herein, we investigated the membrane insertion of RW16 into zwitterionic membranes, a versatile CPP that also presents antibacterial and antitumor activities. Using complementary approaches, including NMR spectroscopy, fluorescence spectroscopy, circular dichroism, and molecular dynamic simulations, we determined the high-resolution structure of RW16 and measured its membrane insertion and orientation properties into zwitterionic membranes. Altogether, these results contribute to explaining the versatile properties of this peptide toward zwitterionic lipids.

Keywords: cell-penetrating peptide; peptide–lipid interaction; lipid model systems; molecular dynamics; NMR; membrane biophysics

---

1. Introduction

The biological membrane is one of the key structural elements of living cells, and constitutes the first barrier that is encountered by molecules and ions that need to be transported into cells. The ability to deliver drugs to the interior of cells is critical for diagnosis and therapeutic applications, and cell-penetrating peptides (CPPs) can overcome this limitation [1]. CPPs constitute a heterogeneous class of small peptides that can translocate through cell membranes and transport cargoes into cells, in a receptor- and energy-independent process. A major advantage of CPPs is their passivity towards cells, i.e., they do not present cytotoxicity. Their mode of action has been considerably studied since their discovery in the 1990s (for a detailed review, see [2]) and it is currently accepted that several parameters in their primary sequence are essential to confer these properties, including a net positive charge (high arginine or sometimes lysine content) and an optimal balance between charged amino acids and hydrophobic ones, i.e., amphipathcity [3–5]. These properties are often shared with antimicrobial peptides (AMPs), whose main role is to kill bacteria, but it is not clear how these similar peptides can exert extremely different functions. Altogether, these membrane-active peptides (MAPs) exert their biological activity by initially interacting with the plasma membrane, and therefore investigating the peptide-membrane interaction of such molecules is essential to understand their mode of action.

As plasma membranes are mainly composed of lipids, the peptide–lipid interactions are crucial for the initial binding of CPPs prior to internalization [6–8]. Polar residues, and especially arginines (Arg), interact with high affinity through their guanidinium group to negatively-charged lipids and lipid phosphate groups, and thus enhance the binding of Arg-rich peptides to membranes [9,10]. In parallel, hydrophobic residues such as tryptophans (Trp) have been shown to establish hydrophobic contacts with lipid acyl chains and play a role on the insertion of MAPs into the membrane [11,12]. Structural plasticity of these peptides during membrane contact may then bring sufficient peptide charge neutralization (e.g., through electrostatic interactions) to help the peptides translocate. However, it has also been suggested that highly hydrophobic residues might prevent peptide internalization, with the peptide trapped in the membrane due to these strong interactions [7,13]. Although CPPs have been broadly reported to have an enhanced affinity for negatively-charged membranes, peptide interaction and insertion in zwitterionic membranes is not fully described. A large amount of studies is performed on anionic systems (negatively-charged membranes) due to the establishment of important electrostatic interactions between the CPP and the cell membrane that are important for their internalization. While lipids in the outer eukaryotic cell membrane leaflet are mainly zwitterionic, with less than 2% anionic lipids, the cell membrane possesses an anionic character due to the glycosaminoglycans. Therefore, study of CPP interaction with zwitterionic lipids is important and most of the studies, to the best of our knowledge, mainly use zwitterionic systems as a comparison model to anionic ones.

In this study, we investigated one CPP, RW16 (RRWRRWWRRWWRRWRR), which possesses cell internalization capacity, but also shows antimicrobial and antitumor activity [14,15]. The design of RW16 was derived through structure–activity relationship (SAR) studies from penetratin (pAntp), and is composed of 10 Arg and 6 Trp to form an idealized amphipathic peptide. RW16 has been successfully shown to be an efficient CPP while exhibiting no cytotoxicity on fibroblast cells [15,16].

The interaction of RW16 with anionic or zwitterionic membranes has been fully described, but reveals atypical and poorly understood behaviors when interacting with zwitterionic liposomes [14, 17]. Only a few studies to date have investigated this behavior at the molecular level. Lamazière et al. have shown that RW16 induced giant unilamellar vesicles (GUVs) via adhesion and aggregation with anionic lipid membrane compositions [17]. They also found that RW16 induced calcein leakage from liposomes, which suggests a link to membrane perturbation but remains non-lethal to cells at comparable concentrations. In a previous study, we observed that RW16 possesses enhanced membrane interaction and perturbation of membranes containing anionic lipids. This property could, in part, explain its antitumor and antibacterial activity, as cancer cells and bacteria contain anionic lipids in the outer leaflet of their membranes [14]. In contrast, RW16 shows weak perturbation of zwitterionic membranes, although this interaction is associated with fast and strong calcein leakage [14]. The polyvalent property displayed by RW16 is not observed for other CPPs and, therefore, it is important to understand it at a molecular level.

Herein, we focus on the first stages of interaction and membrane insertion of RW16 in contact with zwitterionic membranes (i.e., mimicking "healthy" cell membranes). By applying complementary and multidisciplinary biophysical methods, we investigated the molecular behavior of the peptide in contact with zwitterionic liposomes. Using nuclear magnetic resonance (NMR) spectroscopy, we calculated the structure of RW16 in the presence of zwitterionic micelles and used the NMR structure coordinates to perform molecular dynamics (MD) simulations. This approach allowed us to provide a complete molecular view of the peptide structure and orientation while inserted in the membrane, including insertion depth in zwitterionic membranes. We also provide an explanation for its ambiguous effect previously observed on zwitterionic membranes in comparison to anionic membranes.

2. Results

2.1. Ensemble Insertion Analysis of RW16 vs. Penetratin in Zwitterionic Membranes

RW16 is a cell-penetrating peptide (CPP) derived from the well-characterized penetratin peptide (pAntp), and our initial approach was to compare RW16 to penetratin by studying membrane insertion into liposomes (LUVs, large unilamellar vesicles). To this aim, we employed the neutral hydrophilic quencher acrylamide to quench intrinsic Trp fluorescence of the peptides. Acrylamide is unable to penetrate the hydrophobic membrane core, such that only fluorophores not embedded in the bilayer are quenched. However, it should also be noted that the response is an ensemble average response from all Trp residues in the peptide sequence. Tryptophan quenching experiments of penetratin have been previously performed by other groups using different experimental conditions (e.g., different buffer, P/L ratio) [18,19]. It was shown that Trp residues locate at the interface between the polar headgroup region and the hydrophobic core of the lipids [6,20]. Using a Trp fluorescence quenching approach, we measured insertion properties of RW16 and penetratin into a zwitterionic membrane. Fluorescence spectra of the peptides were recorded in the absence and presence of liposomes of dioleoylphosphatidylcholine (DOPC), and with increasing concentrations of acrylamide. Stern–Volmer plots of acrylamide quenching are shown in Figure 1c. In buffer, the Stern–Volmer constant (K_{SV}) of the two peptides were similar and comparable to the values found in the literature for similar concentrations of peptide and under equivalent experimental conditions [19,21]. In the presence of LUVs composed of DOPC, the Stern–Volmer constant of both peptides decreased significantly, demonstrating a strong insertion of the peptides in the membrane. The Stern–Volmer coefficients (K_{SV}) were normalized to the K_{SV} calculated in buffer (NAF for "normalized accessibility factor") to allow for the comparison between both peptides (Table 1 and Figure 1d) [11,21]. A higher NAF value corresponds to higher Trp exposure to solvent, and inversely, a lower NAF value relates to higher insertion in the bilayer. The NAF of RW16 in DOPC was lower than for penetratin (0.18 compared to 0.53; Table 1), which highlights that RW16 is inserted deeper in the membrane than penetratin. Based on the composition of RW16, the higher number of Trp as compared to penetratin might explain this finding, since additional Trp residues are predicted to have stronger hydrophobic contacts with the hydrophobic core of the membrane.

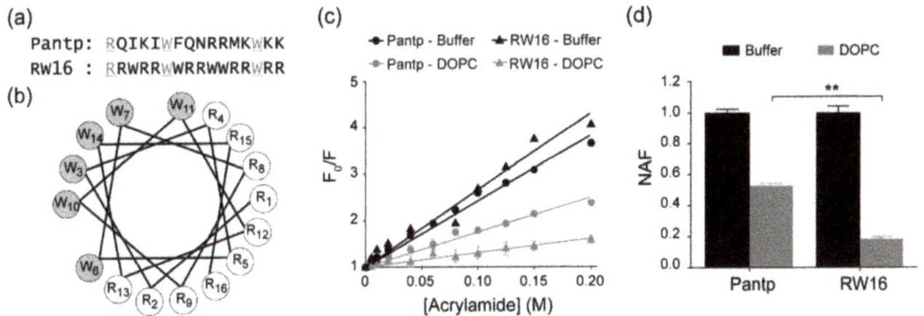

Figure 1. RW16 and penetratin insertion in the membrane. (**a**) Sequence alignment of RW16 and penetratin. (**b**) Edmunson wheel of RW16 along the axis generated from Helixator, http://www.tcdb.org. (**c**) Inhibition rate (F_0/F) of RW16 and penetratin Trp fluorescence in buffer and in the presence of DOPC liposomes (P/L 1:50 mol:mol), with increasing concentrations of acrylamide. (**d**) Normalized accessibility factor (NAF) for penetratin and RW16 in presence of DOPC liposomes. Significance was tested with a Student's *t*-test, where ** $0.001 < p < 0.01$.

Table 1. Stern–Volmer coefficients (K_{SV}) determined by fluorescence quenching by acrylamide and normalized accessibility factors (NAF). The experiment was performed in duplicate.

	K_{SV} (M^{-1})		NAF	
	Buffer	**DOPC**	**Buffer**	**DOPC**
Penetratin	14.2 ± 0.3	7.5 ± 0.2	1	0.53 ± 0.02
RW16	16.6 ± 0.7	3.0 ± 0.2	1	0.18 ± 0.02

2.2. Solution Structure of RW16 in the Presence of Zwitterionic Membranes

To further address details of membrane insertion and to obtain information on RW16 at the atomic level, we next calculated the structure of the peptide in the presence of zwitterionic lipids. We prepared a sample of RW16 with dodecylphosphocholine-d$_{38}$ (DPC-d$_{38}$) micelles and used NMR spectroscopy to obtain distance restraints for structural calculation. DPC was employed instead of DOPC to obtain the high-resolution structure of the peptide, which would not be possible with DOPC liposomes using liquid-state NMR spectroscopy. The use of deuterated lipids in the sample allowed us to remove lipid contributions from the NMR signal and thus only observe peptide resonances. Strikingly, and as illustrated in Figure 2, the crosspeak signals in natural abundance 2D ^{1}H,^{13}C-HSQC (Figure 2a), 2D ^{1}H,^{1}H-TOCSY (Figure 2b), and 2D ^{1}H,^{1}H-NOESY (Figure 2c) were very well resolved and separated, despite the peptide sequence symmetry and only two types of residues (Arg and Trp). All ^{1}H chemical shifts for the 6 Trp can be unambiguously assigned, as can the ^{1}H resonances for the 10 Arg. The chemical shift assignments have been deposited in the Biological Magnetic Resonance Bank (BMRB) under accession number 34400. The 2D ^{1}H,^{1}H-NOESY (Figures 2c and 3a) is indicative of a single population of structures, and we were able to derive 408 distance and 28 backbone dihedral restraints for structure calculations using Aria1.2/CNS1.21 [22] (Table 2). The ensemble of the 10 lowest energy structures of RW16 in the presence of DPC micelles is presented in Figure 3b and has been deposited in the PDB (Protein Data Bank) as entry 6RQS. As predicted, we observe that the Arg and Trp are mainly segregated to each side of the helix, thus creating an amphipathic helix. Arg15 appears to be an exception and seems to be isolated from the others as it is located on the "Trp side" of the helix. This arginine has upfield shifted chemical shift values and displays nuclear Overhauser effect (NOE) crosspeaks to neighboring Trp side chains (Figure 3a). Although not included in the structure calculation, the peptide also contains a N-terminal biotin-aminopentanoic acid tag (Biot-Apa).

A notable feature of the micelle-bound RW16 peptide is the high degree of helicity. The helical αnature of bound RW16 is supported both by NOE crosspeaks (Figures 2 and 3a), as well as the negative secondary chemical shift values of the backbone ^{1}Hα nuclei (Figure 4a). To confirm this observation, we measured the secondary structure content of RW16 by circular dichroism (CD) in buffer and in the presence of DPC micelles (Figure 4b). By deconvoluting the measured CD signal, we observe a predominant α-helical structure for the peptide in buffer, as well as in the presence of zwitterionic lipids (Table 3). We also measure an increase in the α-helix content with zwitterionic membranes compared to buffer. Furthermore, the helical content is slightly higher (76% vs. 60%) with micelles than with LUVs. This trend was previously found in Jobin et al. [14] and demonstrates a stronger structuring of the peptide in presence of micelles (Table 3).

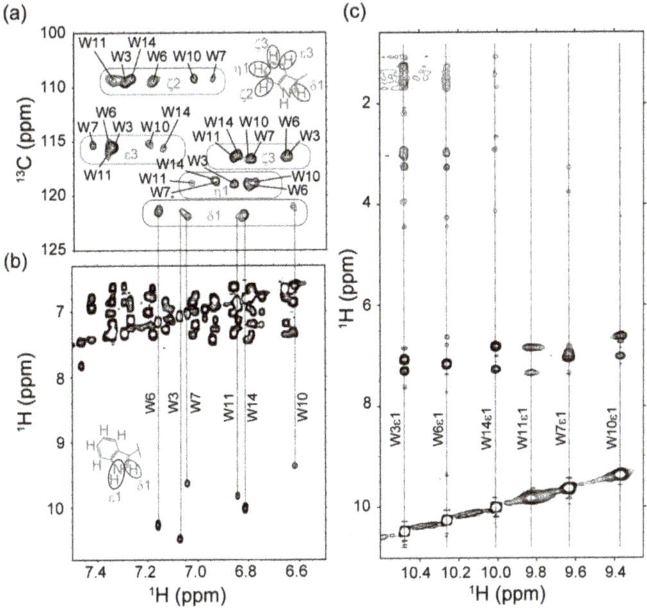

Figure 2. NMR spectroscopy of RW16 in DPC-d$_{38}$ micelles. (**a–c**) Chemical shift assignments of RW16 using a combination of 2D ^{1}H,^{13}C-HSQC, 2D ^{1}H,^{1}H-TOCSY, and 2D ^{1}H,^{1}H-NOESY at 310 K. (**a**) Selected region of 2D ^{1}H,^{13}C-HSQC, illustrating assigned δ1, ζ2, η2, ε3, and ζ3 ^{1}H-^{13}C resonances for the 6 Trp residues. (**b**) Selected region of the 2D ^{1}H,^{1}H-TOCSY, highlighting assignment of the Trp ε1 ^{1}H resonances from the δ1 crosspeaks in the 2D ^{1}H,^{13}C-HSQC. (**c**) Selected regions from the 2D ^{1}H,^{1}H-NOESY spectrum used to obtain distances for structure calculation, with representative NOE strips indicated for Trp ε1 ^{1}H nuclei.

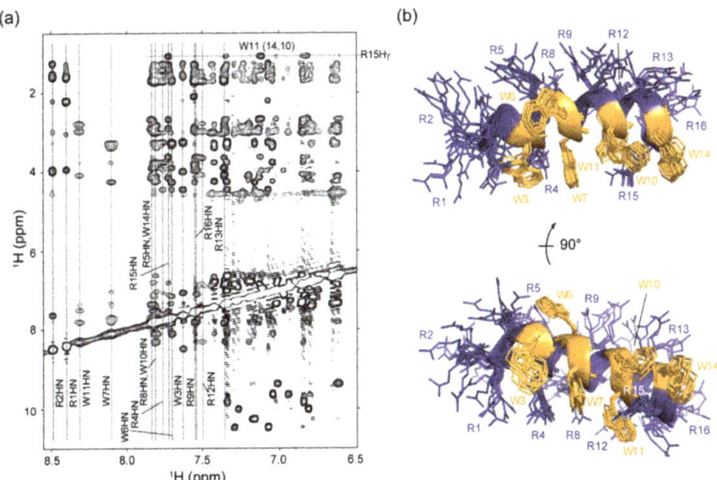

Figure 3. Solution structure of micelle-bound RW16.

(**a**) Selected regions from the 2D ^{1}H,^{1}H-NOESY spectrum used to obtain distances for structure calculation, with representative NOE strips indicated for all backbone amide ^{1}H^N nuclei. The upfield shifted side chain ^{1}Hγ resonance of Arg15 is also indicated, with NOE crosspeaks to Trp11 and Trp14. (**b**) Ensemble of 10 structures calculated for RW16 bound to DPC-d$_{38}$ micelles. The 6 Trp residues (orange) and 10 Arg residues (blue) are labeled. Note that the N-terminal biotin-aminopentanoic acid tag, although present in the sample, was not included in the structural models. The ensemble of structures has been deposited in the Protein Data Bank under accession number 6RQS.

Table 2. NMR and refinement statistics for RW16. PDB entry: 6RQS.

Distance and Dihedral Restraints	
Total Distance Restraints	408
Intraresidue	171
Sequential ($\lvert i - j \rvert = 1$)	44
Short Range ($1 < \lvert i - j \rvert < 5$)	20
Long Range ($\lvert i - j \rvert > 4$)	9
Ambiguous	162
Dihedral Restraints	28
Structure Statistics	
Violations (mean and SD)	
Distance Constraints (Å)	0.020 ± 0.003
Dihedral Angle Constraints (°)	1.1 ± 0.3
Deviations from Idealized Geometry	
Bond Lengths (Å)	0.001 ± 0.000
Bond Angles (°)	0.306 ± 0.008
Impropers (°)	0.21 ± 0.01
Ramachandran plot (%) [a]	
Most favored	86.7
Additionally favored	13.3
Generally allowed	0.0
Disallowed	0.0
Average pairwise rmsd (Å)	
Protein backbone all	0.5 ± 0.1
Protein heavy all	1.8 ± 0.2

[a] Determined by using PROCHECK-NMR [23].

Molecular details of membrane interaction with RW16 could not easily be obtained from our NMR spectroscopy data, due to the necessary use of deuterated lipids such that NOEs between peptide and lipids were not visible. Therefore, we decided to use molecular dynamics (MD) simulations in a zwitterionic membrane, using the solution structure of membrane-bound RW16 as a starting point in the simulations. This method allowed us to get insight into the burying of RW16 in DOPC membranes on a microsecond time scale (Supplementary Movie 1).

Our first analysis of the MD simulations confirmed a persistent helical structure of RW16, with the helix fraction calculated over the three trajectories shown in Figure 4c and a snapshot is shown in Figure 4d (Supplementary Movie 1). The segments 2–9 display a very stable helix fraction (80 to 100%). The segments 10–14 also remain helical but are more labile with some excursions to a turn (T state), explaining the larger error bars. Interestingly, we also observed these segments fluctuating between some turn state and π-helix in the first trajectory (Figure S1). The occurrence of a π-helix corresponds to a "compression" of the helix on these somehow short segments of the peptide. Overall, our MD trajectories show that the peptide remains helical with some possible fluctuations in the backbone hydrogen bonds on the C-terminal part.

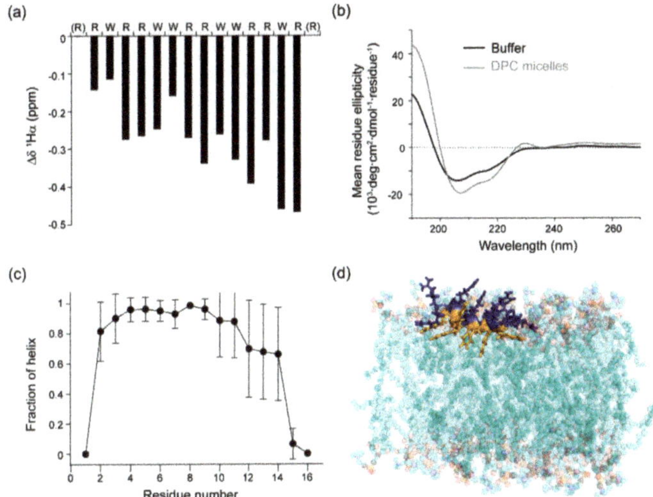

Figure 4. RW16 peptide structure in contact with zwitterionic membranes. (**a**) $^{1}H\alpha$ chemical shifts compared to random coil $^{1}H\alpha$ predictions obtained from NMR data. (**b**) CD spectra of RW16 in phosphate buffer (black line) and in the presence of DPC micelles (gray line). (**c**) Fraction of helix calculated from the molecular dynamics (MD) simulations. The three states H (α-helix), G (3_{10} helix), and I (π-helix) of the DSSP program were considered as part of the helix fraction (see Materials and Methods). After discarding the first 10 ns, each trajectory was cut into two blocks. Each value ± error was calculated as the mean and standard deviation over the six blocks respectively. (**d**) Snapshot of RW16 inserted in DOPC bilayer at t = 633 ns of MD trajectory 1. The C-terminus is on the left and N-terminus on the right. Trp are represented in orange, Arg in blue, the backbone is shown as an orange/blue ribbon and the lipids are drawn as spheres where carbon atoms are in cyan, oxygen in red, and nitrogen in blue.

Table 3. Secondary structure percentages of RW16 calculated from CD spectroscopy in phosphate buffer alone or in the presence of DPC micelles or DOPC LUVs at a P/L ratio 1/50.

	Random Coil	α-Helix	β-Sheet
Buffer	54	46	0
DPC micelles	24	76	0
DOPC LUVs [1]	40	60	0

[1] Data from [14].

Consistent with our NMR data, we observed very small fluctuations, confirming a strong anchoring of RW16 to the membrane thanks to its Trp side chains. Figure 4d shows a snapshot of the peptide being inserted in the membrane of DOPC after 633 ns of one simulation (Supplementary Movie 1). Visual observation of the MD trajectories in the DOPC membrane, moreover, revealed that the azimuthal angle of RW16 (how the peptide rotates about its helix axis) is very constant with the face of Trp residues oriented towards the membrane center and the Arg residues towards the interface membrane/water. These findings, however, also differ from the NMR results, in which a more flexible structure is observed on the N-terminal side (Figure 4a). This variation can arise from different parameters and will be further detailed in the Discussion.

2.3. Membrane Insertion Depth of RW16

To obtain more precise information regarding the membrane partitioning of RW16, and more specifically side chain insertion depth in the lipid bilayer, quenching of Trp fluorescence by brominated lipids was performed. Three different lipids, each containing two bromines covalently attached to the lipid acyl chains at three different positions, were incorporated in DOPC LUVs. These lipids act as an internal probe in the liposome membranes with a quenching radius of 8–9 Å for the brominated probes [24]. It was previously shown that the presence of two bromines on the lipid acyl chain does not modify the physical properties of the lipids (like the phase transition) and preserve the lipid packing properties as in DOPC lipids [25]. Therefore, these are unlikely to influence the peptide-membrane interaction. Trp fluorescence of RW16 was measured in the presence of liposomes composed of DOPC alone or DOPC with a small percentage (30% mol:mol) of the brominated lipids (Figure 5a). The quenching observed in the presence of the different liposomes was normalized to the fluorescence measured in pure DOPC LUVs. The quenching percentage provides the average distribution of Trp from the bilayer center using equations of the distribution analysis (DA) and of the parallax method (PM) (for more details, see Materials and Methods) [21,26–30] (Figure 5b). Our calculation indicates that Trp residues are, on average, inserted at 12–13 Å from the bilayer center (Table 4). According to Wiener et al. [31], DOPC bilayers have a hydrocarbon core of approximately 30 Å and a total bilayer thickness of around 50 Å. We therefore determined the peptide to be located at around 7 Å from the polar headgroup of lipids, at the interface between the hydrophobic core and the polar region. The broad area calculated for Trp insertion depth might arise from different bilayer environments due to heterogeneity of the different Trp residue locations in the bilayer, also indicating that the helix might be tilted. In comparison, penetratin was observed to be located at the glycerol and phosphate levels upon zwitterionic bilayer insertion, but to a lesser extent than observed for RW16 [32–34]. These results, together with the acrylamide quenching experiments, confirm that RW16 inserts stably into zwitterionic bilayers. In addition, these results could explain, at least partially, our previous observation of calcein leakage measured in zwitterionic lipids and the membrane perturbation of zwitterionic vesicles [14]. In the case of anionic membranes, the interaction between Arg and lipid headgroups creates a charge compensation which delays calcein leakage, whereas the net charge of 0 at the membrane surface of zwitterionic liposomes creates an imbalance in the peptide-lipid interaction and higher fluctuations of the membrane [6].

Table 4. Average insertion depths and fitting parameters of RW16 in DOPC LUVs (P/L 1:50 mol:mol) determined by the distribution analysis and the parallax method.

	Distribution Analysis (DA)			Parallax Method (PM)	
	h_m (Å)	δ (Å)	S	h_m (Å)	Rc (Å)
RW16	12.4	8.9	1.7	12.5	13.8

Figure 5c shows the average location of RW16 in DOPC bilayer, calculated from MD simulations at 17 Å from the bilayer center, which is just below the polar headgroups of the lipids. A detailed analysis of the side chain location allowed us to precisely calculate an average insertion depth of Trp residues to be 13 Å from the bilayer center (Figure 5d and Figure S2). This is in agreement with fluorescence spectroscopy results and confirms that Trp residues are located adjacent to the phospholipid glycerols at the interface between the polar and the hydrophobic region of lipids. This placement is not surprising, given the preference of Trp residues for interfacial partitioning [12,35]. In parallel, Arg residues displayed a broad location between 17 and 22 Å from the bilayer center (Figure 5d and Figure S2), suggesting the presence of electrostatic interactions formed by bidentate hydrogen bonds of Arg and negative charges of the phosphate of the DOPC.

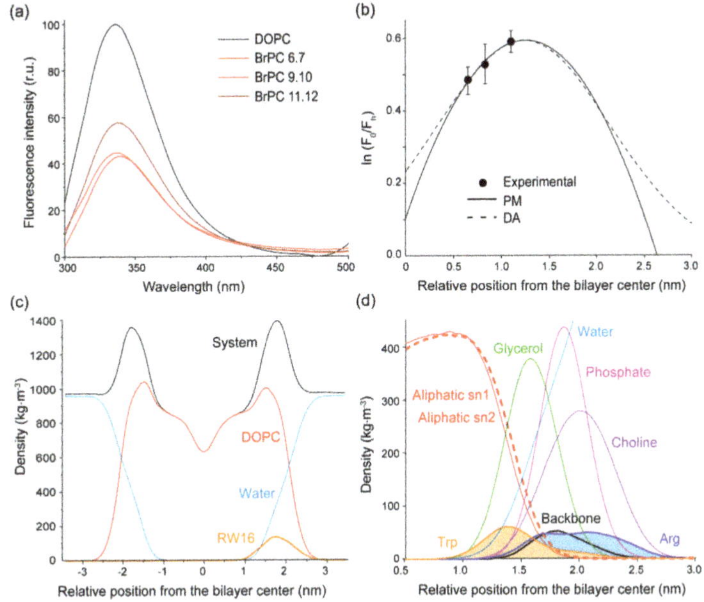

Figure 5. Peptide insertion into the membrane. (**a**) Representative Trp fluorescence spectra of RW16 in the presence of DOPC or DOPC/BrPC liposomes. (**b**) Curve fitting calculated by distribution analysis (DA) or the parallax method (PM) for RW16 in zwitterionic vesicles. The data were averaged over four independent experiments and each value ± error represents the mean and standard deviation. (**c**) Density profiles along the perpendicular axis to the bilayer plane calculated by MD simulations corresponding to water molecules, DOPC molecules, peptides, and the overall system. (**d**) Close up from (c) on the Trp and Arg region of the peptide, and on the lipid subgroups.

The peptide insertion is further illustrated by following the relative exposure of the center of mass (COM) of RW16 side chains in comparison to the lipid atoms (Figure S2). We observed that Trp3, Trp6, Trp7, and Trp10 are buried in the hydrophobic part of the bilayer and located below the central glycerol atom (around 13–14 Å from the bilayer center). In contrast, Trp11 lies at the phosphate level and Trp14 lies at the glycerol level, being therefore both slightly exposed to the solvent. The naf value observed with acrylamide quenching experiments for RW16 might be due to these two amino acids still being exposed to the solvent. As expected, Arg residues are mostly located between the central glycerol atom and the nitrogen of the choline. More specifically, Arg1, Arg2, Arg4, Arg13, and Arg15 lie between the phosphate and the glycerol and are therefore less solvent-exposed. Arg5, Arg8, Arg9, Arg12, and Arg16 are located above the choline nitrogen and are therefore highly solvent-exposed. The total number of hydrogen bonds between the Arg side chains and lipids were further calculated for the three trajectories from MD simulations (Figure S3). We observed an increase in the number of hydrogen bonds during the simulations, starting from 14 (±1, SD) hydrogen bonds reaching a plateau after 500 ns where 19 (±2, SD) hydrogen bonds were calculated (Figure S3a). This correlates well with the presence of 10 bidentate bonds occurring between all 10 Arg side chains and the lipid phosphate groups. Moreover, we calculated the number of Arg–water hydrogen bonds (Figure S3b), and found that they instead slightly decrease from 30 (±2, SD) hydrogen bonds and tend to a plateau at 27 (±2, SD) after a few hundreds of ns. This could indicate the presence of 10 Arg–water hydrogen bonds at the beginning of the simulation and suggests that one Arg might form different types of bonds during the simulations. This is in agreement with our previous results that showed Arg15 underwent important changes in its partitioning during one simulation and the observed shifted chemical shift values and NOE crosspeaks to neighboring Trp side chains observed in NMR.

Overall, the results on solvent accessibility by MD are similar to an insertion of the Trp at around 13 Å from the bilayer center, revealed by Trp fluorescence quenching by acrylamide. This insertion trend is in line with the fact that the peptide stabilizes into the bilayer by creating hydrophobic interactions with the fatty acid chains through Trp residues and electrostatic interactions with the lipid phosphates through Arg residues. A similar behavior has been reported for other cell-penetrating peptides [18,21].

2.4. Tilting of RW16 in the Membrane

We have demonstrated that RW16 is not fully inserted in the bilayer, with some exposure to the aqueous buffer through Arg residues. Previously, Walrant et al. observed that some CPPs are tilted in the bilayer [9], and thus we decided to investigate peptide tilting of RW16 relative to the normal to the bilayer plane. Using the MD simulation data, we were able to precisely calculate a tilt angle of the peptide, defined as the angle between the helix and the normal to the bilayer plane (Figure 6a). Because of the perfectly amphipathic nature of RW16 (Figure 1b), an interfacial partitioning with a horizontal orientation (i.e., tilt of 90°) within the membrane/water interface is energetically more favorable than any low tilt angle [36]. However, we noted that the peptide is not perfectly horizontal in the MD trajectories and presents a slight tilt, with values ranging from 75° to 85°, with an average tilt value around 80°. These results indicate that the N-terminus of the peptide is inserted deeper in the membrane. This trend is consistent with the insertion depth calculations performed for each residue (Figure 5d and Figure S2), with the N-terminal side chains systematically more buried inside the bilayer compared to the C-terminal side chains. Specifically, the Cα of Arg1 is located between the phosphates and the glycerol atoms, whereas the Cα of Arg16 is located above the cholines.

Similar to our results, it was shown that Trp residues of penetratin are facing towards the hydrophobic core of the bilayer [34]. In addition, the results on penetratin show a similar tilting behavior of the peptide. For penetratin, the Trp residues are located at the level of phospholipid glycerols (in a zwitterionic bilayer) with the N-terminus of the amphipathic helix inserted slightly deeper into the bilayer than the C-terminus (tilt angle of 80 to 90°) [20,33]. The small difference in tilt angles between the two peptides may arise from the high density of Trp in RW16, which could induce stronger anchoring of Trp residues to the lipid acyl chains in the hydrophobic core of the membrane.

The orientation of RW16 was also investigated by solution NMR spectroscopy with the sample prepared in the zwitterionic membrane-mimicking DPC micelles. The peptide position in micelles was examined by measuring the accessibility of atoms to the solvent (buffer) as a function of increasing concentrations of the paramagnetic agent Gd(DPTA-BMA). This paramagnetic probe is water-soluble and inert toward peptide-micelle complexes, and leads to faster relaxation towards nuclei in a distance-dependent manner [37]. Therefore, the highest solvent paramagnetic relaxation enhancement (sPRE) is expected for solvent-exposed residues outside of the bilayer, followed by residues at the water/bilayer interface. The sPRE values were measured for several atoms (Figure S4) and the results mapped to the RW16 peptide (Figure 6b). The pattern of accessible atoms reflects a situation in which the N-terminus is more exposed to the buffer as compared to the C-terminus, and therefore indicates a tilt angle greater than 90°. This behavior is opposite to the main tilt observed in the MD simulations. It is likely that this difference arises from the nature of the N-termini: In the NMR spectroscopy measurements, a biotin is linked by the Apa spacer to the N-terminus, whereas a simple acetyl group is used in the MD simulations. The larger biotin is hydrophilic and is not expected to insert deeply into the hydrophobic core of the membrane. In keeping with this hypothesis, the biotin nuclei are highly sensitive to the addition of Gd (DPTA-BMA) (Figure S4) and are therefore exposed to the buffer. The altered tilt preference in the NMR sample may also help explain the fact that RW16 displays a dynamic N-terminus, whereas the N-terminus was structured in the MD simulations and the C-terminus was more labile (with the helix undergoing more fluctuations). Nevertheless, we find by both methods that one terminal side is less partitioned into the membrane, and this side is more flexible than the rest of the helix.

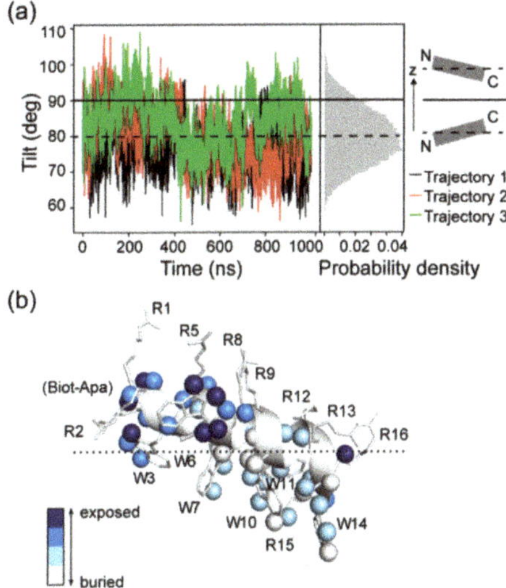

Figure 6. Peptide tilting relative to the normal to the zwitterionic membrane plane calculated by NMR and MD. (**a**) Tilt of the peptide calculated from the difference in z position between the C-alpha of Arg2 and Arg13 calculated over the three trajectories. The histogram (gray) shows the average distribution of z over the three trajectories. The scheme on the right panel shows that a tilt <90° describes a deeper insertion of the N-terminus, while a tilt >90° describes a deeper insertion of the C-terminus. (**b**) Solvent accessibility for RW16 in DPC micelles, as measured by NMR spectroscopy using the paramagnetic probe Gd(DTPA-BMA). The resulting solvent paramagnetic relaxation enhancement values (sPRE; Figure S4) have been measured for several atoms in RW16 and shown as spheres. Atoms that are strongly affected (colored blue) by the added Gd(DTPA-BMA) are more solvent-exposed as compared to atoms that are less affected (colored white). The tilt of the peptide shown represents an estimate based on the observed data.

2.5. Side Chain Contacts of RW16 in the Bilayer

As previously shown by NMR spectroscopy, MD simulations, and fluorescence spectroscopy, the Trp residues in RW16 are strongly anchored in the membrane, which leads to a high stability of the peptide in the bilayer. In contrast, most of the Arg residues extend out of the membrane, with the sole exception of Arg15. As already mentioned, we observed chemical shift values for the Arg15 side chain that are upfield shifted in keeping with a more hydrophobic environment, and we observed clear NOE crosspeaks to side chain atoms of Trp11, as well as Trp10 and Trp14 (Figure 3a). Calculations of the distance between residues in MD simulations also revealed side chain–side chain contacts, mainly between Arg15 and Trp11, but also with Trp10 and Trp14 (Figure 7a). This is consistent with what is observed in NMR and suggests that Arg15 is in a dynamic cavity surrounded by three Trp residues and creates π-cation contacts with them (Figure 7b). The contact to Arg15 may also explain the line-broadening of several nuclei from Trp11, such as the exchangeable side chain hydrogen $^1H\varepsilon$ (Figure 2c). Similar Arg–Trp π-cation contacts were also observed for penetratin and were suggested to help stabilize the peptide inside the bilayer by partially masking its positive charge [33].

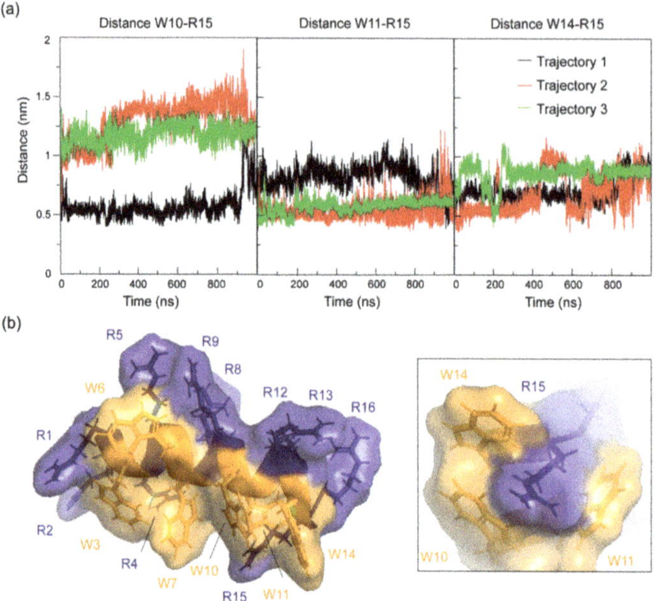

Figure 7. Side chain contacts between Arg15 and residues Trp10, Trp11, and Trp14. (**a**) COM-COM of side chain distances between Trp10 and Arg15 (left, W10–R15), Trp11 and Arg15 (middle, W11–R15), and Trp14 and Arg15 (right, W14–R15) for the three trajectories. (**b**) Surface density of RW16 amino acids showing a special arrangement of Arg15 and Arg–Trp π-cation interactions (left). Close-up on the pocket formed by Trp10, Trp11, and Trp14 around Arg15 (right). Structure generated with Pymol (PDB ID: 6RQS) [38].

3. Discussion

By using a multidisciplinary approach, we have determined the structure and membrane insertion of RW16, an amphipathic and cationic CPP, in the presence of zwitterionic bilayers. Previous data on RW16 highlighted its versatile properties in terms of biological activity and membrane interaction [14–17,39]. While the enhanced interaction and perturbation of anionic membranes by CPPs have been extensively studied, herein we provide a detailed molecular view of the membrane insertion of RW16 into zwitterionic membranes summarized in Figure 8.

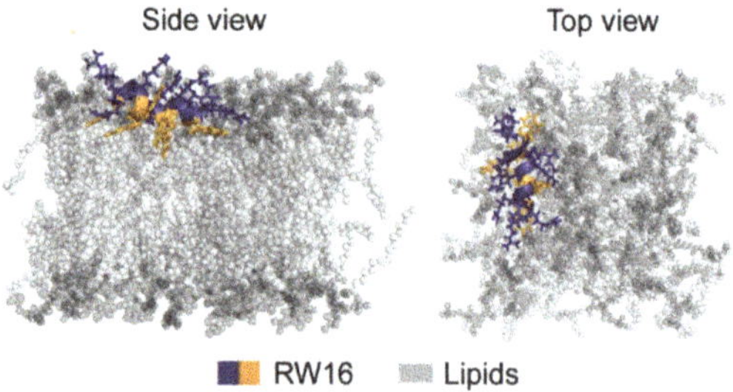

Figure 8. Summary cartoon representing RW16 embedded in a zwitterionic membrane. Shown is the calculated secondary structure and orientation of the peptide with the Trp (orange) and Arg (blue) side chains in a zwitterionic membrane (gray) with the polar headgroups (dark gray) and the aliphatic chains (light gray).

We show that RW16 conformation is mostly α-helical in buffer, and this helicity is slightly enhanced in the presence of zwitterionic liposomes. The use of different membrane models (micelles and LUVs) did not generate significant differences in the secondary structure of the peptide. This is similar to previous observations made on other CPPs [40].

Using fluorescence spectroscopy, NMR, and MD methods, we obtained precise side chain insertion values, i.e., at the interface between the polar and the hydrophobic region of lipids (Figure 8). These values are comparable to calculated insertion depths of penetratin, the parent CPP of RW16 [32–34,41]. The position of RW16 in the membrane is not surprising, given its primary sequence and amphipathicity [42]. As already suggested for penetratin, the Arg residues in RW16 are mostly located at the lipid phosphate groups and form hydrogen bonds and salt bridges to act as an anchor in the membrane for the peptide [43]. Moreover, it was shown with MD simulations that polar side chains can establish long-term contacts with lipids by forming salt bridges and hydrogen bonds, and create local membrane perturbations [44]. Although this effect appears stronger with anionic membranes, where clustering of negatively-charged lipids occurred, it was also observed with zwitterionic membranes.

In our study, we noticed that some Arg residues are inserted deeper into the bilayer and locate at the glycerol region of lipids. We calculated the hydrogen bond number and confirmed that Arg residues establish long-term and stable hydrogen bonds with lipids and water molecules. Although membrane insertion of arginine is energetically unfavorable and associated with a high free energy cost, it has also been described that Arg residues can pull down water molecules in the membrane to stabilize its insertion. This process, known as water defect, was observed for the TAT peptide in zwitterionic membranes by neutron scattering experiments, and was shown to produce local membrane perturbations [45]. Different studies by all-atom MD simulations have indicated that this induces substantial membrane deformations due to the insertion of polar side chains in the hydrophobic core of the membrane [46–49]. This was further suggested as a mechanism for translocation of Arg-rich peptides [50] and might partially explain the membrane perturbation and leakage observed of RW16 in the presence of zwitterionic lipids. Despite strong Arg–water hydrogen bonds observed in our calculations, water insertion was not observed in our simulations, which at 1 μs was unable to be explored for the longer timescales (seconds/microseconds) required for this process. These observations would instead require enhanced sampling techniques, such as umbrella sampling. However, we did observe bidentate hydrogen bonding involving Arg residues with the lipid phosphate groups, and such interactions were reported to be important in cellular uptake mechanisms by creating a small curvature of the membrane, which could induce invagination phenomena [51,52]. These curvatures could contribute to tubulation and internal vesicle formation as induced by RW16 on giant unilamellar vesicles (GUVs) observed in Lamazière et al. [17].

Aside from this process, we have shown that Trp residues create hydrophobic contacts with the lipid acyl chains and therefore stabilize the peptide helix in the membrane hydrophobic core (Figure 8). Strong hydrophobic contacts with the lipid acyl chains were similarly observed using CD and NMR by Czajlik et al. [53].

Our results show that membrane-bound RW16 forms a stable α-helix with limited dynamics restricted to one terminus. The orientation of the helix appears to differ between NMR spectroscopy and MD simulations, and this difference might be explained by an altered helix polarity driven by the nature and size of the N-terminal cap. In NMR spectroscopy, the N-terminal biotin may prevent this end of the helix inserting into the membrane, while in MD simulations the acetyl group does not strongly influence the insertion of the peptide. Moreover, the membrane model used in both methods differs, and it is possible that the membrane curvature might affect the orientation of membrane-active peptides relative to the bilayer [20]. On one side, NMR measurements were performed in DPC micelles, which allows for fast molecular rotation as required for liquid-state NMR spectroscopy. On the other side, MD simulations used DOPC lipids which assemble in a fully flat membrane. The size and curvature of the membranes are therefore different and likely have an impact on how the peptide inserts in the membrane. Our data nevertheless converges on a tilted configuration of the peptide in zwitterionic bilayers (Figure 8).

Moreover, data obtained by NMR spectroscopy and MD simulations both reveal side chain contacts between Arg and Trp residues and, more specifically, π-cation interactions between Arg15 and Trp10, Trp11, and Trp14. A preference was observed between Arg15 and Trp11, which corresponds to the (i, i + 4) positions of the helix, i.e., to one turn of the α-helix. Such side chain contacts, in both buried and solvent-exposed positions, were shown to highly contribute to the conformational stability and the function of biomolecules [54,55]. Herein, a hydrophobic pocket surrounding Arg15 was formed by three tryptophans that help mask the positive charges of the Arg residue. Such pairing of aromatic and polar residues was indicated to decrease the energetic barrier for the motion of cationic side chains through a lower dielectric environment like the bilayer [44]. Similar π-cation interactions were also observed for penetratin between Arg and multiple Trp residues [32,33]. In our case, this process might help RW16 to further stabilize inside the bilayer, and hence create small fluctuations of the lipid membrane. These fluctuations could explain the strong and quick membrane leakage observed previously with zwitterionic liposomes in comparison to anionic liposomes [14].

Overall, the data obtained in this study clarify the membrane interaction and insertion properties of RW16 and connect these findings to observed perturbation of zwitterionic lipids. The stable conformational and insertion behavior observed for RW16 influences its activity toward biological membranes (i.e., cell internalization, antibacterial and antitumoral properties), and this was reported as an important property of CPPs like penetratin. Regardless of the peptide orientation in membranes, we obtained converging results (degree of insertion, tilting, side chain contacts) between the different methods employed in this study and comparable to similar studies on other CPPs [34]. Moreover, by using complementary approaches, we have shown that the choice of cap when protecting the terminal side of peptides may have functional importance and could impact peptide physico-chemical properties and membrane interaction.

4. Materials and Methods

4.1. Materials

All lipids were purchased from Avanti Polar Lipids (Alabaster, AL, USA). Gd(DTPA-BMA) was purified from the commercially available MRI contrast reagent Omniscan (GE Healthcare SAS, Vélizy-Villacoublay, France) by high performance liquid chromatography (HPLC). Omniscan was diluted with water to 250 mM and purified with an Alliance 2695 (Waters, Milford, MA, USA). The isocratic separation was performed on a semi-preparative C18 reversed phase (RP) column (Symmetry300RP-18, 300 mm × 10 mm, particle size 5 μm, Waters, Germany). The mobile phase was 100% water at a flow rate of 3 mL/min. Detection was performed with a variable wavelength detector set at 200 nm. Injection volume was 50 μL. Gd (DTPA-BMA) was eluted as a single peak (tR 9 min). The fractions were combined and lyophilized to obtain a white amorphous powder.

4.2. Liposome Preparation

All liposomes were prepared by initially dissolving the appropriate quantity of phospholipids in chloroform and methanol to ensure the complete mixing of the components and to obtain the desired concentration. A lipid film was then formed by removing the solvent using a stream of N_2 (g) followed by 3 h vacuum. To form MLVs, the dried lipids were dispersed in Tris buffer 10 mM, 150 mM NaCl, 2 mM EDTA, and thoroughly vortexed. To form LUVs, the MLVs dispersion was run through five freeze/thaw cycles and passed 11 times through a mini-extruder equipped with a polycarbonate membrane with a pore diameter of 0.1 μm (Avanti Polar Lipids, Alabaster, AL, USA).

4.3. Molecular Dynamics Simulations

All-atom molecular dynamics (MD) of RW16 within a DOPC membrane were performed in this work. In order to generate the starting conformation for this system, we used the following strategy. First, an all-atom pure DOPC bilayer without peptide was pre-equilibrated for 200 ns. To get an idea of where the peptide partitions in the bilayer, we performed some self-assembly simulations with the MARTINI coarse-grained (CG) force field [56,57]. We then picked the NMR structure with the lowest energy and placed it in the (all-atom) bilayer using a typical CG snapshot as a template. This CG snapshot allowed us to place the NMR structure of RW16 at a probable vertical position, roughly with the center of mass of the peptide between glycerols and phosphates functional groups, and also with a relevant orientation so that Trp residues faced the hydrophobic core of the membrane and Arg residues were oriented towards water. Lipids and water molecules overlapping with RW16 were removed. The number of lipids in the other leaflet was adjusted to get the same number of lipids in both leaflets. Next, lipids were repacked against the peptide using an NPT simulation (at 300 K and 1 atm) of 10 ns with position restraints (PR) on the peptide. At this point, this was our starting conformation.

The system consisted of one RW16 peptide, 100 DOPC lipids (50 per leaflet), 3400 water molecules, and a quantity of Na^+/Cl^- ions so that the system was neutral and the final ion concentration was 150 mM (10 Na^+ and 20 Cl^-). The termini of the peptide were capped with an acetyl at the N-terminus and an amide at the C-terminus. The total number of atoms was 16,233 and the volume of the box was approximately $6.0 \times 6.0 \times 6.7$ nm^3. Using the starting conformation described above, we performed three different runs using different initial velocities. For each run, an equilibration of 100 ps was performed with PR on the peptide, followed by a 1000 ns production (for which the PR were fully released).

All MD simulations were performed using GROMACS 5 [58]. Except for the CG self-assembly simulations described above, the OPLS-AA force field [59] was used for the protein in combination [60] with the Berger lipids for DOPC [61] and the TIP3P model [62] for water. Because OPLS-AA and Berger lipids have different 1–4 combination rules, the half-ε double-pairlist strategy was used to mix both force fields [60,63]. Electrostatic interactions were calculated with the particle-mesh-Ewald (PME) method [64], with a real-space cutoff of 1 nm. Bond lengths were constrained using the LINCS algorithm [65]. The integration time step was set to 2 fs. Water molecules were kept rigid with the SETTLE algorithm [66]. The system was coupled to a Bussi thermostat [67] and to a semi-isotropic Parrinello–Rahman barostat [68] at a temperature of 300 K and a pressure of 1 atm.

All analyses were performed using GROMACS tools. For each analysis on which we computed averages, the first 10 ns of the production runs were systematically discarded. RW16 secondary structures were determined with the DSSP program [69,70], implemented in the GROMACS tool do_dssp. Molecular graphics were generated with Pymol [38].

4.4. Fluorescence Spectroscopy Experiments

4.4.1. Brominated Lipid Quenching Experiments

Depth-dependent fluorescence quenching of tryptophan of RW16 was performed in LUVs composed of DOPC and either (6,7)-, (9,10)-, or (11,12)-BrPC (70:30 mol:mol). Fluorescence intensities in the absence of quencher (F_0) were measured in DOPC vesicles. Spectra were recorded between 300 and 500 nm with an increment of 1 nm, an integration time of 0.1 s, and using an excitation wavelength of 280 nm. The P/L molar ratio was 1/50 and the final peptide concentration was 0.5 µM. Data were corrected for vesicle background. Depth-dependent fluorescence quenching profiles (DFQPs) were fitted to the experimental points in Matlab using the distribution analysis (DA) method [26,27] and the parallax method (PM) [21,28–30] with the following equations:

$$DA : \ln \frac{F_0}{F_{(h)}} = \frac{S}{\sigma \sqrt{2\pi}} e^{\left[-\frac{|h-h_m|^2}{2\sigma^2} \right]} \tag{1}$$

$$\text{PM}: \ln \frac{F_0}{F_{(h)}} = \pi C\left[R_c^2 - [h - h_m]^2\right] \tag{2}$$

where F_0 is the Trp fluorescence intensity in absence of quencher, $F_{(h)}$ is the Trp fluorescence intensity in the presence of quencher at the distance h(Å) from the bilayer center, and h_m corresponds to the average insertion depth of the tryptophan residues. In DA, the DFQP data are fitted with a Gaussian function where σ represents the dispersion, which is related to the in-depth distribution of the tryptophan chromophores, and S is the area under the quenching profile, which is related to the quenching ability of the tryptophan part. The parallax method fits data to a truncated parabola, and R_c is the radius of quenching. Average bromine distances from the bilayer center (h) for (6,7)-BrPC, (9,10)-BrPC and (11,12)-BrPC were taken to be 11.0, 8.3, and 6.5 Å, respectively [21].

4.4.2. Acrylamide Quenching

Following peptide–lipid interactions, the accessibility of the peptides to aqueous quenchers of Trp fluorescence was modified. We used acrylamide as a Trp fluorescence quencher from a stock solution of 4 M. Acrylamide quenching experiments were performed with a 0.5 µM peptide solution in the absence or presence of LUVs and with a titration of acrylamide. The peptide/liposomes mixtures (1:50 mol:mol) were incubated for 15 min at room temperature prior to the measurements. The excitation wavelength was set to 295 nm instead of 280 nm to reduce the absorbance by acrylamide ($\varepsilon^{280} = 4.3\,\text{M}^{-1}\cdot\text{cm}^{-1}$, $\varepsilon^{295} = 0.24\,\text{M}^{-1}\cdot\text{cm}^{-1}$). Fluorescence intensities were then measured after the addition of acrylamide at room temperature. Data were analyzed according to the Stern–Volmer equation for collisional quenching [71]:

$$\frac{F_0}{F} = 1 + K_{SV}\cdot[Q] \tag{3}$$

where F_0 and F correspond to the maximum fluorescence intensities in the absence and presence of quencher respectively, [Q] is the molar concentration of quencher, and Ksv is the Stern–Volmer quenching constant.

4.5. Circular Dichroism

CD spectra were recorded on a Jasco J-815 CD spectrophotometer with a 1 mm cuvette path length. Far-UV spectra were recorded from 180 to 270 nm with a 0.5 nm step resolution and a 2 nm bandwidth at 37 °C. The scan speed was 50 nm/min (0.5 s response time) and the spectra were averaged over 8 scans. CD spectra were collected for all the peptides in phosphate buffer at pH 5.5 with and without micelles at peptide/lipid (P/L) ratio of 1:50 (mol:mol). For each sample, the background (buffer) was automatically subtracted from the signal. Spectra were smoothed using a Savitzky–Golay smoothing filter and were deconvoluted to estimate the secondary structure content using the deconvolution software CDFriend developed in our laboratory (S. Buchoux, not published) [72].

4.6. NMR Spectroscopy Experiments

The RW16 peptide sample was prepared at a concentration of 1 mM in 400 µL of H_2O/D_2O (90:10, v:v) containing 50 mM sodium phosphate buffer at pH 5.5, and 60 mM of DPC-d_{38} forming zwitterionic micelles. NMR experiments were recorded at 310 K on a Bruker 800 MHz Avance III spectrometer equipped with a TCI ^{1}H/^{13}C/^{15}N cryoprobe, or a Bruker 700 MHz spectrometer with a standard TXI triple resonance gradient probe.

For assignment, we used a 2D ^{1}H,^{1}H-TOCSY with a mixing time of 40 ms, collected with 40 scans for each of the 256 increments. Additional assignment and restraints for structure calculation were obtained from a ^{1}H,^{1}H-NOESY spectra with a mixing time of 150 ms, collected with 32 scans for each of the 228 increments. The solvent signal in both experiments was suppressed using two excitation sculpting blocks before the start of the acquisition. Partial assignment of ^{13}C chemical shifts was accomplished with an ^{1}H,^{13}C-HSQC for which 352 scans were acquired for each of the 256 increments.

To measure solvent accessibility by paramagnetic relaxation enhancements, the RW16-DPC-d_{38} sample was titrated with Gd(DTPA-BMA) to final concentrations of 1, 2, 3, 4, 5, 7.5, and 10 mM. Proton T2 relaxation was estimated by crosspeak intensity as a proxy for T2 relaxation in the 2D ^{1}H,^{1}H-TOCSY and ^{1}H,^{1}H-NOESY spectra. All spectra were processed using Bruker Topspin 3.2 and analyzed by Sparky.

Supplementary Materials: Supplementary materials can be found at http://www.mdpi.com/1422-0067/20/18/4441/s1.

Author Contributions: Conceptualization, M.-L.J. and I.D.A.; data curation, M.-L.J., C.D.M., P.F.J.F., and I.D.A.; formal analysis, M.-L.J., C.D.M., and P.F.J.F.; funding acquisition, I.D.A.; investigation, M.-L.J., L.V., R.D., A.G., C.D.M., and P.F.J.F.; methodology, M.-L.J. and I.D.A.; project administration, M.-L.J. and I.D.A.; resources, C.D.M., P.F.J.F., and I.D.A.; software, P.F.J.F.; supervision, I.D.A.; validation, M.-L.J., C.D.M., P.F.J.F., and I.D.A.; visualization, M.-L.J., C.D.M., and P.F.J.F.; writing—original draft, M.-L.J., C.D.M., and P.F.J.F.; writing—review and editing, M.-L.J., C.D.M., P.F.J.F., and I.D.A.

Funding: This work was supported by the French Ministère de l'Enseignement Supérieur et de la Recherche, by La Ligue contre le Cancer and by the French National Research Agency (ANR) (ANR-CROSS ANR17-CE11-0050-01).

Acknowledgments: We thank Rodrigue Marquant for peptide synthesis and purification of RW16. Computational work was performed using HPC resources from GENCI-CINES allocated to P.F.J.F.

Conflicts of Interest: The authors declare no conflicts of interest.

Abbreviations

CPP	Cell-penetrating peptides
AMP	Antimicrobial peptides
MAP	Membrane-active peptide
SAR	Structure–activity relationship
GUV	Giant unilamellar vesicles
NMR	Nuclear magnetic resonance
MD	Molecular dynamics
LUV	Large unilamellar vesicle
DOPC	Dioleoylphosphatidylcholine
DPC-d_{38}	Dodecylphosphocholine-d_{38}
HSQC	Heteronuclear single quantum coherence
TOCSY	Total correlation spectroscopy
NOE	Nuclear Overhauser effect
NOESY	Nuclear Overhauser effect spectroscopy
CD	Circular dichroism

References

1. Guidotti, G.; Brambilla, L.; Rossi, D. Cell-Penetrating Peptides: From Basic Research to Clinics. *Trends Pharm. Sci.* **2017**, *38*, 406–424. [CrossRef] [PubMed]

2. Langel, Ü. *Methods for Structural Studies of CPPs. CPP, Cell-Penetrating Peptides*; Springer Singapore: Singapore, 2019; pp. 289–323.

3. Jiao, C.Y.; Delaroche, D.; Burlina, F.; Alves, I.D.; Chassaing, G.; Sagan, S. Translocation and endocytosis for cell-penetrating peptide internalization. *J. Biol. Chem.* **2009**, *284*, 33957–33965. [CrossRef] [PubMed]

4. Yang, N.J.; Hinner, M.J. Getting across the cell membrane: An overview for small molecules, peptides, and proteins. In *Site-Specific Protein Labeling: Methods and Protocols*; Humana Press: New York, NY, USA, 2015; Volume 1266, pp. 29–53. ISBN 9781493922727. [CrossRef]

5. Kauffman, W.B.; Fuselier, T.; He, J.; Wimley, W.C. Mechanism Matters: A Taxonomy of Cell Penetrating Peptides. *Trends Biochem. Sci.* **2015**, *40*, 749–764. [CrossRef] [PubMed]

6. Christiaens, B.; Grooten, J.; Reusens, M.; Joliot, A.; Goethals, M.; Vandekerckhove, J.; Prochiantz, A.; Rosseneu, M. Membrane interaction and cellular internalization of penetratin peptides. *Eur. J. Biochem.* **2004**, *271*, 1187–1197. [CrossRef] [PubMed]

7.   Walrant, A.; Vogel, A.; Correia, I.; Lequin, O.; Olausson, B.E.S.; Desbat, B.; Sagan, S.; Alves, I.D. Membrane interactions of two arginine-rich peptides with different cell internalization capacities. *Biochim. Biophys. Acta* **2012**, *1818*, 1755–1763. [CrossRef]

8.   Binder, H.; Lindblom, G. Charge-dependent translocation of the Trojan peptide penetratin across lipid membranes. *Biophys. J.* **2003**, *85*, 982–995. [CrossRef]

9.   Walrant, A.; Correia, I.; Jiao, C.Y.; Lequin, O.; Bent, E.H.; Goasdoué, N.; Lacombe, C.; Chassaing, G.; Sagan, S.; Alves, I.D. Different membrane behaviour and cellular uptake of three basic arginine-rich peptides. *Biochim. Biophys. Acta* **2011**, *1808*, 382–393. [CrossRef]

10. Caesar, C.E.B.; Esbjörner, E.K.; Lincoln, P.; Nordén, B. Membrane interactions of cell-penetrating peptides probed by tryptophan fluorescence and dichroism techniques: Correlations of structure to cellular uptake. *Biochemistry* **2006**, *45*, 7682–7692. [CrossRef]

11. De Kroon, A.I.P.M.; Soekarjo, M.W.; De Gier, J.; De Kruijff, B. The role of charge and hydrophobicity in peptide-lipid interaction: A comparative study based on tryptophan fluorescence measurements combined with the use of aqueous and hydrophobic quenchers. *Biochemistry* **1990**, *29*, 8229–8240. [CrossRef]

12. Yau, W.M.; Wimley, W.C.; Gawrisch, K.; White, S.H. The preference of tryptophan for membrane interfaces. *Biochemistry* **1998**, *37*, 14713–14718. [CrossRef]

13. Jobin, M.-L.; Blanchet, M.; Henry, S.; Chaignepain, S.; Manigand, C.; Castano, S.; Lecomte, S.; Burlina, F.; Sagan, S.; Alves, I.D. The role of tryptophans on the cellular uptake and membrane interaction of arginine-rich cell penetrating peptides. *Biochim. Biophys. Acta* **2015**, *1848*. [CrossRef] [PubMed]

14. Jobin, M.-L.; Bonnafous, P.; Temsamani, H.; Dole, F.; Grélard, A.; Dufourc, E.J.; Alves, I.D. The enhanced membrane interaction and perturbation of a cell penetrating peptide in the presence of anionic lipids: Toward an understanding of its selectivity for cancer cells. *Biochim. Biophys. Acta* **2013**, *1828*. [CrossRef] [PubMed]

15. Delaroche, D.; Cantrelle, F.-X.; Subra, F.; Van Heijenoort, C.; Guittet, E.; Jiao, C.-Y.; Blanchoin, L.; Chassaing, G.; Lavielle, S.; Auclair, C.; et al. Cell-penetrating Peptides with Intracellular Actin-remodeling Activity in Malignant Fibroblasts. *J. Biol. Chem.* **2010**, *285*, 7712–7721. [CrossRef] [PubMed]

16. Williams, E.J.; Dunican, D.J.; Green, P.J.; Howell, F.V.; Derossi, D.; Walsh, F.S.; Doherty, P. Selective Inhibition of Growth Factor-stimulated Mitogenesis by a Cell-permeable Grb2-binding Peptide. *J. Biol. Chem.* **1997**, *272*, 22349–22354. [CrossRef] [PubMed]

17. Lamazière, A.; Burlina, F.; Wolf, C.; Rard Chassaing, G.; Trugnan, G.; Ayala-Sanmartin, J. Non-Metabolic Membrane Tubulation and Permeability Induced by Bioactive Peptides. *PLoS ONE* **2007**, e201. [CrossRef] [PubMed]

18. Herbig, M.E.; Fromm, U.; Leuenberger, J.; Krauss, U.; Beck-Sickinger, A.G.; Merkle, H.P. Bilayer interaction and localization of cell penetrating peptides with model membranes: A comparative study of a human calcitonin (hCT)-derived peptide with pVEC and pAntp(43–58). *Biochim. Biophys. Acta Biomembr.* **2005**, *1712*, 197–211. [CrossRef] [PubMed]

19. Christiaens, B.; Symoens, S.; Vanderheyden, S.; Engelborghs, Y.; Joliot, A.; Prochiantz, A.; Vandekerckhove, J.; Rosseneu, M.; Vanloo, B. Tryptophan fluorescence study of the interaction of penetratin peptides with model membranes. *Eur. J. Biochem.* **2002**, *269*, 2918–2926. [CrossRef]

20. Magzoub, M.; Eriksson, L.E.G.; Gräslund, A. Comparison of the interaction, positioning, structure induction and membrane perturbation of cell-penetrating peptides and non-translocating variants with phospholipid vesicles. *Biophys. Chem.* **2003**, *103*, 271–288. [CrossRef]

21. Thoren, P.E.; Persson, D.; Esbjorner, E.K.; Goksor, M.; Lincoln, P.; Norden, B.; Thorén, P.E.G.; Persson, D.; Esbjörner, E.K.; Goksör, M.; et al. Membrane Binding and Translocation of Cell-Penetrating Peptides. *Biochemistry* **2004**, *43*, 3471–3489. [CrossRef]

22. Rieping, W.; Bardiaux, B.; Bernard, A.; Malliavin, T.E.; Nilges, M. ARIA2: Automated NOE assignment and data integration in NMR structure calculation. *Bioinformatics* **2007**, *23*, 381–382. [CrossRef]

23. Laskowski, R.A.C.; Rullmann, J.A.; MacArthur, M.W.; Kaptein, R.; Thornton, J.M. AQUA and PROCHECK-NMR: Programs for checking the quality of protein structures solved by NMR. *J. Biomol. Nmr* **1996**, *8*, 477–486. [CrossRef] [PubMed]

24. Bolen, E.J.; Holloway, P.W. Quenching of tryptophan fluorescence by brominated phospholipid. *Biochemistry* **1990**, *29*, 9638–9643. [CrossRef] [PubMed]

25. East, J.M.; Lee, A.G. Lipid Selectivity of the Calcium and Magnesium Ion Dependent Adenosinetriphosphatase, Studied with Fluorescence Quenching by a Brominated Phospholipid. *Biochemistry* **1982**, *21*, 4144–4151. [CrossRef] [PubMed]

26. Ladokhin, A.S.; Holloway, P.W.; Kostrzhevska, E.G. Distribution Analysis of Membrane Penetration of Proteins by Depth-Dependent Fluorescence Quenching. *J. Fluoresc.* **1993**, *3*, 195–197. [CrossRef] [PubMed]

27. Ladokhin, A.S. Distribution Analysis of Depth-Dependent Fluorescence Quenching in Membranes: A Practical Guide. In *Methods in Enzymology*; Academic Press: Cambridge, MA, USA, 1997; Volume 278, pp. 462–473. [CrossRef]

28. Abrams, F.S.; London, E. Calibration of the parallax fluorescence quenching method for determination of membrane penetration depth: Refinement and comparison of quenching by spin-labeled and brominated lipids. *Biochemistry* **1992**, *31*, 5312–5322. [CrossRef] [PubMed]

29. Abrams, F.S.; London, E. Extension of the parallax analysis of membrane penetration depth to the polar region of model membranes: Use of fluorescence quenching by a spin-label attached to the phospholipid polar headgroup. *Biochemistry* **1993**, *32*, 10826–10831. [CrossRef]

30. Chattopadhyay, A.; London, E. Parallax method for direct measurement of membrane penetration depth utilizing fluorescence quenching by spin-labeled phospholipids. *Biochemistry* **1987**, *26*, 39–45. [CrossRef]

31. Wiener, M.C.; White, S.H. Structure of a fluid dioleoylphosphatidylcholine bilayer determined by joint refinement of x-ray and neutron diffraction data. III. Complete structure. *Biophys. J.* **1992**, *61*, 434–447. [CrossRef]

32. Zhang, W.; Smith, S.O. Mechanism of penetration of Antp(43-58) into membrane bilayers. *Biochemistry* **2005**, *44*, 10110–10118. [CrossRef]

33. Lensink, M.F.; Christiaens, B.; Vandekerckhove, J.; Prochiantz, A.; Rosseneu, M. Penetratin-membrane association: W48/R52/W56 shield the peptide from the aqueous phase. *Biophys. J.* **2005**, *88*, 939–952. [CrossRef]

34. Lindberg, M.; Biverståhl, H.; Gräslund, A.; Mäler, L. Structure and positioning comparison of two variants of penetratin in two different membrane mimicking systems by NMR. *Eur. J. Biochem.* **2003**, *270*, 3055–3063. [CrossRef] [PubMed]

35. Norman, K.E.; Nymeyer, H. Indole localization in lipid membranes revealed by molecular simulation. *Biophys. J.* **2006**, *91*, 2046–2054. [CrossRef] [PubMed]

36. Seelig, J. Thermodynamics of lipid-peptide interactions. *Biochim. Biophys. Acta-Biomembr.* **2004**, *1666*, 40–50. [CrossRef] [PubMed]

37. Respondek, M.; Madl, T.; Göbl, C.; Golser, R.; Zangger, K. Mapping the orientation of helices in micelle-bound peptides by paramagnetic relaxation waves. *J. Am. Chem. Soc.* **2007**, *129*, 5228–5234. [CrossRef] [PubMed]

38. Schrodinger LLC The PyMOL Molecular Graphics System, Version 1.8. 2015. Available online: https://pymol.org/2/ (accessed on 6 September 2019).

39. Derossi, D.; Chassaing, G.; Prochiantz, A. Trojan peptides: The penetratin system for intracellular delivery. *Trends Cell Biol.* **1998**, *8*, 84–87. [CrossRef]

40. Lindberg, M.; Jarvet, J.; Langel, U.; Gräslund, A. Secondary structure and position of the cell-penetrating peptide transportan in SDS micelles as determined by NMR. *Biochemistry* **2001**, *40*, 3141–3149. [CrossRef] [PubMed]

41. Brattwall, C.E.B.; Lincoln, P.; Nordén, B. Orientation and Conformation of Cell-Penetrating Peptide Penetratin in Phospholipid Vesicle Membranes Determined by Polarized-Light Spectroscopy. *J. Am. Chem. Soc.* **2003**, *125*, 14214–14215. [CrossRef]

42. Giménez-Andrés, M.; Čopič, A.; Antonny, B. The many faces of amphipathic helices. *Biomolecules* **2018**, *8*, 45. [CrossRef]

43. Witte, K.; Olausson, B.E.S.; Walrant, A.; Alves, I.D.; Vogel, A. Structure and dynamics of the two amphipathic arginine-rich peptides RW9 and RL9 in a lipid environment investigated by solid-state NMR and MD simulations. *Biochim. Biophys. Acta-Biomembr.* **2013**, *1828*, 824–833. [CrossRef]

44. Polyansky, A.A.; Volynsky, P.E.; Arseniev, A.S.; Efremov, R.G. Adaptation of a membrane-active peptide to heterogeneous environment. I. Structural plasticity of the peptide. *J. Phys. Chem. B* **2009**, *113*, 1107–1119. [CrossRef]

45. Chen, X.; Liu, S.; Deme, B.; Cristiglio, V.; Marquardt, D.; Weller, R.; Rao, P.; Wang, Y.; Bradshaw, J. Efficient internalization of TAT peptide in zwitterionic DOPC phospholipid membrane revealed by neutron diffraction. *Biochim. Biophys. Acta-Biomembr.* **2017**, *1859*, 910–916. [CrossRef] [PubMed]

46. Dorairaj, S.; Allen, T.W. On the thermodynamic stability of a charged arginine side chain in a transmembrane helix. *Proc. Natl. Acad. Sci. USA* **2007**, *104*, 4943–4948. [CrossRef] [PubMed]

47. Lazaridis, T.; Leveritt, J.M.; Pebenito, L. Implicit membrane treatment of buried charged groups: Application to peptide translocation across lipid bilayers. *Biochim. Biophys. Acta-Biomembr.* **2014**, *1838*, 2149–2159. [CrossRef] [PubMed]

48. Freites, J.A.; Tobias, D.J.; von Heijne, G.; White, S.H. Interface connections of a transmembrane voltage sensor. *Proc. Natl. Acad. Sci. USA* **2005**, *102*, 15059–15064. [CrossRef] [PubMed]

49. MacCallum, J.L.; Bennett, W.F.D.; Tieleman, D.P. Partitioning of Amino Acid Side Chains into Lipid Bilayers: Results from Computer Simulations and Comparison to Experiment. *J. Gen. Physiol.* **2007**, *129*, 371–377. [CrossRef] [PubMed]

50. MacCallum, J.L.; Bennett, W.F.F.D.; Tieleman, D.P. Transfer of arginine into lipid bilayers is nonadditive. *Biophys. J.* **2011**, *101*, 110–117. [CrossRef]

51. Rothbard, J.B.; Jessop, T.C.; Wender, P.A. Adaptive translocation: The role of hydrogen bonding and membrane potential in the uptake of guanidinium-rich transporters into cells. *Adv. Drug Deliv. Rev.* **2005**, *57*, 495–504. [CrossRef] [PubMed]

52. Tang, M.; Waring, A.J.; Hong, M. Phosphate-mediated arginine insertion into lipid membranes and pore formation by a cationic membrane peptide from solid-state NMR. *J. Am. Chem. Soc.* **2007**, *129*, 11438–11446. [CrossRef]

53. Czajlik, A.; Meskó, E.; Penke, B.; Perczel, A. Investigation of penetratin peptides. Part 1. The environment dependent conformational properties of penetratin and two of its derivatives. *J. Pept. Sci.* **2002**, *8*, 151–171. [CrossRef]

54. Smith, J.S.; Scholtz, J.M. Energetics of polar side-chain interactions in helical peptides: Salt effects on ion pairs and hydrogen bonds. *Biochemistry* **1998**, *37*, 33–40. [CrossRef]

55. Andrew, C.D.; Penel, S.; Jones, G.R.; Doig, A.J. Stabilizing Nonpolar/Polar Side-Chain Interactions in the a-Helix. *Proteins Struct. Funct. Genet.* **2001**, *45*, 449–455. [CrossRef]

56. Marrink, S.J.; Risselada, H.J.; Yefimov, S.; Tieleman, D.P.; De Vries, A.H. The MARTINI force field: Coarse grained model for biomolecular simulations. *J. Phys. Chem. B* **2007**, *111*, 7812–7824. [CrossRef] [PubMed]

57. Monticelli, L.; Kandasamy, S.K.; Periole, X.; Larson, R.G.; Tieleman, D.P.; Marrink, S.J. The MARTINI coarse-grained force field: Extension to proteins. *J. Chem. Theory Comput.* **2008**, *4*, 819–834. [CrossRef] [PubMed]

58. Abraham, M.J.; Murtola, T.; Schulz, R.; Páll, S.; Smith, J.C.; Hess, B.; Lindahl, E. GROMACS: High performance molecular simulations through multi-level parallelism from laptops to supercomputers. *SoftwareX* **2015**, *1–2*, 19–25. [CrossRef]

59. Jorgensen, W.L.; Maxwell, D.S.; Tirado-Rives, J. Development and testing of the OPLS all-atom force field on conformational energetics and properties of organic liquids. *J. Am. Chem. Soc.* **1996**, *118*, 11225–11236. [CrossRef]

60. Tieleman, D.P.; MacCallum, J.L.; Ash, W.L.; Kandt, C.; Xu, Z.; Monticelli, L. Membrane protein simulations with a united-atom lipid and all-atom protein model: Lipid-protein interactions, side chain transfer free energies and model proteins. *J. Phys. Condens. Matter* **2006**, *18*, S1221. [CrossRef] [PubMed]

61. Berger, O.; Edholm, O.; Jähnig, F. Molecular dynamics simulations of a fluid bilayer of dipalmitoylphosphatidylcholine at full hydration, constant pressure, and constant temperature. *Biophys. J.* **1997**, *72*, 2002–2013. [CrossRef]

62. Jorgensen, W.L.; Chandrasekhar, J.; Madura, J.D.; Impey, R.W.; Klein, M.L. Comparison of simple potential functions for simulating liquid water. *J. Chem. Phys.* **1983**, *79*, 926–935. [CrossRef]

63. Chakrabarti, N.; Neale, C.; Payandeh, J.; Pai, E.F.; Pomès, R. An iris-like mechanism of pore dilation in the CorA magnesium transport system. *Biophys. J.* **2010**, *98*, 784–792. [CrossRef]

64. Essmann, U.; Perera, L.; Berkowitz, M.L.; Darden, T.; Lee, H.; Pedersen, L.G. A smooth particle mesh Ewald method. *J. Chem. Phys.* **1995**, *103*, 8577–8593. [CrossRef]

65. Hess, B.; Bekker, H.; Berendsen, H.J.C.; Fraaije, J.G.E.M. LINCS: A linear constraint solver for molecular simulations. *J. Comput. Chem.* **1997**, *18*, 1463–1472. [CrossRef]

66. Miyamoto, S.; Kollman, P.A. Settle: An analytical version of the SHAKE and RATTLE algorithm for rigid water models. *J. Comput. Chem.* **1992**, *13*, 952–962. [CrossRef]

67. Bussi, G.; Donadio, D.; Parrinello, M. Canonical sampling through velocity rescaling. *J. Chem. Phys.* **2007**, *126*, 014101. [CrossRef] [PubMed]

68. Parrinello, M.; Rahman, A. Polymorphic transitions in single crystals: A new molecular dynamics method. *J. Appl. Phys.* **1981**, *52*, 7182–7190. [CrossRef]

69. Kabsch, W.; Sander, C. Dictionary of protein secondary structure: Pattern recognition of hydrogen-bonded and geometrical features. *Biopolymers* **1983**, *22*, 2577–2637. [CrossRef] [PubMed]

70. Joosten, R.P.; Te Beek, T.A.H.; Krieger, E.; Hekkelman, M.L.; Hooft, R.W.W.; Schneider, R.; Sander, C.; Vriend, G. A series of PDB related databases for everyday needs. *Nucleic Acids Res.* **2011**, *39*. [CrossRef] [PubMed]

71. Lehrer, S.S. Solute Perturbation of Protein Fluorescence. The Quenching of the Tryptophyl Fluorescence of Model Compounds and of Lysozyme by Iodide Ion. *Biochemistry* **1971**, *10*, 3254–3263. [CrossRef]

72. Jean-François, F.; Khemtémourian, L.; Odaert, B.; Castano, S.; Grélard, A.; Manigand, C.; Bathany, K.; Metz-Boutigue, M.H.; Dufourc, E.J. Variability in secondary structure of the antimicrobial peptide Cateslytin in powder, solution, DPC micelles and at the air-water interface. *Eur. Biophys. J.* **2007**, *36*, 1019–1027. [CrossRef]

International Journal of
Molecular Sciences

Article

Drug Conjugation Induced Modulation of Structural and Membrane Interaction Features of Cationic Cell-Permeable Peptides

Edit Pári [1,*], Kata Horváti [2,3], Szilvia Bősze [2], Beáta Biri-Kovács [2,3], Bálint Szeder [4], Ferenc Zsila [5] and Éva Kiss [1,*]

[1] Laboratory of Interfaces and Nanostructures, Institute of Chemistry, Eötvös Loránd University, Budapest 112, P.O. Box 32, 1518 Budapest, Hungary

[2] MTA-ELTE Research Group of Peptide Chemistry, Eötvös Loránd University, Budapest 112, P.O. Box 32, 1518 Budapest, Hungary; khorvati@gmail.com (K.H.); szilvia.bosze@gmail.com (S.B.); beabiri@caesar.elte.hu (B.B.-K.)

[3] Institute of Chemistry, Eötvös Loránd University, Budapest 112, P.O. Box 32, 1518 Budapest, Hungary

[4] Research Centre for Natural Sciences, Signal Transduction and Functional Genomics Research Group, Institute of Enzymology, P.O. Box 286, 1519 Budapest, Hungary; szederbalint@gmail.com

[5] Research Centre for Natural Sciences, Research Group of Biomolecular self-assembly, Institute of Materials and Environmental Chemistry P.O. Box 286, 1519 Budapest, Hungary; zsila.ferenc@ttk.mta.hu

*   Correspondence: pari.edit@gmail.com (E.P.); kisseva@caesar.elte.hu (É.K.)

Received: 3 March 2020; Accepted: 20 March 2020; Published: 22 March 2020

Abstract: Cell-penetrating peptides might have great potential for enhancing the therapeutic effect of drug molecules against such dangerous pathogens as *Mycobacterium tuberculosis* (Mtb), which causes a major health problem worldwide. A set of cationic cell-penetration peptides with various hydrophobicity were selected and synthesized as drug carrier of isoniazid (INH), a first-line antibacterial agent against tuberculosis. Molecular interactions between the peptides and their INH-conjugates with cell-membrane-forming lipid layers composed of DPPC and mycolic acid (a characteristic component of Mtb cell wall) were evaluated, using the Langmuir balance technique. Secondary structure of the INH conjugates was analyzed and compared to that of the native peptides by circular dichroism spectroscopic experiments performed in aqueous and membrane mimetic environment. A correlation was found between the conjugation induced conformational and membrane affinity changes of the INH–peptide conjugates. The degree and mode of interaction were also characterized by AFM imaging of penetrated lipid layers. In vitro biological evaluation was performed with Penetratin and Transportan conjugates. Results showed similar internalization rate into EBC-1 human squamous cell carcinoma, but markedly different subcellular localization and activity on intracellular Mtb.

Keywords: membrane affinity; cell-penetrating peptides; circular dichroism spectroscopy; atomic force microscopy; mycolic acid; Langmuir monolayer; drug–peptide conjugates

---

1. Introduction

In the genus *Mycobacterium*, there are three dangerous pathogens for humans: *Mycobacterium tuberculosis* (Mtb), *M. leprae* and *M. lepromatosis* [1]. These pathogens are causative agents of human diseases such as tuberculosis and leprosy. Tuberculosis is still a major public health problem worldwide; it is one of the top 10 causes of death and the leading cause from a single infectious agent (above HIV/AIDS). In 2017, Tuberculosis caused an estimated 1.3 million deaths among HIV-negative people, and there were an additional 300,000 deaths among HIV-positive people [2].

During the therapeutic treatment of infectious diseases, it is important to consider the structural and physicochemical properties of the cell envelope of mycobacteria, in order to allow the access of drug compounds to the cell, for effective targeted drug delivery and treatments. Processes located at the cell envelope of mycobacteria include the protection of the bacterial cell from the environment, transport of solutes and proteins, adhesion to receptors and mechanical resistance [3]. *Mycobacterium* is considered Gram-positive bacterium which possesses a lipid-rich cell envelope [4] composed of a typical plasma membrane and a three-part complex: mycomembrane, arabinogalactan and peptidoglycan. The mycomembrane is a specific barrier composed of mycolic acids, which are the major chemical components (almost 40%) in the cell envelope [5–7]. Mycolic acids are exceptionally long chain α-alkyl-β-hydroxyl fatty acids that have different conformational behavior covalently linked to polysaccharide arabinogalactans in the cell wall [3,4,8,9]. The folding and unfolding of mycolic acids can influence its biological function and permeability properties for drugs [7,10]. The Langmuir monolayer is an ideal model to investigate the importance of the lateral organization and the specific functions of lipids, use them for cell mimicking, and allows the systematic study of biomolecule–membrane interactions. [11–13]. Hasegawa et al. studied the conformational changes of α-mycolic acid at different surface pressures by using Langmuir–Blodgett technique and atomic force microscopy (AFM) [14]. It was found that α-mycolic acid formed a stable Langmuir monolayer and it is entirely expanded when the surface pressure is high. In experimental model systems, mammalian membranes often mimicked by dipalmitoylphosphatidylcholine (DPPC). This lipid is chosen to represent the most abundant head groups in natural membranes [15]. Chimote et al. used Langmuir monolayers to evaluate the effect of mycolic acid and cord factor (alone and as mixtures) on the surface properties of DPPC layer [7]. It has been found that both mycobacterial lipids inhibit the surface activity of DPPC; furthermore, mycolic acid fluidizes the lung surfactant film, and this can contribute to alveolar collapse in tuberculosis patients.

Various methods have been proposed to inhibit the fission of bacterial cells. Isoniazid (INH) is a first-line antitubercular agent that inhibits the synthesis of mycolic acids. The interaction between antituberculosis drugs (such as INH, rifampicin and ethambutol) and DPPC, which is the main component of the pulmonary surfactant, was investigated by Chimote et al. [16]. Pénzes et al. [17] studied mycolic-acid-containing lipid monolayer models to interpret the membrane affinity of antituberculosis drug conjugates. Change induced in the structure of the phospholipid monolayer by interaction with drug–peptide conjugate was followed by tensiometry and sum-frequency vibrational spectroscopy [18]. Other antibacterial drug candidates and polymers were also applied to determine their membrane affinity and in vitro activity, using lipid mono- and bilayer models [19–21]. These studies show the role of the cell envelope in essential processes, and the investigation of its interactions is still an important research field to develop new drugs or drug–conjugates for more effective treatments.

Recently, cationic amphiphilic and antimicrobial peptides (AMPs) have received considerable attention as a possible solution against resistant pathogens. Some of them also showed an anticancer effect by destabilizing the outer membrane of cancer cells and providing selectivity between normal and malignant cells [22]. Cell-penetrating peptides (CPP) have a great pharmaceutical potential as delivery vectors for a wide variety of bioactive cargos. Penetratin [23] and Transportan [24] are among the most-studied CPPs and have the ability to translocate the plasma membrane, delivering drugs to the cytoplasm. The antimicrobial peptide Buforin II (5–21) [25] and Magainin II [26] were isolated from amphibians. Dhvar4 is a synthetic derivative of Histatin, which is a salivary histidine-rich cationic AMP found in the human parotid secretion [27], Crot(1–9,38–42) [28] and CM15 [29], which are hybrid AMPs composed of venom peptides. OT20 [30] is a synthetic oligomer of Tuftsin, a receptor-binding peptide produced by the enzymatic cleavage of the Fc-domain of the heavy chain of immunoglobulin G.

There is a plethora of reports on the investigation of CPPs and AMPs properties [31–33], but most of them focus on one or two peptides, without comparison of the members of the given family. In our work, we strived for the comparison of various cationic peptides and their investigation under the same measurement conditions with different techniques. This might be useful to reveal the effect of the composition, polarity, molecular mass and conformational behavior on membrane interactions and transport. For this purpose, a representative set of cationic membrane active peptides was selected based on our previous studies on their cell penetrating ability, hemolytic activity, antimicrobial and cytotoxic effects [34]. Our goal was to identify peptides as promising drug carriers for INH. The selected peptides and their INH conjugated derivatives were synthetized and used to study their interactions with DPPC and DPPC + mycolic acid mixed monolayers employing the Langmuir technique. Circular dichroism spectroscopy (CD) was employed to monitor the secondary structure of peptides and their INH-conjugates in water and also in membrane mimetic-trifluoroethanol-containing media. The influence of conjugation was evaluated and compared for the various AMPs. For further analysis, atomic force microscopy (AFM) was used to image the morphology and structure of peptide-penetrated lipid layer. Cellular uptake was studied on EBC-1 squamous cell carcinoma as lung model cells, and the localization of the peptides was imaged by confocal microscopy. Intracellular killing activity of INH–Penetratin and INH–Transportan was assayed on *Mycobacterium tuberculosis* infected human monocytes.

2. Results and Discussion

2.1. Peptide Synthesis and Characterization

Cationic membrane active peptides and their INH-derivatives were synthesized in good yields, using previously described methods [35]. Peptides were characterized with their monoisotopic molecular mass (M), retention time on a C-18 RP-HPLC column (R_t), estimated hydrophilicity values (H) and net charge (z) (see Table 1). Peptides are listed according to their retention times, which are the indicator of their polarity. OT20 was found to be the most hydrophilic peptide, while Transportan proved to be the most hydrophobic one in the given set. This order of peptides' polarity is almost equal to the relative hydrophilicity estimated from the amino acid composition. The conjugation with INH does not change the relative polarity of the peptides, as the R_t values show. It is notable, however, that some increase in the hydrophobicity can be presumed from the slightly increased retention-time values for most of the peptides (except Buforin and Transportan). That might be the result of decreased charge due to INH coupling at the N-terminal, although the addition of INH molecule is expected to increase the polarity of the conjugate.

2.2. Interaction of Peptides and INH–Peptide Conjugates with Lipid Monolayers

The peptide–lipid interactions were investigated by the Langmuir technique, and the results of the penetration measurements are presented in Figures 1 and 2. The lipid layer was formed from DPPC and the mixture of DPPC+mycolic acid at ratio of 3:1. The aqueous solution of the peptide or the INH-conjugated peptide was injected below the compressed (20 mN/m) and conditioned (20 min) lipid monolayer after 10 min relaxation time. The final concentration was 2 μM in the aqueous phase. The changes of the surface pressure as a function time was recorded for 1 h. The increase of the surface pressure ($\Delta\Pi$) gives information of the degree of penetration, which is considered as the membrane affinity of the given peptide to the lipid layer.

The penetration results showed that the membrane affinity to DPPC layer is modest ($\Delta\Pi < 2$) in case of OT20, Crot(1–9,38–42) and Buforin II (5–21), but higher for the Penetratin (Figure 1). The other peptides' membrane affinity then gradually increased. This shows agreement with our previous results, where the membrane affinity order of the peptides almost followed their hydrophobicity [37]. Transportan, CM15 and Magainin are shown to be the most hydrophobic ones in the present set, according to their R_t (>12) and H (< 0) values. Those exhibit significant membrane affinity, exceeding a 4 mN/m increase in surface pressure. The exact order, however, exposes some differences which might indicate that other factors than overall hydrophobicity of the peptide also influence the degree of interaction with the lipid layer.

Table 1. Analytical characterization of the peptides and INH–peptide conjugates studied in this work.

Compound [a]	Sequence	M [b]	R_t [c]	H [d]	Z [e]
OT20	TKPKGTKPKGTKPKGTKPKG	2062.3	6.7	1.1	9+
Crot(1–9,38–42)	YKQCHKKGGKKGSG	1503.8	6.8	0.8	6+
Buforin II (5–21)	RAGLQFPVGRVHRLLRK	2001.2	10.9	0.2	6+
Penetratin	RQIKIWFQNRRMKWKK	2244.3	10.9	0.5	8+
Dhvar4	KRLFKKLLFSLRKY	1838.2	11.8	0.3	7+
Magainin II	GIGKFLHSAKKFGKAFVGEIMNS	2464.3	12.8	−0.1	4+
CM15	KWKLFKKIGAVLKVL	1769.2	13.7	−0.1	6+
Transportan	AGYLLGKINLKALAALAKKIL	2180.4	16.8	−0.3	5+
INH–OT20	INH-TKPKGTKPKGTKPKGTKPKG	2239.3	7.0	-	8+
INH–Crot(1–9,38–42)	INH-YKQCHKKGGKKGSG	1680.9	7.3	-	5+
INH–Buforin II (5–21)	INH-RAGLQFPVGRVHRLLRK	2178.3	11.0	-	5+
INH–Penetratin	INH-RQIKIWFQNRRMKWKK	2421.4	11.5	-	7+
INH–Dhvar4	INH-KRLFKKLLFSLRKY	2015.2	12.8	-	6+
INH–Magainin II	INH-GIGKFLHSAKKFGKAFVGEIMNS	2641.4	13.3	-	3+
INH–CM15	INH-KWKLFKKIGAVLKVL	1946.2	14.3	-	5+
INH–Transportan	INH-AGYLLGKINLKALAALAKKIL	2357.5	16.9	-	4+

[a] All peptides and conjugates were amidated on the C-terminus and isolated as TFA salts. [b] Exact mass measured on a Thermo Scientific Q Exactive™ Focus Hybrid Quadrupole-Orbitrap™ Mass Spectrometer. [c] Retention time on a YMC-Pack ODS-A C18 (5 μm, 120 Å) 150 × 4.6 mm column, gradient: 5% B, 2 min; 5%–100% B, 20 min. [d] Hydrophilicity was calculated by using values of amino acids expressing hydrophilicity of each amino acid [36]. [e] Net charge at neutral pH.

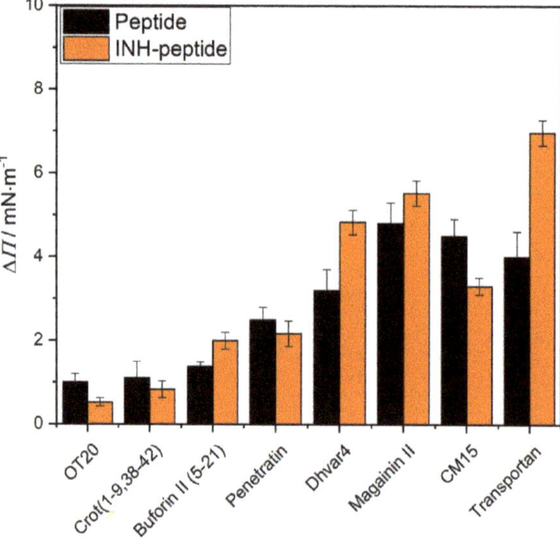

Figure 1. The degree of penetration (ΔΠ) of the original peptides (black) and their INH conjugated derivatives (orange) into DPPC lipid monolayer after 1 h interaction. (The peptides are presented according to their increasing R_t values.)

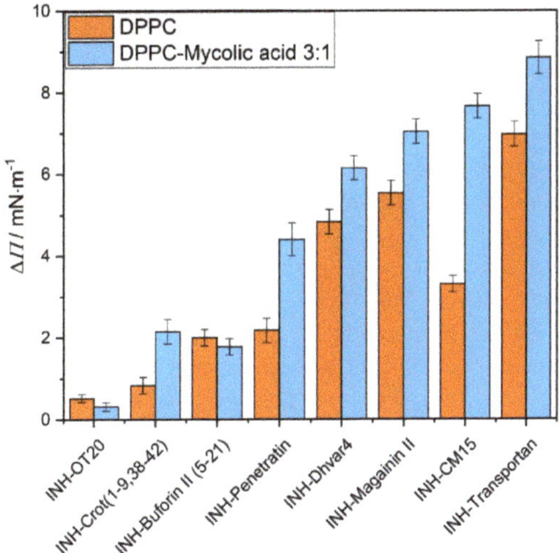

Figure 2. The degree of penetration ($\Delta\Pi$) of isoniazid (INH)-conjugated peptides into DPPC (orange) and DPPC+mycolic acid (3:1 mass ratio) (blue) mixed lipid monolayers after 1 h interaction. (The peptide conjugates are presented according to their increasing R_t values.)

INH conjugation did not change significantly or decrease membrane affinity of peptides with DPPC monolayer, except in the case of Buforin II (5–21), Dhvar4 and Transportan, where the penetration increased. The coupling of INH to the peptides due to the reduction of charge at the N terminal is expected to increase the hydrophobicity and surface activity character of the conjugate. Therefore, the penetration of the most hydrophobic peptide, Transportan, is further increased for INH–Transportan. The explanation might be similar for the case of Dhvar4, where the INH conjugation induced a relative decrease in polarity counts more because of the lower molecular weight. On the contrary, coupling INH to highly charged hydrophilic peptides probably does not induce important changes in the polar character of the derivatives. That is in correspondence with the slightly decreasing membrane affinity of OT20, Crot(1–9,38–42), Penetratin and CM15 conjugates.

Besides the overall polarity, the conjugation with INH is supposed to induce some structural changes that lead to preferred or unfavored interaction with lipid layers.

The membrane affinity of the INH-conjugated peptides was also characterized by using another membrane model, which is formed from DPPC+mycolic acid (3:1) mixture (Figure 2). By mixing mycolic acid, the monolayer provides a negatively charged layer, which gives the possibility of the electrostatic interactions with the cationic CPPs. This negatively charged environment in the lipid membrane favors the insertion of CPPs [31,38]. Moreover, the incorporation of mycolic acid molecules with long alkyl chains into the lipid layer changes the molecular order and reduces the regular close packing of the DPPC molecules, allowing the interaction with and penetration of the peptides. According to the experimental findings, the membrane affinity significantly increased in all cases except for INH–OT20 and INH–Buforin II (5–21). INH–Transportan and INH–CM15 have the highest interactions with the DPPC+mycolic acid mixed lipid monolayer.

2.3. Circular Dichroism Spectroscopic Evaluation of the Secondary Structure of CPPs/AMPs and Their INH Conjugates

Circular dichroism (CD) spectra recorded in the far-UV region (190–260 nm) provides valuable information on the secondary structure of peptides in a particular medium [39]. In many instances, cationic AMPs/CPPs are substantially unstructured in aqueous solution but acquire helical conformation upon contact with negatively charged cell membranes and also in certain organic solvents, like trifluoroethanol (TFE) [40–42]. TFE and related haloalcohols provide a low dielectric, lipomimetic environment which favors and propagates the structuration process of cationic peptides being disordered in water. Membrane-binding-induced folding of cationic peptides into amphipathic helices is a crucial step in their biological actions. Therefore, their structural modifications enhancing helix propensity often result in increased antibacterial and/or hemolytic activity [43–45].

All native peptides studied herein exhibit a main negative CD band in water, with a deep minimum, around 196–200 nm (Figures 3 and 4). Such a CD pattern is consistent with the dynamic equilibrium of the mixture of random conformational states. In the presence of TFE, however, a significant amount of helical structure was induced, as evident from the high-intensity positive–negative CD couplet evolved below 210 nm and the pronounced minimum at 220–222 nm, which represent the contribution of the optically active π-π^* and n-π^* transitions of the peptide bonds arranged along a right-handed helix [39]. Consideration of peak intensities of the CD couplets observed in TFE and water:TFE mixtures permits comparisons of relative α-helical propensities of the peptides. The largest $\Delta\varepsilon$ extrema were observed for CM15, Dhvar4 and Transportan, whereas Penetratin and Magainin II showed somewhat lower values (Figures 3 and 4). Buforin II (5–21) displayed the weakest helical induction with a $\Delta\varepsilon_{max}$ of +10 as measured in TFE (Figure 3d).

It is to be noted that Crot(1–9,38–42) and OT20 behaved differently, since neither of them showed helical conversion upon addition of TFE (Figure 4c,d). For OT20, this result can be explained by the combined occurrence of proline and glycine residues in the sequence that are well-known, potent helix breaker residues [46,47]. Crot(1–9,38–42) lacks proline, but its 14 AA long sequence contains three glycine side chains, creating the highest relative helix breaker frequency among the peptides under study (apart from OT20). Therefore, glycines might be responsible for preventing the canonical α-helical folding of Crot(1–9,38–42).

According to our previous data [48–50], small-molecule-binding-induced disorder-to-helix transitions of cationic AMPs are associated with characteristic CD spectral alterations: In the very first phase of the titration with the folding inducer, the negative CD band of the peptide below 210 nm loses intensity, and its λ_{min} is shifted to longer wavelengths. Similar changes were reported for TFE titration of intrinsically disordered peptides [51] and protein regions [52,53]. Taking these results into consideration, the bathochromic shift, as well as the intensity reduction of the main CD peak observed in water, suggests that INH conjugation affects the aqueous conformational equilibrium of all peptides (Figures 3 and 4). Furthermore, as demonstrated by band-intensity increases for some peptide conjugates, this effect is more pronounced in TFE and H_2O:TFE solution and refers to the moderate enhancement of the α-helical fraction compared to the native form. Such kinds of spectral alterations were obtained for INH–Transportan, INH–Magainin II and INH–Buforin II (Figure 3). The most advanced CD spectral modifications were seen for INH conjugate of Transportan (Figure 3a). At the short-wavelength side of the aqueous spectrum, the initial part of the helix related positive CD peak appeared, and the n-π^* CD intensities increased noticeably above 215 nm. Taking into account the positive correlation between peptide helicity and bilayer-disturbing potency [43–45], the chiroptical data predict that the INH conjugates bind to lipid membranes with greater affinity compared to the native peptides. Based on the measure of the conjugation-induced spectroscopic changes (Figure 3), an estimation of the relative membrane affinity increment can be proposed (in decreasing order): INH–Transportan, INH–Magainin II, and INH–Buforin II (5–21). Importantly, this prediction is in good correlation with the membrane affinity enhancement of these conjugates displayed in Figure 1. However, the reduced $\Delta\varepsilon$ values of INH–Dhvar4 obtained in TFE and H_2O:TFE mixture indicate a lower helical propensity (Figure 4b), but its DPPC affinity is larger than that of the parent peptide (Figure 1). In contrast, the reduced CD values above 200 nm in the TFE spectrum of INH–CM15 (Figure 4a) also suggest a lower helicity that is in line with the decreased membrane binding shown in Figure 1. Beyond the impact of the helical conformation, these examples warn to the role of additional INH-conjugation-related physicochemical modifications which may also affect the peptide-membrane interactions.

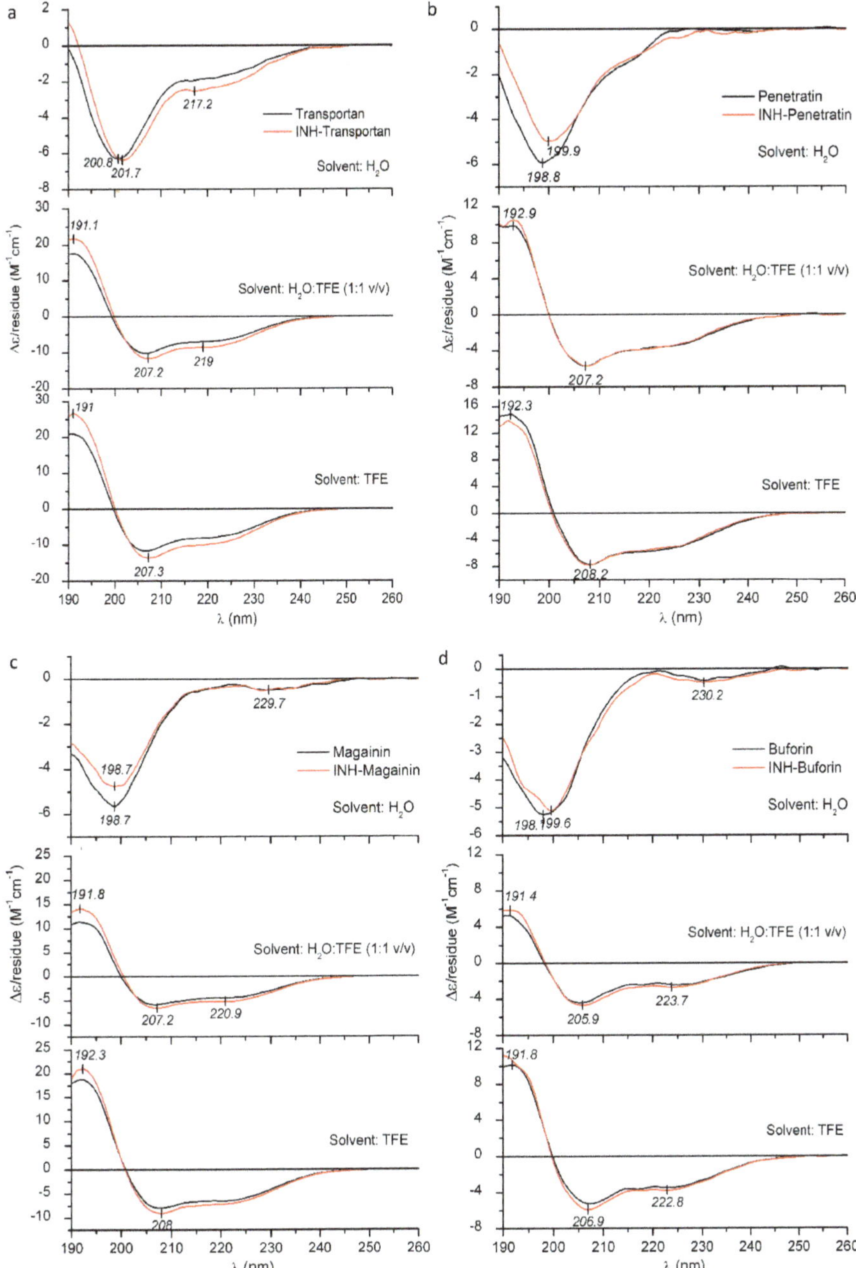

Figure 3. Far-UV CD spectra of Transportan (**a**), Penetratin (**b**), Magainin II (**c**), Buforin II (5–21) (**d**) and their INH conjugates measured in deionized water, TFE and in water:TFE mixture.

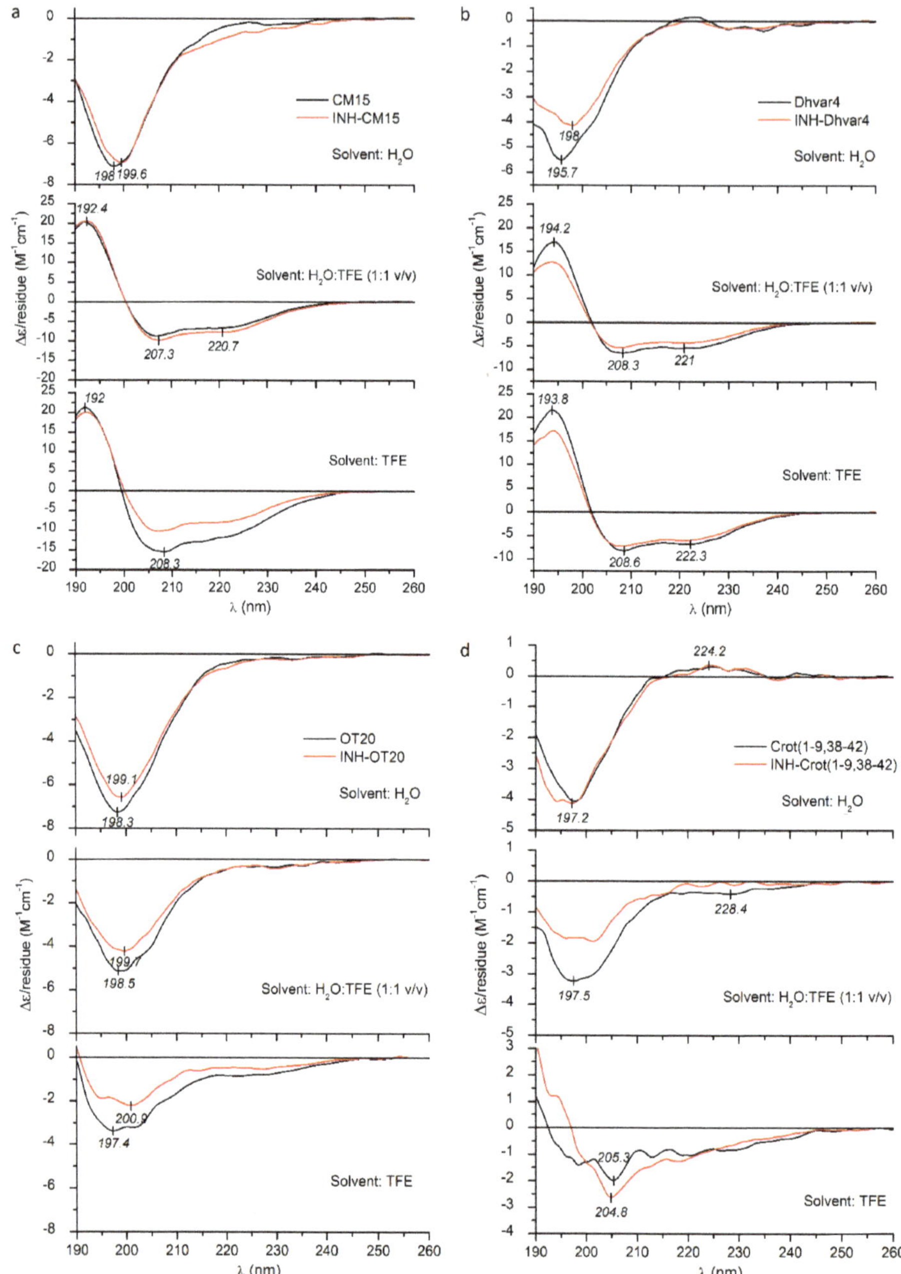

Figure 4. Far-UV CD spectra of CM15 (**a**), Dhvar4 (**b**), OT20 (**c**), Crot(1–9,38–42) (**d**) and their INH conjugates measured in deionized water, TFE and in water:TFE mixture.

The presence of helix breaker residues makes it difficult to assess the conjugation-induced structural impact of OT20 and Crot(1–9,38–42) (Figure 4).

2.4. Atomic Force Microscopy (AFM)

At the end of the membrane affinity measurements, the penetrated monolayers were transferred to a previously cleaned cover glass, to study the structural changes of the membranes by atomic force microscopy (Figures 5 and 6). The surface of the original DPPC and mixed DPPC+mycolic acid monolayer is rather smooth, with a morphology described by typical roughness values presented in Table 2. (R_a is the average deviation of all points from a mean level over the evaluation area. R_q (RMS) is the root mean square roughness, while R_z is the average absolute value of the five highest peaks and the five lowest valleys over the evaluation area, being sensitive to positive or negative extremes of the image.) Well-defined homogenous lipid islands are formed by the ordered lipid molecules in equilibrium with the less-oriented film with smaller surface density as expected from the surface pressure–area isotherm. Comparing to this, the morphology of the penetrated monolayers was changed. The 3D images provide a visual impression on the surface, while the cross-section profiles allow the measurement of the height differences. As examples, for INH–Penetratin and INH–Transportan interaction with DPPC lipid monolayer was found that the homogenous lipid layer was disturbed by the insertion of the peptide-conjugate and several small aggregates, extended protrusions appeared with about 1–3 nm height (Figures 5a and 6a). This morphological change can also be expressed by the roughness factors determined by analyzing the AFM images (Table 2).

Table 2. Roughness values, R_a, R_q and R_z characterizing the morphology of lipid layers and penetrated lipid layers determined by AFM imaging.

Peptide	Lipid System	R_a/nm	R_q/nm	R_z/nm
	DPPC	0.10	0.13	0.36
	DPPC+mycolic acid	0.25	0.33	1.08
INH–Penetratin	DPPC	0.21	0.27	0.92
	DPPC+mycolic acid	0.35	0.45	1.33
INH–Transportan	DPPC	0.23	0.37	0.90
	DPPC+mycolic acid	0.59	0.94	2.40

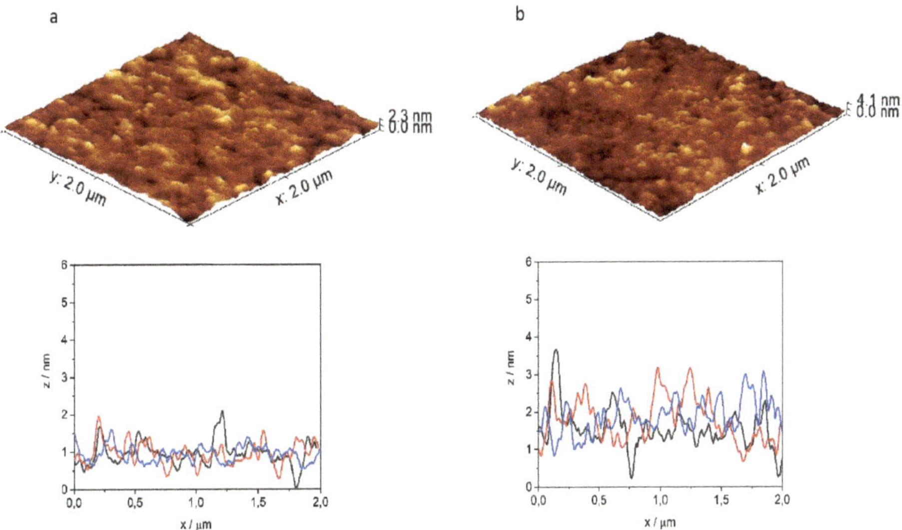

Figure 5. Three-dimensional atomic-force microscopy image of DPPC (**a**) and DPPC+mycolic acid (**b**) lipid layer after 1 h interaction with INH–Penetratin, and characteristic cross-section profiles of the penetrated films.

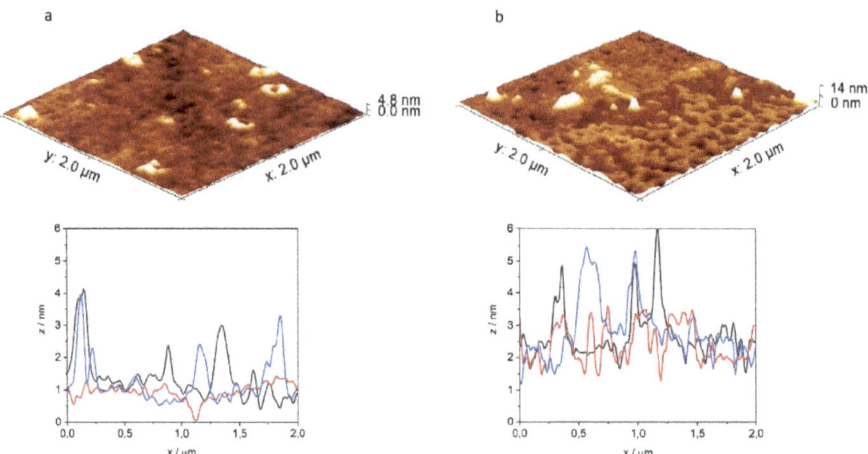

Figure 6. Three-dimensional atomic-force microscopy image of DPPC (**a**) and DPPC+mycolic acid (**b**) lipid layer after 1 h interaction with INH–Transportan, and characteristic cross-section profiles of the penetrated films.

It is noticeable that, considering all three parameters, the roughness of the surface increased as a result of its interaction with the peptide conjugates. Similar membrane morphology change was detected previously in the course of interaction of lipid monolayers with cationic, amphiphilic polymers presenting antibacterial properties [21].

The result of penetration of INH–Penetratin and INH–Transportan into DPPC+mycolic acid film was presented in Figures 5b and 6b. The disturbance of the order of lipid molecules and the deviation from the smooth surface are even higher, in comparison to the corresponding interaction with DPPC layer. The INH–Transportan forms aggregates as a sort of spherical dots and rings between and within the lipid islands with 3–6 nm height, while in the case of INH–Penetratin, a generally disturbed surface appeared with vertical deviation exceeding 3 nm. The intense disturbance of the mixed lipid layer by interaction with INH–Transportan resulted in the fragmentation of the monolayer. The appearance of aggregated structural units is particularly characteristic for the interaction with the negatively charged DPPC+mycolic acid layer. It has already been shown that such aggregation is the consequence of the proximity of the membrane, and this phenomenon can be triggered by a high concentration of the CPPs on negative phospholipids and induced conformational changes [32]. The electrostatic interactions between cationic peptides and negatively charged lipid layer facilitate the conformational change, and agglomeration of CPPs especially for amphipathic peptides such as Penetratin [31].

The significantly enhanced interaction of INH–peptide conjugates with the DPPC+mycolic acid lipid layer, shown by the highly disturbed structure, is in harmony with their high degree of penetration, measured by the increase of surface pressure of the film (Figure 2). It is also in agreement with the CD measurements, since this type of morphology might be the result of enhanced membrane interaction, which involves a membrane-induced increase of the α-helical fraction observed by CD spectroscopy.

2.5. Cellular Uptake, Localization and Efficacy against Intracellular Mycobacterium Tuberculosis

For the in vitro biological evaluation, two peptides were selected, namely Penetratin and Transportan, which are the two most-studied cell-penetrating peptides. Internalization of the peptides was measured on EBC-1 human squamous cell carcinoma, which was chosen as a lung cell model. Squamous-cell aggregates are common within the lung tissue associated with *Mycobacterium tuberculosis* (Mtb) infection. Mtb infected host cells (mainly macrophages and other lung tissue cells) play a pivotal role in chronic-tuberculosis-induced carcinogenesis caused by DNA damage, growth factors, proliferative activities, etc. To eliminate the intracellular bacteria from the host cells is crucial for successful therapy. The infection caused mainly by proliferative changes in the lung tissue (mainly in bronchial and alveolar mucosa) that Mtb leaves behind cannot be ignored [54,55]. Therefore, squamous cell culture can be employed as a suitable model to investigate the main features of the carrier peptides. Flow cytometric measurements revealed a concentration-dependent internalization of the peptides. Penetratin and Transportan showed similar internalization potency based on the detected fluorescent intensity measured by flow cytometry (Figure 7a). However, on the confocal microscopy images, markedly different subcellular localization was observed. Penetratin showed diffused intracellular distribution, while in the case of Transportan, mostly vesicular localization was captured (Figure 7b). Besides cytoplasmic localization, Penetratin showed co-localization with lysosomal and nucleus staining, possibly indicating that this compound internalizes not only through the endo-lysosomal pathway, but also through direct cell penetration (Figure 7b). This observation is in line with previous reports on the nuclear translocation of Penetratin [56]. The vast majority of CPPs can be trapped in endosomes, following delivery into the cells. Subsequently, any conjugated cargo might be trapped, as well [57]. Significantly lower intracellular antitubercular activity of Transportan (Figure 7c) might be the consequence of an endosomal escape problem, while the diffuse intracellular delivery makes INH–Penetratin an efficient candidate against intracellular *Mycobacterium tuberculosis* (Figure 7d).

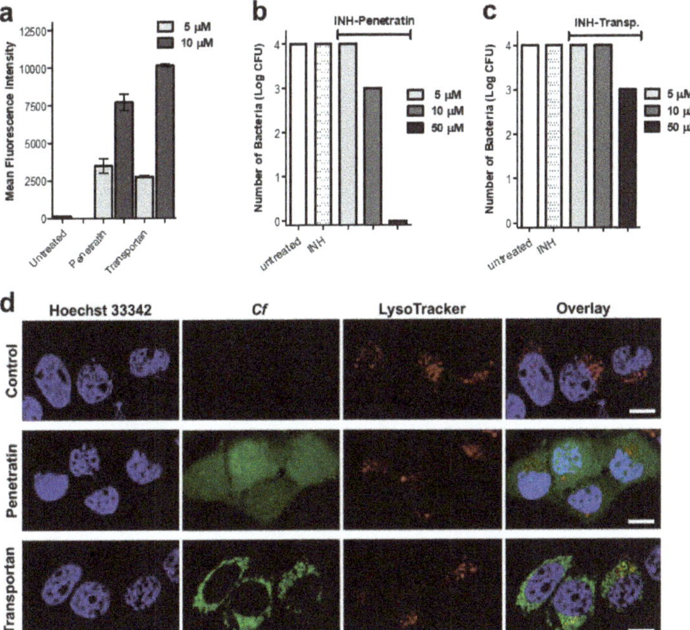

Figure 7. Internalization and intracellular antibacterial activity of the peptides. Cellular uptake of EBC-1 human lung carcinoma cells was measured by flow cytometry after 2 h treatment with 5 and 10 µM peptide concentrations (**a**). Localization of 5(6)-carboxyfluorescein-labeled Penetratin and Transportan in EBC-1 cells was visualized by confocal laser scanning microscopy (**b**). Cells were incubated with *Cf*-labeled peptides (green), lysosomes were stained with LysoTracker Deep Red (red) and nuclei were stained with Hoechst 33342 (blue). The scale bar represents 20 µm. Penetratin and Transportan were conjugated with Isoniazid, and INH–Penetratin and INH–Transportan were assayed against *Mycobacterium tuberculosis*-infected MonoMac-6 human monocytes. After treatment, cells were lysed with SDS, and number of bacteria was enumerated on LJ tubes (**c**,**d**).

3. Materials and Methods

3.1. Peptide Synthesis and Characterization

For peptide synthesis, amino acid derivatives were obtained from Iris Biotech (Germany). The reagents *N,N'*-diisopropylcarbodiimide (DIC), triisopropylsilane (TIS), 1-hydroxybenzotriazole (HOBt), 1,8-diazabicyclo[5.4.0]undec-7-ene (DBU), isoniazid (INH), glyoxylic acid and 5(6)-carboxy fluorescein (*Cf*) were purchased from Sigma-Aldrich (Budapest, Hungary). Fmoc-Rink Amide MBHA resin was from Merck (Budapest, Hungary). Trifluoroacetic acid (TFA), *N,N*-dimethylformamide (DMF), dichloromethane (DCM), diethyl ether and acetonitrile (AcN) were from VWR (Budapest, Hungary).

For the in vitro assays, RPMI-1640 medium, fetal calf serum (FCS) and trypan blue were obtained from Sigma-Aldrich. Trypsin was from Gibco. HPMI buffer (9 mM glucose, 10 mM $NaHCO_3$, 119 mM NaCl, 9 mM HEPES, 5 mM KCl, 0.85 mM $MgCl_2$, 0.053 mM $CaCl_2$, 5 mM $Na_2HPO_4 \times 2H_2O$, pH = 7.4) was prepared in-house, using components obtained from Sigma-Aldrich (Budapest, Hungary).

Synthetic Procedures

Peptides were produced on solid phase (Fmoc-Rink Amide MBHA resin, capacity = 0.67 mmol/g), in an automated peptide synthesizer (Syro-I, Biotage, Uppsala, Sweden), using standard Fmoc/tBu strategy with DIC/HOBt coupling reagents. Isoniazid was conjugated to the *N*-terminus of the peptides by using glyoxylic acid as a linker (the synthesis is detailed in [35]). Fluorescently labeled derivatives were synthesized with the use of 5(6)-carboxyfluorescein (*Cf*) with DIC/HOBt coupling method. Peptides were cleaved from the resin with TFA/H_2O/TIS (9.5:2.5:2.5 *v/v*) mixture (2 h, RT). After filtration, compounds were precipitated in cold diethyl ether, centrifuged (4000 rpm, 5 min) and freeze-dried from water.

RP-HPLC purification was performed on a Thermo Fisher UltiMate 3000 Semiprep HPLC with a Phenomenex Jupiter Proteo C-12 column (250 × 10 mm), using gradient elution, consisting of 0.1% TFA in water (eluent A) and 0.1% TFA in acetonitrile/water = 80/20 (*v/v*) (eluent B).

Purified peptides were analyzed by RP-HPLC, on an analytical C-18 column, using gradient elution with the abovementioned eluent A and B (flow rate was 1 mL/min, UV detection at $\lambda = 220$ nm). Molecular mass of the peptides was determined on a Thermo Scientific Q Exactive™ Focus Hybrid Quadrupole-Orbitrap™ Mass Spectrometer (ThermoFisher Scientific, Waltham, MA, USA).

3.2. Membrane Affinity Measurements

Monomolecular lipid membrane models were used to characterize the interactions between the monolayers and a series of isoniazid (INH)-conjugated peptides. The membrane affinity was investigated by using the Langmuir technique. Surface pressure was recorded with the tensiometric method, with ±0.5 mN/m accuracy, employing previously purified filter papers, such as Wilhelmy plates. The trough was made of Teflon and was equipped with one surface pressure sensor and two polyoxymethylene (POM) barriers. Dichloromethane (VWR Chemicals, analytical purity) and methanol (VWR Chemicals, Reag. Ph. Eur. for HPLC) were used for the cleaning of the troughs and barriers before each measurement. Then, 1,2-Dipalmitoyl-sn-glycero-3-phosphocholine (DPPC) (Avanti Polar Lipids Inc., Alabaster, AL, USA) and its mixture with mycolic acid from *Mycobacterium tuberculosis* (human strain) (98% Sigma-Aldrich) were applied in chloroform (VWR Chemicals, analytical purity), to form the monomolecular lipid layer on the top of the double-distilled water subphase in the Langmuir through. (Molecular structure of isoniazid, DPPC and mycolic acid are presented in Figure 8). The composition of the lipid mixture was a mass ratio of 75w% DPPC and 25w% mycolic acid. Double-distilled water was checked by its conductivity (<5 mS) and surface tension (>72.0 mN/m at 23 ± 0.5 °C) values. After the cleaning procedure, the through was filled with double-distilled water. DPPC or the mixture of DPPC and mycolic acid was dissolved in chloroform to a concentration 0.2 mg/mL. The lipid solution was spread on the subphase. The system was left for 10 min, to evaporate the chloroform, before the compression. Surface pressure–area isotherms were recorded two times before each penetration measurement. The lipid layer was conditioned by barrier adjustment for 20 min, and the aqueous solution of the peptide was injected under the lipid monolayer, after 10 min relaxation time. The peptides were dissolved in double-distilled water, at 200 μM concentration. This peptide solution was injected into the subphase, to achieve 2 μM final concentration. The changes of the surface pressure as a function of time were recorded for 1 h.

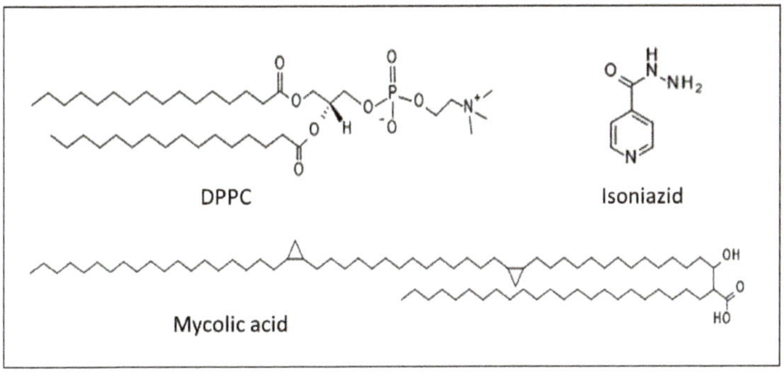

Figure 8. Chemical composition of the drug and lipids applied in Langmuir monolayer experiments

3.3. Circular Dichroism (CD) Spectroscopic Measurements

Peptide samples were dissolved in deionized water and in TFE at 4–6 μM concentrations. Far-UV CD curves were taken on a JASCO J-715 spectropolarimeter, at 25 ± 0.2 °C, in 0.5 cm path-length rectangular quartz cuvette (Hellma, Plainview, NY, USA). Temperature control was provided by a Peltier thermostat. The CD data were monitored in continuous scanning mode, between 190 and 260 nm, at a rate of 100 nm/min, with a step size of 0.1 nm, response time of 2 s, eight accumulations and 2 nm bandwidth. The CD curves were corrected by spectral contribution of the blank solvent. CD spectra were plotted in mean residue molar CD units ($\Delta\varepsilon$/residue) calculated by the following equation:

$$\Delta\varepsilon = \Theta/(32.98cl) \tag{1}$$

where Θ is the measured ellipticity (deg) as a function of wavelength (nm), c is the mean residue molar concentration and l is the optical path length (cm).

3.4. Atomic Force Microscopy Measurements

Penetrated Langmuir monolayers were transferred to a previously cleaned hydrophilic cover glass for atomic microscopy measurements. The cover glasses were cleaned by ultra-cleaning, with the mixture of H_2O_2 and H_2SO_4, for two hours. After washing and drying, they were suitable for the deposition. The structure of the monolayers was investigated by Nanosurf Flex-Axiom (Liestal, Switzerland) atomic force microscope system, operating in contact mode. Silicon cantilever ContGD-G (BudgetSensors) was used, with a tip curvature radius less than 10 nm, an average force constant of 0.2 N/m, resonance frequency of 13 kHz and with gold reflective coating. Each sample was imaged in $2 \times 2\ \mu m^2$ areas, at 10 randomly selected locations. The image analysis was performed by Gwyddion 2.50 program.

3.5. In Vitro Assays

3.5.1. Cellular Uptake and Localization of the Peptides

Internalization of *Cf*-labeled Penetratin and Transportan was measured on EBC-1 (squamous cell carcinoma, origin: bronchi) [58,59]. Cells were maintained as an adherent culture in DMEM medium supplemented with 10% heat-inactivated fetal calf serum (FCS) L-glutamine (2 mM) and gentamicin (35 µM), at 37 °C, in a humidified atmosphere containing 5% CO_2. For the assay, cells were treated with peptides at 5 and 10 µM final concentration and were incubated for 2 h. After centrifugation (1000 rpm, 5 min) and washing with serum-free DMEM medium, supernatant was removed, and 100 µL 0.25% trypsin was added to the cells. After 10 min incubation, 0.8 mL 10% FCS/HPMI medium was added, and then cells were washed and resuspended in 0.25 mL HPMI medium. The intracellular fluorescence intensity of the cells was measured on a BD LSR II flow cytometer (BD Biosciences, San Jose, CA, USA) on channel FITC (emission at λ = 505 nm), and data were analyzed with FACSDiva 5.0 software. All measurements were performed in triplicates, and the mean fluorescent intensity, together with standard error of the mean (SEM), was graphically presented.

Localization of the peptides was visualized by confocal microscopy. EBC-1 cells were seeded (10^5 cells/well) one day prior to treatment, on cover glass containing 24-well plates (Sarstedt, Nümbrecht, Germany). Lysosomes were stained by LysoTracker™ Deep Red (Invitrogen, Carlsbad, CA, USA) for 30 min, followed by incubation with *Cf*-labeled Penetratin and Transportan for 60 min. Subsequently, nuclei were stained by Hoechst 33342 solution (Thermo Scientific, Waltham, MA, USA). After each step, cells were washed three times with serum-free medium. Cells were fixed by 4% paraformaldehyde for 15 min and mounted to microscopy slides with Mowiol® 4–88. Imaging was performed on a Zeiss LSM 710 system with 40× oil immersion objective. Images were processed with ZEN lite software (Zeiss, Oberkochen, Germany).

3.5.2. Determination of Antibacterial Efficacy against Intracellular *Mycobacterium tuberculosis*

MonoMac-6 human monocytes (2×10^5 cells/1 mL medium/well) were cultured in RPMI-1640 medium containing 10 % FCS, in a 24-well plate, 24 h prior to the experiment. Adherent cells were infected with *Mycobacterium tuberculosis* $H_{37}Rv$, at a multiplicity of infection (MOI) of 10. Nonphagocytosed extracellular bacteria were removed by washing the culture three times with serum-free RPMI. After 1 day of incubation, infected cells were treated with INH-conjugated Penetratin and Transportan 5, 10 and 50 µM final concentrations. For comparison, INH was also assayed at 100 µM concentration. As control, infected cells were treated with culture media. After 3 days, the treatment was repeated with fresh solutions of the compounds, and cells were incubated for an additional 3 days. After 3 washing steps, infected cells were lysed with 2.5 % sodium dodecyl sulphate (SDS) solution, and 100 µL lysate was transferred to L-J medium (BBL Löwenstein–Jensen medium, Beckton Dickinson). Colony-forming units (CFU) were enumerated after 4 weeks of incubation, and the number of bacteria was calculated by using a standard dilution series of *M. tuberculosis* [60].

4. Conclusions

INH-conjugated peptides with various composition, hydrophobicity and charge character were compared in an experimental membrane affinity study, using lipid monolayer models. In most cases, increased membrane affinity was observed, especially with mycolic-acid-containing lipid layer induced by the conjugation with INH. Hydrophobic character of the peptides seems to favor the interaction with lipid layers within the set of cationic peptides studied here. CD spectroscopic measurement allowed for the analysis of the correlation between the secondary structural change of the peptides and their INH-conjugates and the membrane affinity. The penetration into the lipid layer was in line with the destruction of ordered structure of lipid layer observed visually by AFM in nanometric range.

In vitro assays supported the cellular uptake and antibacterial effect of the INH–peptide conjugates; furthermore, an important difference in the intracellular localization was also revealed, depending on the carrier peptide. Nuclear translocation of Penetratin and its ability to carry various chemically conjugated cargos within the nucleus has been reported so far [56], and this feature made Penetratin a good candidate to utilize in gene therapy [61,62]. Our recent study added a supplement feature to Penetratin as a carrier molecule, namely the use as vector for peptide delivery of antitubercular agents and utilization of it against intracellular bacteria.

Author Contributions: K.H. performed the peptides synthesis and characterization; E.P., K.H., S.B., B.S., B.B.-K. and F.Z. performed the experiments, analyzed the data and wrote the paper; É.K. accomplished writing—review and editing the paper. All authors have read and agree to the published version of the manuscript.

Funding: This work was financed by the ELTE Institutional Excellence Program (NKFIH-1157-8/2019-DT), which is supported by the Hungarian Ministry of Human Capacities; by the ELTE Thematic Excellence Programme supported by the Hungarian Ministry for Innovation and Technology (SZINT+); and by grants from the European Union and the State of Hungary, co-financed by the European Regional Development Fund (VEKOP-2.3.3-15-2017-00020, VEKOP-2.3.2-16-2017-00014). Project no. 2018-1.2.1-NKP-2018-00005 was implemented with the support provided from the National Research Development and Innovation Fund of Hungary.

Acknowledgments: Authors thank László Buday for the confocal microscopy; Gitta Schlosser for the HRMS measurements; Zsuzsa Senoner, Sándor Dávid, Zoltán Kis, Bernadett Pályi, Judit Henczkó and Nóra Magyar for the mycobacterial work and the BSL-3 laboratory guide. K. Horváti was supported by the Premium Postdoctoral Research Program 2019 of the Hungarian Academy of Sciences. E. Pári was supported by the ÚNKP-19-3 New National Excellence Program of the Ministry for Innovation and Technology. F. Zsila acknowledges the support of the National Competitiveness and Excellence Program (NVKP_16-1-2016-0007) and the BIONANO_GINOP-2.3.2-15-2016-00017b project.

Conflicts of Interest: The authors declare no conflict of interest.

References

1. Han, X.Y.; Seo, Y.H.; Sizer, K.C.; Schoberle, T.; May, G.S.; Spencer, J.S.; Li, W.; Nair, R.G. A new Mycobacterium species causing diffuse lepromatous leprosy. *Am. J. Clin. Pathol.* **2008**, *13*, 856–864. [CrossRef] [PubMed]

2. WHO. *Global Tuberculosis Report 2018*; Report No.; WHO: Geneva, Switzerland, 2018; ISBN 978-92-4-156564-6.

3. Chiaradia, L.; Lefebvre, C.; Parra, J.; Marcoux, J.; Burlet-Schiltz, O.; Etienne, G.; Tropis, M.; Daffé, M. Dissecting the mycobacterial cell envelope and defining the composition of the native mycomembrane. *Nat. Sci. Rep.* **2017**, *12807*, 1–12. [CrossRef] [PubMed]

4. Zuber, B.; Chami, M.; Houssin, C.; Dubochet, J.; Griffiths, G.; Daffé, M. Direct visualization of the outer membrane of mycobacteria and corynebacteria in their native state. *J. Bacteriol.* **2008**, *190*, 5672–5680. [CrossRef] [PubMed]

5. Le Chevalier, F.; Cascioferro, A.; Majlessi, L.; Herrmann, J.L.; Brosch, R. Mycobacterium tuberculosis evolutionary pathogenesis and its putative impact on drug development. *Future Microbiol.* **2014**, *9*, 969–985. [CrossRef]

6. Hasegawa, T.; Amino, S.; Kitamura, S.; Matsumoto, L.; Katada, S.; Nishijo, J. Study of the Molecular Conformation of r- and Keto-Mycolic Acid Monolayers by the Langmuir-Blodgett Technique and Fourier Transform Infrared Reflection-Absorption Spectroscopy. *Langmuir* **2003**, *19*, 105–109. [CrossRef]

7. Chimote, G.; Banerjee, R. Lung surfactant dysfunction in tuberculosis: Effect of mycobacterial tubercular lipids on dipalmitoylphosphatidylcholine surface activity. *Colloid. Surface B.* **2005**, *45*, 215–223. [CrossRef]

8. Groenewald, W.; Baird, M.S.; Verschoor, J.A.; Minnikin, D.E.; Croft, A.K. Differential spontaneous folding of mycolic acids from Mycobacterium tuberculosis. *Chem. Phys. Lipids* **2014**, *180*, 15–22. [CrossRef]

9. Pinheiro, M.; Giner-Casares, J.J.; Lúcio, M.; Caio, J.M.; Moiteiro, C.; Lima, J.L.; Reis, S.; Camacho, L. Interplay of mycolic acids, antimycobacterial compounds and pulmonary surfactant membrane: A biophysical approach to disease. *Biochim. Biophys. Acta* **2013**, *1828*, 896–905. [CrossRef]

10. McNeil, M.; Brennan, P. Structure, function and biogenesis of the cell envelope of mycobacteria in relation to bacterial physiology, pathogenesis and drug resistance; some thoughts and possibilities arising from recent structural information. *Res. Microbiol.* **1991**, *142*, 451–463. [CrossRef]

11. Phan, M.D.; Shin, K.A. Langmuir Monolayer: Ideal Model Membrane to Study Cell. *J. Chem. Biol. Interfaces* **2014**, *2*, 1–5. [CrossRef]

12. Gyulai, G.; Pénzes, C.B.; Mohai, M.; Csempesz, F.; Kiss, É. Influence of surface properties of polymeric drug delivery nanoparticles on their membrane affinity. *Eur. Polym. J.* **2013**, *49*, 2495–2503. [CrossRef]

13. Ábrahám, Á.; Katona, M.; Kasza, G.; Kiss, É. Amphiphilic polymer layer–cell membrane interaction studied by QCM and AFM in model systems. *Eur. Polym. J.* **2017**, *93*, 212–221. [CrossRef]

14. Hasegawa, T.; Nishijo, J.; Watanabe, M. Conformational characterization of r-mycolic acid in a monolayer film by the Langmuir-Blodgett technique and atomic force microscopy. *Langmuir* **2000**, *16*, 7325–7330. [CrossRef]

15. Knobloch, J.; Suhendro, D.K.; Zieleniecki, J.L.; Shapter, J.G.; Köper, I. Membrane–drug interactions studied using model membrane systems. *Saudi. J. Biol. Sci.* **2015**, *22*, 714–718. [CrossRef] [PubMed]

16. Chimote, G.; Banerjee, R. Effect of antitubercular drugs on dipalmitoylphosphatidylcholine monolayers: Implications for drug loaded surfactants. *Respir. Physiol. Neurobiol.* **2005**, *145*, 65–77. [CrossRef]

17. Pénzes, C.B.; Schnöller, D.; Horváti, K.; Bősze, S.; Mező, G.; Kiss, É. Membrane affinity of antituberculotic drug conjugate using lipid monolayer containing mycolic acid. *Colloid. Surface. A.* **2012**, *413*, 142–148. [CrossRef]

18. Hill, K.; Pénzes, C.B.; Schnöller, D.; Horváti, K.; Bősze, S.; Hudecz, F.; Keszthelyi, T.; Kiss, É. Characterization of the membrane affinity of an isoniazide peptide-conjugate by tensiometry, atomic force microscopy and sum-frequency vibrational spectroscopy, using a phospholipid Langmuir monolayer model. *Phys. Chem. Chem. Phys.* **2010**, *12*, 11498–11506. [CrossRef]

19. Ábrahám, Á.; Baranyai, Z.; Gyulai, G.; Pári, E.; Horváti, K.; Bősze, S.; Kiss, É. Comparative analysis of new antitubercular drug peptide conjugates—Model membrane and in vitro studies. *Colloid. Surface B.* **2016**, *147*, 106–115. [CrossRef]

20. Schnöller, D.; Pénzes, C.B.; Horváti, K.; Bősze, S.; Hudecz, F.; Kiss, É. Membrane affinity of new antitubercular drug candidates using a phospholipid Langmuir monolayer model and LB technique. *Prog. Colloid Polym. Sci.* **2011**, *138*, 131–138.

21. Kiss, É.; Heine, E.T.; Hill, K.; He, Y.-C.; Keusgen, N.; Pénzes, C.B.; Schnöller, D.; Gyulai, G.; Mendrek, A.; Keul, H.; et al. Membrane affinity and antimicrobal properties of polyelectrolytes with different hydrophobicity. *Macromol. Biosci.* **2012**, *12*, 1181–1189. [CrossRef]

22. Nyström, L.; Malmsten, M. Membrane interactions and cell selectivity of amphiphilic anticancer peptides. *Curr. Opin. Colloid Interface Sci.* **2018**, *38*, 1–17. [CrossRef]

23. Derossi, D.; Joliot, A.H.; Chassaing, G.; Prochiantz, A. The third helix of the Antennapedia homeodomain translocates through biological membranes. *J. Biol. Chem.* **1994**, *269*, 10444–10450.

24. Langel, U.; Pooga, M.; Kairane, C.; Zilmer, M.; Bartfai, T. A galanin-mastoparan chimeric peptide activates the Na+,K(+)-ATPase and reverses its inhibition by ouabain. *Regul. Pept.* **1996**, *62*, 47–52. [CrossRef]

25. Park, C.B.; Kim, M.S.; Kim, S.C. A novel antimicrobial peptide from Bufo bufo gargarizans. *Biochem. Biophys. Res. Commun.* **1996**, *218*, 408–413. [CrossRef]

26. Zasloff, M. Magainins, a class of antimicrobial peptides from Xenopus skin: Isolation, characterization of two active forms, and partial cDNA sequence of a precursor. *Proc. Natl. Acad. Sci. USA* **1987**, *84*, 5449–5453. [CrossRef]

27. Helmerhorst, E.J.; Van't Hof, W.; Veerman, E.C.; Simoons-Smit, I.; Amerongen, A.V.N. Synthetic histatin analogues with broad-spectrum antimicrobial activity. *Biochem. J.* **1997**, *326*, 39–45. [CrossRef]

28. Andreu, D.; Ubach, J.; Boman, A.; Wahlin, B.; Wade, D.; Merrifield, R.B.; Boman, H.G. Shortened cecropin A-melittin hybrids. Significant size reduction retains potent antibiotic activity. *FEBS Lett.* **1992**, *296*, 190–194. [CrossRef]

29. Radis-Baptista, G.; de la Torre, B.G.; Andreu, D. A novel cell-penetrating peptide sequence derived by structural minimization of a snake toxin exhibits preferential nucleolar localization. *J. Med. Chem.* **2008**, *51*, 7041–7044. [CrossRef]

30. Mezo, G.; Kalászi, A.; Reményi, J.; Majer, Z.; Hilbert, A.; Láng, O.; Köhidai, L.; Barna, K.; Gaál, D.; Hudecz, F. Synthesis, conformation, and immunoreactivity of new carrier molecules based on repeated tuftsin-like sequence. *Biopolymers* **2004**, *73*, 645–656. [CrossRef]

31. Kalafatovic, D.; Giralt, E. Cell-penetrating peptides: Design strategies beyond primary structure and amphipathicity. *Molecules* **2017**, *22*, 1929. [CrossRef]

32. Di Pisa, M.; Chassaing, G.; Swiecicki, J.M. Translocation mechanism(s) of cell-penetrating peptides: Biophysical studies using artificial membrane bilayers. *Biochemistry* **2015**, *54*, 194–207. [CrossRef]

33. Lee, E.Y.; Wong, G.C.L.; Ferguson, A.L. Machine learning-enabled discovery and design of membrane active peptides. *Bioorg. Med. Chem.* **2018**, *26*, 2708–2718. [CrossRef]

34. Horváti, K.; Bacsa, B.; Mlinkó, T.; Szabó, N.; Hudecz, F.; Zsila, F.; Bősze, S. Comparative analysis of internalisation, haemolytic, cytotoxic and antibacterial effect of membrane-active cationic peptides: Aspects of experimental setup. *Amino Acids* **2017**, *49*, 1053–1067. [CrossRef]

35. Horváti, K.; Mező, G.; Szabó, N.; Hudecz, F.; Bősze, S. Peptide conjugates of therapeutically used antitubercular isoniazid-design, synthesis and antimycobacterial effect. *J. Pept. Sci.* **2009**, *15*, 385–391. [CrossRef]

36. Hopp, T.P.; Woods, K.R. Prediction of protein antigenic determinants from amino acid sequences. *Proc. Natl. Acad. Sci. USA* **1981**, *78*, 3824–3828. [CrossRef]

37. Kiss, É.; Gyulai, G.; Pári, E.; Horváti, K.; Bősze, S. Membrane affinity and fluorescent labelling: Comparative study of monolayer interaction, cellular uptake and cytotoxicity profile of carboxyfluorescein-conjugated cationic peptides. *Amino Acids* **2018**, *50*, 1557–1571. [CrossRef]

38. Via, M.A.; Del Pópolo, M.G.; Wilke, N. Negative dipole potentials and carboxylic polar head-groups foster the insertion of cell-penetrating-peptides into lipid monolayers. *Langmuir* **2018**, *34*, 3102–3111. [CrossRef]

39. Toniolo, C.; Formaggio, F.; Woody, R.W. Electronic Circular Dichroism of Peptides. In *Comprehensive Chiroptical Spectroscopy: Applications in Stereochemical Analysis of Synthetic Compounds, Natural Products, and Biomolecules*; Berova, N., Polavarapu, P.L., Nakanishi, K., Woody, R.W., Eds.; John Wiley & Sons: Hoboken, NJ, USA, 2012; pp. 499–544.

40. Gopal, R.; Park, J.S.; Seo, C.H.; Park, Y. Applications of circular dichroism for structural analysis of gelatin and antimicrobial peptides. *Int. J. Mol. Sci.* **2012**, *13*, 3229–3244. [CrossRef]

41. Chen, Y.; Guarnieri, M.T.; Vasil, A.I.; Vasil, M.L.; Mant, C.T.; Hodges, R.S. Role of peptide hydrophobicity in the mechanism of action of α-helical antimicrobial peptides. *Antimicrob. Agents. Chemother.* **2007**, *51*, 1398–1406. [CrossRef]

42. Gong, Z.; Ikonomova, S.P.; Karlsson, A.J. Secondary structure of cell-penetrating peptides during interaction with fungal cells. *Protein. Sci.* **2018**, *27*, 702–713. [CrossRef]

43. Dathe, M.; Wieprecht, T. Structural features of helical antimicrobial peptides: Their potential to modulate activity on model membranes and biological cells. *Biochim. Biophys. Acta* **1999**, *1462*, 71–87. [CrossRef]

44. Arouri, A.; Dathe, M.; Blume, A. The helical propensity of KLA amphipathic peptides enhances their binding to gel-state lipid membranes. *Biophys. Chem.* **2013**, *180*, 10–21. [CrossRef]

45. Cherry, M.A.; Higgins, S.K.; Melroy, H.; Lee, H.S.; Pokorny, A. Peptides with the same composition, hydrophobicity, and hydrophobic moment bind to phospholipid bilayers with different affinities. *J. Phys. Chem. B.* **2014**, *118*, 12462–12470. [CrossRef]

46. Yan, J.; Liang, X.; Liu, C.; Cheng, Y.; Zhou, L.; Wang, K.; Zhao, L. Influence of proline substitution on the bioactivity of mammalian-derived antimicrobial peptide NK-2. *Probiotics Antimicrob. Proteins* **2018**, *10*, 118–127. [CrossRef]

47. Imai, K.; Mitaku, S. Mechanisms of secondary structure breakers in soluble proteins. *Biophysics* **2005**, *1*, 55–65. [CrossRef]

48. Zsila, F.; Bősze, S.; Horváti, K.; Szigyártó, I.C.; Beke-Somfai, T. Drug and dye binding induced folding of the intrinsically disordered antimicrobial peptide CM15. *RSC Adv.* **2017**, *7*, 41091–41097. [CrossRef]

49. Zsila, F.; Juhász, T.; Bősze, S.; Horváti, K.; Beke-Somfai, T. Hemin and bile pigments are the secondary structure regulators of intrinsically disordered antimicrobial peptides. *Chirality* **2018**, *30*, 195–205. [CrossRef]

50. Zsila, F.; Kohut, G.; Beke-Somfai, T. Disorder-to-helix conformational conversion of the human immunomodulatory peptide LL-37 induced by antiinflammatory drugs, food dyes and some metabolites. *Int. J. Biol. Macromol.* **2019**, *129*, 50–60. [CrossRef]

51. Garcin, D.; Marq, J.B.; Iseni, F.; Martin, S.; Kolakofsky, D. A short peptide at the amino terminus of the Sendai virus C protein acts as an independent element that induces STAT1 instability. *J. Virol.* **2004**, *78*, 8799–8811. [CrossRef]

52. Fealey, M.E.; Binder, B.P.; Uversky, V.N.; Hinderliter, A.; Thomas, D.D. Structural impact of phosphorylation and dielectric constant variation on synaptotagmin's IDR. *Biophys. J.* **2018**, *114*, 550–561. [CrossRef]

53. Anderson, V.L.; Ramlall, T.F.; Rospigliosi, C.C.; Webb, W.W.; Eliezer, D. Identification of a helical intermediate in trifluoroethanol-induced alpha-synuclein aggregation. *Proc. Natl. Acad. Sci. USA* **2010**, *107*, 18850–18855. [CrossRef] [PubMed]

54. Nalbandian, A.; Yan, B.S.; Pichugin, A.; Bronson, R.T.; Kramnik, I. Lung carcinogenesis induced by chronic tuberculosis infection: The experimental model and genetic control. *Oncogene* **2009**, *28*, 1928–1938. [CrossRef] [PubMed]

55. Cukic, V. The Association Between Lung Carcinoma and Tuberculosis. *Med. Arch.* **2017**, *71*, 212–214. [CrossRef] [PubMed]

56. Derossi, D.; Chassaing, G.; Prochiantz, A. Trojan peptides: The penetratin system for intracellular delivery. *Trends. Cell. Biol.* **1998**, *8*, 84–87. [CrossRef]

57. LeCher, J.C.; Nowak, S.J.; McMurry, J.L. Breaking in and busting out: Cell-penetrating peptides and the endosomal escape problem. *Biomol. Concepts* **2017**, *8*, 131–141. [CrossRef]

58. Hiraki, S.; Miyai, M.; Seto, T.; Tamura, T.; Watanabe, Y.; Ozawa, S.; Ikeda, H.; Nakata, Y.; Ohnoshi, T.; Kimura, I. Establishment of human continuous cell lines from squamous cell, adeno- and small cell carcinoma of the lung and the results of hetero transplantation. *Lung Cancer* **1982**, *22*, 53–58.

59. Imanishi, K.; Yamaguchi, K.; Suzuki, M.; Honda, S.; Yanaihara, N.; Abe, K. Production of transforming growth factor-alpha in human tumour cell lines. *Br. J. Cancer* **1989**, *59*, 761–765. [CrossRef]

60. Horváti, K.; Bacsa, B.; Szabó, N.; Dávid, S.; Mező, G.; Grolmusz, V.; Vértessy, B.; Hudecz, F.; Bősze, S. Enhanced cellular uptake of a new, in silico identified antitubercular candidate by peptide conjugation. *Bioconjug. Chem.* **2012**, *23*, 900–907.

61. Abes, R.; Arzumanov, A.A.; Moulton, H.M.; Abes, S.; Ivanova, G.D.; Iversen, P.L.; Gait, M.J.; Lebleu, B. Cell-penetrating-peptide-based delivery of oligonucleotides: An overview. *Biochem. Soc. Trans.* **2007**, *35*, 775–779. [CrossRef]

62. Mäe, M.; Langel, U. Cell-penetrating peptides as vectors for peptide, protein and oligonucleotide delivery. *Curr. Opin. Pharmacol.* **2006**, *6*, 509–514. [CrossRef]

 International Journal of
Molecular Sciences

Article

Revisiting the Interaction of Melittin with Phospholipid Bilayers: The Effects of Concentration and Ionic Strength

Thiru Sabapathy [1], Evelyne Deplazes [1,2,†] and Ricardo L. Mancera [1,*,†]

[1] School of Pharmacy and Biomedical Sciences, Curtin Health Innovation Research Institute, Curtin University, GPO Box U1987, Perth, WA 6845, Australia; t.sabapathy@curtin.edu.au (T.S.); evelyne.deplazes@uts.edu.au (E.D.)

[2] School of Life Sciences, University of Technology Sydney, Ultimo, NSW 2007, Australia

* Correspondence: R.Mancera@curtin.edu.au

† Co-corresponding authors.

Received: 19 December 2019; Accepted: 21 January 2020; Published: 23 January 2020

Abstract: Melittin is an anti-microbial peptide (AMP) and one of the most studied membrane-disrupting peptides. There is, however, a lack of accurate measurements of the concentration-dependent kinetics and affinity of binding of melittin to phospholipid membranes. In this study, we used surface plasmon resonance spectroscopy to determine the concentration-dependent effect on the binding of melittin to 1-palmitoyl-2-oleoyl-glycero-3-phosphocholine (POPC) bilayers in vesicles. Three concentration ranges were considered, and when combined, covered two orders of magnitudes (0.04 µM to 8 µM), corresponding to concentrations relevant to the membrane-disrupting and anti-microbial activities of melittin. Binding kinetics data were analysed using a 1:1 Langmuir-binding model and a two-state reaction model. Using in-depth quantitative analysis, we characterised the effect of peptide concentration, the addition of NaCl at physiological ionic strength and the choice of kinetic binding model on the reliability of the calculated kinetics and affinity of binding parameters. The apparent binding affinity of melittin for POPC bilayers was observed to decrease with increasing peptide/lipid (P/L) ratio, primarily due to the marked decrease in the association rate. At all concentration ranges, the two-state reaction model provided a better fit to the data and, thus, a more reliable estimate of binding affinity. Addition of NaCl significantly reduced the signal response during the association phase; however, no substantial effect on the binding affinity of melittin to the POPC bilayers was observed. These findings based on POPC bilayers could have important implications for our understanding of the mechanism of action of melittin on more complex model cell membranes of higher physiological relevance.

Keywords: surface plasmon resonance; melittin; liposomes; peptide–lipid interactions; anti-microbial peptides; pore-forming peptides

---

1. Introduction

Anti-microbial peptides (AMPs) are found throughout the animal and plant kingdom, where they form an important part of the innate immune system [1]. Due to the fact of their potent activity against Gram-positive and Gram-negative bacteria as well as other pathogens, AMPs have been actively pursued as lead molecules for the development of new antimicrobial agents [2–5]. In addition, AMPs have been investigated for their anti-cancer activity to address the issue of chemotherapy resistance [6–8], and there is emerging evidence for their use as adjuvant cancer therapeutics [9,10].

The cytotoxic effect of many AMPs primarily stems from their ability to lyse the phospholipid bilayer of the plasma or mitochondrial cell membranes [11–13]. While AMPs show a wide range of secondary structures and different mechanisms of action, they share certain physico-chemical characteristics. Like other membrane-disruptive peptides, AMPs are positively charged (cationic) and have a relatively large number of hydrophobic residues that are often clustered to give the peptide an overall amphipathic character [5,11–15]. Their cationic nature enables electrostatic interactions with the negatively charged (anionic) lipid head groups found in bacterial membranes, while the amphipathic nature facilitates insertion into the hydrophobic core of the membrane. However, there is still an incomplete understanding of the molecular mechanisms by which AMPs bind to and, subsequently, disrupt cell membranes, as well as the physico-chemical factors affecting the interaction [12]. Characterising the membrane-binding properties of AMPs is particularly important for the development of therapeutic peptides (e.g., antibiotics or anti-cancer agents) with higher specificity for microbes or cancer cells.

One of the most studied AMPs is melittin [16], the major component of honey bee venom [17]. Melittin is a 26 amino acid long cationic peptide and its cytolytic effects on lipid vesicles as well as on bacterial and mammalian cells has been demonstrated in a large number of studies [18–27]. Early work using nuclear magnetic resonance (NMR) and circular dichroism (CD) spectroscopy on the structure of melittin in water and lipid environments showed that the peptide is mostly unstructured in solution but undergoes a transition to α-helix upon binding to a lipid–water or hydrophobic interface [28–36]. Melittin is also known to dimerise and form higher-order oligomers under certain conditions such as high peptide concentrations and high ionic strength [29,37–39].

The membrane-disrupting mechanism of melittin has been studied extensively using a wide range of experimental and molecular simulation approaches [40]. Findings from these studies suggest that membrane-bound melittin can be present in two orientations: parallel, where it lies on the membrane surface interacting mostly with the lipid head groups, and perpendicular, where it is inserted into the hydrophobic core of the membrane, oriented perpendicularly to the membrane surface. A two-step model is often used to describe the mechanism of formation of membrane pores by melittin [26,41,42], wherein the peptide, at low concentration, binds to the bilayer surface in a parallel conformation and then shifts to a perpendicular orientation at higher concentration, leading to membrane pore formation. The existence of two membrane-bound orientations has been demonstrated by molecular simulation studies [43–48] and the retention of amphiphilic peptides in a surface-absorbed state has been explained by elasticity theory [49].

A recent study has suggested a variation to this two-step model, in which the transition from parallel to perpendicular does not occur but, rather, parallel surface binding and direct insertion are competing processes [26]. Numerous studies of melittin make it clear that surface binding, membrane disruption and pore formation constitute an inter-dependent and complex process. In addition, each one of these processes is affected by environmental factors such as membrane lipid composition [23,27,39,50–53], peptide concentration and peptide–lipid ratio [21,54], pH and ionic strength [51,53], and local membrane curvature [55].

Like other AMPs, melittin exhibits preferred binding to negatively charged (anionic) lipids [53,56] due to the strong electrostatic attraction between the C-terminal region of melittin and negatively charged lipid headgroups [53]. It has been suggested that the basis for the strong attraction between melittin and anionic lipids is an "electrostatic arrest" (adsorption) of melittin in its parallel (inactive) orientation on the lipid surface [53]. Indeed, this postulate is used to explain the tolerance of anionic liposomes to membrane disruption by melittin [32,57,58]. This suggests that, compared to neutral (or zwitterionic membranes), the presence of anionic lipids promotes increased surface adsorption but, by favouring the (inactive) parallel conformation of melittin, it also hinders pore formation and, thus, reduces leakage. Likewise, melittin interacts with the headgroups of zwitterionic lipids (e.g., phosphatidylcholine), but the interaction is much weaker than for anionic lipids (e.g., phosphatidylglycerol or phosphatidylinositol) [53,55,59,60].

Among the most commonly used methods to study membrane binding of AMPs, including melittin, are isothermal titration calorimetry (ITC) [61] and surface plasmon resonance (SPR) [62–65]. Other methods used for the study of melittin to lipid membranes include ultrafiltration assays [55,59] and fluorescence measurements [53]. Isothermal titration calorimetry has been used to determine the enthalpy (ΔH), entropy (ΔS) and free energy of binding (ΔG) as well as the associated equilibrium dissociation constant (K_D) of the interaction between melittin and model cell membranes with various lipid compositions [66,67]. Surface plasmon resonance has provided real-time measurements of the association (k_a) and dissociation rates (k_d) of binding as well as estimates of ΔG for melittin–lipid interactions [62,63,65,68]. In one of the earliest SPR studies, Aguilar and co-workers [64] reported on the binding of melittin to dimyristoylphosphocholine (DMPC) and dimyristoylphosphatidylglycerol (DMPG) using peptide concentrations between 10 and 140 µM. While this concentration range is not high enough to induce the formation of the tetrameric aggregate of melittin in solution [28,29,69–71], it may likely cause the accumulation of the peptide on the membrane surface thus affecting both the subsequent binding of melittin and the barrier properties of the lipid bilayer [72]. This is important because, based on the classical carpet model [73], it has been postulated that, at high concentration, cationic amphiphilic peptides (including melittin) can accumulate at the lipid surface creating an asymmetry of mass, charge and surface pressure. Subsequently, this asymmetry is dissipated by the re-organization of the lipid bilayer, leading to a transient increase in the permeability of the peptides across the lipid bilayer until an equilibrium is established between both sides of the bilayer. A similar occurrence of transient permeability and the subsequent appearance of resistance in the lipid bilayer following the action of melittin has been shown in phosphatidylcholine (PC) vesicles at a peptide/lipid ratio of 1/200 [40]. Consequently, experiments conducted at high peptide concentrations could be problematic when studying the interaction of peptides with lipid bilayers. In addition, the concentrations of melittin required to induce leakage [17,18,24,25,55,57] are in the sub-µM to low-µM range, and using much higher peptide concentrations might affect the relevance of the observed binding to the mechanism of membrane disruption. In a more recent SPR study, Aguilar and co-workers [62] re-investigated the interaction of melittin with DMPC at lower concentrations (0.125–12 µM). This study reported a concentration-dependent change in binding (resonance units or RU levels); however, it did not quantify the changes in binding affinity due to the reported poor fits for the 1:1 and two-state reaction binding models.

In this study, we carried out SPR experiments of the binding of melittin to 1-palmitoyl-2-oleoylphosphatidylcholine (POPC) membranes using three sets of concentration ranges that covered two orders of magnitude (0.04–8.0 µM), and also investigated the effect of ionic strength on this interaction. To our knowledge, this is the first study reporting extensive quantitative analysis of the concentration-dependent kinetics and binding affinity of melittin and how this might relate to the various stages of its mechanism of membrane binding and insertion.

2. Results

The POPC SUVs were deposited on the L1 sensor chip at a flow rate of 5 µL/min for 60 min, attaining an immobilization level up to a maximum of ~8400 RU. Following immobilization, multi-cycle or single-cycle kinetics experiments were conducted and the resulting sensorgram data were fitted to either the 1:1 Langmuir model or a two-state reaction model to estimate k_a, k_d and K_D of the melittin–POPC interaction. As noted in Section 2, single-cycle kinetics experiments were used for the low-range and medium-range concentrations, while multi-cycle kinetics experiments were used for the high-range concentrations.

2.1. Effect of Analyte Concentration on Kinetics Analysis

One of the aims of this study was to investigate the effect of melittin (analyte) concentration on the apparent affinity of its interaction with POPC, the quality of fit to the data using different kinetic binding models and the subsequent reliability of the estimated binding constants. Multi-cycle kinetics (MCK) or single-cycle kinetics (SCK) experiments were carried out with three different ranges of analyte concentrations, referred to as high-range, mid-range and low-range. The analyte concentrations were:

0.5, 1, 2, 4 and 8 µM for the high-range; 0.3, 0.6, 0.9, 1.2 and 1.5 µM for the mid-range; and 0.04, 0.06, 0.08, 0.1 and 0.12 µM for the low-range.

Figure 1 shows the sensorgrams obtained from MCK experiments with the high-range concentrations, while Figures 2 and 3 show the sensorgrams from SCK experiments obtained with the mid- and low concentration ranges, respectively. For the high concentration range, MCK were used as attempts with SCK resulted in very high RU levels with the initial injections, which saturated the signal for subsequent additions. For all concentration ranges, data were fitted to a 1:1 Langmuir model and a two-state reaction model, and the corresponding residual plots are shown along with the fit of the model to the raw data in the sensorgrams. In addition to this, a steady-state fit with the maximum binding response at equilibrium for each analyte concentration was conducted for all the concentration ranges (see Figures S1–S3). Tables 1 and 2 report the estimates of the kinetic parameters, resulting binding affinities, R_{max} as well as the Chi^2 and U-values for the 1:1 Langmuir model and the two-state reaction model, respectively. Experiments were conducted three times with freshly prepared POPC liposomes each time, and the data is reported as mean ± SEM.

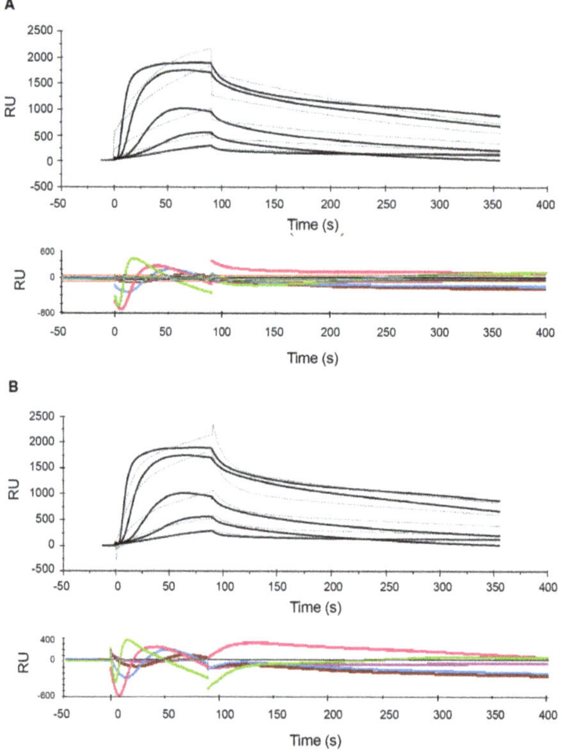

Figure 1. Sensorgrams of the melittin–POPC (1-palmitoyl-2-oleoyl-glycero-3-phosphocholine) interaction using multi-cycle kinetics for the high-range concentrations. Following immobilization of POPC liposomes (1 mM lipid) to a binding response of 8400 RU, increasing concentrations of melittin (0.5, 1, 2, 4 and 8 µM) were serially injected at a flow rate of 30 µL/min for 80 s with intermittent regeneration cycles with 10 mM glycine–HCl (pH 2.5). Binding kinetics analysis of the sensorgrams with a Langmuir 1:1 model (**A**) and a two-state reaction model (**B**) were conducted using BIAevaluation® software. The estimated rate constants and affinity values, based on both these models, are listed in Tables 1 and 2.

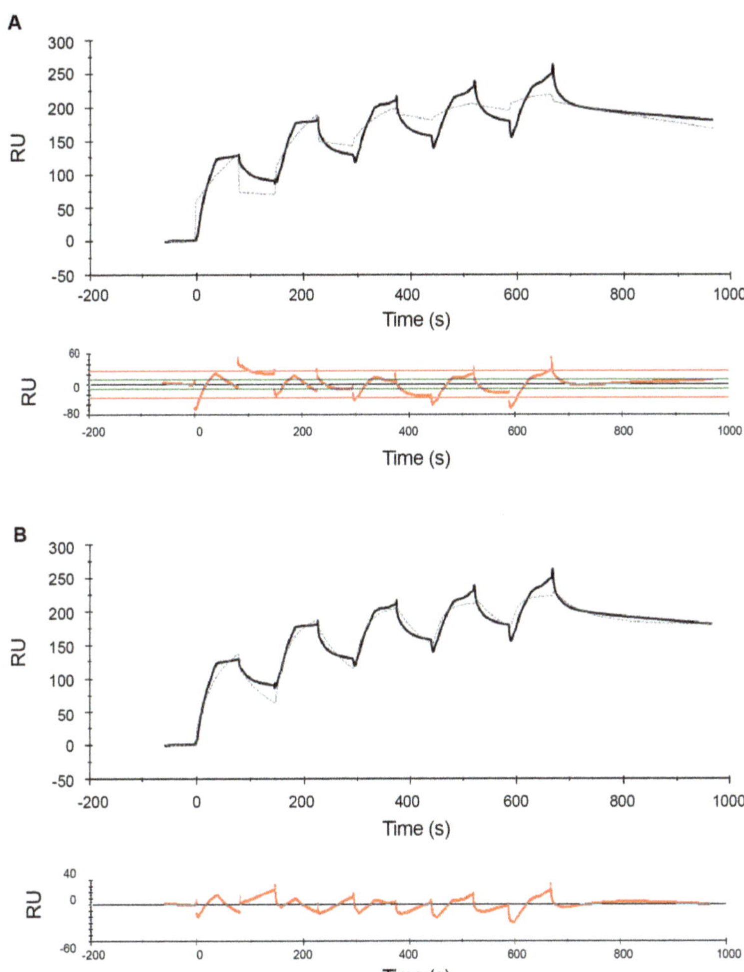

Figure 2. Sensorgrams of the melittin–POPC interaction using single-cycle kinetics for the mid-range concentrations. Following immobilization of POPC liposomes (1 mM lipid) to a binding response of ~4500 RU, increasing concentrations of melittin (0.3, 0.6, 0.9, 1.2 and 1.5 µM) were serially injected in a single-cycle at a flow rate of 30 µL/min for 80 s with a single regeneration step with 10 mM glycine–HCl (pH 2.5) at the end, i.e., after all five concentrations of melittin. Binding kinetic analysis of the sensorgrams with a Langmuir 1:1 model (**A**) and a two-state reaction model (**B**) were conducted using BIAevaluation® software. The estimated rate constants and affinity values, based on both these models, are listed in Tables 1 and 2.

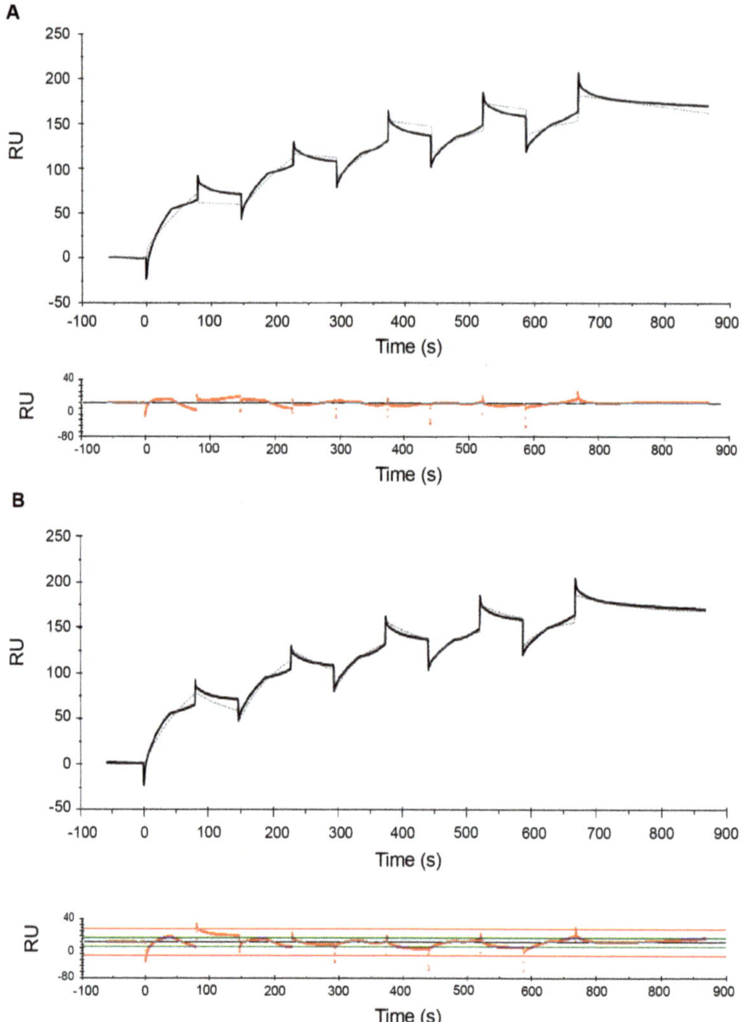

Figure 3. Sensorgrams of the melittin–POPC interaction using single-cycle kinetics for the low-range concentrations. Following immobilization of POPC liposomes (1 mM lipid) to a binding response of ~3500 RU, increasing concentrations of melittin (0.04, 0.06, 0.08, 0.1 and 0.12 μM) were serially injected in a single-cycle at a flow rate of 30 μL/min for 80 s with a single regeneration step with 10 mM glycine–HCl (pH 2.5) at the end, i.e., after all five concentrations of melittin. Binding kinetics analysis of the sensorgrams using a Langmuir 1:1 model (**A**) and a two-state reaction model (**B**) were conducted using BIAevaluation® software. The estimated rate constants and affinity values, based on both these models, are listed in Tables 1 and 2.

Table 1. Kinetic and data-fitting parameters for melittin–POPC interactions based on a 1:1 Langmuir model. Data were measured at the three concentration ranges and obtained from fitting sensorgrams to a 1:1 Langmuir model to obtain estimates of association (k_a) and dissociation rates (k_d) as well as binding constants (K_D), R_{max}, Chi2 and U values. Uncertainties are given as mean $\pm$ SEM obtained from the curve fitting.

Analyte (Melittin) Concentration (µM)	k_a (M^{-1}·s^{-1})	k_d (M^{-1}·s^{-1})	K_D (M)	R_{max} (RU)	Chi^2 (RU)	U-Value
0.5–8.0 (high-range)	$2.2 \pm 0.1 \times 10^3$	$2.7 \pm 0.1 \times 10^{-3}$	$1.2 \pm 0.1 \times 10^{-6}$	2841 ± 24	192	12
0.5–8.0 (high-range) +0.15 M NaCl	$1.1 \pm 0.3 \times 10^3$	$8.2 \pm 1.1 \times 10^{-4}$	$0.6 \pm 0.3 \times 10^{-6}$	2168 ± 28	198	12
0.3–1.5 (mid-range)	$1.6 \pm 0.2 \times 10^4$	$7.4 \pm 0.2 \times 10^{-4}$	$4.1 \pm 0.1 \times 10^{-8}$	216 ± 11	12	9
0.04–0.12 (low-range)	$1.2 \pm 0.3 \times 10^5$	$5.7 \pm 0.8 \times 10^{-4}$	$4.7 \pm 0.6 \times 10^{-9}$	197 ± 8	7	5

Table 2. Kinetic and data-fitting parameters for melittin–POPC interactions based on a two-state reaction model. Data were measured at the three concentration ranges and obtained from fitting sensorgrams to a two-state reaction model to obtain estimates of association (k_{a1} and k_{a2}) and dissociation rates (k_{d1} and k_{d2}) as well as binding constants (K_D), R_{max} and Chi2 values. Uncertainties are given as mean $\pm$ SEM obtained from the curve fitting.

Analyte (Melittin) Concentration (µM)	k_{a1} (M^{-1}·s^{-1})	k_{d1} (s^{-1})	k_{a2} (M^{-1}·s^{-1})	k_{d2} (s^{-1})	K_D (M)	R_{max} (RU)	Chi^2 (RU)
0.5–8.0 (high-range)	$5.2 \pm 0.1 \times 10^3$	$4.7 \pm 0.2 \times 10^{-2}$	$1.3 \pm 0.01 \times 10^{-2}$	$3.1 \pm 0.1 \times 10^{-3}$	$1.7 \pm 0.2 \times 10^{-6}$	3488 ± 48	134
0.5–8.0 (high-range) +0.15 M NaCl	$2.2 \pm 0.6 \times 10^4$	$5.8 \pm 1.2 \times 10^{-2}$	$3.8 \pm 1.1 \times 10^{-3}$	$8.5 \pm 0.7 \times 10^{-4}$	$0.5 \pm 0.2 \times 10^{-6}$	2725 ± 18	122
0.3–1.5 (mid-range)	$4.6 \pm 0.2 \times 10^4$	$1.3 \pm 0.1 \times 10^{-2}$	$3.4 \pm 0.3 \times 10^{-3}$	$1.2 \pm 0.3 \times 10^{-6}$	$1.5 \pm 0.2 \times 10^{-11}$	243 ± 13	7
0.04–0.12 (low-range)	$2.1 \pm 0.1 \times 10^5$	$7.1 \pm 0.1 \times 10^{-3}$	$5.2 \pm 0.4 \times 10^{-3}$	$2.2 \pm 0.8 \times 10^{-6}$	$1.5 \pm 0.3 \times 10^{-11}$	200 ± 9	3

The 1:1 model assumes the formation of an analyte (A)-ligand (L) complex in a single step and with a 1:1 ratio, given by the reaction scheme

$$A + L \underset{k_a}{\overset{k_d}{\rightleftharpoons}} AL \tag{1}$$

where k_a is the association rate constant for formation of AL, and k_d is the dissociation rate constant for complex AL. While this is one of the most commonly used kinetic binding model in SPR analysis, data from various studies indicates that the inherent assumptions of this model do not appropriately explain the complex mechanism involved in the binding of AMPs to membranes [65]. In contrast, the two-state reaction model assumes that the AL complex formed on the membrane surface undergoes a change (either a conformational change, a change in agglomeration or orientation) to form AL* and is given by the reaction scheme

$$A + L \underset{k_{a1}}{\overset{k_{d1}}{\rightleftharpoons}} AL \underset{k_{a2}}{\overset{k_{d2}}{\rightleftharpoons}} AL^* \tag{2}$$

where k_{a1} is the association rate constant for formation of AL, k_{d1} is the dissociation rate constant for complex AL, k_{a2} is the association rate constant for conversion of AL to AL*, and k_{d2} is the dissociation rate constant for conversion of AL* to AL. The model assumes that the formation of AL* (and back) can only go via AL (i.e., A and L cannot form AL* and AL* cannot breakup without going through the state AL). It also assumes that the formation of AL follows first-order kinetics and that the rate of change is equal in both directions [74]. Note that it is not possible by SPR alone to determine the exact nature of the AL complex or determine the type of change that the complex undergoes between AL and AL*.

Comparison of the sensorgrams across the three concentration ranges used shows that one of the striking differences among the experiments at different concentration ranges were the R_{max} values. In the low- and mid concentration ranges, R_{max} levels are ~200 RU (Tables 1 and 2). In contrast, for the high concentration range, R_{max} reached much higher levels (>2800 RU and >3400 RU for the 1:1 and two-state model, respectively). Interestingly, the difference in R_{max} values between mid- and low concentration ranges is substantially smaller to that observed in the high concentration range. This is observed in both the 1:1 and the two-state reaction models (Tables 1 and 2). This is consistent with the widely accepted view that the binding of melittin to lipids is not based on the absolute concentration of the peptide, but rather it is determined by the peptide/lipid (P/L) ratio [26].

Comparison of the residual plots for the three concentration ranges, for both the 1:1 and two-state reaction models, shows that the fits were much better in the mid- and low concentration ranges than in the high concentration range. This is reflected quantitatively in the corresponding Chi^2 and U-values (Tables 1 and 2). Chi^2 measures the average deviation of the experimental data from the model and is an overall measure of the "goodness of fit", with values below 10% of R_{max} being considered as a good fit. U-values are a measure of the "uniqueness" of the predicted K_D value with a lower U-value indicating greater confidence in the predicted value. With the 1:1 model the Chi^2 values range from 3% for the low concentration range to 7% for the high concentration range. With the two-state model the Chi^2 values range from 2 to 4%. Similarly, the U-values are lower for the mid- and low concentration ranges than the high concentration range, indicating greater confidence in the estimated K_D values for mid- and low concentration ranges compared to high concentration range. The Chi^2 and U-values are lower for the two-state model compared to the 1:1 model, indicating that the fit is much better for the former. This is consistent with observations from previous SPR studies of melittin by Aguilar and co-workers [62,65] whose comparison of the 1:1 model, a parallel model and the two-state model suggested that the fit for the latter is comparatively better.

Comparison of kinetic rate constants and the binding affinity from the different concentration ranges of melittin (analyte) used in this study indicate that these parameters are concentration dependent. This is true for both the 1:1 and the two-state model. While the absolute values of the kinetic rates and binding affinities are different between the models, the concentration-dependent trends are the same. Also, it is generally recommended to use a concentration range that the expected K_D value lies within the range of concentrations used to derive that K_D value. The K_D value, estimated as per the steady-state fit, for all three concentration ranges are within the analyte concentration range used; this is shown in the Supplementary Materials (Figures S1–S3). A steady state affinity measurement is based on the plot of responses at equilibrium states against the concentrations of analyte used in the assay. However, it is notable that the K_D values estimated using kinetic measurements (i.e., the k_a (on-rate) and k_d (off-rate) values), especially for the mid and low concentration ranges, were outside the analyte concentration range. This could be due to the fact of several reasons; in principle, the rate constants k_a and k_d and the K_D derived from these rate constants are independent of both analyte and ligand concentrations. Secondly, as can be noted in Tables 1 and 2, the concentration-dependent effect on the change in affinity (K_D) was higher due to the increasing on-rate (k_a) than due to the observed decrease in the off-rate (k_d). Other factors that may affect the K_D, such as the pH, temperature etc., were kept constant across the experiments.

In both models, the association is slower in the high concentration range than in the low concentration range. For example, in the two-state model, the rate for the initial surface binding (k_{a1}) was 2.1×10^5 $M^{-1} \cdot s^{-1}$ for the low concentration range, and it dropped to 4.1×10^4 $M^{-1} \cdot s^{-1}$ for the mid concentration range. The k_{a1} further decreased to 5.2×10^3 $M^{-1} \cdot s^{-1}$ for the high concentration range (Table 2). In other words, a 100-fold increase in concentration resulted in an approximately 100-fold decrease in the association rate. The association rates (k_{a1} and k_{a2}) derived from the two-state reaction model also suggested that the initial binding to form the AL complex was much more rapid than the subsequent transition to AL*, and that k_{a1} was affected more by the concentration than k_{a2}. Nevertheless, in both models all association rates decreased with increasing peptides concentrations. This is inconsistent with previous findings reported by Aguilar and co-workers [62], where the association for a 0.125–1 μM concentration range was found to be slow, while rapid association was observed at higher concentrations (4–12 μM). However, this observation was based on a qualitative assessment of sensorgrams rather than a quantitative analysis, as the authors concluded that the latter was not possible due to the poor fit of their experimental data which may likely be due to the very high RU levels. In addition, the conclusion about the absence of binding at low concentrations was based on a comparison of the RU levels at low peptide concentrations (which were well below 200 RU) to the RU levels at high peptide concentrations (which were in the range of 3000–7000 RU). A semi-quantitative comparison of data with such different RU levels is likely less reliable than the analysis of binding kinetics rates derived from the sensorgrams.

Comparison of the dissociation rates from the different concentration ranges suggests that k_d might depend more on the number of analyte–ligand complexes formed (as given by R_{max} values) rather than the analyte concentration. In the 1:1 model this can be seen by comparing the k_d of the mid and low concentration ranges to the high concentration range. The k_d values from the mid and low concentration ranges were close to each other and both concentration ranges showed low RU levels. For the high concentration range, where the RU level was much higher, k_d deviated by an order of magnitude to that in the other two concentration ranges. A similar effect was seen for k_{d1} in the two-state model.

As a result of these concentration-dependent changes in the association and dissociation rates, the binding constant K_D differed by orders of magnitudes for the low, mid and high concentration ranges. For both binding kinetics models, K_D was in the μM (10^{-6}) range for the high concentration range but shows nM (10^{-9}) and sub-nM affinity for the low concentration range in the 1:1 and two-state reaction models, respectively.

2.2. Effects of NaCl on the Melittin–POPC Interaction

Ionic strength is known to affect the biological activity of AMPs [75] including the oligomerisation of melittin in solution [29,70,71]. To assess the effect of NaCl on the binding of melittin to POPC, experiments with melittin in the high concentration range (0.5 to 8 µM) were conducted in the presence of 0.15 M NaCl in the running/analyte buffer. Figures 4 and 5 show the sensorgrams for MCK and SCK for the melittin–POPC interaction in the presence of 0.15 M NaCl, fitted to 1:1 and two-state reaction models, respectively. Comparison of this figure with the sensorgrams obtained in the absence of NaCl (Figure 1) suggests that addition of NaCl results in a decrease in signal response during the association phase for almost all of the concentrations. Consequently, the sensorgrams did not have the desired curvature optimal for performing fitting analysis. However, the curve-fitting step was conducted for experiments involving the use of NaCl and the estimates are reported, but no major conclusions were made using the estimates. Comparison of the R_{max} levels in Tables 1 and 2 shows that the drop from 2841 RU to 2168 for the 1:1 model and from 3488 to 2725 RU in the two-state model, corresponds to ~20%–25% reduction in R_{max}. This decrease in signal response resulted in only a small decrease in the association rate (k_a) in the 1:1 model (Table 1). In the two-state model, it was k_{a2} that was reduced, while k_{a1} slightly increased in the presence of NaCl. In both models, the dissociation rate (k_d for the 1:1 model and k_{d1} for the two-state model) was also reduced, more so than the reduction in k_a. As a result, there was an overall small increase in affinity upon addition of NaCl.

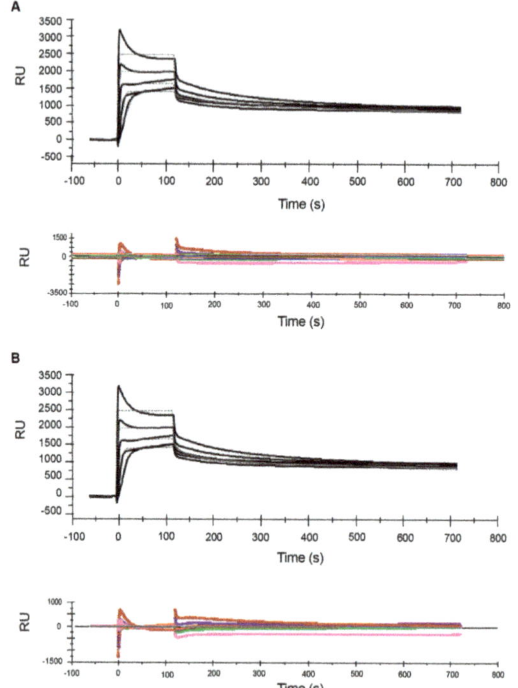

Figure 4. Effect of sodium chloride on the association phase of the melittin–POPC interaction. Following immobilization of POPC liposomes (1 mM lipid), increasing concentrations of melittin (2, 4, 6, 8 and 10 µM) were serially injected with intermittent regeneration cycles with 10 mM glycine–HCl (pH: 2.5) (**A**—multi-cycle) and single regeneration using the same buffer at the end of all cycles (**B**—single-cycle). The estimated rate constants and affinity values, based on both these models, are listed in Tables 1 and 2.

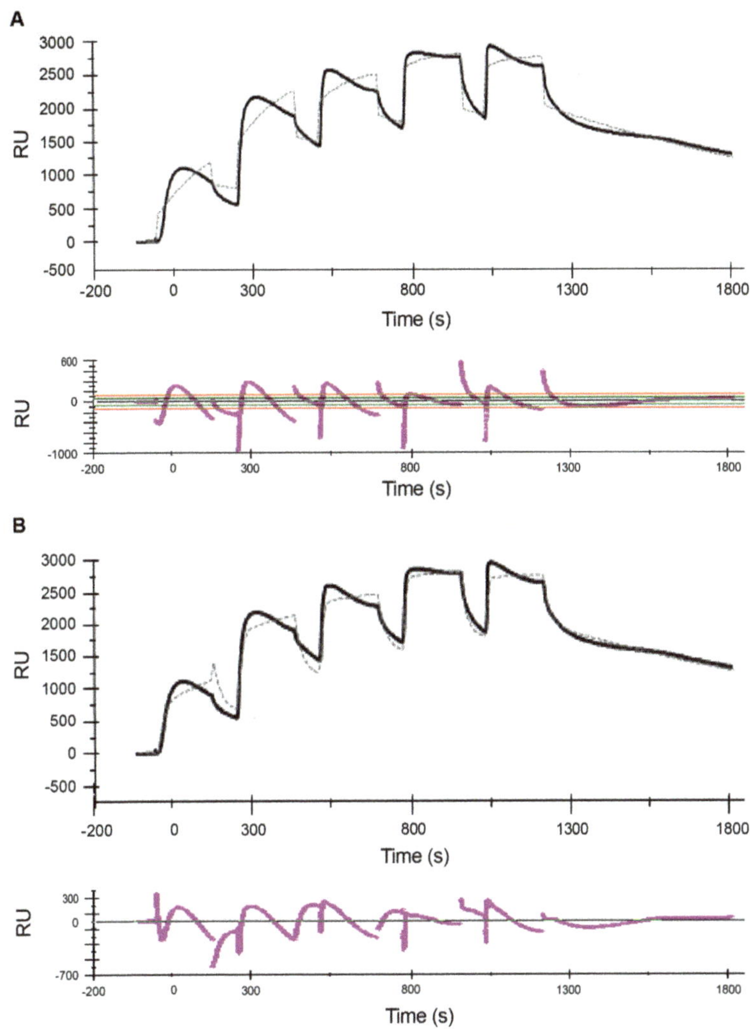

Figure 5. Effect of sodium chloride on melittin–POPC interaction—single-cycle kinetics. Following immobilization of POPC liposomes (1 mM lipid), increasing concentrations of melittin (0.5, 1, 2, 4 and 8 μM), made up in running buffer containing 0.15 M NaCl, were serially injected, following a single-cycle kinetic approach. The sensorgrams were fitted to a 1:1 Langmuir model (**A**) and two-state reaction model (**B**) using BIAevaluation® software. The estimated rate constants and affinity values, based on both these models, are listed in Tables 1 and 2.

3. Discussion

This study aimed to quantify the effect of peptide concentration on the kinetics and affinity of melittin binding to POPC bilayers. The kinetics of the melittin–POPC interaction was studied at three different concentration ranges spanning two orders of magnitude. To understand the mechanism of melittin (and other AMPs), it is not so much the actual concentration of peptide present that is important but the relative amounts of peptide and lipid, i.e., the P/L ratio. The analyte concentrations used in this study correspond to the following P/L ratios: high range concentration (0.5–8 μM) is a P/L of ~1/2000 to 1/125; mid range concentration (0.3–1.5 µM) is a P/L of ~1/3330 to 1/660; low range concentration (0.04–0.12 µM) is a P/L of ~1/20,000 to 1/8300.

As noted in the introduction, there have been previous studies reporting the binding affinity of melittin for phospholipid bilayers. When comparing binding affinities for peptide–membrane systems, it is important to consider the technique or assay used as this often dictates the theoretical model used to analyse the data. In SPR, K_D is calculated using binding kinetics rates and/or Langmuir adsorption isotherms or related binding models. In contrast, data from ITC, NMR or filtration experiments of peptide–membrane systems can be analysed using a surface partition equilibrium model. In this case, a partition coefficient K_p, rather than K_D, is calculated. K_p is given by the ratio of peptides bound to the membrane and peptides remaining in solution, while electrostatic effects are corrected for by using the Gouy–Chapman theory [76]. A direct comparison of the raw data between these types of experiments or analysis using both models is often not possible. In both cases, the free energy associated with the binding affinity is given by $\Delta G = -RT \ln(K)$, where R is the gas constant, T is the temperature in Kelvin and K is either K_p or $1/K_D$. Nevertheless, the underlying assumptions in these models are different and affinities for the same system under the same conditions can vary [77]. In the following discussion, the $\Delta G = -RT \ln(K)$ relation is used for comparison of data from ITC and NMR experiments from previous studies and the SPR data from this study.

In one of the earliest studies of melittin binding to membranes, Vogel [35] used circular dichroism (CD) and fluorescence spectroscopy to estimate the binding affinity of melittin to DMPC vesicles. The lipid and peptide concentration used in that study correspond to P/L ratios between 1/10 to 1/250 which is comparable to the upper end of the high concentration range used in this study. The K_D from the CD titration experiments was calculated using a binding isotherm and estimated to be 2 µM which is in very good agreement with the binding affinities of 1.2 µM and 1.7 µM of our experiments. Allende et al. [55,59] used an ultrafiltration assay to study the interaction of melittin with membranes of a wide range of different lipid compositions including egg phosphatidylcholine (EPC) and DOPC vesicles. The P/L ratios ranged from 1/300 to 1/600, corresponding to our high concentration ranges. The K_p values for EPC and DOPC were ~3.6 µM, slightly higher but still in good agreement with our results. In two NMR studies, Beschiaschvili et al. [60,78] estimated the binding affinity of melittin to POPC vesicles using P/L ratios from ~1/600 to 1/1400. This corresponds to the lower part of the high range and upper part of the mid range from our experiments. The K_p was estimated between 1.5 to 2 µM. This agrees with our K_D values for the high concentration range but not for the values in the mid concentration range, which are estimated to be much lower (sub-µM and nM range). However, a direct comparison of the K_D values from our SPR experiments to the K_p values from the NMR is complicated by the fact that, based on our findings, the concentration range used in the NMR experiments is where the effect of concentration is the largest. Another difference is that in the aforementioned studies the size of the vesicles used ranges from 1–10 µM which is 20–200 times larger than the vesicles used in this study. This difference in size implies differences in the curvature of the membrane surface, which can affect the peptide-lipid interaction.

The only other SPR study that used quantitative analysis to report K_D values of melittin binding to PC-only membranes is that by Lee et al. [64], where the binding of melittin to DMPC was studied. The K_D value was estimated to be 33 µM for P/L ratios between 1/3 and 1/100 which is higher than the highest concentration used in our study. This could be the result of the very high P/L ratio or the fact that supported lipid monolayers formed on an HPA chip instead of the intact bilayer vesicles used in this study. In our experiments, we assumed that the vesicles (ligand) stay intact after immobilization [79]. The tighter lipid packing of a supported bilayer compared to SUVs might cause slightly reduced binding [80].

In another SPR study, Papo et al. [68] quantified the kinetics and affinity of binding of melittin to POPC/chol (10:1 w/w) vesicles using P/L ratios between ~1/1600 and ~1/3300. This corresponds roughly to the low range concentrations used in our study. Data were analysed using a two-state model, and the K_D estimated based on this model was 1.6 µM. This is significantly lower than the K_D determined in the sub-nM range for our low concentration. While the study by Allende et al. [55] showed that cholesterol can lower the binding affinity by 1–2 orders of magnitude, there is still a large difference between the K_D reported by Papo et al. and our estimates in this study. We note, however, that the value of k_{a1} reported by Papo et al. was very low and could possibly be below the acceptable detection limit of the Biacore X instrument used in their experiments. We speculate this based on the acceptable limits for the kinetic binding constants for the Biacore T200 instrument used in this study which is a more sensitive instrument with a wider detection limit compared to the Biacore X instrument.

Overall, the results of our SPR experiments for the high concentration range are in good agreement with the binding affinities from earlier studies. Combined with the fact that our estimates of the binding affinities in the mid- and low-range concentrations are more reliable, it increases our confidence in the predicted effect of concentration on the kinetics and binding affinity in the mid- and low concentration ranges. We can now interpret our finding that the apparent binding affinity of melittin for POPC decreases with increasing P/L ratio in the context of what is known about the mechanism of action of melittin.

As mentioned in the introduction, a number of studies have shown that melittin can orient itself into different orientations with respect to the bilayer surface: in a parallel orientation, where the peptide lies flat on the membrane surface, and a perpendicular orientation where the peptide inserts into the hydrophobic core region of the bilayer. The parallel orientation is often referred to as the 'inactive' state, as there is evidence suggesting that this orientation is non-pore forming and non-lytic. In contrast, the perpendicular orientation is the 'active' state that results in the formation of the pores [26,81,82]. In the original two-state model [41] the peptide binds to the surface and then transitions from the parallel to the perpendicular orientation. In an alternative model, parallel surface binding and direct insertion are competing processes [26]. Independent of the mechanism, there is evidence that suggests that the most important factor controlling the amount of peptides found in the parallel or perpendicular orientation is the P/L ratio [26,81]. Yang et al. [81] used oriented CD spectroscopy and neutron scattering to demonstrate that at melittin concentrations corresponding to a P/L ratio of ~1/40, 85% of peptides were orientated parallel to the membrane surface. At peptide concentrations corresponding to a P/L ratio of ~1/15, 68% of peptides were orientated perpendicular to the membrane. Another study demonstrated that at higher P/L ratio, melittin inserts perpendicularly into the hydrophobic core of the bilayer, to translocate and redistribute on either side of the bilayer [54]. This study suggests that the redistribution of melittin on either side of the bilayer at a higher P/L ratio (~1/100) occurs due to the formation of transient but not stable pores, and that the latter occur only above a critical P/L ratio (~1/45). Interestingly, the formation of such transient pores in phospholipid bilayers by the action of melittin has also been observed at concentrations as low as a P/L of 1/5000. For SUVs, this P/L ratio corresponds to the low- and mid-range concentrations used in this study. Similarly, the existence of transient pores induced by melittin and the subsequent 're-sealing' has been shown in bacterial membranes [72]. Based on these findings, a variant of the binding kinetics model of the classical carpet model has been proposed, in which permeabilisation of membranes by cationic amphipathic peptides is transient and stochastic on the scale of a single vesicle or bacteria [40,72,83,84]. Peptides accumulate on the bilayer surface, causing an asymmetry in peptide concentration across the bilayer, i.e., an asymmetry in mass, charge and surface pressure on the bilayer. Subsequently, this asymmetry is abolished by a set of stochastic membrane re-organization events, resulting in the influx of peptide molecules across the bilayer until equilibrium in concentration is attained. Finally, simulations show that while the energy barrier for the reorientation of single peptide is large, the barrier is reduced by a higher P/L as well as by the presence of transient water-filled pores in the membrane [45,85].

Based on the above findings, we can assume the following conditions at the low- and mid concentration ranges used in our experiments: the vast majority of peptides are found in parallel orientation and no stable pores are present, but transient pores might stochastically form and 're-seal'. This means that at a low P/L ratio, any additional increase in peptide concentration results in an accumulation of peptides on the membrane surface creating an accumulation of positive charge, leading to an increase in the repulsion for subsequent peptide binding. This could explain the continuous decrease in the association rates (k_{a1} and k_a) with increasing concentration of melittin observed in this study. The largest effect of concentration on K_D occurs when going from the mid- to the high concentration range, which goes from supra-nM to μM range. The upper range of the high concentration range in this study reaches a P/L ratio where the presence of peptides found in the perpendicular orientation is increasingly likely. This is consistent with the fact that concentration ranges used in this study were relevant to the concentration of melittin required to induce leakage [17,18,24,25,55,57] and the minimum inhibitory concentration of melittin on bacteria.

Our experiments also showed that binding kinetics parameters were also affected by the choice of binding model. At all concentrations, a two-state reaction model gives the best fit to the data, suggesting that it is better at describing the process of binding than a 1:1 model. At first this appears to confirm the classical two-state model of melittin binding and pore formation [41] which involves the transition from a parallel to a perpendicular orientation, excluding the model where parallel adsorption and perpendicular insertion are described as competing processes [26]. However, our findings only confirm that a two-state model provides a better description of binding than the 1:1 model, but cannot exclude the parallel–perpendicular competition model [26]. This would only be possible by a direct comparison of data fitted to both of these models. We note that the conventional binding models used in this study do not account for any electrostatic effects. The CD spectroscopy experiments by Beschiaschvili et al. [60] demonstrated that melittin binding to lipids (SUVs) decreased as the membrane surface charge was neutralized by previously bound melittin. This effect was observed to be higher in zwitterionic lipids (POPC) than hybrid SUVs (80/20 POPC/POPG). If taken into account, the influence of electrostatic effects could potentially correspond to the transition from positive to negative cooperativity in melittin–membrane interactions which, to the best of our knowledge, cannot be qualitatively assessed by SPR.

In addition to the effect of peptide concentration on the kinetics and affinity of binding, we also investigated the effect of physiological ionic strength (0.15 M NaCl) on the interaction of melittin with POPC bilayers. At the pH, ionic strengths and peptide concentrations used in this study, melittin is be expected to remain in its monomeric form [28,29,69–71]. In the presence of NaCl, an apparent decrease in the signal response during the association phase was observed in both single and multi-cycle kinetics experiments, leading to a somewhat reduced association rate (k_a). This could be caused by a screening effect of the NaCl between the cationic peptide and the zwitterionic membrane. Another reason could be that the binding of Na^+ ions to the membrane surface changes the electrostatics of the water-lipid interfacial region. Several studies have shown that Na^+ ions can bind to phospholipid bilayers, where they interact with the phosphate and carbonyl oxygen of the lipid headgroups [86,87]. The binding of Na^+ shields the negative charge of the phosphate group and, thus, indirectly increases the electrostatic charge of the membrane (as there are now more unbound positively charged choline groups than negatively charged phosphate groups). As the initial association is likely driven by electrostatics, this could account for the reduced k_a in the presence of NaCl. Further characterisation of the melittin–POPC interaction with a range of NaCl concentrations will be needed to better understand this effect. Also, future studies should focus on the use of complex cell membrane models (e.g., hybrid lipid systems or virus-like particles derived from mammalian cell models) that could enable interaction studies in a more physiologically relevant context.

4. Materials and Methods

4.1. Materials

The POPC (1-palmitoyl-2-oleoyl-glycero-3-phosphocholine) was purchased from Avanti Polar Lipids (Birmingham, USA) through Sigma–Aldrich (Castle Hill, Australia). Mechanical extrusion for the preparation of lipid vesicles was carried out using an extruder from Avanti Polar Lipids (Birmingham, USA). Melittin was purchased from AnaSpec (Fremont, USA). Reagents for buffer solutions, glycine-HCl and CHAPS (3-((3-cholamidopropyl) dimethylammonio)-1-propanesulfonate) solutions were purchased from Sigma–Aldrich (Castle Hill, Australia). Running buffer for SPR experiments consisted of 0.05 M HEPES (Sigma–Aldrich, Australia) at a pH of 7.4 and also included 150 mM NaCl for experiments investigating the effect of ionic strength. The SPR experiments were conducted in a BiacoreT200 instrument using a L1 sensor chip S series (GE Healthcare Life Sciences, Paramatta, Australia).

4.2. Liposome Preparation

The POPC vesicles at a 1.0 mM lipid concentration were used for all experiments in this study and were prepared as follows. First, the required amount of POPC was weighed in a glass vial, dissolved in chloroform and dried under a narrow jet of nitrogen gas. Complete dehydration of the lipid film was ensured by placing the vial in a vacuum desiccator attached to a pump for at least 12 h. The resulting lipid film was rehydrated with running buffer to a final lipid concentration of 1.0 mM. The solution was vortexed for 1 h and allowed to stand for at least 2 h or to a maximum overnight period to enable osmotic swelling. A uniform population of small unilamellar vesicles (SUVs) was obtained by mechanical extrusion through a polycarbonate filter with pore size of ~50 nm for at least 21 times. This approach was previously shown to be effective in attaining a higher degree of homogeneity and reproducibility in nanosizing liposomes compared to other commonly used approaches for generating SUVs [88].

4.3. Surface Plasmon Resonance Experiments

4.3.1. Immobilization of POPC Vesicles on a L1 Chip

The L1 sensor chip was equilibrated at room temperature and then docked into a BiacoreT200 instrument. After priming the system with running buffer, a manual run was set-up for three consecutive injections with running buffer containing 20 mM CHAPS at a flow rate of 30 µL/min. This washing step was conducted in every subsequent use of the chip before and after the binding experiments. The POPC SUVs of ~50 nm diameter at a 1.0 mM lipid concentration were then steadily immobilized on to the L1 chip surface through a manual run at a flow rate of 5 µL/min. The target immobilization levels varied from ~4000 to 8000 RU, based on the concentration of melittin used in the binding assay. Following liposome immobilization, a short pulse with running buffer was injected at a flow rate of 2 µL/min for 5 min to ensure the stability of the baseline in the flow cell containing the immobilized vesicles.

4.3.2. Binding experiments

Melittin solutions at the desired range of concentrations were prepared in running buffer by dilution from a stock solution (2.2 mM). Melittin solutions were also prepared in a running buffer containing 0.15 M NaCl, the POPC SUVs were prepared in a running buffer without NaCl, while all other experimental conditions were the maintained as same as in the experiments conducted in the absence of NaCl. The kinetics of the melittin–POPC interactions were estimated using either single or multi-cycle kinetics. Melittin solutions were serially injected at increasing concentrations at 30 µL/min, with (multi-cycle) or without (single-cycle) intermittent regeneration cycles among various analyte injections, using 10 mM glycine–HCl, pH 2.5.

In a single-cycle kinetics experiment increasing concentrations of analyte are serially injected with a single dissociation phase after the injection of the highest concentration. It should be noted that, although this approach includes only a single dissociation phase at the end, there is a short pulse of running buffer after each analyte injection period during when some dissociation can be noted. Depending on the rate of dissociation, it is possible that some analyte remains bound to the immobilized ligands before the next injection. However, the fitting models used for the analysis of single-cycle kinetic experiments account for the varying concentrations of analyte injections in time and different amounts of preformed ligand–analyte complexes on the surface [89]. The fact that in a single-cycle kinetic experiment there is no intermittent regeneration cycles, it can be advantageous as it prevents the detrimental effects of a strong acidic regeneration buffer on the ligand. Single-cycle and multi-cycle approaches differ in their experimental procedure but have been shown to yield comparable results. Nevertheless, the single-cycle approach can be challenging for high analyte concentrations, because the number of ligand–analyte complexes being formed after initial analyte injections can substantially impair analyte binding in subsequent injections. Therefore, a multi-cycle approach, which includes separate association, dissociation and regeneration phase for each analyte injection, is used only for the high-range analyte concentration in this study.

4.4. Curve Fitting

All sensorgrams obtained from single- and multi-cycle kinetic analyses were fitted using either the 1:1 Langmuir model or the two-state reaction model available in the BIAevaluation® software version 1.0 (GE Healthcare Life Sciences, Paramatta, Australia) to estimate k_a, k_d and K_D. The curve fitting procedure is an iterative numerical process that determines the best fit of experimental data to the set of equations specific for a binding kinetics model (e.g., the 1:1 Langmuir or the two-state reaction model) thus defining the interaction and yielding the kinetic rate constants (k_a and k_d) from which the binding constants (K_D) can be determined. The "goodness of fit" for the fitted curve is given by Chi^2. In addition, residual plots show the magnitude of the deviation (in units of RU) of the experimental data from the fitted data as a function of time. The experimental data are shown as solid lines and the fitted data as dotted lines in all of the sensorgrams. The quality of fit and reliability of the estimated binding parameters were assessed by the Chi^2 and U-values (only for the Langmuir 1:1 model), respectively.

Supplementary Materials: Supplementary Materials can be found at http://www.mdpi.com/1422-0067/21/3/746/s1.

Author Contributions: Conceptualization, R.L.M. and E.D.; methodology, T.S., E.D. and R.L.M.; formal analysis, T.S.; investigation, T.S., E.D. and RLM.; writing—original draft preparation, T.S. and E.D.; writing—review and editing, T.S., E.D. and R.L.M.; visualization, T.S. and E.D.; supervision, R.L.M. and E.D.; project administration, R.L.M. and E.D.; funding acquisition, R.L.M. and E.D. All authors have read and agreed to the published version of the manuscript.

Funding: Part of this research was carried out during the tenure of a Suzanne Cavanagh Early Career Investigator grant from Cancer Council Western Australia to E.D. and R.L.M.

Conflicts of Interest: The authors declare no conflict of interest. The funders had no role in the design of the study; in the collection, analyses, or interpretation of data; in the writing of the manuscript, or in the decision to publish the results.

References

1. Breen, S.; Solomon, P.S.; Bedon, F.; Vincent, D. Surveying the potential of secreted antimicrobial peptides to enhance plant disease resistance. *Front. Plant Sci.* **2015**, *6*, 900. [CrossRef]

2. Anaya-López, J.L.; López-Meza, J.E.; Ochoa-Zarzosa, A. Bacterial resistance to cationic antimicrobial peptides. *Crit. Rev. Microbiol.* **2013**, *39*, 180–195. [CrossRef] [PubMed]

3. Brogden, N.K.; Brogden, K.A. Will new generations of modified antimicrobial peptides improve their potential as pharmaceuticals? *Int. J. Antimicrob. Agents* **2011**, *38*, 217–225. [CrossRef] [PubMed]

4. Guaní-Guerra, E.; Santos-Mendoza, T.; Lugo-Reyes, S.O.; Terán, L.M. Antimicrobial peptides: General overview and clinical implications in human health and disease. *Clin. Immunol.* **2010**, *135*, 1–11. [CrossRef] [PubMed]

5. Fjell, C.D.; Hiss, J.A.; Hancock, R.E.; Schneider, G. Designing antimicrobial peptides: form follows function. *Nat. Rev. Drug Discov.* **2011**, *11*, 37–51. [CrossRef]

6. Schweizer, F. Cationic amphiphilic peptides with cancer-selective toxicity. *Eur. J. Pharmacol.* **2009**, *625*, 190–194. [CrossRef]

7. Felício, M.R.; Silva, O.N.; Gonçalves, S.; Santos, N.C.; Franco, O.L. Peptides with Dual Antimicrobial and Anticancer Activities. *Front. Chem.* **2017**, *5*, 5. [CrossRef]

8.   Gaspar, D.; Veiga, A.S.; Castanho, M.A. From antimicrobial to anticancer peptides. A review. *Front. Microbiol.* **2013**, *4*, 294. [CrossRef]

9.   Leuschner, C.; Hansel, W. Membrane disrupting lytic peptides for cancer treatments. *Curr. Pharm. Des.* **2004**, *10*, 2299–2310. [CrossRef]

10.   Fernandez-Rojo, M.A.; Deplazes, E.; Pineda, S.S.; Brust, A.; Marth, T.; Wilhelm, P.; Martel, N.; Ramm, G.A.; Mancera, R.L.; Alewood, P.F.; et al. Gomesin peptides prevent proliferation and lead to the cell death of devil facial tumour disease cells. *Cell Death Discov.* **2018**, *4*, 1–11. [CrossRef]

11.   Brogden, K.A. Antimicrobial peptides: pore formers or metabolic inhibitors in bacteria? *Nat. Rev. Microbiol.* **2005**, *3*, 238–250. [CrossRef] [PubMed]

12.   Sani, M.-A.; Separovic, F. How Membrane-Active Peptides Get into Lipid Membranes. *Acc. Chem. Res.* **2016**, *49*, 1130–1138. [CrossRef] [PubMed]

13.   Zasloff, M. Antimicrobial peptides of multicellular organisms. *Nature* **2002**, *415*, 389–395. [CrossRef] [PubMed]

14.   Wang, Z.; Wang, G. APD: the Antimicrobial Peptide Database. *Nucleic Acids Res.* **2004**, *32*, D590–D592. [CrossRef] [PubMed]

15.   Zhang, R.; Eckert, T.; Lutteke, T.; Hanstein, S.; Scheidig, A.; MJJBonvin, A.; ENifantiev, N.; Kozar, T.; Schauer, R.; Abdulaziz Enani, M.; et al. Structure-Function Relationships of Antimicrobial Peptides and Proteins with Respect to Contact Molecules on Pathogen Surfaces. *Curr. Top. Med. Chem.* **2016**, *16*, 89–98. [CrossRef] [PubMed]

16.   Raghuraman, H.; Chattopadhyay, A. Melittin: a Membrane-active Peptide with Diverse Functions. *Biosci. Rep.* **2007**, *27*, 189–223. [CrossRef]

17.   Habermann, E. Bee and Wasp Venoms. *Science* **1972**, *177*, 314–322. [CrossRef]

18.   Andersson, A.; Danielsson, J.; Gräslund, A.; Mäler, L. Kinetic models for peptide-induced leakage from vesicles and cells. *Eur. Biophys. J.* **2007**, *36*, 621–635. [CrossRef]

19.   Burton, M.G.; Huang, Q.M.; Hossain, M.A.; Wade, J.D.; Palombo, E.A.; Gee, M.L.; Clayton, A.H. Direct Measurement of Pore Dynamics and Leakage Induced by a Model Antimicrobial Peptide in Single Vesicles and Cells. *Langmuir* **2016**, *32*, 6496–6505. [CrossRef]

20.   Hur, J.; Kim, K.; Lee, S.; Park, H.; Park, Y. Melittin-induced alterations in morphology and deformability of human red blood cells using quantitative phase imaging techniques. *Sci. Rep.* **2017**, *7*, 1–10. [CrossRef]

21.   Matsuzaki, K.; Yoneyama, S.; Miyajima, K. Pore formation and translocation of melittin. *Biophys. J.* **1997**, *73*, 831–838. [CrossRef]

22.   Oren, Z.; Shai, Y. Selective lysis of bacteria but not mammalian cells by diastereomers of melittin: structure-function study. *Biochemistry* **1997**, *36*, 1826–1835. [CrossRef] [PubMed]

23.   Pan, J.; Khadka, N.K. Kinetic Defects Induced by Melittin in Model Lipid Membranes: A Solution Atomic Force Microscopy Study. *J. Phys. Chem. B* **2016**, *120*, 4625–4634. [CrossRef] [PubMed]

24.   Rex, S. Pore formation induced by the peptide melittin in different lipid vesicle membranes. *Biophys. Chem.* **1996**, *58*, 75–85. [CrossRef]

25.   Rex, S.; Schwarz, G. Quantitative Studies on the Melittin-Induced Leakage Mechanism of Lipid Vesicles. *Biochemistry* **1998**, *37*, 2336–2345. [CrossRef] [PubMed]

26.   Van Den Bogaart, G.; Guzmán, J.V.; Mika, J.T.; Poolman, B. On the mechanism of pore formation by melittin. *J. Biol. Chem.* **2008**, *283*, 33854–33857. [CrossRef]

27.   Van Den Bogaart, G.; Mika, J.T.; Krasnikov, V.; Poolman, B. The Lipid Dependence of Melittin Action Investigated by Dual-Color Fluorescence Burst Analysis. *Biophys. J.* **2007**, *93*, 154–163. [CrossRef]

28.   Brown, L.R.; Braun, W.; Kumar, A.; Wüthrich, K. High resolution nuclear magnetic resonance studies of the conformation and orientation of melittin bound to a lipid-water interface. *Biophys. J.* **1982**, *37*, 319–328. [CrossRef]

29.   Brown, L.R.; Lauterwein, J.; Wüthrich, K. High-resolution 1H-NMR studies of self-aggregation of melittin in aqueous solution. *Biochim. Biophys. Acta (BBA) Protein Struct.* **1980**, *622*, 231–244. [CrossRef]

30.   Dawson, C.R.; Drake, A.F.; Helliwell, J.; Hider, R.C. The interaction of bee melittin with lipid bilayer membranes. *Biochim. Biophys. Acta (BBA) Biomembr.* **1978**, *510*, 75–86. [CrossRef]

31.   Ikura, T.; Gō, N.; Inagaki, F. Refined structure of melittin bound to perdeuterated dodeclylphoscholine micelles as studied by 2D-NMR and distance geometry calculation. *Proteins Struct. Funct. Bioinform.* **1991**, *9*, 81–89. [CrossRef] [PubMed]

32. Ladokhin, A.S.; White, S.H. Folding of amphipathic α-helices on membranes: energetics of helix formation by melittin11Edited by D. Rees. *J. Mol. Biol.* **1999**, *285*, 1363–1369. [CrossRef] [PubMed]

33. Lam, Y.H.; Wassall, S.R.; Morton, C.J.; Smith, R.; Separovic, F. Solid-State NMR Structure Determination of Melittin in a Lipid Environment. *Biophys. J.* **2001**, *81*, 2752–2761. [CrossRef]

34. Lauterwein, J.; Brown, L.R.; Wüthrich, K. High-resolution 1H-NMR studies of monomeric melittin in aqueous solution. *Biochim. Biophys. Acta (BBA) Protein Struct.* **1980**, *622*, 219–230. [CrossRef]

35. Vogel, H. Incorporation of melittin into phosphatidylcholine bilayers. Study of binding and conformational changes. *FEBS Lett.* **1981**, *134*, 37–42. [CrossRef]

36. Vogel, H. Comparison of the conformation and orientation of alamethicin and melittin in lipid membranes. *Biochemistry* **1987**, *26*, 4562–4572. [CrossRef]

37. Hristova, K.; Dempsey, C.E.; White, S.H. Structure, location, and lipid perturbations of melittin at the membrane interface. *Biophys. J.* **2001**, *80*, 801–811. [CrossRef]

38. Iwadate, M.; Asakura, T.; Williamson, M.P. The structure of the melittin tetramer at different temperatures. *Eur. J. Biochem.* **1998**, *257*, 479–487. [CrossRef]

39. Park, S.C.; Kim, J.Y.; Shin, S.O.; Jeong, C.Y.; Kim, M.H.; Shin, S.Y.; Cheong, G.W.; Park, Y.; Hahm, K.S. Investigation of toroidal pore and oligomerization by melittin using transmission electron microscopy. *Biochem. Biophys. Res. Commun.* **2006**, *343*, 222–228. [CrossRef]

40. Wimley, W.C. How Does Melittin Permeabilize Membranes? *Biophys. J.* **2018**, *114*, 251–253. [CrossRef]

41. Huang, H.W. Action of antimicrobial peptides: two-state model. *Biochemistry* **2000**, *39*, 8347–8352. [CrossRef] [PubMed]

42. Terwilliger, T.C.; Eisenberg, D. The structure of melittin. *J. Biol. Chem.* **1982**, *257*, L6010–L6015.

43. Berneche, S.; Nina, M.; Roux, B. Molecular dynamics simulation of melittin in a dimyristoylphosphatidylcholine bilayer membrane. *Biophys. J.* **1998**, *75*, 1603–1618. [CrossRef]

44. Deplazes, E. Molecular simulations of venom peptide-membrane interactions: Progress and challenges. *Pept. Sci.* **2018**, *110*, e24060. [CrossRef]

45. Irudayam, S.J.; Berkowitz, M.L. Binding and reorientation of melittin in a POPC bilayer: computer simulations. *Biochim. Biophys. Acta* **2012**, *1818*, 2975–2981. [CrossRef] [PubMed]

46. Leveritt, J.M., III; Pino-Angeles, A.; Lazaridis, T. The Structure of a Melittin-Stabilized Pore. *Biophys. J.* **2015**, *108*, 2424–2426. [CrossRef]

47. Lyu, Y.; Xiang, N.; Zhu, X.; Narsimhan, G. Potential of mean force for insertion of antimicrobial peptide melittin into a pore in mixed DOPC/DOPG lipid bilayer by molecular dynamics simulation. *J. Chem. Phys.* **2017**, *146*, 155101. [CrossRef]

48. Upadhyay, S.K.; Wang, Y.; Zhao, T.; Ulmschneider, J.P. Insights from Micro-second Atomistic Simulations of Melittin in Thin Lipid Bilayers. *J. Membr. Biol.* **2015**, *248*, 497–503. [CrossRef]

49. Huang, H.W. Elasticity of lipid bilayer interacting with amphiphilic helical peptides. *J. Phys. II* **1995**, *5*, 1427–1431. [CrossRef]

50. Chen, L.Y.; Cheng, C.W.; Lin, J.J.; Chen, W.Y. Exploring the effect of cholesterol in lipid bilayer membrane on the melittin penetration mechanism. *Anal. Biochem.* **2007**, *367*, 49–55. [CrossRef]

51. Dufourcq, J.; Faucon, J.F. Intrinsic fluorescence study of lipid-protein interactions in membrane models. Binding of melittin, an amphipathic peptide, to phospholipid vesicles. *Biochim. Biophys. Acta (BBA) Biomembr.* **1977**, *467*, 1–11. [CrossRef]

52. Ghosh, A.K.; Rukmini, R.; Chattopadhyay, A. Modulation of tryptophan environment in membrane-bound melittin by negatively charged phospholipids: implications in membrane organization and function. *Biochemistry* **1997**, *36*, 14291–14305. [CrossRef] [PubMed]

53. Strömstedt, A.A.; Wessman, P.; Ringstad, L.; Edwards, K.; Malmsten, M. Effect of lipid headgroup composition on the interaction between melittin and lipid bilayers. *J. Colloid Interface Sci.* **2007**, *311*, 59–69. [CrossRef] [PubMed]

54. Lee, M.T.; Sun, T.L.; Hung, W.C.; Huang, H.W. Process of inducing pores in membranes by melittin. *Proc. Natl. Acad. Sci. USA* **2013**, *110*, 14243–14248. [CrossRef] [PubMed]

55. Allende, D.; Simon, S.; McIntosh, T.J. Melittin-induced bilayer leakage depends on lipid material properties: evidence for toroidal pores. *Biophys. J.* **2005**, *88*, 1828–1837. [CrossRef] [PubMed]

56. Ladokhin, A.S.; White, S.H. 'Detergent-like'permeabilization of anionic lipid vesicles by melittin. *Biochim. Biophys. Acta (BBA) Biomembr.* **2001**, *1514*, 253–260. [CrossRef]

57. Benachir, T.; Lafleur, M. Study of vesicle leakage induced by melittin. *Biochim. Biophys. Acta* **1995**, *1235*, 452–460. [CrossRef]

58. Hincha, D.K.; Crowe, J.H. The lytic activity of the bee venom peptide melittin is strongly reduced by the presence of negatively charged phospholipids or chloroplast galactolipids in the membranes of phosphatidylcholine large unilamellar vesicles. *Biochim. Biophys. Acta (BBA) Biomembr.* **1996**, *1284*, 162–170. [CrossRef]

59. Allende, D.; McIntosh, T.J. Lipopolysaccharides in Bacterial Membranes Act like Cholesterol in Eukaryotic Plasma Membranes in Providing Protection against Melittin-Induced Bilayer Lysis. *Biochemistry* **2003**, *42*, 1101–1108. [CrossRef]

60. Beschiaschvili, G.; Seelig, J. Melittin binding to mixed phosphatidylglycerol/phosphatidylcholine membranes. *Biochemistry* **1990**, *29*, 52–58. [CrossRef]

61. Klocek, G.; Schulthess, T.; Shai, Y.; Seelig, J. Thermodynamics of melittin binding to lipid bilayers. Aggregation and pore formation. *Biochemistry* **2009**, *48*, 2586–2596. [CrossRef] [PubMed]

62. Hall, K.; Lee, T.H.; Aguilar, M.I. The role of electrostatic interactions in the membrane binding of melittin. *J. Mol. Recognit.* **2011**, *24*, 108–118. [CrossRef] [PubMed]

63. Hall, K.; Mozsolits, H.; Aguilar, M.I. Surface plasmon resonance analysis of antimicrobial peptide-membrane interactions: affinity & mechanism of action. *Lett. Pept. Sci.* **2003**, *10*, 475–485.

64. Lee, T.H.; Mozsolits, H.; Aguilar, M.I. Measurement of the affinity of melittin for zwitterionic and anionic membranes using immobilized lipid biosensors. *J. Pept. Res.* **2001**, *58*, 464–476. [CrossRef]

65. Mozsolits, H.; Wirth, H.J.; Werkmeister, J.; Aguilar, M.I. Analysis of antimicrobial peptide interactions with hybrid bilayer membrane systems using surface plasmon resonance. *Biochim. Biophys. Acta (BBA) Biomembr.* **2001**, *1512*, 64–76. [CrossRef]

66. Bhunia, A.; Domadia, P.N.; Bhattacharjya, S. Structural and thermodynamic analyses of the interaction between melittin and lipopolysaccharide. *Biochim. Biophys. Acta (BBA) Biomembr.* **2007**, *1768*, 3282–3291. [CrossRef]

67. Henriksen, J.R.; Andresen, T.L. Thermodynamic profiling of peptide membrane interactions by isothermal titration calorimetry: a search for pores and micelles. *Biophys. J.* **2011**, *101*, 100–109. [CrossRef]

68. Papo, N.; Shai, Y. Exploring peptide membrane interaction using surface plasmon resonance: differentiation between pore formation versus membrane disruption by lytic peptides. *Biochemistry* **2003**, *42*, 458–466. [CrossRef]

69. Bello, J.; Bello, H.R.; Granados, E. Conformation and aggregation of melittin: dependence of pH and concentration. *Biochemistry* **1982**, *21*, 461–465. [CrossRef]

70. Raghuraman, H.; Chattopadhyay, A. Effect of ionic strength on folding and aggregation of the hemolytic peptide melittin in solution. *Biopolym. Orig. Res. Biomol.* **2006**, *83*, 111–121. [CrossRef]

71. Talbot, J.C.; Dufourcq, J.; De Bony, J.; Faucon, J.F.; Lussan, C. Conformational change and self association of monomeric melittin. *FEBS Lett.* **1979**, *102*, 191–193. [CrossRef]

72. Yang, Z.; Choi, H.; Weisshaar, J.C. Melittin-induced permeabilization, re-sealing, and re-permeabilization of E. coli membranes. *Biophys. J.* **2018**, *114*, 368–379. [CrossRef] [PubMed]

73. Shai, Y.; Oren, Z. From "carpet" mechanism to de-novo designed diastereomeric cell-selective antimicrobial peptides. *Peptides* **2001**, *22*, 1629–1641. [CrossRef]

74. Crescenzo, D.; Grothe, S.; Lortie, R.; Debanne, M.T.; O'Connor-McCourt, M. Real-time kinetic studies on the interaction of transforming growth factor alpha with the epidermal growth factor receptor extracellular domain reveal a conformational change model. *Biochemistry* **2000**, *39*, 9466–9476. [CrossRef] [PubMed]

75. Walkenhorst, W.F. Using adjuvants and environmental factors to modulate the activity of antimicrobial peptides. *Biochim. Biophys. Acta (BBA) Biomembr.* **2016**, *1858*, 926–935. [CrossRef] [PubMed]

76. Santos, N.C.; Prieto, M.; Castanho, M.A.R.B. Quantifying molecular partition into model systems of biomembranes: an emphasis on optical spectroscopic methods. *Biochim. Biophys. Acta (BBA) Biomembr.* **2003**, *1612*, 123–135. [CrossRef]

77. Domingues, T.M.; Mattei, B.; Seelig, J.; Perez, K.R.; Miranda, A.; Riske, K.A. Interaction of the antimicrobial peptide gomesin with model membranes: a calorimetric study. *Langmuir* **2013**, *29*, 8609–8618. [CrossRef]

78. Beschiaschvili, G.; Baeuerle, H.-D. Effective charge of melittin upon interaction with POPC vesicles. *Biochim. Biophys. Acta (BBA) Biomembr.* **1991**, *1068*, 195–200. [CrossRef]

79. Yuan, J.; Wu, Y.; Aguilar, M.-I. Surface Plasmon Resonance Spectroscopy in the Biosciences. In *Amino Acids, Peptides and Proteins in Organic Chemistry*; Wiley-VCH Verlag GmbH & Co. KGaA: Weinheim, Germany, 2011; pp. 225–247.

80. Kuchinka, E.; Seelig, J. Interaction of melittin with phosphatidylcholine membranes. Binding isotherm and lipid head-group conformation. *Biochemistry* **1989**, *28*, 4216–4221. [CrossRef]

81. Yang, L.; Harroun, T.A.; Weiss, T.M.; Ding, L.; Huang, H.W. Barrel-stave model or toroidal model? A case study on melittin pores. *Biophys. J.* **2001**, *81*, 1475–1485. [CrossRef]

82. Woo, S.Y.; Lee, H. Aggregation and insertion of melittin and its analogue MelP5 into lipid bilayers at different concentrations: effects on pore size, bilayer thickness and dynamics. *Phys. Chem. Chem. Phys.* **2017**, *19*, 7195–7203. [CrossRef] [PubMed]

83. Wheaten, S.A.; Ablan, F.D.; Spaller, B.L.; Trieu, J.M.; Almeida, P.F. Translocation of Cationic Amphipathic Peptides across the Membranes of Pure Phospholipid Giant Vesicles. *J. Am. Chem. Soc.* **2013**, *135*, 16517–16525. [CrossRef] [PubMed]

84. Wiedman, G.; Herman, K.; Searson, P.; Wimley, W.C.; Hristova, K. The electrical response of bilayers to the bee venom toxin melittin: Evidence for transient bilayer permeabilization. *Biochim. Biophys. Acta (BBA) Biomembr.* **2013**, *1828*, 1357–1364. [CrossRef] [PubMed]

85. Irudayam, S.J.; Pobandt, T.; Berkowitz, M.L. Free energy barrier for melittin reorientation from a membrane-bound state to a transmembrane state. *J. Phys. Chem. B* **2013**, *117*, 13457–13463. [CrossRef] [PubMed]

86. Deplazes, E.; White, J.; Murphy, C.; Cranfield, C.G.; Garcia, A. Competing for the same space: protons and alkali ions at the interface of phospholipid bilayers. *Biophys. Rev.* **2019**, *11*, 483–490. [CrossRef]

87. Tatulian, S.A. Binding of alkaline-earth metal cations and some anions to phosphatidylcholine liposomes. *Eur. J. Biochem.* **1987**, *170*, 413–420. [CrossRef]

88. Ong, S.G.M.; Chitneni, M.; Lee, K.S.; Ming, L.C.; Yuen, K.H. Evaluation of extrusion technique for nanosizing liposomes. *Pharmaceutics* **2016**, *8*, 36. [CrossRef]

89. Karlsson, R.; Katsamba, P.S.; Nordin, H.; Pol, E.; Myszka, D.G. Analyzing a kinetic titration series using affinity biosensors. *Anal. Biochem.* **2006**, *349*, 136–147. [CrossRef]

International Journal of
Molecular Sciences

Article

Lipidated Analogs of the LL-37-Derived Peptide Fragment KR12—Structural Analysis, Surface-Active Properties and Antimicrobial Activity

Elżbieta Kamysz [1,*], Emilia Sikorska [2], Maciej Jaśkiewicz [3], Marta Bauer [3], Damian Neubauer [3], Sylwia Bartoszewska [3], Wioletta Barańska-Rybak [4] and Wojciech Kamysz [3]

1   Laboratory of Chemistry of Biological Macromolecules, Department of Molecular Biotechnology, Faculty of Chemistry, University of Gdańsk, 80-308 Gdańsk, Poland

2   Laboratory of Structural Studies of Biopolymers, Department of Organic Chemistry, Faculty of Chemistry, University of Gdańsk, 80-308 Gdańsk, Poland; emilia.sikorska@ug.edu.pl

3   Department of Inorganic Chemistry, Faculty of Pharmacy, Medical University of Gdańsk, 80-416 Gdańsk, Poland; mj@gumed.edu.pl (M.J.); marta.bauer@gumed.edu.pl (M.B.); damian.neubauer@gumed.edu.pl (D.N.); sylwia.bartoszewska@gumed.edu.pl (S.B.); wojciech.kamysz@gumed.edu.pl (W.K.)

4   Department of Dermatology, Venereology and Allergology, Faculty of Medicine, Medical University of Gdańsk, 80-214 Gdańsk, Poland; wioletta.baranska-rybak@gumed.edu.pl

*   Correspondence: elzbieta.kamysz@ug.edu.pl; Tel.: +48-58-523-5011

Received: 21 December 2019; Accepted: 28 January 2020; Published: 30 January 2020

Abstract: An increasing number of multidrug-resistant pathogens is a serious problem of modern medicine and new antibiotics are highly demanded. In this study, different n-alkyl acids (C_2-C_{14}) and aromatic acids (benzoic and *trans*-cinnamic) were conjugated to the N-terminus of KR12 amide. The effect of this modification on antimicrobial activity (ESKAPE bacteria and biofilm of *Staphylococcus aureus*) and cytotoxicity (human red blood cells and HaCaT cell line) was examined. The effect of lipophilic modifications on helicity was studied by CD spectroscopy, whereas peptide self-assembly was studied by surface tension measurements and NMR spectroscopy. As shown, conjugation of the KR12-NH_2 peptide with C_4-C_{14} fatty acid chains enhanced the antimicrobial activity with an optimum demonstrated by C_8-KR12-NH_2 (MIC 1–4 µg/mL against ESKAPE strains; MBEC of *S. aureus* 4–16 µg/mL). Correlation between antimicrobial activity and self-assembly behavior of C_{14}-KR12-NH_2 and C_8-KR12-NH_2 has shown that the former self-assembled into larger aggregated structures, which reduced its antimicrobial activity. In conclusion, N-terminal modification can enhance antimicrobial activity of KR12-NH_2; however, at the same time, the cytotoxicity increases. It seems that the selectivity against pathogens over human cells can be achieved through conjugation of peptide N-terminus with appropriate n-alkyl fatty and aromatic acids.

Keywords: ESKAPE pathogens; *Staphylococcus aureus*; KR12; LL-37; lipopeptide; critical aggregation concentration; CD spectroscopy; NMR; biofilm; cytotoxicity

---

1. Introduction

The occurrence of multidrug-resistant (MDR) bacterial strains faces many difficulties in the therapy of some infections due to prolonged treatment and frequent relapses. An increasing number of MDR pathogens is mainly associated with persistent and abused use of antibiotics and just those strains are mostly associated with hospital flora. The ESKAPE pathogens (*Enterococcus faecium, Staphylococcus aureus, Klebsiella pneumoniae, Acinetobacter baumannii, Pseudomonas aeruginosa* and *Enterobacter* spp.) are bacterial species responsible for most of nosocomial infections [1]. Moreover, recent epidemiological data have shown that the therapy of infections caused by those bacteria is also associated with the highest risk of mortality [2]. According to the latest reports of World Health Organization

(WHO), all of the ESKAPE pathogens are listed in the group of bacteria for which new antibiotics are highly demanded [3]. It should be noted that one of the species in the high priority group are methicillin-resistant *S. aureus* (MRSA), which are prevalent species in environment making them the major source of hospital-acquired infections (HAIs) [4]. It has been estimated that almost 44% of all HAIs are caused by those bacteria, with indication of being responsible for over 20% of excessive mortality [5,6]. The therapy of infections caused by MRSA is even more challenging as these strains produce a number of mechanisms allowing them to invade into the organisms, including avoidance of opsonization by antibodies and complement system, disruption of chemotaxis and lysis of neutrophils. Because of their ability to survive inside leukocytes, the infections tend to move into a chronic stage and recur after recovering. Furthermore, the therapy often needs prolonged hospitalization and commonly tends to be ineffective. An additional complication of the therapy is the ability of bacteria to form biofilms—an organized three-dimensional structure characterized by enhanced resistance to antibiotics [7]. It has been estimated that approximately 80% of chronic and recurrent infections are associated with the biofilm occurrence [8]. Low effectiveness of the current approaches to the therapy of HAIs together with accompanying side-effects adversely affect the patient's health. A multitude of antibiotics often fail to be effective in the treatment because of MDR strains. Therapeutic difficulties accompanying the majority of infections escalates the need to search for new effective drugs. Antimicrobial peptides (AMPs) are a promising class of antimicrobial compounds which have a chance to fight resistant pathogens owing to their rapid membrane-targeting bactericidal mode of action and the predicted low propensity for development of resistance [9–11]. One of the AMPs is a linear, cationic, α-helical and amphipathic peptide LL-37 (LLGDFFRKSKEKIGKEFKRIVQRIKDFLRNLVPRTES), the member of the human cathelicidin family [12–14]. This peptide is released from its precursor, hCAP-18, through proteolytic processing by proteinase 3, a serine proteinase secreted from neutrophils [14]. Interestingly, the hCAP-18 found in seminal plasma can also be hydrolyzed by vaginal gastricsin. As a result, instead of LL-37 another peptide (ALL-38) can be generated. Although this compound contains additional alanine at the *N*-terminus, exhibits comparable antimicrobial activity [15]. Furthermore, LL-37 is found in variety of cells, tissues and body fluids such as leukocytes, bone marrow, milk, salivary glands, skin, respiratory tract, epithelium cells and leukocytes within the digestive tract, urinary tract as well as squamous epithelium of the mouth and tongue [12,13,16,17]. This compound exhibits a broad spectrum of antibacterial activity against both planktonic cells and biofilms of Gram-positive and Gram-negative bacteria, which promotes it as a candidate for a new antibiotic [18,19]. However, LL-37 is a relatively long peptide, which makes it to be expensive for manufacturing. Thus, the search for a novel, shorter analogs of LL-37 is desired. Some fragments of LL-37 have been evaluated to identify improved antimicrobial derivatives (for instance KR12, FK-13, GF17, 17BIPHE2) [20–24]. Both LL-37 and its shorter active analogs adopt a helical structure in the presence of membrane lipids [23–25]. The shortest α-helical fragment of LL-37 with documented antimicrobial activity is KR12 amide (KRIVQRIKDFLR-NH$_2$) [20,21,26]. This peptide, a truncated form of LL-37, shares two common features of antimicrobial peptides: a positive net charge and an amphipathic structure, which determine their antimicrobial activity.

In this article, we report the synthesis of a series of lipopeptides derivatized with variable length fatty acids or aromatic acids covalently attached to the *N*-terminus of KR12-NH$_2$, and their antimicrobial activity against planktonic cells of ESKAPE bacteria. The fatty acid tail was introduced to KR12-NH$_2$, because in the literature it was found that addition a fatty acid residue to AMPs may improve effectiveness of peptides as antimicrobial agents by enhancing their ability to form either secondary structures or oligomerize upon interacting with bacterial membranes [27–29]. We also demonstrate the activity of the tested peptides against biofilm formed by reference strains of *S. aureus* (including *MRSA*), because one of the obstacles complicating therapy of staphylococcal infections is the growth of biofilm. Relationships between antimicrobial activity and hemolytic activity as well as cytotoxicity of the peptides were also determined. The effect of *N*-terminal modifications on helicity of the KR12-NH$_2$ peptide was studied by

CD spectroscopy. The ability of the selected lipopeptides to spontaneous self-assembly in solution was evaluated with surface tension measurements and NMR spectroscopy.

2. Results and Discussion

This section describes and discusses results MS and RP-HPLC analyses of peptides (2.1), evaluation of their antimicrobial activity against planktonic *S. aureus* and ESKAPE strains and biofilm of *S. aureus* reference strains (2.2), as well as studies on hemolysis (2.3) and cytotoxicity (2.4). Moreover, CD spectroscopy (2.5), critical aggregation concentration (CAC) and NMR spectroscopy (2.6) were included to learn how *N*-terminal modification affects secondary-structure and peptide self-assembly. As found, the activity of the studied peptides was determined by many concurrent parameters, including hydrophobicity, conformation or tendency to self-assembly.

2.1. Peptide Synthesis and Purification

Peptides **I-X** were synthesized by solid-phase method using Fmoc chemistry. Their purity was higher than 95% as shown by analytical reversed-phase high-performance liquid chromatography (RP-HPLC). The electrospray ionization mass spectrometry (ESI MS) in positive ion mode confirmed identity of the purified peptides. Physicochemical characteristics of the peptides are shown in Table 1.

Table 1. Characteristics of the peptides.

Peptide	Name	Net Charge	HPLC t'_R (min)	Average Mass (Da)	z	m/z calc.	m/z found
I	Ac-KR12-NH$_2$ (C$_2$-KR12-NH$_2$)	+4	3.58	1612.96	2	807.49	807.29
					3	538.66	538.69
					4	404.25	404.47
II	C$_4$-KR12-NH$_2$	+4	3.99	1641.02	2	821.52	821.38
					3	548.01	547.73
					4	411.26	411.36
III	C$_6$-KR12-NH$_2$	+4	4.42	1669.07	2	835.54	835.16
					3	557.36	556.84
					4	418.28	418.35
IV	C$_8$-KR12-NH$_2$	+4	4.88	1697.12	2	849.57	849.46
					3	566.71	566.65
					4	425.29	425.31
					5	340.43	340.56
V	C$_{10}$-KR12-NH$_2$	+4	5.38	1725.18	2	863.60	863.43
					3	576.07	575.86
					4	432.30	432.20
VI	C$_{12}$-KR12-NH$_2$	+4	5.86	1753.23	2	877.62	877.03
					3	585.42	585.04
					4	439.32	439.18
VII	C$_{14}$-KR12-NH$_2$	+4	6.39	1781.28	2	891.65	891.42
					3	594.77	594.59
					4	446.33	446.04
VIII	Benzoic acid-KR12-NH$_2$	+4	4.23	1675.03	2	838.52	838.14
					3	559.35	559.00
					4	419.77	419.76
IX	*trans*-Cinnamic acid-KR12-NH$_2$	+4	4.44	1701.07	2	851.54	851.34
					3	568.03	567.58
					4	426.28	425.85
X	KR12-NH$_2$ (KRIVQRIKDFLR-NH$_2$)	+5	2.68	1570.93	2	786.47	786.23
					3	524.65	524.69
					4	393.74	393.94
					5	315.19	315.18

z—ion charge, m/z—mass to charge ratio; adjusted retention time (t'_R) is an analyte's retention time (t_R) minus the elution time of an unretained peak (t_m).

Conjugation of KR12-NH$_2$ *N*-terminal amino group with aliphatic or aromatic acids result in compounds with a reduced net charge (+4 vs. +5) and enhanced hydrophobicity, as shown by RP-HPLC. Results of this evaluation are presented in Table 1. Moreover, the number of carbon atoms in the *N*-terminal acid was plotted against the adjusted retention time (Figure 1).

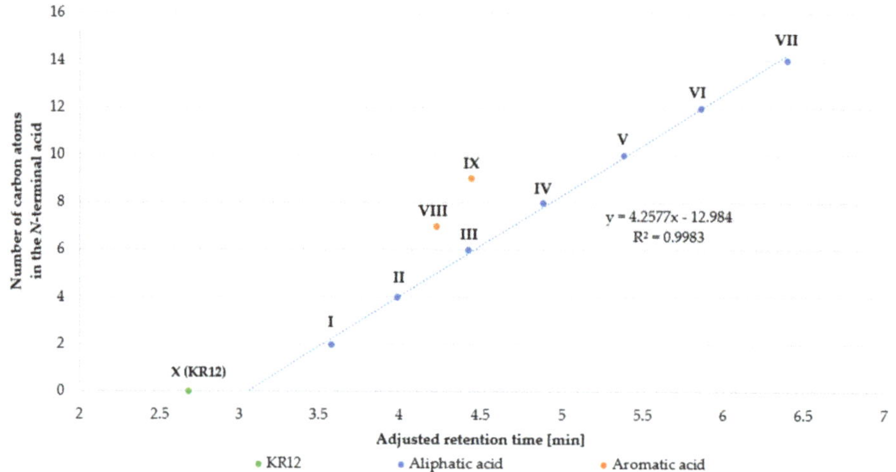

Figure 1. The number of carbon atoms in the *N*-terminal acid residue *versus* adjusted retention time.

Peptides with *N*-terminal aliphatic acid to generate a homologous series differ in the number of carbon atoms (methylene groups). It can be seen that retention time (hydrophobicity) of peptides **I–VII** increased proportionally to the number of carbon atoms (linear regression, $R^2 = 0.9983$). Retention time of KR12-NH$_2$ (+5) is shifted to lower values than those predicted using regression equation of **I–VII** (calculated 3.05 vs. measured 2.68 min), mainly due to its higher net charge (+5). Analogs with an aromatic acid at the *N*-terminus did not follow this trend. The calculated retention times of analogs with identical number of carbon atoms in aliphatic acids as compared to those with aromatic **VIII** (C$_7$) and **IX** (C$_9$) acids were distinctly higher (more hydrophobic) than the experimentally determined ones (**VIII**: 4.23 vs. 4.69 min calc.; **IX**: 4.44 vs. 5.16 min calc.). This phenomenon is the result of different carbon hybridization. In aromatic acids, carbon atoms are sp^2 hybridized and in the aliphatic ones they are sp^3 hybridized (excluding carbon atom of carboxylic group), which influences polarity, shape and planarity (aromatic ring) of aromatic compounds, and both can affect retention time [30–32].

2.2. Antimicrobial Assay

In our preliminary research, we tested LL-37 and KR12-NH$_2$ (**X**) against a reference strain of *S. aureus* ATCC 25923. Minimal inhibitory concentrations (MICs) of *S. aureus* strain were 256 µg/mL for peptide KR12-NH$_2$ and >512 µg/mL for LL-37 in analysis performed in the Mueller-Hinton medium. MICs for *S. aureus* strain cultivated in 1% Bacto Peptone medium were 64 µg/mL for peptide KR12-NH$_2$ and >512 for LL-37. We also tested antimicrobial activity of LL-37 and KR12-NH$_2$ against clinical strains of *S. aureus* acquired from the skin and nose and it strongly depended on the bacterial strains of *S. aureus* (MICs values ranged between 1 and >512 µg/mL) [33]. Because antistaphylococcal activities of KR12-NH$_2$ and LL-37 were comparable, we decided to introduce a lipophilic residue to peptide KR12-NH$_2$ (**X**). Peptide **X** and its nine analogs (**I–IX**) were tested against selected reference strains of ESKAPE bacteria (Table 2—*E. faecium*, *K. pneumoniae*, *A. baumannii*, *P. aeruginosa*, *K. aerogenes*; Table 3—several reference strains of *S. aureus* including *MRSA* ATCC 33591) and staphylococcal biofilm (Table 4). The antimicrobial activity of the synthesized peptides was dependent on the number of carbon atoms in the *N*-acyl substituent. A high activity against planktonic forms of the bacteria and the staphylococcal biofilm was found for peptides **III–V** and **IX** (Tables 2–4). The most effective was the analog of KR12-NH$_2$ modified in the *N*-terminal part of the molecule with octanoic acid residue (C$_8$-KR12-NH$_2$, peptide **IV**) for which the minimal inhibitory concentration (MIC) values ranged between 1 and 4 µg/mL (Tables 2 and 3), while the minimal biofilm eradication concentrations (MBECs) of *S. aureus* strains were four-fold higher than the MIC values and ranged between 4 and 16 µg/mL. Generally, the conjugation of the KR12-NH$_2$ with both longer and shorter hydrocarbon acyl chains than that of C$_8$ resulted in a decrease in antimicrobial activity. The next active compound was analog KR12-NH$_2$ modified with *trans*-cinnamic acid residue (peptide **IX**). For this particular compound, the MIC values ranged between 1 and 8 µg/mL (Tables 2 and 3). The MBEC values were four-fold higher than those of MIC values and ranged between 4 and 32 µg/mL. As a rule, modification of the KR12 amide with fatty acid residues (C$_4$-C$_{14}$) intensified antimicrobial potency against the tested bacteria. An exception was found only for analog modified with C$_4$ (**II**), which was inactive against *A. baumannii*. In general, antimicrobial activity of the analogs depended on their hydrophobicity (*N*-terminal acid). The relation between antimicrobial activity of the peptides against Gram-positive (*S. aureus* ATCC 25923) and Gram-negative (*P. aeruginosa* ATCC 9027) strains and their hydrophobicity (adjusted retention time) is presented on the Figure 2.

Table 2. The minimal inhibitory concentration (MIC) values (µg/mL) of the peptides against reference strains of ESKAPE pathogens.

Peptide	*E. faecium* ATCC 700221	*K. pneumoniae* ATCC 700603	*A. baumannii* ATCC BAA-1605	*P. aeruginosa* ATCC 9027	*K. aerogenes* ATCC 13048
I	16	>256	>256	64	>256
II	4	128	>256	32	128
III	2	16	16	8	16
IV	1	2	2	2	2
V	2	16	8	8	16
VI	4	32	16	32	32
VII	8	64	32	128	128
VIII	1	16	16	8	16
IX	1	4	4	4	4
X	8	>256	256	16	>256

Table 3. The MIC values (µg/mL) of the test peptides against reference strains of *S. aureus*.

Peptide	*S. aureus* ATCC 25923	*S. aureus* ATCC 6538	*S. aureus* ATCC 33591	*S. aureus* ATCC 9144	*S. aureus* ATCC 12598
I	>256	256	>256	>256	>256
II	128	32	128	64	64
III	16	4	16	8	8
IV	2	2	2	2	4
V	4	4	4	4	4
VI	32	32	16	32	32
VII	128	64	16	64	128
VIII	32	8	32	16	16
IX	8	2	4	4	4
X	>256	256	>256	>256	>256

Table 4. The MBEC values (µg/mL) of the test peptides against reference strains of *S. aureus*.

Peptide	*S. aureus* ATCC 25923	*S. aureus* ATCC 6538	*S. aureus* ATCC 33591	*S. aureus* ATCC 9144	*S. aureus* ATCC 12598
I	>256	>256	>256	>256	>256
II	256	128	128	128	128
III	32	16	16	16	16
IV	8	16	4	4	4
V	32	32	32	16	8
VI	256	256	128	64	32
VII	256	256	256	128	64
VIII	64	32	32	16	16
IX	32	8	16	8	4
X	>256	>256	>256	>256	>256

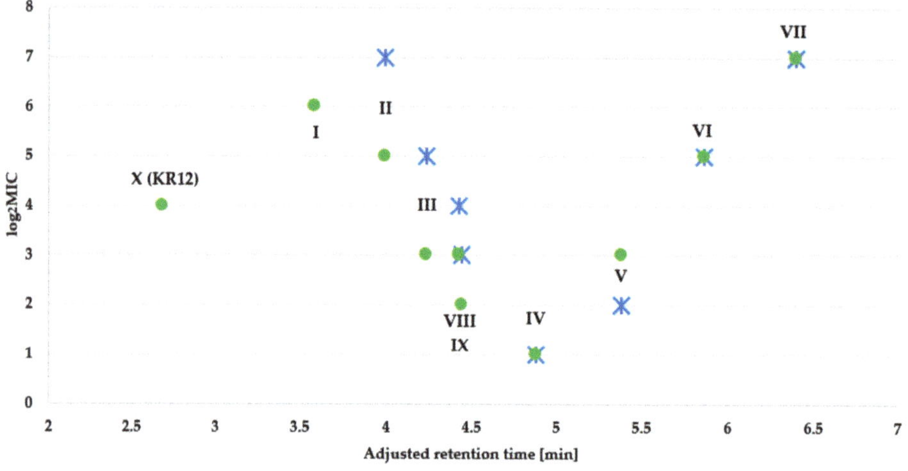

Figure 2. Antimicrobial activity of KR12 amide and its analogs ($\log_2$MIC) against *S. aureus* ATCC 25923 and *P. aeruginosa* ATCC 9027 *versus* adjusted retention time.

Positive charge of the peptide is essential for its antimicrobial activity due to interactions with negatively charged pathogen cells. A gradual reduction of positive charge usually results in decrease or loss of antimicrobial activity [34,35]. However, as observed, there was no simple correlation between charge and activity. In case of the studied analogs the *N*-terminal modification reduced net charge from +5 to +4. Despite the reduction of total positive charge, most of the analogs displayed improved antimicrobial activity when compared to KR12-NH$_2$. This finding emphasizes that *N*-terminal modification also modifies other structural parameters of the peptide that are crucial for activity. The antimicrobial activity of lipopeptides depended on the length of the acyl substituent, which is compatible with earlier reports [27,36,37]. However, in the literature, different fatty acids have been suggested as the optimum modification to provide enhanced antimicrobial activity. For instance, Laverty et al. conjugated a tetrapeptide amide H-Orn-Orn-Trp-Trp-NH$_2$ with saturated fatty acids (C$_6$-C$_{16}$) and demonstrated that *N*-acyl substituents of 12–14 carbon atoms in length exhibited the strongest antimicrobial and antibiofilm activities [36]. As a result, the hydrophobicity of the *N*-acyl substituent was pointed as a key determinant of antimicrobial activity for the peptides [36]. Albada et al. also studied the influence of lipidation (C$_2$-C$_{14}$) on antimicrobial potency of short active unnatural AMPs. In this case, the highest activities against a broad spectrum of pathogens were found for compounds modified with C$_8$ and C$_{10}$ residues. The authors suggested that the lowered activity of peptides lipidated with C_{12} and C_{14} could be associated with their poor solubility in media used for microbiological assays [37]. In our study, the optimum modification was found for lipidation with octanoic acid (C_8).

2.3. Hemolysis Assay

The hemolytic activity of peptides **I–X** was assessed for human red blood cells (hRBCs) to verify their toxicity (Figure 3 and Table 5). Our results indicate that hemolysis of the erythrocytes depended on the number of carbon atoms of the conjugated acid. For longer hydrocarbon acyl chains (C_2 to C_{14}), the hemolytic activity increased. For instance, for peptides **I** (C_2—ethanoic acid) and **II** (C_4—butyric acid) the hemolysis of hRBCs was not detected within the studied concentration range (0.5–256 µg/mL). Analogs **III** and **IV** (with hexanoic acid – C_6 and octanoic acid – C_8) exhibited a higher hemolytic activity, but it was still below their MIC and MBEC values. The most effective compound against both planktonic cells and the biofilm of *S. aureus* was peptide **IV**, which caused 5% hemolysis of hRBCs at a concentration of 64 µg/mL. Peptide **IV** had a high selectivity index amounting to almost 28 (Table 5). In the case of peptides, **V–VII** containing hydrocarbon acyl chains from C_{10} to C_{14} the hemolytic activity was comparable and exceeded the values of MIC and MBEC. For these compounds, hemoglobin release was found, beginning from 2–4 µg/mL. Lysis of hRBCs for those peptides was noticed in the range of 64–128 µg/mL. As a result, an analog with a tetradecanoic acid (**VII**) turned out to be the most toxic one. Noteworthy is the fact that conjugation of KR12-NH$_2$ with aromatic carboxylic acids, such as benzoic (C_7) and *trans*-cinnamic (C_9) acids led to analogs **VIII** and **IX** that were less hemolytic than analogs **III** (C_6) and **IV** (C_8). However, selectivity index of peptide **IX** was lower than that of peptide **IV** (Table 5). Peptide **VIII** did not cause hemolysis of hRBCs (>5%) over the whole concentration range. SI value for this peptide was not calculated because minimal hemolytic concentration (MHC) exceeded 256 µg/mL. It should be emphasized that peptide KR12-NH$_2$ (**X**), which was used as a base for all modifications, did not cause any significant hemolysis (>5%) over the whole concentration range (0.5–256 µg/mL).

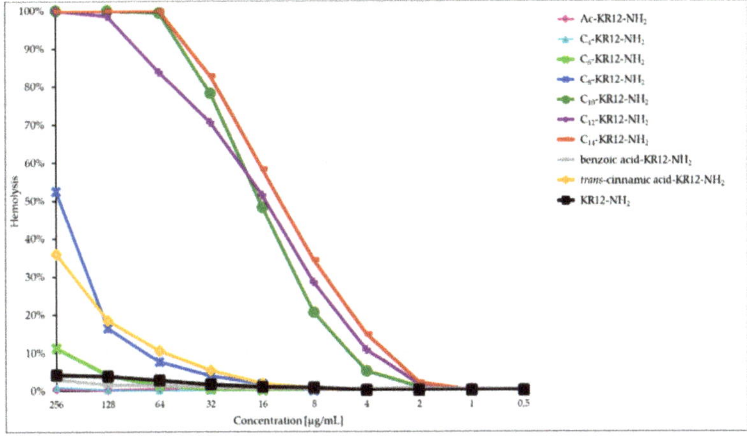

Figure 3. Percentage of hemolysis of erythrocytes *versus* peptide concentration.

2.4. MTT Assay

All the tested peptides exhibited cytotoxicity against human keratinocytes cell line (HaCaT). A 65-fold difference in half maximal inhibitory concentration (IC$_{50}$) values was observed between peptides displaying the highest and lowest degrees of cytotoxicity (Table 5). Only for peptides **III–V**, **VIII** and **IX**, the IC$_{50}$ values were higher than the mean MIC ones (GM) for *S. aureus*. Moreover, the highest toxicity was found for peptide **VII** (C_{14}), but its activity against *S. aureus* was relatively poor. On the other hand, the least toxic was peptide **I**. Therefore, for this compound the antimicrobial activity was determined only against *E. faecium* and *P. aeruginosa* (16 and 64 µg/mL, respectively). Selectivity indices IC_{50}/GM against reference strains of *S. aureus* were low. The most selective were analogs modified with aromatic acid residues **VIII** (C_7) and **IX** (C_9) with selectivity indices of 2.20 and 2.50, respectively. Several articles on the safety-profile of KR-12 analogs have already been published, indicating high antimicrobial activity and low cytotoxicity of the aforementioned analogs. For instance, in the study of Jacob et al. a series of KR12 analogs were designed and synthesized in order to optimize the α-helical structure (KR12-a1 to a7) [21]. As a result, all of the analogs showed insignificant cytotoxicity against macrophages of RAW264.7 cell line and anti-inflammatory activity. Moreover, on the basis of these results, Kim et al. conducted a research on D-amino acid substituted analogs of KR-12-a5 (KRIVKLILKWLR-NH₂) which appeared to be non-toxic against macrophages (RAW264.7) and fibroblasts (NIH-3T3) at whole concentration range [38]. It is worth noticing that Rajasekaran et al. conducted research on alanine scan of FK-13 peptide (FKRIVQRIKDFLR-NH₂), which is also considered to be an antimicrobial region of LL-37 [22]. In this study, the cytotoxicity was determined against both RAW264.7 and HaCaT cell lines. As a result, for majority of peptides the viability of test cells at MIC concentrations were not significantly affected. Furthermore, in the last two cases, the cell selectivity (therapeutic index) was determined in the relation to hemolysis. Respectively, in the study of Kim et al. the most selective was compound with 6-DL, with selectivity of 61.2, while in the article of Rajasekaran et al. the most selective was FK13-a4 with a selectivity equal to 138.4 [22,38]. However, the obtained results should not be compared as in one assay the MHC that caused 10% of hemolysis was taken into calculation while in another considered concentration was HC50 (50% hemolysis).

Table 5. MHC, IC_{50}, GM and selectivity indices (SI) of peptides determined for reference strains of *S. aureus*.

Peptide	MHC [1] (µg/mL)	IC_{50} (µg/mL)	GM [2] (µg/mL)	Selectivity Index (SI) [3] MHC/GM	IC_{50}/GM
I	>256.00	84.20	>256.00	NA	NA
II	> 256.00	38.73	73.52	NA	0.53
III	256.00	10.13	9.19	27.86	1.10
IV	64.00	3.23	2.30	27.83	1.41
V	4.00	5.97	4.00	1.00	1.49
VI	4.00	7.60	27.86	0.14	0.27
VII	4.00	1.29	64.00	0.06	0.02
VIII	>256.00	40.50	18.38	NA	2.20
IX	32.00	10.00	4.00	8.00	2.50
X	>256.00	74.60	>256.00	NA	NA

[1] MHC is the minimal hemolytic concentration that caused 5% hemolysis of human red blood cells. [2] The geometric mean (GM) of the MIC values against *S. aureus* was calculated. [3] SI is the ratio of MHC/IC_{50} to GM. More selective compounds are characterized by the highest values of SI [39]. NA: not applicable; SI values were not calculated for compounds with MHC and/or GM values higher than 256 µg/mL.

Peptide net charge and hydrophobicity can affect biological activity, including antimicrobial potency and cytotoxicity as well as also selectivity. Data in Tables 1 and 5 were used to find a relationship between selectivity and hydrophobicity of KR12-NH₂ analogs (Figures 4 and 5). As seen, longer acyl chain and higher hydrophobicity influenced hemolytic activity to a greater extent than antimicrobial activity, resulting in reduced selectivity indices MHC/GM (Figure 4). Similar tendency was observed in case of selectivity indices IC_{50}/GM; however, the differences between the peptides are not so spectacular (Figure 5). Interestingly, KR12-NH₂ modified with aromatic acids (analogs **VIII** and **IX**) had highest selectivity indicies (IC_{50}/GM). It has been shown that conjugation of the *N*-terminal amino group of the cationic peptide (Orn-Orn-Trp-Trp-NH₂) with cinnamic acid (and its derivatives) lead to compounds with promising antimicrobial activity against Gram-positive bacteria (*S. aureus*) and low cytotoxicity (HaCaT cell line) and hemolytic activity [40].

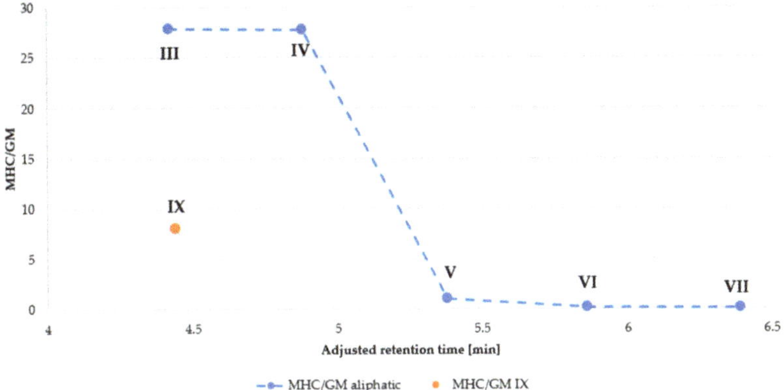

Figure 4. Selectivity for *S. aureus* over erythrocytes *versus* adjusted retention time.

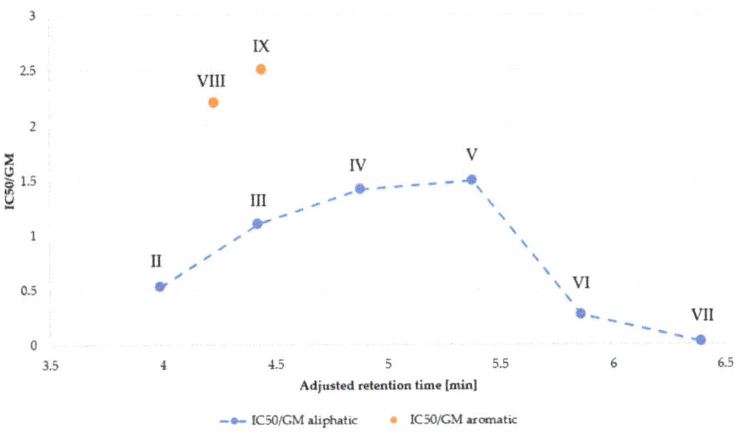

Figure 5. Selectivity for *S. aureus* over human cells (HaCaT) *versus* adjusted retention time.

2.5. Conformational Studies

The CD spectra (Figure 6) revealed the peptides to be generally devoid of a stable conformation in water and phosphate buffered saline (PBS) solutions, and only addition of membrane-mimicking surfactants such as sodium dodecyl sulfate (SDS) and dodecylphosphocholine (DPC) as well as liposomes such as 1-palmitoyl-2-oleoyl-sn-glycero-3-phosphocholine (POPC) and 1-palmitoyl-2-oleoyl-sn-glycero-3-phosphoglycerol (POPG) induced an α-helical structure, but lipopeptides C_{10}-KR12-NH_2 (**V**), C_{12}-KR12-NH_2 (**VI**) and C_{14}-KR12-NH_2 (**VII**) clearly deviated from this trend. As seen, the CD spectra of compounds **V**, **VI** and **VII** in PBS, as well as of analog **VII** in water, displayed typical features of peptides with an α-helix folding with two well-defined minimums at 208 and 222 nm. A straightforward explanation of this fact is a self-assembly of the lipopeptides, which seems to be a sufficient factor to provide hydrophobic environment stabilizing the helical conformation. In the case of C_{14}-KR12-NH_2 (**VII**), hydrophobic interactions between the tetradecanoic acyl chains were sufficient to overcome the electrostatic and steric repulsion between the peptide residues in both non-buffered and buffered aqueous solutions. In turn, with C_{10}-KR12-NH_2 (**V**) and C_{12}-KR12-NH_2 (**VI**), only an increase in the solution ionic strength resulted in an effective screening of the electrostatic peptide repulsion leading to peptide's self-assembly. Interestingly, a $\Theta222/\Theta208$ ratio greater than 1 noticed in PBS indicated a coiled-coil formation and was a further proof for self-assembly. A similar tendency has previously been observed for conjugates of magainin with lipophilic acids [41].

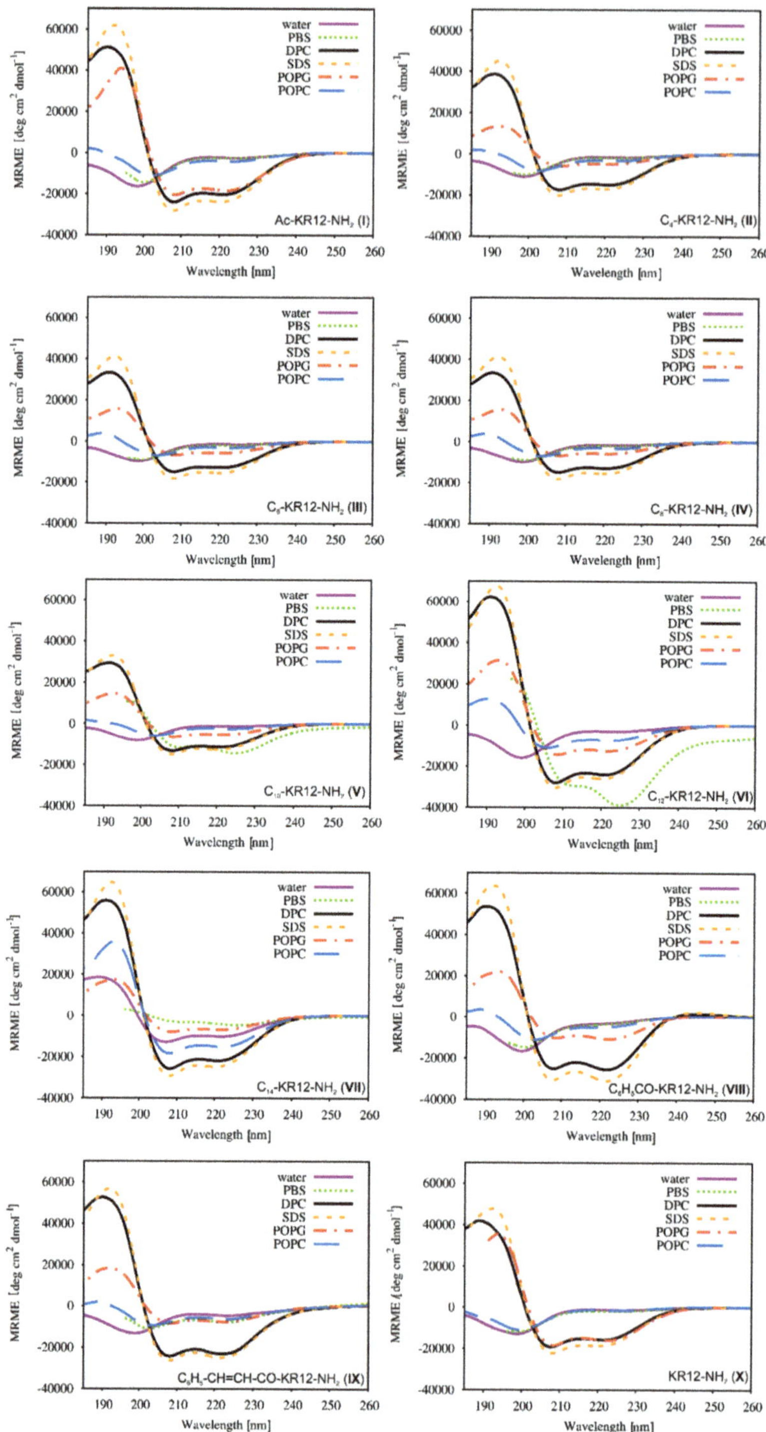

Figure 6. Far-UV CD spectra of the peptides.

As seen in Figure 6, there were only insignificant differences in the CD spectra of the peptides in SDS and DPC micelles. In both detergents, a high helical content was found for all the peptides studied (45–86%, Table 6). In turn, POPG and POPC liposomes had a different effect on conformation of the peptides. In general, the presence of POPC liposomes, a model of eukaryotic membranes, induced an increase in helicity with elongation of the attached acyl chain, which correlated well with a rise in hemolytic activity. The C_{14}-KR12-NH_2 (**VII**) analog with the highest helical percentage in POPC liposomes exhibited also the highest cytotoxicity against human keratinocytes. On the other hand, analogs Ac-KR12-NH_2 (**I**) and C_8-KR12-NH_2 (**IV**) displayed the same helical fraction (16%) but extremely different cytotoxicity, because the former was found to be the least cytotoxic, whereas the latter one of the most cytotoxic peptides. This indicated no linear correlation between helicity and toxicity of the peptides studied. In turn, the CD spectra in POPG liposomes, representing a negatively charged bacterial membrane, showed that modifications of KR12-NH_2 with acyl chains longer than that of C_4 as well as aromatic substituents (C_7 and C_9) reduced the helical fraction, but without any clear correlation with antimicrobial activity. This result was not surprising at all because numerous studies to date have shown different relationships between helicity and antimicrobial activity. In particular, Shai and Oren have demonstrated that reducing helicity by incorporating D-amino acids decreased hemolytic activity but did not affect most of the potent antimicrobial activity of the diastereomeric analogs as compared to that of the parent peptides [42,43]. Comparable results have been reported for Temporin L analogs [44]. In turn, a study on enantiomers of Pleurocidin [45] has shown all the D-amino acid-containing peptides exhibited a decreased antibacterial activity and a dramatically decreased hemolytic activity as that of compared to L-amino acid-containing counterpart despite a higher percentage of helical structure. All this suggests that conformation of the peptides is not the only factor affecting biological activity.

Table 6. Helical content determined based on CD spectra.

Peptide	Helical Content %					
	Water	PBS	DPC	SDS	POPG	POPC
I	8	8	72	82	69	16
II	7	6	69	71	67	9
III	7	7	65	67	24	16
IV	6	6	53	62	27	16
V	6	51	45	54	26	15
VI	8	74	81	86	55	32
VII	35	17	77	84	29	63
VIII	8	5	73	82	41	15
IX	7	19	72	77	30	18
X	7	7	66	70	66	7

2.6. Self-Assembly Studies

The critical aggregation concentrations (CACs) were determined for the analogs with *N*-terminal fatty acids C_8-C_{14} (Figure 7). The surface tension measurements were carried out in pure water, because the peptides aggregated in PBS solution. This is probably the consequence of self-assembly at much lower CAC values according to the rule that an increase in ionic strength of the solution decreases the CAC value of ionic surfactants [46–48].

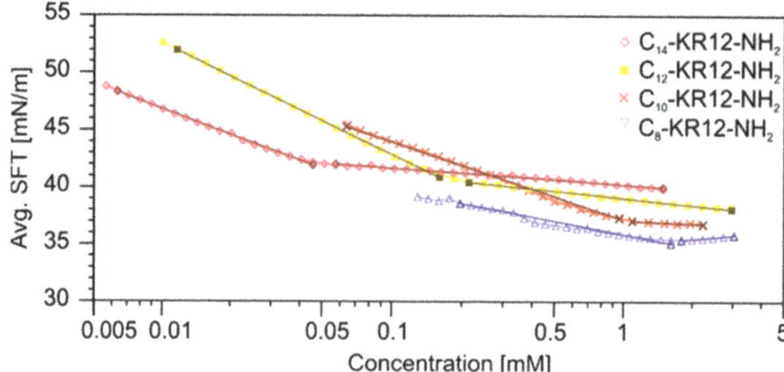

Figure 7. Relationship between the surface tension and peptide concentration.

As expected, the longer was the lipophilic acyl chain, the more effective self-assembly became, owing to an increase in intermolecular hydrophobic interactions. The CAC values decreased with increasing alkyl chain length following the order: 1.46 mM (2475 µg/mL), 1.05 mM (1807 µg/mL), 0.17 mM (297 µg/mL) and 0.042 mM (74 µg/mL) for **IV**, **V**, **VI** and **VII**, respectively. In the ^{1}H NMR spectra, the self-assembly induced broadening of the resonance lines (Figure 8A), which is related to a decrease in system's tumbling rate and shortening of the T_2 relaxation times. Interestingly, the NMR spectra of C$_8$-KR12-NH$_2$ (**IV**) and C$_{14}$-KR12-NH$_2$ (**VII**), both recorded at a concentration above CAC, differed from each other despite identical peptide sequences, reflecting different conformations of the peptides. In the latter case, the amide proton resonances were spread out over a wider range of chemical shifts as compared to that of the former, this being characteristic of formation of helical structure [49]. The translation diffusion coefficient (D_{tr}) determined at a concentration higher than CAC of C$_{14}$-KR12-NH$_2$ (**VII**) and the corresponding hydrodynamic radius (R$_H$), derived from the Stokes-Einstein's equation were 7.76 × 10^{-11} m^2/s and 32 Å, respectively. Due to the low CAC value, it was difficult to measure the self-diffusion coefficient for the monomer of peptide **VII**. Hence, the length of a single molecule was established to be ~35 Å assuming a tetradecanoic acyl tail and the peptide moiety to exist in full-extended and helical conformations, respectively. This value corresponded well with R$_H$ extracted from the translation diffusion coefficient. Therefore, we concluded that C$_{14}$-KR12-NH$_2$ self-assembled into micelles. In the case of C$_8$-KR12-NH$_2$ (**IV**), the translation diffusion coefficients, D_{tr}, extracted from the NMR experiments at concentrations three-fold lower and three-fold higher than CAC were 1.95 ×·10^{-10} and 1.74 × 10^{-10} m^2/s, respectively, and corresponded to the $D_{tr,\ oligomer}/D_{tr,\ monomer}$ ratio of roughly 0.89. Based on the previous study, the $D_{tr,\ oligomer}/D_{tr,\ monomer}$ ratio of ~0.8 is related to dimer formation by assuming that both the monomer, and the dimer adopt compact (spherical) structures. With the peptides of elongated shapes, this ratio may increase [50,51]. For comparison, in the case of common surfactants, SDS and 1,2-diheptanoyl-sn-grycero-3-phosphocholine (DHPC), as well as other antimicrobial lipopeptides, which self-assemble into spherical micelles, the $D_{tr,\ micelle}/D_{tr,\ monomer}$ ratio is lower than 0.5 [52–54]. Taking all this into account, we speculate that C$_8$-KR12-NH$_2$ self-assembled into dimers and the oligomerization over the tested concentration range does not favor helix formation.

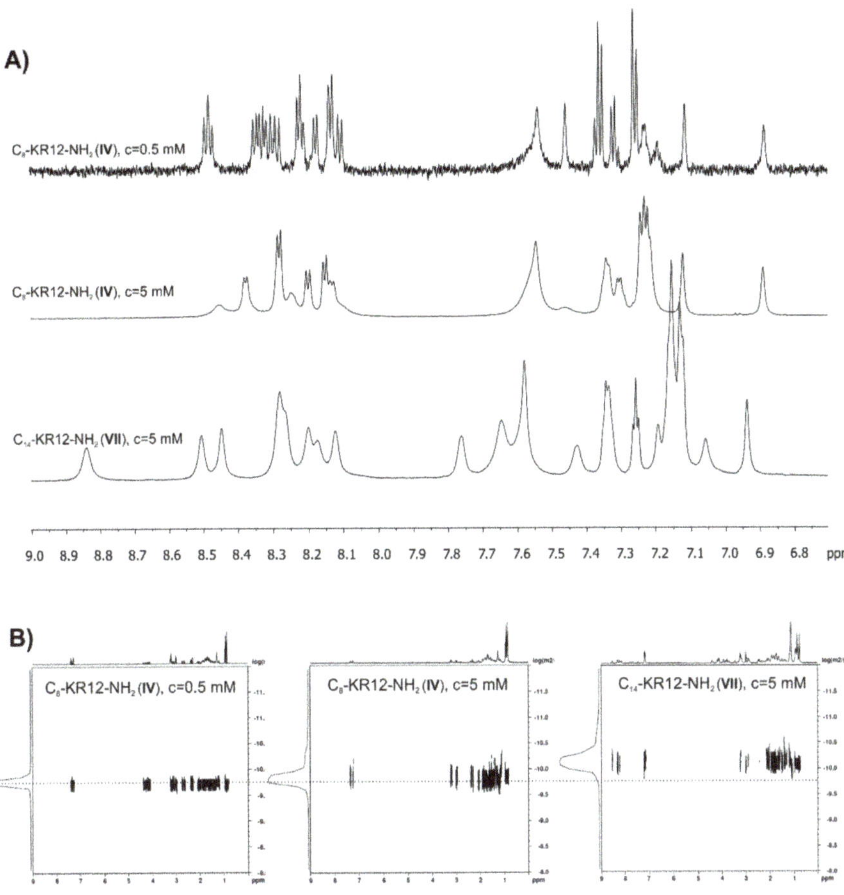

Figure 8. ^{1}H NMR (**A**) and DOSY (**B**) spectra for C_8-KR12-NH$_2$ and C_{14}-KR12-NH$_2$ recorded at concentrations below and/or above critical aggregation concentration (CAC).

3. Materials and Methods

3.1. Peptide Synthesis

The peptides (Table 1) were synthesized by solid-phase method using Fmoc chemistry on a resin modified by a Rink amide linker with a loading of 1.0 mmol/g (Orpegen Peptide Chemicals GmbH, Heidelberg, Germany) [55,56]. N^{α}-Fmoc-protected amino acids and the coupling reagents were obtained from Iris Biotech GmbH (Marktredwitz, Germany). The following amino acids side-chain-protecting groups were used: Trt (for Gln), OtBu (Asp), Boc (Lys), Pbf (Arg). Peptide synthesis was carried out manually. Single deprotection of the Fmoc group was performed in a 20% (*v/v*) piperidine (Iris Biotech GmbH, Marktredwitz, Germany) solution in *N,N*-dimethylformamide (DMF) for 15 min. Acylation with a protected amino acid was conducted in a dichloromethane (DCM)/DMF (Merck, Poland) solution with coupling agents for 1.5 h using a 3-fold molar excess of *N,N'*-diisopropylcarbodiimide (DIC; Peptideweb, Zblewo, Poland) and OxymaPure (Iris Biotech GmbH, Marktredwitz, Germany). Every step was preceded by rinsing the resin and running the chloranil test. Coupling reactions of lipophilic residues (fatty acids, aromatic acids) were performed by the same method as that used for protected amino acids. After the synthesis, the peptide resins were dried under vacuum. The peptides were cleaved from the resin using a mixture of trifluoroacetic acid (TFA; Apollo Scientific, Denton, UK), triisopropylsilane (TIS; Sigma-Aldrich, St. Louise, MO, USA), and water (95:2.5:2.5 *v/v/v*). The cleaved peptides were precipitated with cold diethyl ether and lyophilized. All peptides were purified using the RP-HPLC on a Knauer system controlled by an LPchrom data system (Lipopharm.pl, Zblewo, Poland) with a Knauer Kromasil C8 column (8 × 250 mm, 100Å pore size, 5 μm particle size). The eluates were fractionated and analyzed by analytical RP-HPLC. The purity of the peptides was determined on a Varian ProStar HPLC system controlled by a Galaxie Chromatography Data System with Phenomenex Gemini-NX C18 column (4.6 × 150 mm, 110 Å pore size, 5 μm particle size). The solvent systems used were: 0.1% aqueous TFA (A) and 0.1% TFA in acetonitrile (ACN) (B). UV detection at 214 nm was used, and the peptides were eluted with a linear gradient 10–100% B in A over 10 min at 25 ± 0.1 °C. The mobile phase flow rate was 2.0 mL/min. The ESI MS (Waters Alliance e2695 system with Acquity QDA detector, Waters, Milford, MA, USA) was used to identify the masses of the obtained peptides.

3.2. Antimicrobial Assays

3.2.1. Microbial Strains and Antimicrobial Assay

Examination of antimicrobial activity of the test compounds was conducted on reference strains of bacteria assigned to ESKAPE group of pathogens: *Enterococcus faecium* ATCC 700221, *Klebsiella pneumoniae* ATCC 700603, *Acinetobacter baumannii* ATCC BAA-1605, *Pseudomonas aeruginosa* ATCC 9027, *Klebsiella aerogenes* ATCC 13048 (formerly *Enterobacter aerogenes*) and reference strains of *Staphylococcus aureus*, namely: *S. aureus* ATCC 25923, *S. aureus* ATCC 6538, *S. aureus* ATCC 33591 (MRSA), *S. aureus* ATCC 9144 and *S. aureus* ATCC 12598. All the strains were stored at −80 °C in Roti®-Store cryo vials (Carl Roth GmbH, Karlsruhe, Germany) and before the tests were transferred into fresh Mueller-Hinton Medium (BioMaxima, Lublin, Poland) and incubated for 24 h at 37 °C. Subsequently, each bacterial inoculum was seeded on Mueller-Hinton Agar plates (BioMaxima) and incubated again for 24 h. The cultures prepared in this way were used in further antimicrobial assays and prepared as described above. The MIC values were determined by the broth microdilution method according to Clinical and Laboratory Standards Institute Protocol [57]. For this purpose, initial inoculums of bacteria (0.5×10^5 colony forming unit (CFU)/mL) in Mueller–Hinton Broth were exposed to the ranging concentrations of compounds (0.5–256 μg/mL) and incubated for 18 h at 37 °C. The experiments were conducted on 96-well microtiter plates, with a final volume of 100 μL. Cell densities were adjusted spectrophotometrically (Multiskan™ GO Microplate Spectrophotometer, Thermo Scientific) at 600 nm. The MICs were taken as the lowest drug concentration at which a visible growth of microorganisms was inhibited [58]. All experiments were conducted in triplicate and included positive (growth) and negative (sterility) controls.

3.2.2. Activity Against Staphylococcal Biofilm

The MBECs values were determined according to the method reported previously [59,60]. Briefly, 24-h cultures of *S. aureus*, namely: *S. aureus* ATCC 25923, *S. aureus* ATCC 6538, *S. aureus* ATCC 33591 (MRSA), *S. aureus* ATCC 9144 and *S. aureus* ATCC 12598 were diluted to a concentration of 0.5×10^7 CFU/mL and added to the test wells of polystyrene microdilution flat-bottom plates. After 24-h of incubation at 37 °C, the wells were rinsed three times with PBS to remove non-adherent cells. Subsequently, 100 μL of the test compounds over a concentration range 0.5–256 μg/mL were added to each well and incubated again for 24 h at 37 °C. After this period, 20 μL of the resazurin (7-hydroxy-3H-phenoxazin-3-one-10-oxide, 4 mg/mL) solution was added to each well. After 1 h of incubation, the MBECs were read out. The determined values were recorded as the lowest concentration at which the reduction of resazurin (from blue to pink) was lower or equal to 10 ± 0.5% as compared to the positive (100%) and negative (0%) controls. All experiments were performed in triplicate.

3.3. The Hemolysis Assay

The assay was conducted according to the procedure described previously by Avrahami and Shai [28]. Briefly, the fresh human RBCs with ethylenediaminetetraacetic acid (EDTA) as anticoagulant were rinsed three times with PBS by centrifugation at $800\times g$ for 10 min and resuspended in PBS. Serial dilution of peptides (0.5–256 µg/mL) was conducted in PBS on 96-well plates. Then the stock RBCs solution was added up to a final volume of 100 µL with a 4% concentration of erythrocytes (v/v). The control wells for 0 and 100% hemolysis were also prepared. They consisted of RBCs suspended in PBS and 1% of Triton-X 100, respectively. Then, the plates were incubated for 60 min at 37 °C and centrifuged at $800\times g$ for 10 min at 4 °C (Sorvall ST 16R Centrifuge, Thermo Scientific). After centrifugation, the supernatant was carefully transferred to new microtiter plates and the release of hemoglobin was monitored by measurement of absorbance at 540 nm (Multiskan™ GO Microplate Spectrophotometer). Percentage of hemolysis was calculated based on wells with 100% hemolysis.

3.4. MTT Assay

The cytotoxicity of test compounds (IC_{50}) was evaluated for human keratinocytes (HaCaT, ATCC) using classic MTT assay on 96-well plates [61]. In this assay, a colorimetric determination of the cell metabolic activity was carried out. Specifically, the color intensity reflects the number of live cells that can be measured spectrophotometrically. Briefly, the cell line was cultured in a Dulbecco's modified Eagle Medium (Sigma-Aldrich) supplemented with 10% fetal bovine serum (v/v), 100 units/mL of penicillin, 100 µg/mL of streptomycin, and 2 mM L-glutamine and was kept at 37 °C in a humidified 5% CO_2 incubator. A day after plating 500 cells per well, a series of concentrations (0.5–500 µg/mL) of the test compounds were added. Dimethyl sulfoxide (DMSO) was used as a control in cells at a final concentration of 1.0% (*v/v*), which was related to the maximal concentration of the solvent compounds used in the experiment. After 24 h of incubation with test compounds, a medium containing 1 mg/mL of MTT was added up to a final concentration of 0.5 mg/mL and subsequently incubated at 37 °C for 4 h. Then, the medium was aspirated and the formazan product was solubilized with DMSO. The background absorbance at 630 nm was subtracted from that at 570 nm for each well (Epoch, BioTek Instruments, Winooski, VT, USA). Six replicates were conducted for each concentration. All experiments were repeated at least twice and the resulting IC_{50} values were calculated with a GraFit 7 software (v. 7.0, Erithacus, Berkley, CA, USA).

3.5. CD Measurements

Circular dichroism studies were performed in water, 10 mM PBS buffer (pH 7.4), 20 mM SDS micelles, 20 mM DPC micelles, and 1.3 mM LUVs POPG and POPC liposomes. Large unilamellar vesicles (LUVs) were prepared according to the previously described procedure [62]. The CD spectra were recorded on a JASCO J-815 spectropolarimeter at 25 °C in the 185–260 nm range. The peptide concentration was 0.15 mg/mL. Every spectrum was scanned three times to amplify the signal-to-noise ratio. The spectra were plotted as a function of the mean residue molar ellipticity (MRME, degree $cm^2 dmol^{-1}$) vs. wavelength (nm). Deconvolution of the CD spectra were carried out using CDPro software with CONTINILL algorithm and SMP56 database set [63].

3.6. Surface Tension Measurements

Surface tension measurements were performed to determine CAC of selected lipidated KR12-NH$_2$ analogs. The measurements were carried out using a Wilhelmy plate method on a K100 tensiometer equipped with two micro-dispensers (Krüss GmbH, Hamburg, Germany). The average value of the surface tension for every concentration was obtained on the basis of 10 measurements. The standard deviations did not exceed 0.1 mN/m. The CAC was determined by plotting the surface tension against the logarithm of compound concentration and was found as the intersection of two lines that best fit through the pre- and post-CAC data.

3.7. NMR Measurements

The NMR spectra were acquired on a Bruker Avance III 700 MHz spectrometer running Topspin 3.2 software in D_2O and H_2O:D_2O (9:1 *v/v*) solution at 298 K. The ^1H NMR spectra with excitation sculpting water suppression were recorded with 16 k data points in F2 dimension. The translation diffusion coefficients (D_{tr}) were measured by the standard Bruker pulse program (stebpgp1s19) with WATERGATE solvent suppression, 4k data points in the F2 dimension, 32 data points (gradient strengths) in the F1 dimension and with 2 s relaxation delay. The diffusion time (Δ) and the maximum duration of gradient distance (δ) were 200 ms and 4 ms in all experiments, respectively. The spectra were processed and analyzed using Topspin 3.2 (BrukerBiospins, Rheinstetten, Germany)).

4. Conclusions

The modification of KR12 amide (**X**) with a lipophilic residue in the *N*-terminal part of the molecule has been found to be an effective way to fortify its antimicrobial activity. For each of the synthesized lipopeptides, the activity against *S. aureus* as well as against bacteria of the ESKAPE group depended on the number of carbon atoms in the substituent. For example, the analog of KR12-NH$_2$ (**IV**) containing octanoic acid residue (C_8) exhibited the highest potency against all organisms tested in planktonic form (MIC 1–4 μg/mL). Moreover, it was able to eradicate biofilms of *S. aureus* strains at relatively low concentrations (MBEC 4–16 μg/mL). Furthermore, this peptide was characterized by low toxicity against hRBCs (MHC 64 μg/mL). For HaCaT, the IC_{50} value was 3.23 μg/mL, but the highest SI values were found for peptides **III** and **IV** (MHC/GM amounting to almost 28) and peptide **IX** with IC_{50}/GM ratio of 2.50.

As has previously been argued, fatty acid conjugation enhances the peptide-membrane interactions [64]. On the other hand, it can either induce or enhance ability to self-assemble in solution, which in turn can perturb the water-membrane partition equilibrium by protecting hydrocarbon chains from water phase, thereby reducing the possibility of peptide membrane insertion. However, aggregation can also increase selectivity of membrane-active anticancer and antimicrobial peptides by reducing effective peptide hydrophobicity and thus affinity towards membranes composed of neutral lipids, such as the outer leaflet of healthy eukaryotic cell membranes [65]. In the case of the peptides studied, an increase in the length of the attached alkyl chain enhanced propensity for self-assembly, promoted formation of larger aggregates and decreased antimicrobial activity, but not cytotoxicity of KR12-NH$_2$ analogs. Interestingly, self-assembly induced also α-helix formation in analogs with C_{10}-C_{14} lipophilic residues. The remaining peptides underwent a conformational switch typical for most antimicrobial peptides only in the presence of surfactants or lipids mimicking membrane environment. No correlation was found between helicity and activity of the peptides, which shows that the antimicrobial activity is the result of many factors. Those affecting activity include conformation, hydrophobicity, hydrophobic moment, charge and its distribution, size of the hydrophobic/hydrophilic domain or aggregation state in solution [35,65]. Conjugation of KR12-NH$_2$ peptide with lipophilic acids affected all of them to clearly demonstrate the complexity of lipopeptide-membrane interactions with multiple interconnected phenomena contributing in the final activity.

Analog KR12-NH$_2$ (**IV**), containing octanoic acid, has a strong potential to eliminate both planktonic cells of ESKAPE pathogens and the staphylococcal biofilm, as demonstrated in this study. After characterizing its proteolytic stability, this compound might be a useful peptide template for developing novel antimicrobial agents. We do not exclude the possibility of changes in the peptide sequence, because both LL-37 and its fragments can be degraded by proteases [66]. The literature describes LL-37 derivatives that displayed antistaphylococcal activity in vitro but also maintained their activity in the presence of physiological salts and human serum (analogs FK-13-a1 and FK-13-a7) and were active in vivo and/or ex vivo (17BIPHE2, SAAP-148) [22,24,67,68]. A supplementary examination of improvement of peptide **IV** selectivity index and its ability to prevent the biofilm formation should also be considered. In addition, the promising antimicrobial activity and low toxicity of peptide **IX** modified with *trans*-cinnamic acid residue is noteworthy, supporting further studies on improving selectivity index and potential application in staphylococcal infections.

Importantly, lipopeptides are already used in the therapy of bacterial infections. Daptomycin is applied in the treatment of systemic bacterial infections. Moreover, polymyxin B is administered parenterally in patients with bacteremia and urinary-tract infections. Unfortunately, the major disadvantage of polymyxin B treatment is its relatively high nephrotoxicity and neurotoxicity [69,70]. On the other hand, daptomycin therapy is associated with dose-dependent myopathy [71]. Lipoglycopeptides are another class of drugs available on the market. Dalbavancin is used in patients with acute bacterial skin and skin structure infections (ABSSSI). This drug is considered to be safe and well-tolerated in the treatment of ABSSSI [72]. Telavancin is another FDA approved lipoglycopeptide for treatment of complicated skin and skin structure infections (cSSSI). Both dalbavancin and telavancin disrupt membrane integrity and cell-wall synthesis [73,74]. Conjugation of a peptide with a fatty acid can increase its stability in serum, tissues and organs. It has been shown that lipidated peptides bind to serum albumin. Moreover, chain length plays pivotal role in peptide stability [41,75–78]. Presumably, conjugation of KR12-NH$_2$ with a fatty acid at its *N*-terminus may lead to increased enzymatic stability. The most active and selective peptides in this study may be useful peptide templates for novel antimicrobial agents. Further studies should estimate peptides proteolytic stability, activity in animal infection models and the influence of the position of the lipophilic moiety within KR12-NH$_2$ on both antimicrobial activity and toxicity. As is known from literature, changing the location of the fatty moiety from the *N*-terminus of the molecule to its *C*-terminus can lead to a decrease in hemolytic activity of the molecule while not adversely affecting its antibacterial activity [37].

Author Contributions: Conceptualization, E.K., E.S. and W.K.; methodology, E.K., E.S., M.J., M.B. and S.B.; investigation, E.K., E.S., M.J., D.N., M.B., W.B.-R. and S.B.; resources, W.K., E.S. and E.K.; writing—original draft preparation, E.K., E.S, M.J., M.B. and W.K.; writing—review and editing, E.K.; visualization, E.K, E.S., D.N. and M.B.; supervision, E.K.; project administration, E.K. and W.K.; funding acquisition, W.K. All authors have read and agree to the published version of the manuscript.

Funding: This study was supported by a Grant from the Polish National Science Centre (Project No. 2016/23/B/NZ7/02919).

Conflicts of Interest: The authors declare no conflict of interest. The funders had no role in the design of the study; in the collection, analyses or interpretation of data; in the writing of the manuscript, or in the decision to publish the results.

Abbreviations

AMPs	antimicrobial peptides
Ac	acetyl group
Boc	*tert*-butyloxycarbonyl
C$_4$	butanoic acid residue
C$_6$	hexanoic acid residue
C$_8$	octanoic acid residue
C$_{10}$	decanoic acid residue
C$_{12}$	dodecanoic acid residue
C$_{14}$	tetradecanoic acid residue
CAC	critical aggregation concentration
CD	circular dichroism
CFU	colony forming unit
DCM	dichloromethane
DHPC	1,2-diheptanoyl-sn-glycero-3-phosphocholine
DIC	*N*,*N*′-diisopropylcarbodiimide
DMF	*N*,*N*-dimethylformamide

DMSO	dimethyl sulfoxide
DOSY	Diffusion-Ordered Spectroscopy
DPC	dodecylphosphocholine
EDTA	ethylenediaminetetraacetic acid
ESI MS	electrospray ionization mass spectrometry
ESKAPE	*Enterococcus faecium, Klebsiella pneumoniae, Acinetobacter baumannii,*
bacteria	*Pseudomonas aeruginosa, Klebsiella aerogenes, Staphylococcus aureus*
Fmoc	9-fluorenylmethoxycarbonyl
GM	geometric mean
HAIs	hospital-acquired infections
hRBCs	human red blood cells
IC_{50}	half maximal inhibitory concentration
MBEC	minimal biofilm eradication concentration
MDR	multidrug-resistant
MHC	minimal hemolytic concentration
MIC	minimal inhibitory concentration
MRSA	methicillin-resistant *S. aureus*
MTT	3-(4,5-dimethylthiazol-2-yl)-2,5-diphenyltetrazolium bromide
Pbf	2,2,4,6,7-pentamethyl-dihydrobenzofuran-5-sulfonyl residue
PBS	phosphate buffered saline
POPC	1-palmitoyl-2-oleoyl-sn-glycero-3-phosphocholine
POPG	1-palmitoyl-2-oleoyl-sn-glycero-3-phosphoglycerol
RP-HPLC	reversed-phase high-performance liquid chromatography
SDS	sodium dodecyl sulfate
SI	selectivity index
SPPS	solid-phase peptide synthesis
TFA	trifluoroacetic acid
TIS	triisopropylsilane
WHO	World Health Organization

References

1. Mulani, M.S.; Kamble, E.E.; Kumkar, S.N.; Tawre, M.S.; Pardesi, K.R. Emerging Strategies to Combat ESKAPE Pathogens in the Era of Antimicrobial Resistance: A Review. *Front. Microbiol.* **2019**, *10*, 539. [CrossRef]
2. Founou, R.C.; Founou, L.L.; Essack, S.Y. Clinical and economic impact of antibiotic resistance in developing countries: A systematic review and meta-analysis. *PLoS ONE* **2017**, *12*, e0189621. [CrossRef] [PubMed]
3. Tacconelli, E.; Carrara, E.; Savoldi, A.; Harbarth, S.; Mendelson, M.; Monnet, D.L.; Pulcini, C.; Kahlmeter, G.; Kluytmans, J.; Carmeli, Y.; et al. Discovery, research, and development of new antibiotics: The WHO priority list of antibiotic-resistant bacteria and tuberculosis. *Lancet Infect. Dis.* **2018**, *18*, 318–327. [CrossRef]
4. Messina, G.; Ceriale, E.; Lenzi, D.; Burgassi, S.; Azzolini, E.; Manzi, P. Environmental contaminants in hospital settings and progress in disinfecting techniques. *Biomed. Res. Int* **2013**, *2013*, 429780. [CrossRef] [PubMed]
5. Di Ruscio, F.; Guzzetta, G.; Bjørnholt, J.V.; Leegaard, T.M.; Moen, A.E.F.; Merler, S.; Freiesleben de Blasio, B. Quantifying the transmission dynamics of MRSA in the community and healthcare settings in a low-prevalence country. *Proc. Natl. Acad. Sci. USA* **2019**, *116*, 14599–14605. [CrossRef]
6. Köck, R.; Becker, K.; Cookson, B.; van Gemert-Pijnen, J.E.; Harbarth, S.; Kluytmans, J.; Mielke, M.; Peters, G.; Skov, R.L.; Struelens, M.J.; et al. Methicillin-resistant Staphylococcus aureus (MRSA): Burden of disease and control challenges in Europe. *Eurosurveillance* **2010**, *15*, 19688. [CrossRef]
7. Mah, T.-F. Biofilm-specific antibiotic resistance. *Future Microbiol.* **2012**, *7*, 1061–1072. [CrossRef]
8. Sharma, D.; Misba, L.; Khan, A.U. Antibiotics versus biofilm: An emerging battleground in microbial communities. *Antimicrob. Resist. Infect. Control* **2019**, *8*, 76. [CrossRef]
9. Haney, E.F.; Hancock, R.E. Peptide design for antimicrobial and immunomodulatory applications. *Biopolymers* **2013**, *100*, 572–583. [CrossRef]
10. Hancock, R.E.; Haney, E.F.; Gill, E.E. The immunology of host defence peptides: Beyond antimicrobial activity. *Nat. Rev. Immunol.* **2016**, *16*, 321–334. [CrossRef]

11. Li, W.; Tailhades, J.; O'Brien-Simpson, N.M.; Separovic, F.; Otvos, L., Jr.; Hossain, M.A.; Wade, J.D. Proline-rich antimicrobial peptides: Potential therapeutics against antibiotic-resistant bacteria. *Amino Acids* **2014**, *46*, 2287–2294. [CrossRef] [PubMed]

12. Gudmundsson, G.H.; Agerberth, B.; Odeberg, J.; Bergman, T.; Olsson, B.; Salcedo, R. The human gene *FALL39* and processing of the cathelin precursor to the antibacterial peptide LL-37 in granulocytes. *Eur. J. Biochem.* **1996**, *238*, 325–332. [CrossRef] [PubMed]

13. Dürr, U.H.; Sudheendra, U.S.; Ramamoorthy, A. LL-37, the only human member of the cathelicidin family of antimicrobial peptides. *Biochim. Biophys. Acta* **2006**, *1758*, 1408–1425. [CrossRef]

14. Sørensen, O.E.; Follin, P.; Johnsen, A.H.; Calafat, J.; Tjabringa, G.S.; Hiemstra, P.S.; Borregaard, N. Human cathelicidin, hCAP-18, is processed to the antimicrobial peptide LL-37 by extracellular cleavage with proteinase 3. *Blood* **2001**, *97*, 3951–3959. [CrossRef]

15. Sørensen, O.E.; Gram, L.; Johnsen, A.H.; Andersson, E.; Bangsbøll, S.; Tjabringa, G.S.; Hiemstra, P.S.; Malm, J.; Egesten, A.; Borregaard, N. Processing of seminal plasma hCAP-18 to ALL-38 by gastricsin: A novel mechanism of generating antimicrobial peptides in vagina. *J. Biol. Chem.* **2003**, *278*, 28540–28546. [CrossRef]

16. Cowland, J.B.; Johnsen, A.H.; Borregaard, N. hCAP-18, a cathelin/pro-bactenecin-like protein of human neutrophil specific granules. *FEBS Lett.* **1995**, *368*, 173–176. [CrossRef]

17. Burton, M.F.; Steel, P.G. The chemistry and biology of LL37. *Nat. Prod. Rep.* **2009**, *26*, 1572–1584. [CrossRef] [PubMed]

18. Noore, J.; Noore, A.; Li, B. Cationic antimicrobial peptide LL-37 is effective against both extra- and intracellular Staphylococcus aureus. *Antimicrob. Agents Chemother.* **2013**, *57*, 1283–1290. [CrossRef] [PubMed]

19. Haisma, E.M.; de Breij, A.; Chan, H.; van Dissel, J.T.; Drijfhout, J.W.; Hiemstra, P.S.; El Ghalbzouri, A.; Nibbering, P.H. LL-37-Derived Peptides Eradicate Multidrug-Resistant *Staphylococcus aureus* from Thermally Wounded Human Skin Equivalents. *Antimicrob. Agents Chemother.* **2014**, *58*, 4411–4419. [CrossRef] [PubMed]

20. Saporito, P.; Vang Mouritzen, M.; Løbner-Olesen, A.; Jenssen, H. LL-37 fragments have antimicrobial activity against Staphylococcus epidermidis biofilms and wound healing potential in HaCaT cell line. *J. Pept. Sci.* **2018**, *24*, e3080. [CrossRef] [PubMed]

21. Jacob, B.; Park, I.S.; Bang, J.K.; Shin, S.Y. Short KR-12 analogs designed from human cathelicidin LL-37 possessing both antimicrobial and antiendotoxic activities without mammalian cell toxicity. *J. Pept. Sci.* **2013**, *19*, 700–707. [CrossRef] [PubMed]

22. Rajasekaran, G.; Kim, E.Y.; Shin, S.Y. LL-37-derived membrane-active FK-13 analogs possessing cell selectivity, anti-biofilm activity and synergy with chloramphenicol and anti-inflammatory activity. *Biochim. Biophys. Acta Biomembr.* **2017**, *1859*, 722–733. [CrossRef] [PubMed]

23. Epand, R.F.; Wang, G.; Berno, B.; Epand, R.M. Lipid Segregation Explains Selective Toxicity of a Series of Fragments Derived from the Human Cathelicidin LL-37. *Antimicrob. Agents Chemother.* **2009**, *53*, 3705–3714. [CrossRef] [PubMed]

24. Wang, G.; Hanke, M.; Mishra, B.; Lushnikova, T.; Heim, C.E.; Thomas, V.C.; Bayles, K.W.; Kielian, T. Transformation of Human Cathelicidin LL-37 into Selective, Stable, and Potent Antimicrobial Compounds. *ACS Chem. Biol.* **2014**, *9*, 1997–2002. [CrossRef]

25. Wang, G. Structures of human host defense cathelicidin LL-37 and its smallest antimicrobial peptide KR-12 in lipid micelles. *J. Biol. Chem.* **2008**, *283*, 32637–32643. [CrossRef]

26. Mishra, B.; Epand, R.F.; Epand, R.M.; Wang, G. Structural location determines functional roles of the basic amino acids of KR-12, the smallest antimicrobial peptide from human cathelicidin LL-37. *RSC Adv.* **2013**, *3*, 19560–19571. [CrossRef]

27. Li, Z.; Yuan, P.; Xing, M.; He, Z.; Dong, C.; Cao, Y.; Liu, Q. Fatty acid conjugation enhances the activities of antimicrobial peptides. *Recent Pat. Food Nutr. Agric.* **2013**, *5*, 52–56. [CrossRef]

28. Avrahami, D.; Shai, Y. A New Group of Antifungal and Antibacterial Lipopeptides Derived from Non-membrane Active Peptides Conjugated to Palmitic Acid. *J. Biol. Chem.* **2004**, *279*, 12277–12285. [CrossRef]

29. Albada, B. Tuning Activity of Antimicrobial Peptides by Lipidation. In *Health Consequence of Microbial Interactions with Hydrocarbons, Oils, and Lipids, Handbook of Hydrocarbon and Lipid Microbiology*; Goldfine, H., Ed.; Springer International Publishing: Cham, Switzerland, 2019; pp. 1–18.

30. Neubauer, D.; Jaśkiewicz, M.; Bauer, M.; Gołacki, K.; Kamysz, W. Ultrashort Cationic Lipopeptides–Effect of *N*-Terminal Amino Acid and Fatty Acid Type on Antimicrobial Activity and Hemolysis. *Molecules* **2020**, *25*, 257. [CrossRef]
31. Cruz, E.; Euerby, M.R.; Johnson, C.M.; Hackett, C.A. Chromatographic classification of commercially available reverse-phase HPLC columns. *Chromatographia* **1997**, *44*, 151–161. [CrossRef]
32. Kimata, K.; Iwaguchi, K.; Onishi, S.; Jinno, K.; Eksteen, R.; Hosoya, K.; Araki, M.; Tanaka, N. Chromatographic characterization of silica c18 packing materials. Correlation between a preparation method and retention behavior of stationary phase. *J. Chromatogr. Sci.* **1989**, *27*, 721–728. [CrossRef]
33. Jaśkiewicz, M.; Bauer, M.; Sadowska, K.; Barańska-Rybak, W.; Kamysz, E.; Kamysz, W. Antistaphylococcal activity of the KR-12 alanine scan. In Proceedings of the 35th European Peptide Symposium, Dublin, Ireland, 26–31 August 2018. Poster, Unpublished Work.
34. Dathe, M.; Nikolenko, H.; Meyer, J.; Beyermann, M.; Bienert, M. Optimization of the antimicrobial activity of magainin peptides by modification of charge. *FEBS Lett.* **2001**, *501*, 146–150. [CrossRef]
35. Dathe, M.; Wieprecht, T. Structural features of helical antimicrobial peptides: Their potential to modulate activity on model membranes and biological cells. *Biochim. Biophys. Acta* **1999**, *1462*, 71–87. [CrossRef]
36. Laverty, G.; McLaughlin, M.; Shaw, C.; Gorman, S.P.; Gilmore, B.F. Antimicrobial activity of short, synthetic cationic lipopeptides. *Chem. Biol. Drug Des.* **2010**, *75*, 563–569. [CrossRef] [PubMed]
37. Albada, H.B.; Prochnow, P.; Bobersky, S.; Langklotz, S.; Schriek, P.; Bandow, J.E.; Metzler-Nolte, N. Tuning the activity of a short Arg-Trp antimicrobial peptide by lipidation of a C- or N-terminal lysine side-chain. *ACS Med. Chem. Lett.* **2012**, *3*, 980–984. [CrossRef] [PubMed]
38. Kim, E.Y.; Rajasekaran, G.; Shin, S.Y. LL-37-derived short antimicrobial peptide KR-12-a5 and its D-amino acid substituted analogs with cell selectivity, anti-biofilm activity, synergistic effect with conventional antibiotics, and anti-inflammatory activity. *Eur. J. Med. Chem.* **2017**, *136*, 428–441. [CrossRef]
39. Lyu, Y.; Yang, Y.; Lyu, X.; Dong, N.; Shana, A. Antimicrobial activity, improved cell selectivity and mode of action of short PMAP-36-derived peptides against bacteria and Candida. *Sci. Rep.* **2016**, *6*, 27258. [CrossRef]
40. Laverty, G.; McCloskey, A.P.; Gorman, S.P.; Gilmore, B.F. Anti-biofilm activity of ultrashort cinnamic acid peptide derivatives against medical device-related pathogens. *J. Pept. Sci.* **2015**, *21*, 770–778. [CrossRef]
41. Avrahami, D.; Shai, Y. Conjugation of a magainin analogue with lipophilic acids controls hydrophobicity, solution assembly, and cell selectivity. *Biochemistry* **2002**, *41*, 2254–2263. [CrossRef]
42. Oren, Z.; Shai, Y. Selective lysis of bacteria but not mammalian cells by diastereomers of melittin: Structure–function study. *Biochemistry* **1997**, *36*, 1826–1835. [CrossRef]
43. Shai, Y.; Oren, Z. Diastereoisomers of cytolysins, a novel class of potent antibacterial peptides. *J. Biol. Chem.* **1996**, *271*, 7305–7308. [CrossRef] [PubMed]
44. Mangoni, M.L.; Carotenuto, A.; Auriemma, L.; Saviello, M.R.; Campiglia, P.; Gomez-Monterrey, I.; Malfi, S.; Marcellini, L.; Barra, D.; Novellino, E. Structure–activity relationship, conformational and biological studies of temporin L analogues. *J. Med. Chem.* **2011**, *54*, 1298–1307. [CrossRef] [PubMed]
45. Lee, J.; Lee, D.G. Structure-antimicrobial activity relationship between pleurocidin and its enantiomer. *Exp. Mol. Med.* **2008**, *40*, 370–376. [CrossRef] [PubMed]
46. Thongngam, M.; McClements, D.J. Influence of pH, ionic strength, and temperature on self-association and interactions of sodium dodecyl sulfate in the absence and presence of chitosan. *Langmuir* **2005**, *21*, 79–86. [CrossRef] [PubMed]
47. Manzo, G.; Carboni, M.; Rinaldi, A.C.; Casu, M.; Scorciapino, M.A. Characterization of sodium dodecylsulphate and dodecylphosphocholine mixed micelles through NMR and dynamic light scattering. *Magn. Reson. Chem.* **2013**, *51*, 176–183. [CrossRef] [PubMed]
48. Sikorska, E.; Wyrzykowski, D.; Szutkowski, K.; Greber, K.; Lubecka, E.A.; Zhukov, I. Thermodynamics, size, and dynamics of zwitterionic dodecylphosphocholine and anionic sodium dodecyl sulfate mixed micelles. *J. Thermal Anal. Calor.* **2016**, *123*, 511–523. [CrossRef]
49. Jeong, J.-H.; Kim, J.-S.; Choi, S.-S.; Kim, Y. NMR structural studies of antimicrobial peptides: LPcin analogs. *Biophys. J.* **2016**, *110*, 423–430. [CrossRef]
50. Khakshoor, O.; Demeler, B.; Nowick, J.S. Macrocyclic beta-sheet peptides that mimic protein quaternary structure through intermolecular beta-sheet interactions. *J. Am. Chem. Soc.* **2007**, *129*, 5558–5569. [CrossRef]
51. Chang, X.; Keller, D.; O'Donoghue, S.I.; Led, J.J. NMR studies of the aggregation of glucagon-like peptide-1: Formation of a symmetric helical dimer. *FEBS Lett.* **2002**, *515*, 165–170. [CrossRef]

52. Jarvet, J.; Danielsson, J.; Damberg, P.; Oleszczuk, M.; Gräslund, A. Positioning of the Alzheimer Abeta(1-40) peptide in SDS micelles using NMR and paramagnetic probes. *J. Biomol. NMR* **2007**, *39*, 63–72. [CrossRef]

53. Chou, J.J.; Baber, J.L.; Bax, A. Characterization of phospholipid mixed micelles by translational diffusion. *J. Biomol. NMR* **2004**, *29*, 299–308. [CrossRef] [PubMed]

54. Sikorska, E.; Stachurski, O.; Neubauer, D.; Małuch, I.; Wyrzykowski, D.; Bauer, M.; Brzozowski, K.; Kamysz, W. Short arginine-rich lipopeptides: From self-assembly to antimicrobial activity. *Biochim. Biophys. Acta Biomembr.* **2018**, *1860*, 2242–2251. [CrossRef] [PubMed]

55. Chan, W.C.; White, P.D. *Fmoc Solid Phase Peptide Synthesis: A Practical Approach*; Oxford University Press: New York, NY, USA, 2000; pp. 11–74.

56. Atherton, E.; Sheppard, R.C. *Solid Phase Peptide Synthesis: A Practical Approach*; IRL Press: Oxford, UK, 1989; pp. 28–38, 75–86, 149–162.

57. Wayne, P.A. *Methods for Dilution Antimicrobial Susceptibility Tests for Bacteria That Grow Aerobically; Approved Standard—Ninth Edition*; Document M07-A8; Clinical and Laboratory Standards Institute (CLSI): Wayne, PA, USA, 2012; Available online: www.clsi.org (accessed on 20 January 2013).

58. Andrews, J.M. Determination of minimum inhibitory concentrations. *J. Antimicrob. Chemother.* **2002**, *49*, 1049. [CrossRef]

59. Maciejewska, M.; Bauer, M.; Neubauer, D.; Kamysz, W.; Dawgul, M. Influence of amphibian antimicrobial peptides and short lipopeptides on bacterial biofilms formed on contact lenses. *Materials (Basel)* **2016**, *9*, 873. [CrossRef] [PubMed]

60. Migoń, D.; Jaśkiewicz, M.; Neubauer, D.; Bauer, M.; Sikorska, E.; Kamysz, E.; Kamysz, W. Alanine Scanning Studies of the Antimicrobial Peptide Aurein 1.2. *Probiotics Antimicrob. Proteins* **2019**, *11*, 1042–1054. [CrossRef] [PubMed]

61. Stachurski, O.; Neubauer, D.; Małuch, I.; Wyrzykowski, D.; Bauer, M.; Bartoszewska, S.; Kamysz, W.; Sikorska, E. Effect of self-assembly on antimicrobial activity of double-chain short cationic lipopeptides. *Bioorg. Med. Chem.* **2019**, *27*, 115129. [CrossRef]

62. Sikorska, E.; Dawgul, M.; Greber, K.; Iłowska, E.; Pogorzelska, A.; Kamysz, W. Self-assembly and interactions of short antimicrobial cationic lipopeptides with membrane lipids: ITC, FTIR and molecular dynamics studies. *BBA-Biomembranes* **2014**, *1838*, 2625–2634. [CrossRef]

63. Sreerama, N.; Woody, R.W. Estimation of Protein Secondary Structure from Circular Dichroism Spectra: Comparison of CONTIN, SELCON, and CDSSTR Methods with an Expanded Reference Set. *Anal. Biochem.* **2000**, *287*, 252–260. [CrossRef]

64. Chu-Kung, A.F.; Bozzelli, K.N.; Lockwood, N.A.; Haseman, J.R.; Mayo, K.H.; Tirrell, M.V. Promotion of peptide antimicrobial activity by fatty acid conjugation. *Bioconjug. Chem.* **2004**, *15*, 530–535. [CrossRef]

65. Vaezi, Z.; Bortolotti, A.; Luca, V.; Perilli, G.; Mangoni, M.L.; Khosravi-Far, R.; Bobone, S.; Stella, L. Aggregation determines the selectivity of membrane-active anticancer and antimicrobial peptides: The case of killerFLIP. *Biochim Biophys Acta Biomembr.* **2020**, *1862*, 183107. [CrossRef]

66. McCrudden, M.T.C.; McLean, D.T.F.; Zhou, M.; Shaw, J.; Linden, G.J.; Irwin, C.H.R.; Lundy, F.T. The Host Defence Peptide LL-37 is Susceptible to ProteolyticDegradation by Wound Fluid Isolated from Foot Ulcersof Diabetic Patients. *Int. J. Pept. Res. Ther.* **2014**, *20*, 457–464. [CrossRef]

67. Mishra, B.; Golla, R.M.; Lau, K.; Lushnikova, T.; Wang, G. Anti-Staphylococcal Biofilm Effects of Human Cathelicidin Peptides. *ACS Med. Chem. Lett.* **2016**, *7*, 117–121. [CrossRef] [PubMed]

68. Riool, M. Novel Antibacterial Strategies to Combat Biomaterial-Associated Infections. Ph.D. Thesis, University of Amsterdam, Amsterdam, The Netherlands. Available online: https://hdl.handle.net/11245.1/db3811ec-92ce-4831-b7c7-f1c0bcc76bc9 (accessed on 27 January 2020).

69. Falagas, M.E.; Kasiakou, S.K. Toxicity of polymyxins: A systematic review of the evidence from old and recent studies. *Crit Care.* **2006**, *10*, R27. [CrossRef] [PubMed]

70. Pastewski, A.A.; Caruso, P.; Parris, A.R.; Dizon, R.; Kopec, R.; Sharma, S.; Mayer, S.; Ghitan, M.; Chapnick, E.K. Parenteral polymyxin B use in patients with multidrug-resistant gram-negative bacteremia and urinary tract infections: A retrospective case series. *Ann. Pharmacother.* **2008**, *42*, 1177–1187. [CrossRef]

71. Steenbergen, J.N.; Alder, J.; Thorne, G.M.; Tally, F.P. Daptomycin, a New Cyclic Lipopeptide Antibiotic, for the Treatment of Resistant Gram-Positive Organisms. *J. Antimicrob. Chemother.* **2005**, *55*, 283–288. [CrossRef]

72. Dunne, M.W.; Talbot, G.H.; Boucher, H.W.; Wilcox, M.; Puttagunta, S. Safety of Dalbavancin in the Treatment of Skin and Skin Structure Infections: A Pooled Analysis of Randomized, Comparative Studies. *Drug Saf.* **2016**, *39*, 147–157. [CrossRef]

73. Pushkin, R.; Barriere, S.L.; Wang, W.; Corey, G.R.; Stryjewski, M.E. Telavancin for Acute Bacterial Skin and Skin Structure Infections, a Post Hoc Analysis of the Phase 3 ATLAS Trials in Light of the 2013 FDA Guidance. *Antimicrob. Agents Chemother.* **2015**, *59*, 6170–6174. [CrossRef]

74. Roberts, K.D.; Sulaiman, R.M.; Rybak, M.J. Dalbavancin and Oritavancin: An Innovative Approach to the Treatment of Gram-Positive Infections. *Pharmacotherapy* **2015**, *35*, 935–948. [CrossRef]

75. Zhang, L.; Bulaj, G. Converting Peptides into Drug Leads by Lipidation. *Curr. Med. Chem.* **2012**, *19*, 1602–1618. [CrossRef]

76. Poupart, J.; Hou, X.; Chemtob, S.; Lubell, W.D. Application of N-Dodecyl l-Peptide to Enhance Serum Stability while Maintaining Inhibitory Effects on Myometrial Contractions Ex Vivo. *Molecules* **2019**, *24*, 4141. [CrossRef]

77. Toth, I.; Flinna, N.; Hillery, A.; Gibbons, W.A.; Artursson, P. Lipidic conjugates of luteinizing hormone releasing hormone (LHRH)+ and thyrotropin releasing hormone (TRH)+ that release and protect the native hormones in homogenates of human intestinal epithelial (Caco-2) cells. *Int. J. Pharm.* **1994**, *105*, 241–247. [CrossRef]

78. Zhang, L.; Robertson, C.R.; Green, B.R.; Pruess, T.H.; White, H.S.; Bulaj, G. Structural requirements for a lipoamino acid in modulating the anticonvulsant activities of systemically active galanin analogues. *J. Med. Chem.* **2009**, *52*, 1310–1316. [CrossRef] [PubMed]

International Journal of
Molecular Sciences

Article

Synthetic Human β Defensin-3-C15 Peptide in Endodontics: Potential Therapeutic Agent in *Streptococcus gordonii* Lipoprotein-Stimulated Human Dental Pulp-Derived Cells

Yeon-Jee Yoo [1], Hiran Perinpanayagam [2], Jue-Yeon Lee [3], Soram Oh [4], Yu Gu [5], A-Reum Kim [6], Seok-Woo Chang [4], Seung-Ho Baek [1] and Kee-Yeon Kum [1,7,*]

[1] Department of Conservative Dentistry, School of Dentistry, Dental Research Institute, Seoul National University, Daehak-ro 101, Chongno-Ku, Seoul 03080, Korea; duswl32@snu.ac.kr (Y.-J.Y.); shbaek@snu.ac.kr (S.-H.B.)
[2] Department of Dentistry, Schulich School of Medicine & Dentistry, University of Western Ontario, London, ON N6A 5C1, Canada; hperinpa@uwo.ca
[3] Central Research Institute, Nano intelligent Biomedical Engineering Corporation (NIBEC), Seoul 03130, Korea; yeon0417@nibec.co.kr
[4] Department of Conservative Dentistry, School of Dentistry, Kyung Hee University, Seoul 02447, Korea; soram0123@naver.com (S.O.); swc2007smc@gmail.com (S.-W.C.)
[5] Department of Conservative Dentistry, School of Stomatology, Shandong University, Jinan 250012, China; guyu618@126.com
[6] Department of Oral Microbiology and Immunology, School of Dentistry, Dental Research Institute and BK21 Plus Program, Seoul National University, Seoul 08826, Korea; kimareum@snu.ac.kr
[7] National Dental Care Center for Persons with Special Cares, Seoul National University Dental Hospital for Persons with Special Needs, Seoul 03080, Korea
*   Correspondence: kum6139@snu.ac.kr; Tel.: +(82)-2-2072-2656; Fax: +(82)-2-2072-3849

Received: 25 October 2019; Accepted: 17 December 2019; Published: 20 December 2019

Abstract: Human β defensin-3-C15, an epithelium-derived cationic peptide that has antibacterial/antifungal and immuno-regulatory properties, is getting attention as potential therapeutic agent in endodontics. This study aimed to investigate if synthetic human β defensin-3-C15 (HBD3-C15) peptides could inhibit inflammatory responses in human dental pulp cells (hDPCs), which had been induced by gram-positive endodontic pathogen. hDPC explant cultures were stimulated with *Streptococcus gordonii* lipoprotein extracts for 24 h to induce expression of pro-inflammatory mediators. The cells were then treated with either HBD3-C15 (50 μg/mL) or calcium hydroxide (CH, 100 μg/mL) as control for seven days, to assess their anti-inflammatory effects. Quantitative RT-PCR analyses and multiplex assays showed that *S. gordonii* lipoprotein induced the inflammatory reaction in hDPCs. There was a significant reduction of IL-8 and MCP-1 within 24 h of treatment with either CH or HBD3-C15 ($p < 0.05$), which was sustained over 1 week of treatment. Alleviation of inflammation in both medications was related to COX-2 expression and PGE2 secretion ($p < 0.05$), rather than TLR2 changes ($p > 0.05$). These findings demonstrate comparable effects of CH and HDB3-C15 as therapeutic agents for inflamed hDPCs.

Keywords: calcium hydroxide; chemokine; human beta defensin-3-C15; human dental pulp cell; *Streptococcus gordonii* lipoprotein

---

1. Introduction

Streptococcus gordonii are Gram-positive facultative anaerobes that are frequently isolated from cases of recurrent apical periodontitis, due to their propensity to form bacterial biofilms on root canal surfaces [1]. Additionally, they can exchange genes that encode cytotoxins, adhesins, and antibiotic resistance, with *Enterococcus faecalis* [2]. These virulence features enable *S. gordonii* to play an important role in the pathogenesis of apical periodontitis.

Virulence factors in gram-positive bacteria include cell wall-associated lipoteichoic acid (LTA) and lipoproteins, which are recognized by Toll-like receptor 2 (TLR2) [3,4]. These stimulate a variety of host cells to induce pro-inflammatory cytokines and chemokines. For instance, LTA from *Staphylococcus aureus* or *Streptococcus pyogenes* were shown to induce IL-8 in human peripheral blood monocytes. *S. aureus* lipoproteins were shown to induce IL-8 in human intestinal epithelial cells. Likewise, LTA, lipoproteins, and peptidoglycan from *S. gordonii* induced the pro-inflammatory cytokines IL-6 and TNFα in dendritic cells, via TLR2 [5]. Notably, the *S. gordonii* lipoprotein is reported to be a key virulence factor in inducing inflammatory responses [6,7].

To inactivate virulence factors and eliminate bacteria, intracanal medicaments are often applied to root canal systems for treating apical periodontitis. The most widely used medicament is calcium hydroxide (CH), which has antimicrobial effects that are largely due to its high pH [8]. It has been shown to suppress both lipopolysaccharide (LPS) [9] and LTA [10,11], which are virulence factors critical to Gram-negative and -positive bacteria respectively. More recently, antimicrobial peptides have been developed as potential therapeutic agents against microbial biofilms [12]. These include human β-defensin-3 (HBD3), which is a cysteine-rich cationic peptide that has strong antibacterial, antifungal and immuno-regulatory properties [13–16]. The disulfide topology that maintains tertiary structure in HBD3 is dispensable for its antimicrobial functions [17], and the C-terminal end of HBD3 contains 15 amino acids that effectively elicit antimicrobial activity [18]. A synthetic HBD3 peptide that consists of only 15 amino acids from the C-terminus (HBD3-C15, GKCSTRGRKCCRRKK), is being developed as an alternative antibiofilm agent (Figure 1) [12]. This 15-mer peptide derived from the C-terminus of HBD3 (HBD3-C15) has been shown to have considerable antimicrobial activity that is comparable to that of the full-length protein [18–21]. However, little is known about its suppressive effects on gram-positive bacterial virulence factors.

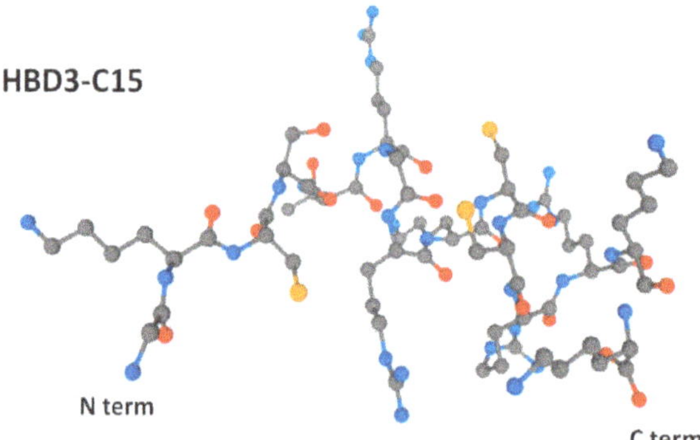

Figure 1. Molecular structure of synthetic human β defensin-3-C15 (HBD3-C15) peptide, prepared by F-moc-based chemical solid phase synthesis from 15 amino acids (GKCSTRGRKCCRRKK). Grey, carbon; blue, nitrogen; red, oxygen; yellow, sulfur.

Where the inactivation or elimination of virulence factors are inadequate, the induction of pro-inflammatory mediators ensure a persistence of pulpitis and apical periodontitis [22,23]. Chemokines such as interleukin-8 (IL-8) and monocyte chemoattractant protein-1 (MCP-1) display potent chemotactic activities for human neutrophils and T lymphocytes, which in turn trigger a series of inflammatory events [24] that release enzymes causing tissue destruction [25]. Accompanying inflammatory mediators such as cyclooxygenase 2 (COX-2) stimulate vasodilation and microvascular permeability by cytoskeletal rearrangement or contraction of vascular smooth muscle [26]. Therefore, the purpose of this study was to examine the anti-inflammatory properties of HBD3-C15 in human dental pulp cells (hDPCs) that had been stimulated by *S. gordonii* derived lipoprotein, by comparing the effects of HBD3-C15 and CH.

2. Results

2.1. S. gordoni Lipoproteins Induced Inflammatory Mediators

S. gordonii lipoprotein induced the expression of chemokines IL-8 and MCP-1 by hDPC explant cultures (Figure 2). When hDPCs were treated with Triton X-114 extracts from *S. gordonii* that contained the bacterial lipoproteins [27], there was an immediate increase in the relative mRNA expressions of IL-8 and MCP-1, as measured by real-time RT-qPCR. The relative levels of IL-8 and MCP-1 were significantly higher ($p < 0.05$) in the *S. gordonii* lipoprotein-treated hDPCs, than in the untreated cultures.

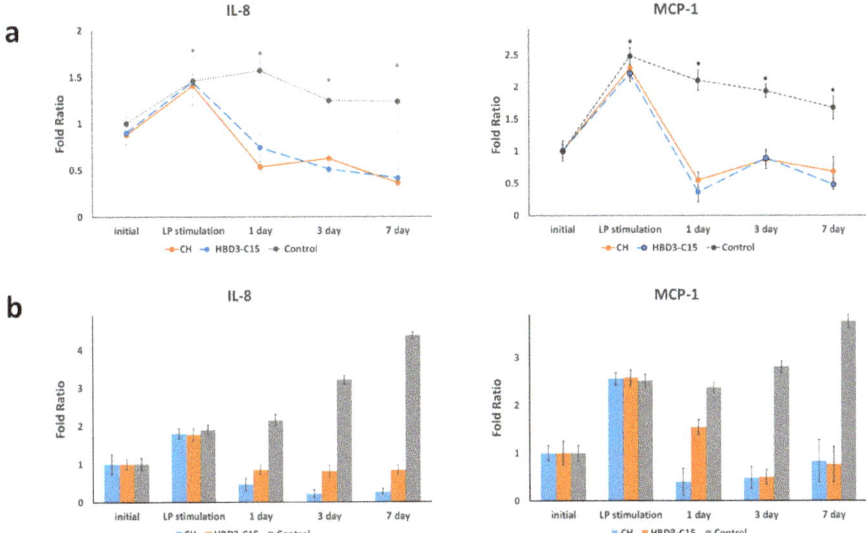

Figure 2. Cultured human dental pulp cells (hDPCs) experienced an inflammatory reaction after Gram-positive *S. gordonii* lipoprotein (LP)-stimulation, and then inflamed hDPCs were treated with either calcium hydroxide (CH) or HBD3-C15. Cell lysates were collected at the indicated time points, and the expressions of IL-8 and MCP-1 were analyzed by (**a**) qRT-PCR and (**b**) multiplex assay. The relative expression levels of each control group are presented.

2.2. HBD3-C15 and CH Suppress Inflammatory Mediators

The stimulated cells were then treated with either HBD3-C15 or CH. The medicaments were not simultaneously treated with *S. gordonii* lipoprotein, to prevent the interactions and denaturation of *S. gordonii* lipoprotein with positively charged HBD3-C15 or highly alkaline CH, and also to reenact clinical relevance in endodontics. Both HBD3-C15 and CH treatments for 1, 3, and 7 days, suppressed the expression of IL-8 and MCP-1 mRNA measured by qRT-PCR (Figure 2a), and their protein level quantified by multiplex assay (Figure 2b). There were significant and sustained reductions in IL-8 and MCP-1 mRNA gene expression and their quantified protein level, to the levels those were at or below their baselines values ($p < 0.05$). There were no significant differences between these two medicaments ($p > 0.05$). Non-medicated cells showed sustained inflammatory status, with increased protein level over time.

2.3. Anti-Inflammatory Effects of HBD3-C15 and CH Were not Mediated by TLR2

To investigate the underlying anti-inflammatory mechanism of HBD3-C15 and CH, TLR2 mRNA expression level was assessed in hDPCs. *S. gordonii* lipoproteins induced upregulated TLR2 expression in hDPCs ($p < 0.05$). However, TLR2 gene expression level were not decreased after medication (Figure 3a). To validate the result, we examined HEK292-TLR2 cell transfection and found that *S. gordonii* lipoproteins potently stimulated NF-κB in HEK292-TLR2 cells as in hDPCs, but there were no significant differences after medication (Figure 3b, $p > 0.05$).

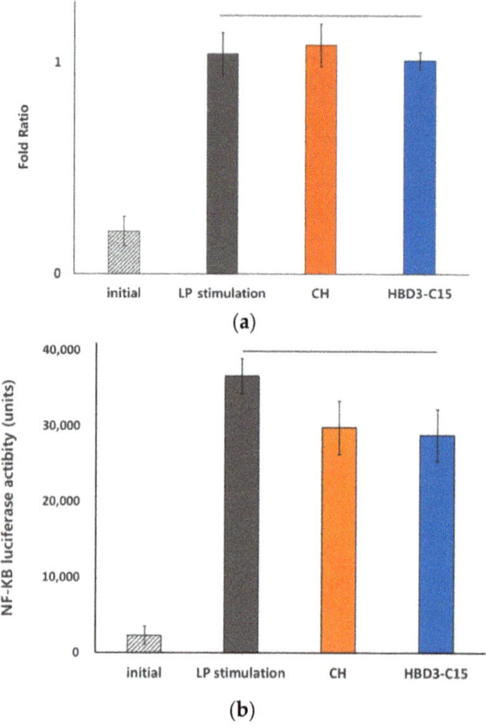

Figure 3. (**a**) Exogenous stimulation of *S. gordonii* lipoproteins augment TLR2 mRNA expressions in hDPCs, and were maintained after calcium hydroxide (CH) or HBD3-C15 treatment. (**b**) *S. gordonii* lipoprotein-stimulated NF-κB are mediated by TLR2. HEK-TLR2 cells (2.5×10^5 cells/mL) were transfected with an NF-κB luciferase reporter plasmid using Attractene transfection reagent. After 16 h, the cells were stimulated with lipoprotein purified from *S. gordonii*, and then treated with either of CH or HBD3-C15. After the cells were lysed, a luciferease assay was conducted.

2.4. Anti-Inflammatory Effects Were Related to COX-2 Expression and PGE2 Level

To extend the findings to speculate on the mechanism of inflammatory status changes after treatment, we investigated the inflammatory marker COX-2 changes of inflamed hDPCs, which is known to be upregulated in pulpal inflammations [28]. It was found that the COX-2 gene expression and PGE2 level were significantly upregulated after *S. gordonii* stimulation of hDPCs ($p < 0.05$), and were significantly alleviated after 24 h of CH and HBD3-C15 treatments ($p < 0.05$), and maintained at low level (Figure 4). There was no significant differences between the two medicaments ($p > 0.05$).

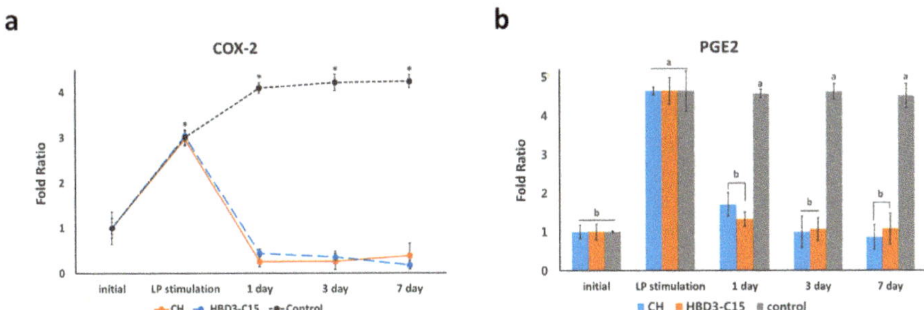

Figure 4. (**a**) Expression of COX-2 gene in *S. gordonii* lipoprotein-stimulated hDPCs after treatment with calcium hydroxide (CH) or HBD3-C15 at determined time points and (**b**) PGE2 secretion levels accordingly. The relative expression levels to each control group are presented.

3. Discussion

Pulp and periapical inflammation are initiated and propagated by chemokines. The pro-inflammatory chemokines IL-8 and MCP-1 are associated with the recruitment of cells to sites of inflammation. A predominant cellular source of IL-8 and MCP-1 are mononuclear phagocytes. However, IL-8 and MCP-1 can also be produced by nonimmune cells, such as fibroblasts, keratinocytes, and endothelial cells in response to either endogenous or exogenous stimuli. The cytokines produced by mononuclear cells are important mediators for chemokine production by cells such as dental pulp fibroblasts, which respond to invading microorganisms and produce chemokines. In dental pulp, the elevation of MCP-1 that is a chemo-attractant, gathers immune cells (to resist infection), which release more pro-inflammatory mediators and inflammatory cells within the tissues [29,30].

Recently, some lipoproteins from gram-positive bacteria were found to be triacylated, which are native ligands of TLR2 [31]. TLR2, along with other pattern recognition receptors (PRRs), are functionally the predominant receptors that stimulate production of the inflammatory mediators, IL-8, IL-6, MCP-1, and PGE2. Therefore, these receptors play important roles in the immune response of the dental pulp, and the progression of pulpitis [32]. Their expression in hDPCs alone appears to upregulate pro-inflammatory mediators. Likewise, our prior study reported that *S. gordonii* induces nitric oxide production through the TLR2 signaling pathway, and lipoprotein is responsible for the induction [7].

Now, this study has shown that the upregulated expression of inflammatory mediators in hDPCs that have been stimulated with *S. gordonii* lipoprotein can be attenuated by either CH or HBD3-C15. Such a capacity to reduce pulpal inflammation, especially in regenerative endodontic procedures on immature teeth may be particularly beneficial. The anti-inflammatory effects of CH are less understood than its known antimicrobial activity. CH inactivates the LPS of gram-negative bacteria, by hydrolyzing fatty acids in the lipid A moiety [9,33–35]. Additionally, it inactivates the LTA of gram-positive bacteria by deacylation, which inhibits their binding of TLR2 [10,11]. However in this study CH and HBD3-C15 suppressed lipoprotein-induced inflammation without blocking TLR2 signaling. It is speculated that the inflammation control of CH and HBD3-C15 as an endodontic medicament did not affect immune related signals in bacterial virulence factor-induced inflammation.

In this study, definite alleviation of the upregulated COX-2 as well as increased PGE2 secretion were observed in after CH- or HBD3-C15-treated hDPCs. Considering that PGE2 could enhance pain transmission in neuro-inflammation [36], the medication-dictated modulation of PGE2 release may provide a plausible background for why CH has been advocated for wide use in endodontics, and also adduce the possibility of HBD3-C15 as an alternative medicament. COX-2 is a membrane associated enzyme that produces PGE2 at sites of pulpal injury and inflammation, which leads to tissue swelling, redness, and pain [37]. As COX-2 participates in the regulation of prostanoid formation in the pathogenesis of pulpal inflammation, it was suggested that COX-2 may play a pivotal role in generating high levels of PGE2 locally resulting in pulpal tissue destruction, which formed the first investigation into using COX-2 inhibitors to control pulpal inflammation [38]. It is necessary to further investigate the effects of CH and HBD3-C15 on differential signal transduction pathways that mediate COX-2 stimulation and PGE2 production in hDPCs, instead of direct effect on TLR2 pathway. Another important clinical factor of the two chemicals would be a topical action in the control of inflammation, which could reduce the side effects of systemic anti-inflammatory drugs (i.e., NSAIDs). However, further studies are necessary to decipher the details of this mechanism.

The anti-inflammatory effects of HBD3-C15 peptide that were found in TLR-2 mediated inflammation in this study, were not in accordance with prior reports on the full length HBD3 protein pathways [16,39,40]. Just as other AMPs that permeabilize microbial membranes and neutralize or disaggregate LPS [16], full length HBD3 can bind directly to LPS and prevent the binding of LPS to host cell receptors [16,39,40] through TLR-4 mediated signaling pathways and the subsequent transcriptional inhibition of inflammatory genes [41]. It has also been suggested that full length HBD3 has anti-inflammatory properties that do not involve direct peptide binding to LPS, in macrophages isolated from human bone marrow [40]. However, this therapeutic application of full length HBD3 is limited by its molecular size, the complexity of disulfide pairing, and attenuated activity at elevated ionic strength. To overcome these limitations and identify the active peptide fragments within HBD3, the C-terminal HBD3 peptide was modified by substituting serine for cysteine residues, and shown to have retained its anti-microbial activity [42]. Recent reports showed that HBD3-C15 peptide could attenuate LPS-induced bone resorption, by disrupting podosome belt formation in osteoclasts and suppressing their differentiation [43]. In anti-inflammatory effect, HBD3-C15 did not affect upregulated TLR-2 level and its following signaling pathways in this study. This may have been due to the study design where CH paste and HBD3-C15 treatments were not applied simultaneously with *S. gordonii* lipoprotein. This was important to avoid the possible impairment or denaturation of lipoprotein and subsequent inflammatory mediators by CH or HBD3-C15, and to focus solely on the changes in lipoprotein-stimulated hDPCs.

Collectively, the results of this study support the use of CH and HBD3-C15 as intracanal medicaments, which may be particularly important for regenerative endodontic procedures in immature necrotic teeth. The synthetic HBD3-C15 peptide has multiple properties that are beneficial for its therapeutic application as an intracanal medicament in recurrent apical periodontitis and also in regenerative endodontic procedures for immature necrotic teeth.

4. Materials and Methods

4.1. Human Dental Pulp Cell Explant Cultures

Human dental pulp cell (hDPCs) explants were grown from the pulps of extracted teeth, which had been approved for collection by the Institutional Review Board of Seoul National University (S-D2014007, 27 March 2014). Crowns of freshly extracted teeth were split open and their pulps carefully harvested. The pulp tissues were diced into fine fragments in culture dishes (Nalge Nunc International, Rochester, NY, USA), and incubated in DMEM (Dulbecco's modified Eagle's medium) supplemented with 10% fetal bovine serum (FBS) and 1% penicillin-streptomycin, with fresh media replenished every 3 days. After 3 weeks of growth, the cells that had extended out from the tissue fragments were carefully harvested and sub-cultured. Following 4–6 passages, cells were utilized in experiments such that each independent experiment was performed with cells from the same passages.

The viability of hDPCs was quantitatively assessed in the presence of CH and HBD3-C15 using a cell counting kit (CellCountEZ^{TM} Cell Survival assay kit, Rockland Immunochemicals, Posttown, PA, USA). Briefly, the hDPCs were seeded in a 96-well plate at a density of 2×10^3 cells/well and cultured with different concentrations of reagents for 24 h. The absorbance was measured with a microplate reader (Bio-Rad Laboratories, Hercules, CA, USA) at 490 nm (Figure 5).

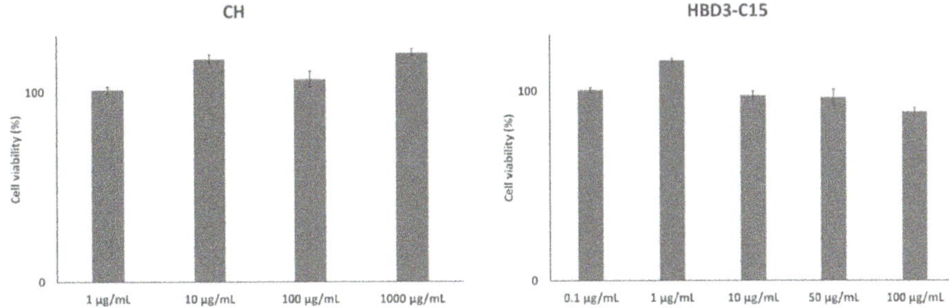

Figure 5. The survival rate of hDPCs was assessed in the presence of calcium hydroxide (CH) and HBD3-C15 using a cell counting kit (CellCountEZ™ Cell Survival assay kit, Rockland Immunochemicals, Posttown, PA, USA). Briefly, the hDPCs were seeded in a 96-well plate at a density of 2×10^3 cells/well and cultured with different concentrations of reagents for 24 h. The absorbance was measured with a microplate reader (Bio-Rad Laboratories, Hercules, CA, USA) at 490 nm.

4.2. *S. gordonii Lipoprotein Extracts*

Bacterial lipoproteins were isolated and purified from *S. gordonii*, as previously reported [44]. Briefly, bacterial pellets were collected and suspended in Tris-buffered saline (TBS) with protease inhibitors. They were sonicated, the lysates suspended in a final concentration of 2% Triton X-114 at 4 °C for 2.5 h, and then centrifuged to discard cell debris. The supernatant was incubated at 37 °C for 15 min and centrifuged again to separate the Triton X-114 phase from the aqueous phase which was discarded. An equal volume of TBS was added to the Triton X-114 phase, incubated at 37 °C for 15 min, and centrifuged again to discard the aqueous phase. Finally, the Triton X-114 phase was incubated overnight with methanol at −20 °C, and lipoprotein precipitates collected and dissolved in 10 mM octyl-beta-D-glucopyranoside in PBS.

4.3. *Cell Transfection*

HEK293-TLR2 cells were transfected as previously described [45]. Briefly, HEK293-TLR2 cells (2.5×10^5 cells/mL, 5 mL, 60 mm dishes) were transfected with an NF-κB (nuclear factor-κB) reporter gene construct (pNF-κB-Luc, Clontech, Mountain View, CA, USA). Transfections were performed in Opti-MEM using Attractene transfection reagent for 16 h (Qiagen, Germantown, MD, USA). After collection, the cells (2.5×10^5 cells/mL, 200 μL, 96-well plates) were plated in complete DMEM. The cells were then stimulated with *S. gordonii* lipoprotein and lysed with GloLysis Buffer (Promega, Madison, WI, USA). The cytoplasmic extracts were assayed with a luminometer (Molecular Devices, Sunnyvale, CA, USA) for luciferase activity.

4.4. *Peptide Preparation*

Peptides were synthesized at the central research institute of NIBEC using a peptide synthesizer (Prelude, Protein Technologies Inc., Tucson, AZ, USA) to produce the C-terminal amide form using standard 9-fluorenylmethoxycarbonyl (F-moc) chemistry and purified using preparative reverse-phase high-performance liquid chromatography (RP-HPLC; Waters, Milford, MA, USA) with a Vydac C18 column and a 50-min gradient from 90% to 10% water/acetonitrile containing 0.1% trifluoroacetic acid (TFA). The purity of the peptides was determined using HPLC (Shimadzu, Kyoto, Japan) and liquid chromatography-mass spectrometry (LC-MS, Shimadzu, Kyoto, Japan). The purity of the peptide was measured as 99.3% and the molecular weight of the peptide was identified as 1766.17 (Da) using LC-MS.

4.5. *Inflammatory Chemokine Analysis*

The hDPCs (2.5×10^5 cells/mL) were plated in six well plates and stimulated with 10 μg/mL of lipoprotein extracts for 24 hrs. The stimulated cells were then treated with either CH (DC Chemical

Co Ltd., Seoul, Korea) mixed with distilled water (100 µg/mL) as the control, or synthetic HBD3-C15 peptide gel (50 µg/mL). The synthetic HBD3-C15 peptide (NIBEC, Seoul, Korea) was prepared by F-moc-based chemical solid phase synthesis from 15 amino acids (GKCSTRGRKCCRRKK). It was used at a concentration (50 µg/mL) that was found to be effective in a previous study [14].

Time points were chosen that included before and after 24 h of lipoprotein stimulation, as well as after 1, 3, and 7 days of the treatments with either CH or HBD3-C15. At these predetermined times, replicate (N = 3) hDPC cultures were harvested and their total cell RNA extracted with RNAiso Plus reagents (Takara, Otsu, Japan). The cDNA was synthesized with gene-specific reverse primers for humans (Table 1) by using the PrimeScript RT reagent kit (Takara). Reverse transcription-quantitative polymerase chain reaction (RT-qPCR) was done with a C1000 Real-time PCR system thermal cycler (Bio-Rad, Hercules, CA, USA). The level of the target gene transcripts (ΔCT) was normalized to that of glyceraldehyde-3-phosphate dehydrogenase (Gapdh). The relative levels of expression in the experimental group were calculated by $\Delta(\Delta$CT) (ΔCT$_{Control} - \Delta$CT$_{Experiment}$). The findings were verified by multiplex analysis according to the manufacturers' instructions. Data was collected using the Luminex-100 system (ver. 1.7, Luminex, Austin, TX, USA) and analyzed by using the Milliplex Analyst (Viagene Tech, Carlisle, MA, USA). A five-parameter regression formula was used to calculate the concentrations from the standard curves. Reported data as normalized pg/mL was assessed. All reactions were run in triplicate.

Table 1. Sequences of primers used in RT-qPCR.

Gene (Human)		Primer (5′→3′)
IL-8	Forward	AGGGTTGCCAGATGCAATAC
	Reverse	CCTTGGCCTCAATTTTGCTA
MCP-1	Forward	GCAGCAAGTGTCCCAAAGA
	Reverse	ACAGGGTGTCTGGGGAAAG
COX2	Forward	TTCAAATGAGATTGTGGGAAAAATTGCT
	Reverse	AGATCATCTCTGCCTGAGTATCTT
TLR2	Forward	CCCATTGCTCTTTCACTGCT
	Reverse	CTTCCTTGGAGAGGCTGATG
GAPDH	Forward	GGCTGAGAACGGGAAGCTT
	Reverse	TCCATGGTGGTGAAGACGC

4.6. Statistical Analysis

Replicate cultures were compared both before and after lipoprotein-stimulation, and then after treatment with either calcium hydroxide or HBD3-C15 for 1, 3, and 7 days. Data were analyzed statistically by a one-way analysis of variance (ANOVA) and Tukey's post-hoc test to a significance of $P < 0.05$.

Author Contributions: Conceptualization: K.-Y.K.; data curation: Y.-J.Y., J.-Y.L., A.-R.K., K.-Y.K.; formal analysis: J.-Y.L., S.O., A.-R.K., Y.G.; investigation: H.P., K.-Y.K.; methodology: J.-Y.L., K.-Y.K.; resources: S.-H.B., S.-W.C., K.-Y.K.; funding: S.-H.B., K.-Y.K.; supervision: K.-Y.K.; validation: H.P.; writing—original draft preparation: Y.-J.Y., K.-Y.K.; writing—review and editing: H.P., K.-Y.K. All authors read and approved the final manuscript. All authors have read and agreed to the published version of the manuscript.

Funding: This study was supported by Seoul National University Dental Hospital Research Fund [04–2018–0102, 02-2019-0008], and the Korea Health Technology R&D Project through the Korea Health Industry Development Institute (KHIDI) and the Ministry of Health & Welfare (HI17C1377).

Conflicts of Interest: All authors declare that they have no conflicts of interest.

Ethical Approval: Study approval was obtained from the Institutional Review Board of Seoul National University Dental Hospital (IRB125/05-16).

References

1. Love, R.M.; Jenkinson, H.F. Invasion of dentinal tubules by oral bacteria. *Crit. Rev. Oral. Biol. Med.* **2002**, *13*, 171–183. [CrossRef] [PubMed]
2. Sedgley, C.M.; Lee, E.H.; Martin, M.J.; Flannagan, S.E. Antibiotic resistance gene transfer between *Streptococcus gordonii* and *Enterococcus faecalis* in root canals of teeth ex vivo. *J. Endod.* **2008**, *34*, 570–574. [CrossRef] [PubMed]
3. Brightbill, H.D.; Libraty, D.H.; Krutzik, S.R.; Yang, R.B.; Belisle, J.T.; Bleharski, J.R.; Maitland, M.; Norgard, M.V.; Plevy, S.E.; Smale, S.T.; et al. Host defense mechanisms triggered by microbial lipoproteins through toll-like receptors. *Science* **1999**, *285*, 732–736. [CrossRef] [PubMed]
4. Nguyen, M.T.; Gotz, F. Lipoproteins of Gram-Positive Bacteria: Key Players in the Immune Response and Virulence. *Microbiol. Mol. Biol. Rev.* **2016**, *80*, 891–903. [CrossRef] [PubMed]
5. Mayer, M.L.; Phillips, C.M.; Townsend, R.A.; Halperin, S.A.; Lee, S.F. Differential activation of dendritic cells by toll-like receptor agonists isolated from the gram-positive vaccine vector *Streptococcus gordonii*. *Scand. J. Immunol.* **2009**, *69*, 351–356. [CrossRef]
6. Kim, A.R.; Ahn, K.B.; Kim, H.Y.; Seo, H.S.; Kum, K.Y.; Yun, C.H.; Han, S.H. *Streptococcus gordonii* lipoproteins induce IL-8 in human periodontal ligament cells. *Mol. Immunol.* **2017**, *91*, 218–224. [CrossRef]
7. Kim, H.Y.; Baik, J.E.; Ahn, K.B.; Seo, H.S.; Yun, C.H.; Han, S.H. *Streptococcus gordonii* induces nitric oxide production through its lipoproteins stimulating Toll-like receptor 2 in murine macrophages. *Mol. Immunol.* **2017**, *82*, 75–83. [CrossRef]
8. Siqueira, J.F., Jr.; Lopes, H.P. Mechanisms of antimicrobial activity of calcium hydroxide: A critical review. *Int. Endod. J.* **1999**, *32*, 361–369. [CrossRef]
9. Safavi, K.E.; Nichols, F.C. Effect of calcium hydroxide on bacterial lipopolysaccharide. *J. Endod.* **1993**, *19*, 76–78. [CrossRef]
10. Baik, J.E.; Jang, K.S.; Kang, S.S.; Yun, C.H.; Lee, K.; Kim, B.G.; Kum, K.Y.; Han, S.H. Calcium hydroxide inactivates lipoteichoic acid from *Enterococcus faecalis* through deacylation of the lipid moiety. *J. Endod.* **2011**, *37*, 191–196. [CrossRef]
11. Baik, J.E.; Kum, K.Y.; Yun, C.H.; Lee, J.K.; Lee, K.; Kim, K.K.; Han, S.H. Calcium hydroxide inactivates lipoteichoic acid from *Enterococcus faecalis*. *J. Endod.* **2008**, *34*, 1355–1359. [CrossRef] [PubMed]
12. Wang, Z.; Shen, Y.; Haapasalo, M. Antibiofilm peptides against oral biofilms. *J. Oral Microbiol.* **2017**, *9*, 1327308. [CrossRef] [PubMed]
13. Schneider, J.J.; Unholzer, A.; Schaller, M.; Schafer-Korting, M.; Korting, H.C. Human defensins. *J. Mol. Med.* **2005**, *83*, 587–595. [CrossRef] [PubMed]
14. Lee, J.K.; Chang, S.W.; Perinpanayagam, H.; Lim, S.M.; Park, Y.J.; Han, S.H.; Baek, S.H.; Zhu, Q.; Bae, K.S.; Kum, K.Y. Antibacterial efficacy of a human beta-defensin-3 peptide on multispecies biofilms. *J. Endod.* **2013**, *39*, 1625–1629. [CrossRef]
15. Lee, J.K.; Park, Y.J.; Kum, K.Y.; Han, S.H.; Chang, S.W.; Kaufman, B.; Jiang, J.; Zhu, Q.; Safavi, K.; Spangberg, L. Antimicrobial efficacy of a human beta-defensin-3 peptide using an *Enterococcus faecalis* dentine infection model. *Int. Endod. J.* **2013**, *46*, 406–412. [CrossRef]
16. Zhang, L.J.; Gallo, R.L. Antimicrobial peptides. *Curr. Biol.* **2016**, *26*, R14–R19. [CrossRef]
17. Wu, Z.; Hoover, D.M.; Yang, D.; Boulegue, C.; Santamaria, F.; Oppenheim, J.J.; Lubkowski, J.; Lu, W. Engineering disulfide bridges to dissect antimicrobial and chemotactic activities of human beta-defensin 3. *Proc. Natl. Acad. Sci. USA* **2003**, *100*, 8880–8885. [CrossRef]
18. Hoover, D.M.; Wu, Z.; Tucker, K.; Lu, W.; Lubkowski, J. Antimicrobial characterization of human beta-defensin 3 derivatives. *Antimicrob. Agents Chemother.* **2003**, *47*, 2804–2809. [CrossRef]
19. Krishnakumari, V.; Rangaraj, N.; Nagaraj, R. Antifungal activities of human beta-defensins HBD-1 to HBD-3 and their C-terminal analogs Phd1 to Phd3. *Antimicrob. Agents Chemother.* **2009**, *53*, 256–260. [CrossRef]
20. Casalinuovo, I.A.; Sorge, R.; Bonelli, G.; Di Francesco, P. Evaluation of the antifungal effect of EDTA, a metal chelator agent, on *Candida albicans* biofilm. *Eur. Rev. Med. Pharmacol. Sci.* **2017**, *21*, 1413–1420.
21. Lim, S.M.; Ahn, K.B.; Kim, C.; Kum, J.W.; Perinpanayagam, H.; Gu, Y.; Yoo, Y.J.; Chang, S.W.; Han, S.H.; Shon, W.J.; et al. Antifungal effects of synthetic human beta-defensin 3-C15 peptide. *Restor. Dent. Endod.* **2016**, *41*, 91–97. [CrossRef] [PubMed]

22. Tsai, C.H.; Chen, Y.J.; Huang, F.M.; Su, Y.F.; Chang, Y.C. The upregulation of matrix metalloproteinase-9 in inflamed human dental pulps. *J. Endod.* **2005**, *31*, 860–862. [CrossRef] [PubMed]

23. Huang, G.T.; Potente, A.P.; Kim, J.W.; Chugal, N.; Zhang, X. Increased interleukin-8 expression in inflamed human dental pulps. *Oral Surg. Oral Med. Oral Pathol. Oral Radiol. Endod.* **1999**, *88*, 214–220. [CrossRef]

24. Rechenberg, D.K.; Galicia, J.C.; Peters, O.A. Biological Markers for Pulpal Inflammation: A Systematic Review. *PLoS ONE* **2016**, *11*, e0167289. [CrossRef] [PubMed]

25. Lacy, P. Mechanisms of degranulation in neutrophils. *Allergy Asthma Clin. Immunol.* **2006**, *2*, 98–108. [CrossRef] [PubMed]

26. Omori, K.; Kida, T.; Hori, M.; Ozaki, H.; Murata, T. Multiple roles of the PGE2 -EP receptor signal in vascular permeability. *Br. J. Pharmacol.* **2014**, *171*, 4879–4889. [CrossRef]

27. Bordier, C. Phase separation of integral membrane proteins in Triton X-114 solution. *J. Biol. Chem.* **1981**, *256*, 1604–1607.

28. Chen, C.L.; Kao, C.T.; Ding, S.J.; Shie, M.Y.; Huang, T.H. Expression of the inflammatory marker cyclooxygenase-2 in dental pulp cells cultured with mineral trioxide aggregate or calcium silicate cements. *J. Endod.* **2010**, *36*, 465–468. [CrossRef]

29. Park, J.; Choi, K.; Jeong, E.; Kwon, D.; Benveniste, E.N.; Choi, C. Reactive oxygen species mediate chloroquine-induced expression of chemokines by human astroglial cells. *Glia* **2004**, *47*, 9–20. [CrossRef]

30. Park, S.H.; Hsiao, G.Y.; Huang, G.T. Role of substance P and calcitonin gene-related peptide in the regulation of interleukin-8 and monocyte chemotactic protein-1 expression in human dental pulp. *Int. Endod. J.* **2004**, *37*, 185–192. [CrossRef]

31. Kurokawa, K.; Lee, H.; Roh, K.B.; Asanuma, M.; Kim, Y.S.; Nakayama, H.; Shiratsuchi, A.; Choi, Y.; Takeuchi, O.; Kang, H.J.; et al. The triacylated ATP binding cluster transporter substrate-binding lipoprotein of *Staphylococcus aureus* functions as a native ligand for toll-like receptor 2. *J. Biol. Chem.* **2009**, *284*, 8406–8411. [CrossRef] [PubMed]

32. Hirao, K.; Yumoto, H.; Takahashi, K.; Mukai, K.; Nakanishi, T.; Matsuo, T. Roles of TLR2, TLR4, NOD2, and NOD1 in pulp fibroblasts. *J. Dent. Res.* **2009**, *88*, 762–767. [CrossRef] [PubMed]

33. Barthel, C.R.; Levin, L.G.; Reisner, H.M.; Trope, M. TNF-alpha release in monocytes after exposure to calcium hydroxide treated Escherichia coli LPS. *Int. Endod. J.* **1997**, *30*, 155–159. [CrossRef] [PubMed]

34. Buck, R.A.; Cai, J.; Eleazer, P.D.; Staat, R.H.; Hurst, H.E. Detoxification of endotoxin by endodontic irrigants and calcium hydroxide. *J. Endod.* **2001**, *27*, 325–327. [CrossRef]

35. Jiang, J.; Zuo, J.; Chen, S.H.; Holliday, L.S. Calcium hydroxide reduces lipopolysaccharide-stimulated osteoclast formation. *Oral Surg. Oral Med. Oral Pathol. Oral Radiol. Endod.* **2003**, *95*, 348–354. [CrossRef]

36. Jiang, J.; Yu, Y.; Kinjo, E.R.; Du, Y.; Nguyen, H.P.; Dingledine, R. Suppressing pro-inflammatory prostaglandin signaling attenuates excitotoxicity-associated neuronal inflammation and injury. *Neuropharmacology* **2019**, *149*, 149–160. [CrossRef]

37. Nakanishi, T.; Shimizu, H.; Hosokawa, Y.; Matsuo, T. An immunohistological study on cyclooxygenase-2 in human dental pulp. *J. Endod.* **2001**, *27*, 385–388. [CrossRef]

38. Chang, Y.C.; Yang, S.F.; Huang, F.M.; Liu, C.M.; Tai, K.W.; Hsieh, Y.S. Proinflammatory cytokines induce cyclooxygenase-2 mRNA and protein expression in human pulp cell cultures. *J. Endod.* **2003**, *29*, 201–204. [CrossRef]

39. Chung, P.Y.; Khanum, R. Antimicrobial peptides as potential anti-biofilm agents against multi-drug resistant bacteria. *J. Microbiol. Immunol.* **2017**, *50*, 405–410. [CrossRef]

40. Semple, F.; Webb, S.; Li, H.N.; Patel, H.B.; Perretti, M.; Jackson, I.J.; Gray, M.; Davidson, D.J.; Dorin, J.R. Human beta-defensin 3 has immunosuppressive activity in vitro and in vivo. *Eur. J. Immunol.* **2010**, *40*, 1073–1078. [CrossRef]

41. Semple, F.; MacPhereson, H.; Webb, S.; Cox, S.L.; Mallin, J.L.; Tyrrell, C.; Grimes, G.R.; Semple, C.A.; Nix, M.A.; Millhouser, G.L.; et al. Human β-defensin 3 affects the activity of pro-inflammatory pathways associated with MyD88 and TRIF. *Eur. J. Immunol.* **2011**, *41*, 3291–3330. [CrossRef] [PubMed]

42. Lee, J.Y.; Suh, J.S.; Kim, J.M.; Kim, J.H.; Park, H.J.; Park, Y.J.; Chung, C.P. Identification of a cell-penetrating peptide domain from human beta-defensin 3 and characterization of its anti-inflammatory activity. *Int. J. Nanomed.* **2015**, *10*, 5423–5434. [CrossRef]

43. Park, O.J.; Kim, J.; Ahn, K.B.; Lee, J.Y.; Park, Y.J.; Kum, K.Y.; Yun, C.H.; Han, S.H. A 15-amino acid C-terminal peptide of beta-defensin-3 inhibits bone resorption by inhibiting the osteoclast differentiation and disrupting podosome belt formation. *J. Mol. Med.* **2017**, *95*, 1315–1325. [CrossRef] [PubMed]

44. Li, Q.; Kumar, A.; Gui, J.F.; Yu, F.S. *Staphylococcus aureus* lipoproteins trigger human corneal epithelial innate response through toll-like receptor-2. *Microb. Pathog.* **2008**, *44*, 426–434. [CrossRef] [PubMed]
45. Lee, J.Y.; Lee, B.H.; Lee, J.Y. Gambogic Acid Disrupts Toll-like Receptor4 Activation by Blocking Lipopolysaccharides Binding to Myeloid Differentiation Factor 2. *Toxicol. Res.* **2015**, *31*, 11–16. [CrossRef] [PubMed]

Article

Characterization of Tachyplesin Peptides and Their Cyclized Analogues to Improve Antimicrobial and Anticancer Properties

Felicitas Vernen [1], Peta J. Harvey [1], Susana A. Dias [2], Ana Salomé Veiga [2], Yen-Hua Huang [1], David J. Craik [1], Nicole Lawrence [1] and Sónia Troeira Henriques [1,3,*]

[1] Institute for Molecular Bioscience, The University of Queensland, Brisbane, Queensland 4072, Australia
[2] Instituto de Medicina Molecular João Lobo Antunes, Faculdade de Medicina, Universidade de Lisboa, 1649-028 Lisboa, Portugal
[3] School of Biomedical Sciences, Faculty of Health, Institute of Health & Biomedical Innovation, Queensland University of Technology, Translational Research Institute, Brisbane, Queensland 4102, Australia
* Correspondence: sonia.henriques@qut.edu.au; Tel.: +61-7-34437342

Received: 26 July 2019; Accepted: 21 August 2019; Published: 26 August 2019

Abstract: Tachyplesin I, II and III are host defense peptides from horseshoe crab species with antimicrobial and anticancer activities. They have an amphipathic β-hairpin structure, are highly positively-charged and differ by only one or two amino acid residues. In this study, we compared the structure and activity of the three tachyplesin peptides alongside their backbone cyclized analogues. We assessed the peptide structures using nuclear magnetic resonance (NMR) spectroscopy, then compared the activity against bacteria (both in the planktonic and biofilm forms) and a panel of cancerous cells. The importance of peptide-lipid interactions was examined using surface plasmon resonance and fluorescence spectroscopy methodologies. Our studies showed that tachyplesin peptides and their cyclic analogues were most potent against Gram-negative bacteria and melanoma cell lines, and showed a preference for binding to negatively-charged lipid membranes. Backbone cyclization did not improve potency, but improved peptide stability in human serum and reduced toxicity toward human red blood cells. Peptide-lipid binding affinity, orientation within the membrane, and ability to disrupt lipid bilayers differed between the cyclized peptide and the parent counterpart. We show that tachyplesin peptides and cyclized analogues have similarly potent antimicrobial and anticancer properties, but that backbone cyclization improves their stability and therapeutic potential.

Keywords: tachyplesin; host defense peptide; anticancer; antimicrobial; antibiofilm; peptide-membrane interaction; structure-activity; model membranes; nuclear magnetic resonance solution structure

---

1. Introduction

The host defense peptides (HDPs) tachyplesin I, II and III (TI, TII and TIII) are active against a broad range of Gram-negative and Gram-positive bacteria and fungi [1–4] and possess anticancer properties [5–15]. Each analogue was isolated from a different species of horseshoe crab, but they share high sequence homology (Table 1). TI, TII and TIII possess 17 amino acid residues, two disulfide bonds, a C-terminal α-amidation, and their structure is organized in a β-hairpin [2,3,16]. Like other HDPs [17], TI, TII and TIII possess an amphipathic secondary structure (i.e., positively charged and hydrophobic amino acids segregate into distinct clusters), thought to be essential for their antimicrobial activity.

Cationic amphipathic HDPs selectively target the anionic surfaces of microbes, rather than the neutral surface of host cells, and kill them by a mechanism that involves binding to and insertion into cell membranes. The initial binding is mediated by electrostatic attractions between the positively-charged residues of HDPs and the anionic microbial surface [18], and is followed by the insertion of hydrophobic residues into lipid membranes in a process that involves van-der-Waal's interactions with the phospholipids [18,19].

Similar to bacterial cells, the surface of cancer cells is negatively-charged due to the increased expression and exposure of phospholipids containing the anionic phosphatidylserine (PS) headgroup [20–23]. In contrast, cell membranes of healthy mammalian cells are asymmetric and phospholipids containing PS-headgroups are found exclusively in the inner leaflet. Variations in the overall cell surface charge [24] in phospholipids with exposed [25,26] membrane fluidity and curvature [18,27], have been shown to modulate the selective toxicity of HDPs towards cancer cells. These differences regulate the affinity of peptides for cell membranes, the effective peptide-to-lipid ratio and the ability for HDP to kill cancerous cells or pathogens rather than healthy cells [28–30].

Several studies have investigated the activity, structure and mechanism of the action of TI, but few have examined the activity of TII and TIII. Early studies suggested that TI kills bacterial cells by a mechanism involving inner membrane permeabilization and the rapid efflux of K^+ [31–33]. Later, TI was shown to translocate across lipid bilayers, cause a phospholipid flip-flop and form toroidal pores [33,34]. In cancerous cells, TI was reported to induce cell disruption and late apoptosis/necrosis [12,35]. Paredes-Gamero et al. proposed that the cancer cell death mechanism was dependent on the peptide dose: at high concentrations, the peptide-induced direct cell membrane disruption, and at lower concentrations, it activated intracellular cell death mechanisms [12].

Because of the expression of TI, TII and TIII in distinct species of horseshoe crab, we were specifically interested in comparing their structure, activity and mode-of-action. As these peptides have a potential application as anticancer and/or antimicrobial agents, we investigated whether backbone cyclization would increase the stability and maintain activity. Our studies show that TI, TII and TIII, and their cyclic analogues cTI, cTII and cTIII, have similar structures and activities against bacteria and cancerous cells. Backbone cyclization reduced the hemolytic activity and increased peptide stability while maintaining potent anticancer and antimicrobial activities. cTI and cTIII especially showed potential to be considered for the development of anticancer peptide-based drugs.

2. Results

2.1. Properties of Tachyplesin I–III and Their Cyclic Analogues

The amino acid sequences of TI, TII and TIII differ in positions 1 or 15, which can be a lysine or an arginine residue (Table 1). So far, TI is the most studied and the only analogue with reported structure calculations [36–38]. We were interested in comparing TI, TII and TIII and their backbone-cyclized analogues (cTI–III) to identify similarities and differences in their three-dimensional structure and stability; and to determine whether these characteristics affect the membrane interactions and biological activity of the peptides.

All peptides were synthesized using solid-phase peptide synthesis, oxidized and correctly folded, as suggested by the observed masses using electrospray ionization mass spectroscopy (ESI-MS; Supplementary Figure S1 and Table 1) and confirmed through clearly dispersed peaks in the amide region of their respective One-dimensional (1D) ^{1}H NMR spectra [39]. The peptides were purified to >95%, as confirmed by analytical reverse-phase high-performance liquid chromatography (RP-HPLC; see chromatograms of pure peptides in Supplementary Figure S1a).

Despite minor differences, their overall hydrophobicity follows the trend cTI > cTII > cTIII > TI > TII > TIII (Table 1), as indicated by their retention time (RT) on analytical RP-HPLC (Supplementary Figure S1). The cyclic analogues appear to be overall more hydrophobic (less polar) than the parent peptides, which is consistent with the loss of the N-terminal charge and the C-terminal amidation resulting from cyclization.

Table 1. The sequence and physicochemical properties of tachyplesin I–III (TI-TIII) and their cyclic analogues (cTI-cTIII).

Peptide	Sequence [1]	Mass (Da) [2]		RT (min) [3]	Charge [4]
		Calc.	Obs.		
TI	**KWCFRVCYRGICYRRCR** *	**2263.8**	2263.5	18.04	+7
TII	RWCFRVCYRGICYRKCR *	2263.8	2263.5	17.79	+7
TIII	KWCFRVCYRGICYRKCR *	2235.8	2235.6	17.68	+7
cTI	KWCFRVCYRGICYRRCRG	2303.8	2303.7	18.69	+6
cTII	RWCFRVCYRGICYRKCRG	2303.8	2303.7	18.53	+6
cTIII	KWCFRVCYRGICYRKCRG	2275.8	2275.5	18.44	+6

[1] Peptide sequences and amino acid residues differing from TI are in bold. * denotes C-terminal amidation. [2] Average mass calculated (Calc.) from the amino acid sequence and experimentally observed (Obs.) using synthetic peptide and determined from m/z 3+ in ESI-MS. [3] Retention time (RT) of peptides on an analytical RP-HPLC; chromatograms shown in Supplementary Figure S1a were obtained with a 2%/min gradient of 0–40% solvent B (90% acetonitrile; 0.05% trifluoroacetic acid (TFA) (*v/v*)) in solvent A (H_2O, 0.05% TFA (*v/v*)) at a flow rate of 0.3 mL/min. [4] Charge of the peptides at pH 7.4.

2.2. Structure of Tachyplesins and Backbone-Cyclised Analogues

The three-dimensional (3D) structures of TII and TIII, and of the backbone-cyclized analogues cTI–cTIII were determined with solution NMR spectroscopy. All backbone resonances were fully assigned apart from the N-terminal amides of the two linear peptides, and the R1/G18 amides of cTII, reflective of some degree of flexibility in these regions. The secondary αH chemical shifts of the native peptides and the cyclic analogues were highly similar, indicating a negligible change in the backbone structure and the β-strands (W2-Y8; I11-R17) (Figure 1a,b).

The 3D solution structures of each peptide were calculated from distance restraints, ranging from 154 for the linear peptides, to 172–284 for the cyclic analogues, along with dihedral angle restraints totaling 34 to 36. The final family of structures for each of the peptides has good structural and energy statistics, as indicated by an overall MolProbity score of less than 1.6, shown in Table S1. Analysis of the structures by PROMOTIF [40] defines antiparallel β-strands being formed by residues W2-Y8 and I11-R17 in all but one of the peptides. The exception is TIII which has slightly shorter strands formed between residues C3-Y8 and I11-C16. All disulfide bonds are defined by PROMOTIF as adopting the short right-hand hook configuration. Structures of the tachyplesin peptides and the cyclized analogues differed primarily in the flexibility of the N- and C-termini of the parent peptides (Figure 1b–d). The reduction of the amino acid side-chain flexibility in the terminal regions due to backbone cyclization is emphasized in Figure 1d with the side chain of the residues K/R1 and K/R15, which differ between TI/cTI, TII/cTII and TIII/cTIII; the side chain of W2 is also shown, as this residue is used to monitor the peptide partitioning into lipid bilayers (see Section 2.5.2). No significant cation-π interactions between the R/K1 and W2 were noted for any of the tachyplesin peptides. Such an interaction might be revealed by NMR in the form of substantial deviations of the chemical shifts of arginine/lysine residues but none were observed (see Figure 1a). The lack of cation-π interactions is also supported by the NMR solution structures that reveal a high degree of flexibility in the termini of the linear peptides, and by the different orientations that these residues acquire in the cyclic peptides, as shown in Figure 1d.

The peptide structures were deposited with the Protein Data Bank (PDB) and the Biological Magnetic Resonance Data Bank (BMRB): TII—PDB ID: 6PI2, BMRB ID: 30617; TIII—PDB ID: 6PI3, BMRB ID: 30618; cTI—PDB ID: 6PIN; BMRB ID: 30619; cTII—PDB ID: 6PIO, BMRB ID: 30620; cTIII—PDB ID: 6PIP, BMRB ID: 30621.

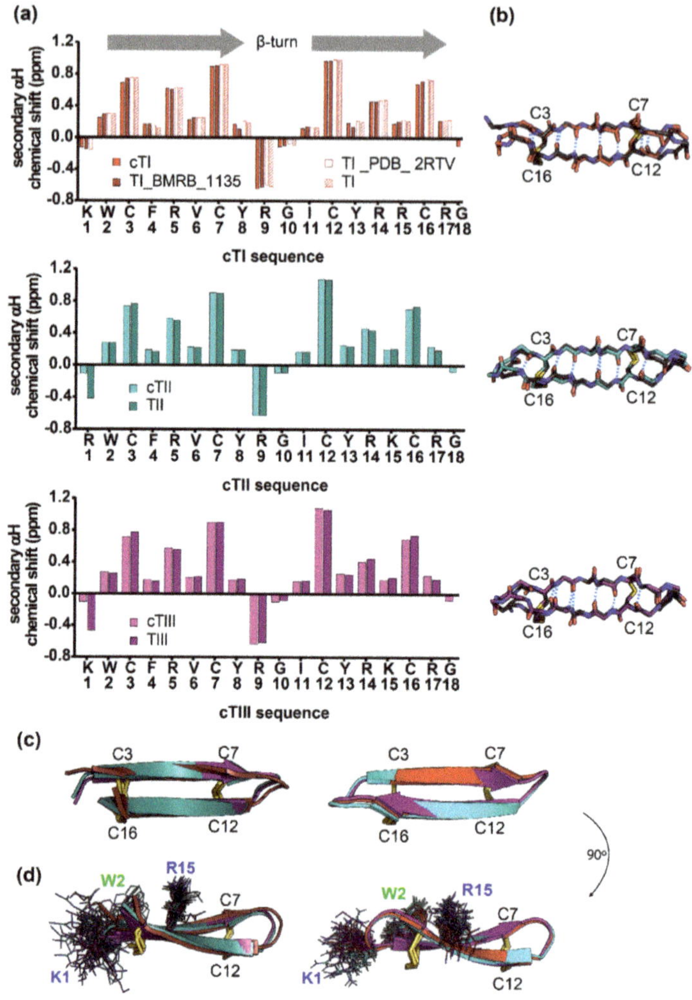

Figure 1. The NMR structures of TI and cTI (red), TII and cTII (cyan), TIII and cTIII (magenta). (**a**) Secondary αH chemical shift at 298 K determined from ^{1}H NMR spectra. Shifts were calculated by subtraction of random coil ^{1}H NMR shifts [41] from the experimental values. For tachyplesin I (TI), the shifts of the peptide synthesized in our lab were compared to shifts obtained from data banks: TI—PDB ID 2RTV [37], and TI—BMRB ID 1135 [38]. Positive shifts greater than 0.1 ppm suggest β-strands, indicated by grey arrows. (**b**) Overlay of the cyclic analogues (colored backbone) with their respective parent peptide (grey backbone). Hydrogen bonds between β-strands are indicated in blue. The structure of TI was obtained from the PDB (ID 1WO0). (**c**) Overlay of TI–TIII and cTI–cTIII. (**d**) Mobility of the residues (K or R) at position 1 and 15, which differ between the peptides, and of W2 of the native sequences and cyclic analogues, respectively.

2.3. *Improved Stability and Reduced Hemolytic Activity of Cyclized Tachyplesin Peptides*

Cyclization has been shown to increase stability [42,43] and reduce the hemolytic activity of some peptides [44,45]. Comparison of resistance to human proteases showed that the cTI analogue did not degrade after treatment for 24 h in 25% (*v/v*) human serum, whereas only 25% of TI remained in the solution (see analytical RP-HPLC chromatograms in Supplementary Figure S2). A peptide (linear and without disulfide bonds) used as a control was fully degraded under the same conditions (Figure 2a). TI has previously been shown to be completely stable for 2 h in mouse or human serum [46].

The percentage of hemolysis of human red blood cells (RBCs) followed the trend TI > TII > TIII > cTII > cTI > cTIII when compared at 128 µM, the highest concentration of peptide tested. TI was the most hemolytic of the native sequences. Backbone cyclization reduced the hemolytic activity of the cTI and cTIII compared to their parent counterparts, but led to no clear improvement for cTII. The C-amidated peptides TI–TIII lysed 66%, 56% and 41% of RBCs at 64 µM respectively (Figure 2b and Table 2). A similar trend but lower hemolytic activity had been reported for TI, TII, and TIII lacking C-terminal amidation at higher peptide concentrations [47], which is known to impact peptide activity [48,49].

The positive control melittin, a hemolytic peptide from honeybee venom, induces 100% hemolysis in human RBCs at concentrations above 2 µM. By comparison, at this concentration, all tachyplesin peptides have a low hemolytic activity of around 10%.

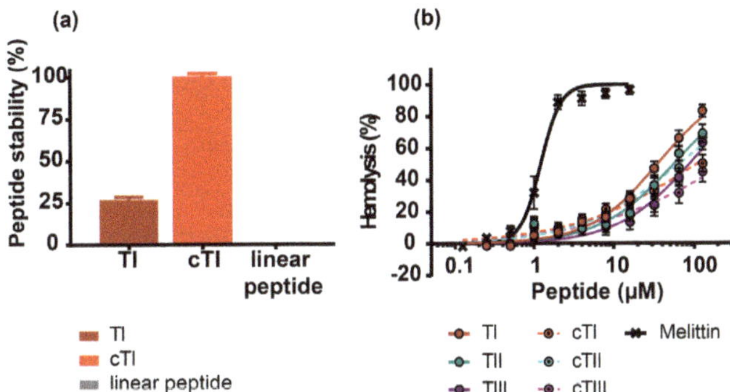

Figure 2. The peptide stability in human serum and activity against human red blood cells (RBCs). (**a**) Stability of TI and cyclic analogue cTI in 25% (*v/v*) human serum for 24 h. The samples were analyzed by analytical RP-HPLC and the percentage of peptide remaining was determined from the area under the peptide peak in the chromatogram in comparison to peak area at time zero. A linear 18 amino acid peptide (KGGGGSGQLIDSMANSFV) was included as the positive control. (**b**) Hemolytic activity of tachyplesins and their cyclic analogues were tested up to 128 µM against human RBCs (0.25% (*v/v*), 1 h incubation at 37 °C). Melittin, a highly hemolytic peptide, was included as the positive control.

Table 2. Concentration of tachyplesin I–III (TI-TIII) and of their cyclic analogues (cTI-cTIII) required to induce 50% lysis in RBCs (HC50). [a.]

Peptide	HC50 (µM)
TI	34.9 ± 2.8
TII	55.4 ± 6.6
TIII	86.4 ± 12.2
cTI	106.9 ± 21.0
cTII	64.1 ± 9.4
cTIII	>128

[a] Values were determined from a minimum of three independent replicates and depicted as mean ± SEM.

2.4. Biological Activity of Tachyplesin I–III and of Their Cyclic Analogues

2.4.1. Activity against Bacteria

The Gram-negative *Escherichia coli* strains ATCC 25922 and DC2 CGSC 7139 and the Gram-positive *Staphylococcus aureus* strains ATCC 25923 and ATCC 6538 were used to determine the antimicrobial activity of the parent tachyplesin peptides and their cyclic analogues against bacteria with differences in the physical and chemical properties of their cell wall. *E. coli* ATCC 25922 ("smooth" LPS) and *S. aureus* ATCC 25923 are common control strains for antimicrobial susceptibility testing [50,51]. *E. coli* DC2 CGSC 7139 is hypersensitive to antibacterial agents and more permeable to dyes [52], while *S. aureus* ATCC 6538 is known to form biofilms [53]. Tachyplesin peptides and their cyclic analogues were tested against planktonic cultures of all strains and against *S. aureus* ATCC 6538 in the biofilm form for direct comparison of their antimicrobial activity.

Tachyplesin I–III were two-to-four times more potent than their cyclic analogues against all bacterial strains in their planktonic growth form (Table 3). Generally, the Gram-negative strains were more susceptible than the Gram-positive strains. Similar MICs have been reported previously for TI against *E. coli* ATCC 25922, [1] and TI and TII against *S. aureus* ATCC 25923 [2,3]. Tachyplesin peptides were most active against *E. coli* DC2 CGSC 7139 and *E. coli* ATCC 25922. The activity against these bacterial strains was reduced when the peptides were cyclized. The peptides were least active against *S. aureus* ATCC 6538 in the planktonic form, and no difference in activity was observed between tachyplesin peptides and cyclic analogues (Table 3).

Table 3. The minimal inhibitory concentrations (MICs) of tachyplesin I–III (TI-TIII) and of their cyclic analogues (cTI-cTIII) against planktonic bacteria.

Peptide	MIC (µM)			
	E. coli DC2 CGSC 7139	*E. coli* ATCC 25922	*S. aureus* ATCC 25923	*S. aureus* ATCC 6538
TI	0.5–1	0.0625–0.5	1–4	4–8
TII	0.5–1	0.125–0.5	1–4	4–8
TIII	0.25–0.5	0.0625–0.5	1–4	4–8
cTI	4	1	2–8	8
cTII	4–8	1–2	2–8	8
cTIII	4–8	1–2	2–8	8

To determine whether the tachyplesin peptides could act against bacterial biofilms, the cells metabolic activity was measured for biofilms formed by *S. aureus* ATCC 6538 (Figure 3 and Table 4). All peptides exhibited similar activity against the bacteria in the biofilm form, with a 50% loss of cell metabolic activity observed at ~20 µM. However, approximately 40% of the biofilm was metabolically active at 32 µM, the highest concentration tested, suggesting a reduced activity against biofilm compared to the planktonic *S. aureus* ATCC 6538 bacteria.

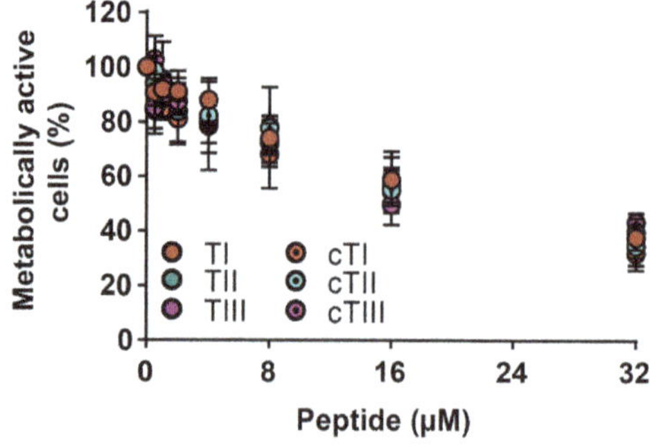

Figure 3. The activity of tachyplesin peptides and their cyclic analogues against an *S. aureus* ATCC 6538 biofilm. The response curve of biofilm treated with increasing the concentrations of peptide ($n = 4$, ± SEM).

Table 4. Activity of tachyplesin I–III (TI–TIII) and cyclic analogues (cTI–cTIII) against *S. aureus* ATCC 6538 in the biofilm growth form.

Peptide	CC50 (µM) [1]	Metabolically Active Cells (%) [2]
TI	21.5 ± 2.2	37.7 ± 4.3
TII	23.1 ± 2.9	39.9 ± 2.9
TIII	24.2 ± 4.1	43.3 ± 4.1
cTI	17.8 ± 3.6	31.9 ± 6.0
cTII	19.4 ± 1.5	34.6 ± 7.1
cTIII	18.0 ± 1.4	37.5 ± 7.2

[1] Peptide concentration required to induce the reduction of 50% of the metabolically active cell population (CC50) determined using one-site specific binding with the Hill slope. [2] Percentage of cells remaining metabolically active at 32 µM, the highest peptide concentration tested.

2.4.2. Activity against Cancerous Cells

TI, TII, TIII, and their cyclic analogues were tested against three melanoma (MM96L, HT144 and WM164) and one cervical cancer (HeLa) cell line (Supplementary Tables S2 and S3) to determine whether the peptides exhibit different cytotoxic activities. The aneuploid immortal keratinocyte cell line HaCaT was included as a non-cancerous control. Cytotoxicity toward cancer cell lines was compared by determining peptide concentrations required to achieve 50% of cell death (CC50) from dose-response curves (Table 5).

Table 5. The cytotoxicity of tachyplesin I–III (TI-TIII) and cyclic analogues (cTI-cTIII) against cultured cells.

	CC50 (µM) [1]					Melanoma Selectivity [5]
Peptide	MM96L [2]	HT144 [2]	WM164 [2]	HeLa [3]	HaCaT [4]	
TI	1.5 ± 0.1	1.7 ± 0.2	2.5 ± 0.1	13.1 ± 1.2	11.6 ± 1.6	2–21
TII	1.6 ± 0.1	2.0 ± 0.1	1.6 ± 0.1	18.0 ± 3.9	3.7 ± 0.2	3–35
TIII	1.8 ± 0.1	2.0 ± 0.1	1.7 ± 0.1	21.7 ± 1.1	7.3 ± 0.5	5–48
cTI	1.3 ± 0.1	1.4 ± 0.1	2.7 ± 0.1	6.7 ± 0.6	7.9 ± 0.5	2–76
cTII	1.1 ± 0.1	0.8 ± 0.04	2.4 ± 0.3	7.2 ± 0.4	2.4 ± 0.3	3–58
cTIII	1.7 ± 0.1	0.9 ± 0.03	1.3 ± 0.1	9.3 ± 0.4	7.5 ± 0.3	3–98

[1] The concentration necessary to kill 50% of cells was calculated from dose-response curves (n ≥ 3, ± SEM). [2] Melanoma cell lines: MM96L, HT144, WM164, [3] cervical cancer cell line: HeLa, [4] healthy epithelial control cell line: HaCaT (aneuploid immortal keratinocyte). Description and verification of each cell line are detailed in Supplementary Tables S2 and S3. [5] selectivity for melanoma cell lines was estimated through the activity-toxicity index (ATI). ATI = MHC/MCC50 (modified from Reference [46]), with MHC being the minimal concentration necessary to induce 10% (lower value) or 50% (higher value) cell death in human RBCs and MCC50 being the median of CC50 values of all melanoma cell lines. Values above 1 indicate a higher selectivity for the cancerous cells over RBCs.

The cytotoxic activities of the tachyplesin peptides and their cyclic analogues are dependent on the cell lines (Table 5). All peptides were most effective against the melanoma cell lines MM96L, HT144 and WM164 (cytotoxic activities ranged between CC50 0.8–2.7 µM). Compared to the control cell line HaCaT, the cytotoxic activities of the peptides against melanoma were significantly different ($p < 0.05$) with the exception of cTII against WM164. The cervical cancer cell line HeLa was more resistant towards the tachyplesin peptides compared to the melanoma cell lines and higher peptide concentrations were necessary to reach 50% cell death. Differences in cytotoxic activity against HeLa, compared to the control cell line HaCaT, can be observed for all peptides except TI and cTI. The cyclic analogues cTI–cTIII were approximately 2x more potent against HeLa than the parent peptides TI–TIII ($p < 0.05$).

The peptides have a similar selectivity for the melanoma cell lines at lower peptide concentrations which would induce ≤10% hemolysis in RBCs. At concentrations inducing ≤50% hemolysis in RBCs, the cyclic analogues cTI–cTIII were more selective. Overall, the most promising therapeutic range was observed for TIII, cTI and cTIII (Table 5).

2.5. Mechanistic Studies

Peptides with an amphipathic arrangement of charged and hydrophobic residues are known to act against bacterial and cancer cells via selective membrane targeting, penetration and/or lysis. To characterize how peptide-membrane interactions affect biological activity, we undertook detailed peptide-membrane binding studies that compare parent and cyclic tachyplesin peptides.

2.5.1. Peptide Binding to Model Membranes

The ability of TI to bind to model membranes has been previously shown and a preference for negatively-charged membranes was found [31,46]. We were interested in comparing the membrane-binding properties of the parent tachyplesins versus the cyclic analogues to determine whether differences in membrane binding could explain the relative biological activities (see Table 5). Given the similar potencies among the three tachyplesins, and among the three cyclic analogues, we compared the interaction of TI and cTI, as representative of parent and cyclic tachyplesin peptides, respectively, with model membranes using surface plasmon resonance (SPR). Phospholipids containing PC-headgroups are the most common in the outer leaflet of the mammalian plasma membrane [54,55]; thus, we prepared model membranes composed of the zwitterionic POPC (1-palmitoyl-2-oleoyl-sn-glycero-3-phosphocholine), which forms fluid bilayers at 25 °C and mimics the overall fluidity and neutral surface of healthy eukaryotic cells [56]. Phospholipids containing the negatively-charged PS-headgroups are normally restricted to the inner leaflet in eukaryotic cell membranes but are exposed at the cell surface of cancerous cells [20,22,23]. Therefore, we used model membranes with 20% of POPS (1-palmitoyl-2-oleoyl-sn-glycero-3-phosphoserine), POPC/POPS (4:1 molar ratio), to represent the negatively-charged surface of cancer cells.

Both TI and cTI bound with higher affinity to negatively-charged POPC/POPS (4:1) than to zwitterionic POPC lipid bilayers. cTI had stronger affinity to both lipid systems than the parent peptide, as shown by a higher peptide-to-lipid (P/L) ratio during association, a slower dissociation rate (k_{off}) from the lipids, and a higher amount of peptide remaining associated to the membrane at the end of dissociation (P/L$_{off}$) (Figure 4a,b and Table 6). Additionally, TI and cTI disrupted large unilamellar vesicles (LUVs) of POPC/POPS (4:1) with higher efficacy than LUVs of POPC, which agrees with their preference for negatively-charged over neutral membranes. However, TI disrupted LUVs of POPC and of POPC/POPS (4:1) more efficiently than the cyclic analogue cTI (Figure 4c, Table 6). Thus, the higher binding saturation and affinity of cTI (see P/L$_{max}$ and kinetic parameters in Table 6) to the membranes did not correlate with its ability to disrupt membranes. The higher efficacy of TI in disrupting membranes, compared to the cyclic analogue, might explain the higher activity of the parent TI–TIII for planktonic bacterial cells (see Table 3), compared to their cyclic analogues.

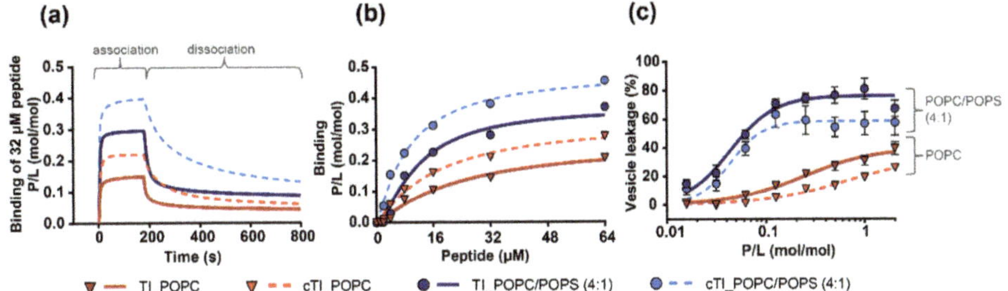

Figure 4. Membrane binding and disruption induced by tachyplesin I (TI) and cyclic tachyplesin I (cTI). Model membranes composed of 1-palmitoyl-2-oleoyl-sn-glycero-3-phosphocholine (POPC) and POPC/1-palmitoyl-2-oleoyl-sn-glycero-3-phosphoserine (POPS) (4:1) were compared. (**a**) Surface plasmon resonance sensorgrams obtained with 32 µM peptide injected over lipid bilayers deposited on an L1 chip surface for 180 s (association); dissociation was monitored for 600 s. Response units (RU) were converted into a peptide-to-lipid ratio (P/L (mol/mol)) to take into consideration the differences in lipid packing resulting in different amounts being deposited to cover the chip surface. (**b**) The dose-response curves show P/L obtained at the end of the association phase ($t = 170$ s) and plotted as a function of peptide concentration injected. (**c**) Percentage of vesicle leakage determined by fluorescence emission intensity of 5-carboxyfluorescein (λ_{ex} = 490 nm, λ_{em} = 513 nm) leaking from LUVs at increasing concentrations of peptide. Lipid concentration used was 5 µM, and the peptide was tested up to 10 µM. Dose-response curves were fitted with one-site specific binding with Hill slope equation in GraphPad Prism.

Table 6. The kinetic and affinity parameters from surface plasmon resonance analysis of the interaction of 32 µM tachyplesin I (TI) and cyclic tachyplesin I (cTI) with neutral (POPC) and negatively-charged model membranes (POPC/POPS (4:1)) and leakage induced by the same peptides and lipid systems.

Peptide	Lipid System	P/L_{max} (mol/mol) [1]	K_D (µM) [1]	k_{off} (x 10^{-2} s^{-1}) [2]	P/L_{off} (mol/mol) [2]	LC_{max} (%) [3]
TI	POPC	0.26 ± 0.06	22.4 ± 10.9	1.50 ± 0.11	0.046 ± 0.001	39.5 ± 4.6
cTI		0.33 ± 0.07	16.7 ± 8.4	0.91 ± 0.03	0.065 ± 0.001	32.9 ± 9.4
TI	POPC/POPS	0.37 ± 0.04	11.8 ± 2.4	2.75 ± 0.22	0.096 ± 0.001	76.0 ± 2.8
cTI	(4:1)	0.48 ± 0.08	9.2 ± 3.4	0.70 ± 0.03	0.137 ± 0.002	58.5 ± 3.6

[1] P/L_{max} and K_D were calculated from the dose-response curves (one-site specific binding with Hill slope equation, GraphPad Prism) in Figure 4b. The P/L_{max} value represents the peptide-to-lipid ratio (mol/mol) when peptide-lipid binding reaches saturation, K_D is the peptide concentration necessary to reach the half-maximal binding response. [2] k_{off} is the dissociation constant and P/L_{off} is the peptide-lipid ratio at the end of association phase calculated from the sensorgrams obtained with 32 µM peptide in Figure 4a. k_{off} and P/L_{off} were fitted in GraphPad Prism, assuming a Langmuir kinetic. [3] Percentage of leakage achieved when incubating 10 µM peptide with 5 µM LUVs. POPC is 1-palmitoyl-2-oleoyl-sn-glycero-3-phosphocholine; POPS is 1-palmitoyl-2-oleoyl-sn-glycero-3-phosphoserine.

To examine the ability of cTI to bind model membranes that mimic bacterial cell membranes, we prepared vesicles with an *E. coli* polar lipid extract composed of zwitterionic phosphatidylethanolamine (PE)-phospholipids, negatively-charged phosphatidylglycerol (PG)-phospholipids, and cardiolipin (CA) in the proportion 67:23.2:9.8 (*wt/wt%*). The SPR sensorgram and dose-response curves (Supplementary Figure S3) show that cTI has a high affinity for *E. coli* lipids. Comparison of the dose-response curves and fitted parameters show that the maximum amount of cTI bound to *E. coli* lipids (P/L_{max}, Supplementary Table S4) is not as high as for the other tested negatively-charged membranes or for the zwitterionic POPC, but the P/L_{off} is higher than from the other tested membranes (see Supplementary Table S4), suggesting that a large amount of peptide remains bound to the bilayers that mimic bacterial membranes. To investigate whether cTI distinguishes the negatively-charged headgroups present in bacteria (i.e., PG) from those in cancer cells (i.e., PS), we compared the binding of cTI to POPC/POPG (1-palmitoyl-2-oleoyl-sn-glycero-3-phosphoglycerol;

4:1)) and to POPC/POPS (4:1). cTI has a slightly higher affinity for model membranes containing PS-phospholipids than to those containing PG-phospholipids, as shown by the higher maximum amount of peptide bound to POPC/POPS (4:1), and a slower dissociation rate (see P/L_{max}, k_{off} and P/L_{off} in Supplementary Figure S3).

2.5.2. Partitioning of Trp Residue into Model Membranes

The fluorescence emission of Trp is sensitive to the local environment. In an aqueous environment, the fluorescence emission spectrum of Trp has a maximum at ~350 nm (λ_{ex} = 280 nm), in a hydrophobic environment the fluorescence emission spectrum is blue-shifted and an increase in the fluorescence quantum yield is usually observed [57–59]. Tachyplesin peptides have one Trp residue: the fluorescence emission properties of this residue can be used to examine the environment surrounding it and inform on the partitioning and orientation of the peptides bound to model lipid bilayers. In the current study, we followed the changes in Trp fluorescence emission spectra of TI–TIII and of the cyclic analogues cTI–cTIII upon titration with LUVs composed of zwitterionic (POPC) or of negatively-charged membranes with two distinct negatively-charged headgroups, i.e., PS and PG (Figure 5).

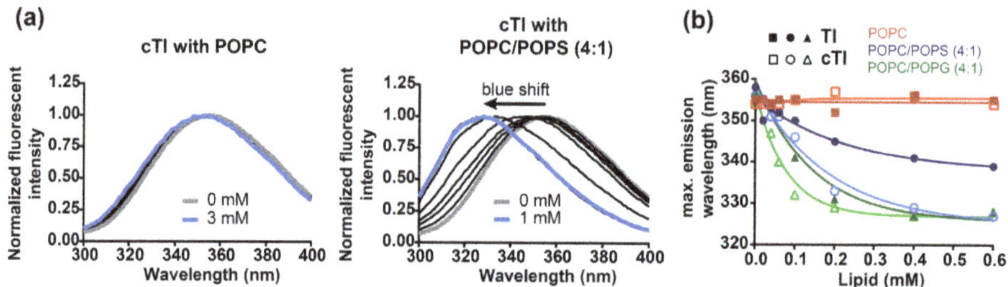

Figure 5. The partitioning of the 12.5 μM of tachyplesin I–III (TI-TIII) and cyclic analogues (cTI-cTIII) into lipid membranes. (**a**) Normalized fluorescence emission spectra of cTI in aqueous solution (grey) and upon titration with LUVs composed of POPC (1-palmitoyl-2-oleoyl-sn-glycero-3-phosphocholine; up to 3 mM) or of POPC/POPS (1-palmitoyl-2-oleoyl-sn-glycero-3-phosphoserine; 4:1) up to 1 mM). Fluorescence emission spectra were obtained with excitation at 280 nm (**b**) Wavelength at which TI (solid symbol) and cTI (open symbol) have their maximum fluorescence emission in buffer and with increasing concentrations of LUVs composed of different lipid mixtures (POPC: red; POPC/POPS (4:1): blue; POPC/POPG (1-palmitoyl-2-oleoyl-sn-glycero-3-phosphoglycerol; 4:1): green).

TI and cTI bound to both zwitterionic POPC and negatively-charged POPC/POPS (4:1) bilayers, as shown with SPR (see Figure 4a,b); however, a blue shift in the Trp fluorescence emission spectra was only observed for model membranes containing negatively-charged phospholipids (Figure 5a, Table 7). Even at the highest concentration of 3 mM POPC LUVs, no change in the Trp fluorescence emission spectra was observed for any of the peptides. In contrast to POPC, a blue shift was detected when peptides were incubated with negatively-charged model membranes POPC/POPS (4:1) and POPC/POPG (4:1) at lipid concentrations as low as 0.1 mM (Figure 5).

In the presence of negatively-charged membranes, the Trp fluorescence emission maxima of cTI–cTIII had a larger shift than the respective parent peptides TI–TIII (Table 7), as illustrated with TI and cTI (Figure 5b). The differences between parent and cyclic peptides were most pronounced with POPC/POPS (4:1) bilayers: TI–TIII required more than twice the amount of POPC/POPS (4:1) LUVs compared to cTI–cTIII (0.19–0.30 mM vs. 0.08–0.12 mM) to induce half of the maximal shifts of the Trp fluorescence emission spectra (Figure 5b, Table 7). Comparison among parent peptides shows similar spectral shifts for a given model membrane (see for instance Trp fluorescence emission shifts of TI–TIII when titrated with POPC/POPS); the same is true among cyclic analogues. These results suggest that the mutations R/K1 or R/K15 have a weak influence, whereas backbone cyclization between R/K1 and G18 impacts the insertion of the Trp residue into lipid membranes.

Table 7. Shift in the fluorescence emission maximum wavelength of tachyplesin I–III (TI-TIII) and cyclic analogues (cTI-cTIII) in buffer and in the presence of model membranes.

	POPC	POPC/POPS (4:1)		POPC/POPG (4:1)	
Peptide	shift (nm) [1]	shift (nm)	0.5 [L] (mM) [2]	shift (nm)	0.5 [L] (mM)
TI	2	22	0.19	30	0.09
TII	0	19	0.30	29	0.09
TIII	−3	18	0.30	23	0.10
cTI	0	28	0.11	28	0.06
cTII	1	27	0.08	29	0.05
cTIII	3	27	0.12	28	0.08

[1] blue shifts observed with 3 mM POPC (1-palmitoyl-2-oleoyl-sn-glycero-3-phosphocholine); 1 mM POPC/POPS (1-palmitoyl-2-oleoyl-sn-glycero-3-phosphoserine; 4:1), or 1 mM POPC/POPG (1-palmitoyl-2-oleoyl-sn-glycero-3-phosphoglycerol; 4:1) LUVs. [2] lipid concentrations required to achieve half of the maximum blue shift observed (0.5 [L] in mM).

The Trp fluorescence emission spectra of all tachyplesin analogues displayed larger blue shifts in the presence of POPC/POPG (4:1) compared to POPC/POPS (4:1) LUVs, which suggests that the Trp residue partitions better and/or inserts further into PG-containing membranes than into PS-containing membranes (Table 7). Despite the blue shift in the fluorescence emission spectra of all the peptides when in the presence of negatively-charged membranes, no increase in quantum yield was detected (data not shown). This could be explained by fluorescence quenching induced by neighboring amino acid side chains, local carbonyl groups [58], or through photon re-absorption between Trp residues of neighbor peptides molecules due to their closer proximity once inserted into lipidic membranes.

2.5.3. Insertion of Trp Residue into Model Membranes

The in-depth location of the Trp residue of tachyplesin peptides bound to lipid membranes was investigated using fluorescence quenching methodologies. Acrylamide, an aqueous quencher unable to partition into lipid bilayers, was used to quench the fluorescence of Trp residues exposed to the aqueous solution. If the Trp residue inserts into the lipid bilayer, it becomes inaccessible to quenching by the aqueous-soluble acrylamide. In addition to acrylamide quenching, we used the lipidic quenchers 5- and 16-doxyl stearic acids (5DS and 16DS) to gain information on the in-depth location of TI and cTI within the membrane. The acids 5- and 16DS are fatty acids that insert into the lipid bilayer and possess the quencher moiety, nitroxide, located at carbon 5 and 16 of the fatty acid chain, respectively. The proximity of the nitroxide moiety to the Trp residue is required for quenching of its fluorescence emission; thus, comparison between the 5- and 16DS quenching efficacies gives information on the location of the Trp residue when the peptide is partitioned into membranes [60,61]. Changes in the Trp fluorescence emission of individual peptides were followed upon titration with each of the quenchers. The fluorescence emission intensity in the absence and presence of the quencher (I_0/I) were plotted as a function of the quencher concentration and used to determine the Stern-Volmer constants (K_{SV}) (Figure 6), which are proportional to the accessibility of the quencher to the fluorophore.

A reduction in the slope (K_{SV}) of the fitted data points in the presence of lipids, compared to the aqueous solution, indicates the reduced quenching efficacy by acrylamide. The percentage of Trp accessible to acrylamide was estimated by the ratio of K_{SV} obtained in the presence and absence of LUVs, assuming that the Trp residue is fully exposed to acrylamide when the peptide is in an aqueous solution. Acrylamide quenched the Trp fluorescence emission of both TI and cTI with a similar efficacy in the buffer and in the presence of 1 mM POPC LUVs (Figure 6a, Table 8). These results show that when TI and cTI are bound to POPC, their Trp residue is accessible to acrylamide and likely to be exposed to the aqueous environment. In contrast, when in the presence of 1 mM POPC/POPS (4:1), the Trp fluorescence emission of TI and cTI was not efficiently quenched by acrylamide, suggesting that their Trp residue is not accessible to acrylamide. A similar result is expected for all the analogues given the lack of insertion of the Trp residue of TI–TIII and of cTI–cTIII when bound to POPC membranes and the large blue shift in the fluorescence emission spectra of the Trp residue of all the analogues when bound to POPC/POPS (4:1) membranes (see Table 7).

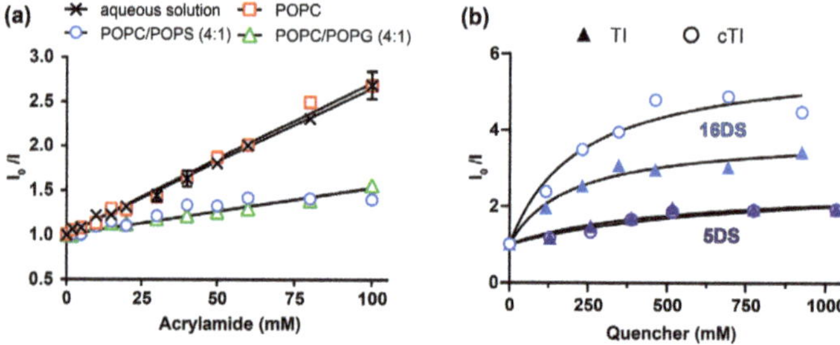

Figure 6. The effect of aqueous (acrylamide) and lipidic (5DS and 16DS) quenching on the tryptophan fluorescence intensity to investigate peptide in-depth location within model membranes. (**a**) The Trp fluorescence emission of 12.5 μM cTI in aqueous solution or in the presence of LUVs (1 mM POPC (1-palmitoyl-2-oleoyl-sn-glycero-3-phosphocholine), 1 mM POPC/POPS (1-palmitoyl-2-oleoyl-sn-glycero-3-phosphoserine; 4:1), 0.1 mM POPC/POPS (4:1) or 0.1 mM POPC/POPG (1-palmitoyl-2-oleoyl-sn-glycero-3-phosphoglycerol; 4:1)) in the absence (I_0) and upon titration with increasing concentrations of acrylamide (I). Data are shown as Stern-Volmer plots (i.e., I_0/I versus concentration of quencher). (**b**) Quenching of Trp fluorescence emission of 12.5 μM TI and cTI incorporated in 1 mM POPC/POPS (4:1) membranes by 5DS or 16DS.

Table 8. Fluorescence quenching of 12.5 μM tachyplesin I-III (TI-TIII) or cylic tachyplesin I-III (cTI-cTIII) by acrylamide.

	Acrylamide Accessibility (%) [1]			
	1 mM	1 mM	0.1 mM	0.1 mM
Peptide	POPC	POPC/POPS (4:1)	POPC/POPS (4:1)	POPC/POPG (4:1)
TI	90.5 ± 4.3	16.2 ± 21.9	72.3 ± 3.9	54.2 ± 5.8
TII			76.2 ± 3.2	56.8 ± 6.9
TIII			82.1 ± 4.7	75.4 ± 4.6
cTI	103.8 ± 2.3	22.5 ± 14.3	32.3 ± 8.3	32.0 ± 4.1
cTII			61.4 ± 5.5	36.1 ± 7.7
cTIII			48.7 ± 6.5	46.2 ± 5.7

[1] Tryptophan accessibility of tachyplesin I-III (TI–TIII) and cyclic tachyplesin (cTI–cTIII) in the presence of LUVs (1 mM POPC (1-palmitoyl-2-oleoyl-sn-glycero-3-phosphocholine), 1 mM POPC/POPS (1-palmitoyl-2-oleoyl-sn-glycero-3-phosphoserine; 4:1), 0.1 mM POPC/POPS (4:1) or 0.1 mM POPC/POPG (1-palmitoyl-2-oleoyl-sn-glycero-3-phosphoglycerol; 4:1)) to the aqueous quencher acrylamide. The accessibility of the Trp residue to the quencher was calculated from the Stern-Volmer constants, K_{SV}, which were derived from the linear fit of the data points (Equation (1)) of the peptide in buffer and peptide in the presence of lipids (see Figure 6a). Full exposition (100%) of the Trp residue to an aqueous environment was assumed in the buffer.

To identify potential differences between the six analogues, we monitored the quenching of Trp fluorescence emission by acrylamide in the presence of 0.1 mM POPC/POPS (4:1) or of 0.1 mM POPC/POPG (4:1) membranes. In the presence of 0.1 mM POPC/POPS (4:1) LUVs, acrylamide quenched the Trp fluorescence emission of TI–TIII with a higher efficacy than that of the cyclic analogues, suggesting that the Trp residue of cTI–cTIII is more protected from acrylamide. The Trp of the native tachyplesin peptides was less accessible for quenching in the presence of 0.1 mM POPC/POPG (4:1) LUVs compared to 0.1 mM POPC/POPS (4:1) LUVs through insertion into the membrane. The accessibility of the acrylamide to the Trp residues of the cyclic analogues remained approximately the same with both lipid systems (Table 8). These results suggest that when compared at the same lipid concentrations, more TI–TIII are inserted into POPC/POPG (4:1), than into POPC/POPS (4:1) membranes, whereas the cyclic analogues did not distinguish between the two negatively-charged lipid mixtures.

Comparison of the quenching efficacy by 5DS and 16DS tested with POPC/POPS (4:1) bilayers, showed that 16DS quenched the Trp fluorescence emission of TI and cTI with a higher efficacy than 5DS, which was particularly evident for the cyclic analogue (see Stern-Volmer plots and K_{SV} values in Figure 6b, Table 9). The higher quenching efficacy induced by 16DS suggests that the Trp residue of TI and of cTI is located within the hydrophobic core of the POPC/POPS (4:1) bilayers. The negative deviation to the linearity (see Figure 6b and f_b values in Table 9) suggests that the Trp residue of a fraction of peptide molecules is not accessible to the quenchers. This can be an indication of peptide molecules adopting different orientations within the membrane; nevertheless, the average location of the Trp residue of cTI and TI confirms that both peptides have their Trp residue deeply inserted and, in particular, the cyclic analogue. Overall, these results suggest that the Trp residue of cyclic analogues partitioned with a higher efficacy and/or were inserted deeper into anionic lipid bilayers, than parent tachyplesin peptides (Figure 7).

Table 9. Fluorescence quenching of tachyplesin I (TI) and and cyclic tachyplesin I (cTI) partitioned into 1 mM POPC/POPS (4:1) LUVs by the quenchers 5DS and 16DS. [1]

	5DS		16DS		
Peptide	K_{SV} (M^{-1})	f_b	K_{SV} (M^{-1})	f_b	Z (Å) [2]
TI	3.7 ± 1.3	0.63 ± 0.57	19.6 ± 4.3	0.74 ± 0.37	8.8
cTI	4.9 ± 1.8	0.61 ± 0.54	26.3 ± 7.1	0.83 ± 0.52	5.5

[1] The K_{SV} and f_b (the fraction of light emitted by the peptide accessible to the quencher) were determined using the Lehrer equation (Equation (2)). [2] The calculated average distance, Z in Å, of the Trp residue from the bilayer center was determined using the parallax method (Equations (3) and (4)) [62].

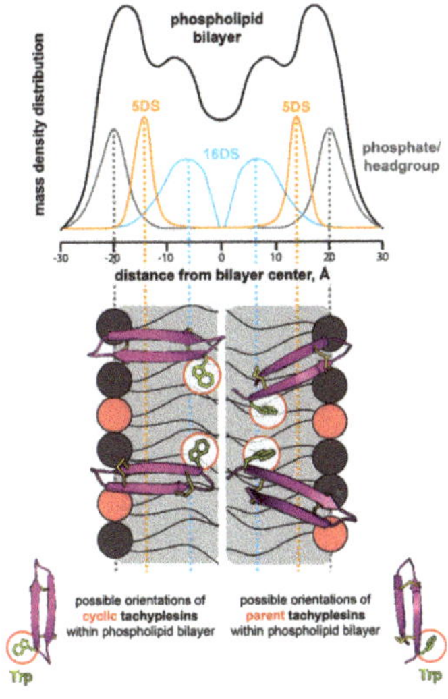

Figure 7. Schematic representation of the proposed orientations of tachyplesin peptides and cyclic analogues, when inserted into negatively-charged membranes based on the tryptophan fluorescence quenching, results with 5- and 16DS. The phospholipids in grey represent PC headgroups and the phospholipids in red represent PS headgroups. The peptides have a backbone length of approximately 25 Å (measured in PyMol) and are able to span across a lipid monolayer. Other orientations are possible. The distribution of the 5DS and 16DS quencher within a phospholipid bilayer was illustrated based on previous publications [60,63].

3. Discussion

In this study we demonstrated the high structural homology of the three tachyplesin peptides TI, TII and TIII. The primary sequence of these peptides differs only between Lys or Arg residues at positions 1 or 15, and both of these positions are located close to the flexible peptide termini. The cyclic analogues cTI, cTII and cTIII share a high structural homology with each other and with their respective parent peptide, but the backbone cyclization reduced the range of motion of the amino acid side chains located in the region of the termini (see Figure 1d).

TI–TIII and cTI–cTIII were active against representative Gram-positive and Gram-negative bacteria at low micromolar MICs (see Table 3). The antimicrobial activities of the three tachyplesin peptides were similar, as were the activities of the three cyclic analogues; however, differences were observed between the two groups. Overall, the parent tachyplesin peptides exhibited stronger antimicrobial activities than the cyclic analogues against all bacteria tested in the planktonic growth form. The Gram-negative *E. coli* strains were more sensitive to the peptides than the Gram-positive *S. aureus* strains. Against the *S. aureus* biofilm (see Figure 3 and Table 4), the tachyplesin peptides were >5-fold less active compared to the planktonic cells, likely due to the reduced accessibility of the peptides to the bacterial cell in biofilm form or the low ability of the peptides to penetrate the extracellular matrix of exopolysaccharides and access the cells [64]. The activities of the tachyplesin peptides against biofilms are comparable to other tested antimicrobial peptides [53]. The efficacy against the biofilms could be increased by using higher concentrations of the tachyplesin peptides or by using the peptide in combination with other antibiofilm or extracellular matrix disrupting agents [65]. Interestingly, TIII had been previously shown to prevent the formation of a *Pseudomonas aeruginosa* biofilm in a rat model of ureteral stent infection [66]; thus, even if the peptides cannot be used to completely disrupt the biofilm at the tested concentration, they could be useful for preventing biofilm formation on biomedical devices [67].

TI–TIII and cTI–cTIII peptides had a higher efficacy against the melanoma cell lines MM96L, HT144 and WM164 than to non-cancerous cells or the HeLa cervical cancer cell line (see Table 5). Interestingly, the cyclic tachyplesin analogues were more active against the cervical cancer cell line HeLa, than the parent tachyplesins. Differences in membrane composition between the different cells might explain the selectivity of the six analogues for melanoma over other cell lines [43], whereas a variation in the lipid-binding affinity and orientation within the membrane between parent versus cyclic analogues might explain more subtle differences observed against individual cell lines.

The experiments with model membranes confirmed the preference of all peptides for negatively-charged membranes (see Figure 4). TI, and its cyclized analogue cTI had a higher affinity for the negatively-charged POPC/POPS (4:1) than for neutral POPC bilayers, as shown in SPR experiments. cTI bond stronger and dissociated slower from POPC/POPS (4:1) and POPC bilayers than TI; however, TI disrupted both lipid bilayers with a higher efficacy than cTI. The ability of tachyplesin peptides to disrupt negatively-charged and neutral membranes with higher efficacy than their respective cyclic analogues, corroborates with their higher antimicrobial and hemolytic activities.

The fluorescence spectroscopy experiments show that the Trp residue of all six tachyplesin analogues partition and/or insert deeper into POPC/POPG (4:1), than into POPC/POPS (4:1) lipid bilayers (see Figure 5); in contrast, SPR studies show that cTI has a slightly higher affinity for POPC/POPS (4:1) than for POPC/POPG (4:1) membranes. This apparent contradiction emphasizes that specific phospholipid headgroups and membrane composition play a role in regulating and fine-tuning peptide-lipid binding interactions and the location of specific residues within the membrane.

The orientation of the peptide bound to membranes might also be affected by the phospholipid headgroup. When TI–TIII and cTI–cTIII were bound to neutral model membranes, their tryptophan residue did not insert into the bilayer. These results support an orientation in which both parent and cyclic tachyplesin peptides are parallel to the water-lipid interface, or with an orientation in which the β-turn (Y8-I11) inserts into the hydrophobic region. In the presence of negatively-charged membranes, the Trp residue of both parent and cyclic tachyplesin peptides partitioned into the membrane. However, the Trp residue in cyclic peptides cTI–cTIII seemed to be more deeply inserted into the bilayer and less exposed to the aqueous environment (see Figure 6, Tables 8 and 9).

Previous studies detected minor alterations to the peptide backbone of TI: the bending of the termini when the peptide was in contact with lipid membranes [36,68]. Doherty et al. [69] suggested that large amplitude motions of TI in the plane of the lipid membrane are essential for translocation, pore formation and membrane disruption. Since backbone cyclization impacts the flexibility of the termini region, cyclic tachyplesin analogues could be more restricted in the conformational changes of their backbone upon interaction with the membrane. The lower flexibility in the termini region might explain the lower ability of the cyclic analogue cTI to disrupt lipid bilayers compared to the parent tachyplesin. Furthermore, backbone cyclization created a second turn (Y17-W2) at which the Trp residue is located (see Figure 1d). The cyclic tachyplesin peptides are likely to insert deeply into the lipid membrane with the second turn (Y17-W2) inserting close to the center of the bilayer (see Figure 6 and Table 9), whereas in the parent peptides, the Trp residue is slightly closer to the membrane interface. A shallower location of the Trp residue could result from the parent peptides adopting a tilted orientation within the membrane (see Figure 7). Differences in the orientation of the peptides when inserted into lipid membranes could explain the higher membrane-disruptive properties and hemolytic activity of TI, compared to cTI (see Figure 4c and Table 6).

Backbone cyclization was shown to increase stability to proteolytic degradation, reduce membrane disruption and decrease hemolysis (see Figure 2). Interestingly, cTII has a slightly higher hemolytic activity compared to cTI and cTIII. cTII has an Arg residue at position 1 and a Lys at position 15, whereas cTI and cTIII have Lys residues at position 1. Arg and Lys residues establish different interactions with the phospholipid headgroups. The side chain of Arg residues can establish hydrogen bonds (H-bond) with two phospholipid headgroups, whereas the side chain in Lys residues can only form one H-bond. In contrast to Lys, Arg residues can form an H-bond through their side chains while being involved in cation-π-interactions [70]. Taking into consideration the increased toxicity of cTII due to an Arg residue, a Lys residue at position 1 favors the selectivity towards cancer cells.

In conclusion, TI–TIII and their cyclized analogues cTI–cTIII have high structural homology. Backbone cyclization increased peptide stability against proteolytic degradation, reduced the flexibility of amino acid side chains located in the terminal regions (i.e., R/K1, W2), and hemolytic activity while maintaining potent anticancer and antimicrobial activities. Since high resistance to proteolysis and low toxicity for host cells are preferred properties for peptide-based drug templates, the cyclic analogues have a higher potential than the parent counterparts, despite a possible loss of potency depending on the target. The Arg at position 1 of the amino acid sequence of TII and cTII proved to be non-beneficial for reducing hemolytic activity and improving selectivity for cancerous cells. Thus, cTI and cTIII have a higher potential than cTII as peptide-based anticancer drug templates. We also found that melanoma cell lines are more susceptible to the treatment with the tachyplesin peptides than the cervical cancer cell line HeLa. Systematic changes of the amino acid sequence might be applied to further increase the selectivity and/or anticancer activity of cyclic tachyplesin analogues (i.e. cTI or cTIII) to target specific cancers.

4. Material and Methods

4.1. Peptide Synthesis, Folding and Purification

The synthesis, folding and purification of synthetic peptides were carried out as previously described [42,71]. Briefly, the peptides were synthesized using 9-fluorenylmethoxycarbonyl (Fmoc) solid-phase peptide synthesis (SPPS) on an automatic peptide synthesizer (Symphony, Protein Technologies Inc., Tucson, USA). Rink amide resin was used for the synthesis of parent tachyplesin peptides and 2-chlorotrityl (2-CTC) resin for the backbone cyclized analogues. The peptides were oxidized overnight in 0.1 M ammonium bicarbonate buffer at pH 8.5 and purified using reverse-phase HPLC (solvent A: H_2O, 0.05% (*v/v*) trifluoroacetic acid (TFA), solvent B: 90% (*v/v*) acetonitrile,

0.05% (*v/v*) TFA) until the desired purity of >95%. The correct peptide mass was confirmed with ESI-MS, while native disulfide connectivity was inferred from the dispersion of peaks in the 1D NMR spectra using a Bruker Avance 600 MHz spectrometer (Billerica, USA). The peptide concentration was determined from the absorbance at 280 nm (ε_{280} = 8730 $M^{-1}.cm^{-1}$ as estimated extinction coefficient based on the contribution of Tyr and Trp residues, and disulfide bonds).

4.2. NMR Spectroscopy

For the structural analysis of TI–III and cTI–III, peptide (1 mg/mL) was dissolved in H_2O/D_2O (10:1, *v/v*) and the pH adjusted to pH 4–5. 1D ^1H spectra, two-dimensional total correlated spectroscopy (TOCSY) and nuclear Overhauser effect spectroscopy (NOESY) were acquired with a Bruker Avance 600 MHz NMR spectrometer (Billerica, USA) at a temperature of 298 K. Additional spectra for ^1H-^{13}C HSQC and ^1H-^{15}N HSQC in H_2O/D_2O (10:1, *v/v*) and exclusive correlation spectra (E.COSY) in D_2O were acquired. Spectra were referenced to an internal standard 2,2-dimethyl-2-silapentone-5-sulfonate (DSS) at 0 ppm. CYANA 3.97 was used to automatically calculate and refine structures based on distance restraints derived from the NOESY spectra [72], and torsion angles (ϕ and φ) generated using TALOS-N and Hα, Cα, Cβ, HN chemical shifts derived from NOESY, ^1H-^{13}C HSQC and ^1H-^{15}N HSQC spectra [73]. Several χ1 side-chain angle restraints were added based on E.COSY and NOESY data. A final set of structures was generated with CNS [74] using torsion angle dynamics, refinement and energy minimization in explicit solvent. Final structures were assessed for stereochemical quality using MolProbity [75].

4.3. Serum Stability

The serum stability assay was carried out as previously described [42] with some modifications. Briefly, tachyplesin variants were incubated in 25% (*v/v*) human serum diluted in phosphate-buffered saline (PBS) at a final concentration of 50 µM at 37 °C. Triplicates were collected at time 0 h and 24 h and the serum proteins were precipitated with acetonitrile (1:3 ratio) supplemented with 3% TFA. Samples were kept on ice for 10 min before centrifugation at 17,000× *g* for 10 min at 4 °C. The peptide containing supernatant of each sample was harvested and quantified using RP-HPLC (10 to 45% solvent B, 1%/min gradient). The percentage of peptide stability in human serum was calculated by comparing the area of the peptide peak obtained at 24 h to that at time 0 h. A linear peptide containing 18 amino acid residues (KGGGGSGQLIDSMANSFV) was included as a control susceptible to proteolytic degradation.

4.4. Hemolytic Studies

A small amount of blood was collected from three healthy human donors. The blood was immediately diluted in PBS and centrifuged 4–5 times for 1 min at 4000 rpm to wash and separate the human red blood cells (RBCs). RBCs suspension (0.25% (*v/v*) in PBS) was incubated with peptides with two-fold serial dilutions of the peptide (highest concentration tested was 128 µM, and the lowest was 0.25 µM) in a 96-well plate. Melittin, a membrane disruptive peptide, was used as control. The plates were incubated for 1 h at 37 °C. After incubation, the plates were centrifuged for 5 min at 1000 rpm to pellet any non-lysed RBCs. A total of 100 µL of the supernatant were transferred to a new 96-well plate [76]. The hemoglobin released into the supernatant from lysed cells was measured by absorbance at 415 nm using the Tecan infinite M1000Pro multiplate reader (Männedorf, Switzerland).

4.5. Cell Culture

Cells were grown in cell culture flasks and incubated in a humidified atmosphere (5% CO_2, 37 °C). The cancer cell lines HeLa and the control cell line HaCaT were grown in a DMEM medium supplemented with 1% (*v/v*) penicillin/streptomycin and 10% (*v/v*) fetal bovine serum (FBS). The cancer cell lines MM96L, HT144 and WM164 were grown in a RPMI medium supplemented with 1% (*v/v*) penicillin/streptavidin, 10% (*v/v*) FBS, 20 mM L-glutamine, and 10 mM sodium pyruvate. Cell cultures were maintained by dilution upon reaching confluence, each 48–72 h. More information about the cell lines [77] is available in Tables S2 and S3.

4.6. Cytotoxicity Assays

Cells were seeded into 96-well flat-bottom plates at 5×10^3 cells/well and incubated overnight. The medium was removed and replaced with 90 µL serum-free medium, 10 µL of 10x concentrated peptide solutions in PBS were added. PBS was added as blank and 0.1% (*v/v*) Triton X-100 was used to establish 100% of cell death. After 2 h incubation at 37 °C, 10 µL of filtered 0.05% (*w/v*) resazurin solution was added to each well [70]. Resazurin is converted to the pink and fluorescent compound resorufin by viable cells [78]. After incubation overnight, the fluorescence intensity (λ_{ex} = 565 nm and λ_{em} = 584 nm) was measured with the Tecan infinite M1000Pro multiplate reader (Männedorf, Switzerland).

The selectivity was calculated through the activity-toxicity index (ATI) [46]. ATI = MHC/MCC50, with MHC being the minimal concentration necessary to induce 10% or 50% cell death in human red blood cells and MCC50 the median of cytotoxic concentrations (CC50) of all tested melanoma cell lines.

4.7. Antimicrobial Studies

S. aureus ATCC 25923, *S. aureus* ATCC 6538, *E. coli* ATCC 25922, and *E. coli* DC2 CGSC 7139 were grown in Mueller Hinton Broth (Sigma Aldrich, St. Luis, USA). Bacterial cultures in the exponential growth phase were diluted to an OD600nm of 0.001 and seeded into 96-well plates. Peptides at different concentrations, starting at 64 µM and with two-fold serial dilutions (i.e., 64, 32, 16, 8, 4, 2, 1, 0.5, 0.25, 0.125 and 0.0625 µM), were incubated with cells [43]; 0.05% (*v/v*) resazurin was added to the cultures the next day and the conversion to resorufin was measured after 1 h of incubation at 37 °C using a 565 nm excitation and 584 nm emission wavelength, as above.

4.8. Biofilm Studies

S. aureus ATCC 6538 (1×10^6 cfu/mL) was cultured in Tryptic Soy Broth, containing 0.25% (*w/v*) glucose and incubated in 96-well microtiter flat-bottomed polystyrene plates for 24 h at 37 °C. Preformed biofilms were then washed with Mueller Hinton Broth to remove non-adherent cells. Two-fold serial dilutions of each peptide (highest concentration tested was 32 µM and lowest was 0.25 µM) were added to the biofilms for 4 h. Untreated 24 h preformed biofilms were used as a control. The metabolic activity of biofilm-embedded cells was determined using a resazurin reduction fluorometric assay as previously described [53].

4.9. Lipid Vesicle Preparation

Mixtures with synthetic lipids (POPC, POPS, POPG, Avanti Polar Lipids) or *E. coli* lipid extract (Avanti Polar Lipids, Alabaster. USA) were extruded in HEPES buffer (10 mM HEPES, 150 mM NaCl, pH 7.4) to produce lipid vesicles, as previously described [43,79]. LUVs (Ø ≤ 100 nm) were used in fluorescence spectroscopy assays and small unilamellar vesicles (SUVs, Ø ≤ 50 nm) for SPR.

4.10. Fluorescence Spectroscopy Assays

Fluorescence emission spectra (300–400 nm, excitation at 280 nm, slits 3/3 mm) of 12.5 µM peptide and ʟ-Trp in HEPES buffer (in quartz cuvettes, path length of 0.5 cm) were scanned upon titration with LUVs composed of various lipid compositions (up to 3 mM POPC, and up to 1.5 mM POPC/POPS (4:1) or POPC/POPG (4:1)) [80] using a FluoroMax-4 spectrofluorometer (Horiba, Kyoto, Japan). Integrated areas of the fluorescence emission spectra were corrected for fluorophore dilution and light dispersion due to titration with LUVs suspension; the blank was discounted.

4.11. Quenching of Tryptophan Fluorescence

The membrane in-depth location of the Trp residue within the peptides was followed using Acrylamide (Sigma Aldrich, St. Luis, USA) and 5- and 16-DS (Sigma Aldrich, St. Luis, USA). Quenching induced by acrylamide was monitored with 12.5 µM peptide in a HEPES buffer, in the presence of 1 mM of POPC, 0.1 mM POPC/POPS (4:1), 1 mM POPC/POPS (4:1), or 0.1 mM POPC/POPG (4:1) titrated with increasing concentration of acrylamide [81]. The fluorescence emission spectra were determined with an excitation wavelength of 290 nm to reduce the quencher/fluorophore light absorption ratio. The fluorescence emission spectra area was corrected for the inner filter effect [82] due to increased absorbance of acrylamide. Data points were analyzed using the Stern-Volmer representation (Equation (1)), the K_{SV} is determined from the slope.

$$\frac{I_0}{I} = 1 + K_{SV}[Q] \tag{1}$$

A total of 12.5 µM of TI, or of cTI, in HEPES buffer and 1 mM POPC/POPS (4:1) were titrated with increasing concentration of 5DS, or of 16DS. The fluorescence emission spectra area was corrected for the inner filter effect [82] due to increased absorbance of 5-,16DS. The effective concentration of 5DS and 16DS in the lipid bilayer was calculated: partition coefficients of 5DS and 16DS into fluid membranes are 89,000 and 9730, respectively [59,81]. The data points had a negative deviation to linearity and the K_{SV} was determined by fitting the data with the Lehrer equation (Equation (2)) [59,81].

$$\frac{I_0}{I} = \frac{1 + K_{SV}[Q]}{(1 + K_{SV}[Q])(1 - f_B) + f_B} \tag{2}$$

I_0 = fluorescent intensity in buffer
I = fluorescent intensity in presence of quencher
K_{SV} = Stern-Volmer constant
[Q] = quencher concentration
f_B = fraction of light accessible to the quencher = $\frac{I_{0,B}}{I_0}$
$I_{0,B}$ = fluorescent intensity of the accessible population of the quencher when [Q] = 0

The average distance of the Trp residue from the bilayer center (Å) was determined using the parallax method [62] following the equation,

$$z_{1F} = \frac{\left(\frac{1}{-\pi C} \ln \frac{F_1}{F_2} - L_{21}^2\right)}{2L_{21}} \tag{3}$$

$$z_{cF} = z_{1F} + L_{c1} \tag{4}$$

in which the z_{1F} is the distance between the Trp residue and the quencher moiety in 5DS, L_{21} is the distance between the quencher groups in 5 and 16DS (5 Å), F_1 is the fluorescence of 12.5 µM of TI, or of cTI, in the presence of 1 mM POPC/POPS (4:1) and 0.4 mM 5DS, F_2 is the fluorescence of 12.5 µM of TI, or of cTI, in the presence of 1 mM POPC/POPS (4:1) and 0.4 mM 16DS. C is the molar fraction of quencher within the total lipid concentration per unit area (assuming the surface of lipid to be 70 Å^2) [62]. z_{cF} is the distance between the Trp residue and the center of the bilayer, L_{c1} is the distance between the center of the bilayers and the quencher group of 5DS (15 Å).

4.12. Surface Plasmon Resonance (SPR)

SPR was used to investigate the affinity and binding kinetics of peptides to membranes of different compositions. The experiments were conducted with an L1 biosensor chip (GE Healthcare) at 25 °C using a BIAcore 3000 instrument (GE Healthcare, Chicago, USA) [83,84]. The HEPES buffer was used for sample preparation and as a running buffer. Lipid bilayers were immobilized onto L1 chip by injection of SUVs at a flow rate of 2 μL/min. Peptide samples with two-fold serial dilutions (the highest concentration tested was 64 μM and the lowest was 1 μM) were injected over the lipid bilayer at a flow rate of 5 μL/min. Association of the peptides onto the lipid bilayer was followed for 180 s and the dissociation from the lipid for 600 s. The BIAeval software was used to analyze the sensorgrams. The response units (RU) were normalized to the peptide-to-lipid ratio (P/L); the P/L obtained at a fixed time point at the end of the association curve (at 170 s), at which the response has reached a plateau and the binding is close to equilibrium, was used to compare the affinity of the peptides to the different lipid systems [84].

4.13. Vesicle Leakage Assay

LUVs (Ø ≤ 100 nm) were prepared with the HEPES buffer containing 40 mM of the fluorescent 5-carboxyfluorescein (CF, Sigma Aldrich, St. Luis, USA). LUVs filled with CF at self-quenching concentrations were separated from the non-encapsulated dye on a Sephadex G-50 column equilibrated with HEPES buffer [85]. The concentration of CF-LUVs was determined through a calibration curve prepared from the original lipid mixture using Stewart's assay (absorbance at 485 nm) [86]. Peptides were incubated (25 °C, 20 min) at two-fold dilutions (starting at 10 μM) with LUVs (5 μM) in the HEPES buffer in a 96-well flat-bottom black optiplates (Perkin Elmer, Waltham, USA). Vesicles and peptides were incubated for 20 min in dark and the release of CF was measured in a Tecan infinite M1000Pro multiplate reader using 490 nm as the excitation and 513 nm as emission wavelengths.

4.14. Statistical Analysis

Values (mean or fit ± SEM) were analyzed in GraphPad Prism 7 to test for significant differences in the cytotoxic activity of the peptides. The multiple t-test function with the Holm–Sidak method was applied when indicated. P values below 0.05 were considered to be significant.

Supplementary Materials: Supplementary materials can be found at http://www.mdpi.com/1422-0067/20/17/4184/s1.

Author Contributions: F.V. conceptualized this study with supervision and input from S.T.H. and N.L. F.V., S.A.D., and P.J.H. conducted experimental work: F.V. folded and purified the peptides, conducted toxicity assays, model membranes, surface plasmon resonance and fluorescence spectroscopy assays; S.A.D. conducted the biofilm assay and helped F.V. with the antimicrobial assays; P.J.H. performed NMR spectroscopy and structural analysis of the peptides; Y.-H.H. established the serum stability assay. F.V. analyzed and interpreted data with assistance from S.T.H. and expert contributions from all the co-authors. S.T.H., N.L., P.J.H., D.J.C., Y.-H.H. and A.S.V. revised and edited the manuscript. S.T.H. and D.J.C. provided the resources and acquired the funding.

Funding: This project was funded by a National Health Medical Research Council (NHMRC) project grant (APP1084965). F.V. was supported by the UQ Research Scholarship, S.T.H. is an Australian Research Council (ARC) Future Fellow (FT150100398), D.J.C. is an ARC Australian Laureate Fellow (FL150100146). Marie Skłodowska-Curie Research and Innovation Staff Exchange grant (RISE; call: H2020-MSCA-RISE-2014, grant agreement 644167) funded secondments of S.A.D. and of A.S.V. to the University of Queensland. The Translational Research Institute is supported by a grant from the Australian Government.

Acknowledgments: Many thanks to QUEDDI, the Matt Cooper lab and the Helmut Schaider lab for the cell lines HaCaT, HT144 and WM164 respectively. The authors thank Olivier Cheneval and Joachim Weidmann (IMB, UQ) for their assistance with peptide synthesis.

Conflicts of Interest: The authors declare no conflict of interest. The funders had no role in the design of the study; in the collection, analyses, or interpretation of data; in the writing of the manuscript, or in the decision to publish the results.

Abbreviations

1D	One-dimensional
2-CTC	2-chlorotrityl resin
3D	Three-dimensional
ACN	Acetonitrile
ATI	Activity/toxicity index
ATTC	American Type Culture Collection

BMRB	Biological magnetic resonance data bank
CC	Cytotoxic concentration
CF	5-carboxyfluorescein
cTI	Cyclic Tachyplesin I
cTII	Cyclic Tachyplesin II
cTIII	Cyclic Tachyplesin III
cTI–cTIII	Cyclic Tachyplesin I, II and III
DS	Doxyl-stearic acid
E.COSY	Exclusive correlation spectra
ESI-MS	Electrospray ionization mass spectroscopy
FBS	Fetal bovine serum
Fmoc	9-fluorenylmethoxycarbonyl
HC50	Hemolytic concentration necessary for 50% lysis of RBCs
HDP	Host defense peptide
K_{SV}	Stern – Volmer constant
LPS	Lipopolysaccharide
LUV	Large unilamellar vesicle
MCC	Minimal cytotoxic concentration
MHC	Minimal hemolytic concentration
MIC	Minimal inhibitory concentration
NMR	Nuclear magnetic resonance
NOESY	Nuclear Overhauser effect spectroscopy
PBS	Phosphate buffered saline
PDB	Protein data bank
PC	Phosphatidylcholine
PS	Phosphatidylserine
PG	Phosphatidylglycerol
P/L	Peptide-to-lipid ratio
POPC	1-palmitoyl-2-oleoyl-glycero-3-phosphocholine
POPS	1-palmitoyl-2-oleoyl-sn-glycero-3-phospho-L-serine
POPG	1-palmitoyl-2-oleoyl-sn-glycero-3-phospho-(1′-rac-glycerol)
RBCs	Red blood cells
RP-HPLC	Reverse-phase high-performance liquid chromatography
RT	Retention time
RU	Response unit
SEM	Standard error of mean
SPPS	Solid phase peptide synthesis
SPR	Surface plasmon resonance
SUV	Small unilamellar vesicle
TI	Tachyplesin I
TII	Tachyplesin II
TIII	Tachyplesin III
TI–TIII	Tachyplesin I, II and III
TFA	Trifluoroacetic acid
TOCSY	Total correlated spectroscopy

References

1. Edwards, I.A.; Elliott, A.G.; Kavanagh, A.M.; Zuegg, J.; Blaskovich, M.A.T.; Cooper, M.A. Contribution of Amphipathicity and Hydrophobicity to the Antimicrobial Activity and Cytotoxicity of β-Hairpin Peptides. *ACS Infect. Dis.* **2016**, *2*, 442–450. [CrossRef] [PubMed]
2. Miyata, T.; Tokunaga, F.; Yoneya, T.; Yoshikawa, K.; Iwanaga, S.; Niwa, M.; Takao, T.; Shimonishi, Y. Antimicrobial peptides, isolated from horseshoe crab hemocytes, tachyplesin II, and polyphemusins I and II: chemical structures and biological activity. *J. Biochem.* **1989**, *106*, 663–668. [CrossRef] [PubMed]

3.   Nakamura, T.; Furunaka, H.; Miyata, T.; Tokunaga, F.; Muta, T.; Iwanaga, S.; Niwa, M.; Takao, T.; Shimonishi, Y. Tachyplesin, a class of antimicrobial peptide from the hemocytes of the horseshoe crab (Tachypleus tridentatus). Isolation and chemical structure. *J. Biol. Chem.* **1988**, *263*, 16709–16713. [PubMed]

4.   Ohta, M.; Ito, H.; Masuda, K.; Tanaka, S.; Arakawa, Y.; Wacharotayankun, R.; Kato, N. Mechanisms of antibacterial action of tachyplesins and polyphemusins, a group of antimicrobial peptides isolated from horseshoe crab hemocytes. *Antimicrob. Agents Chemother.* **1992**, *36*, 1460–1465. [CrossRef] [PubMed]

5.   Buri, M.V.; Torquato, H.F.V.; Barros, C.C.; Ide, J.S.; Miranda, A.; Paredes-Gamero, E.J. Comparison of Cytotoxic Activity in Leukemic Lineages Reveals Important Features of beta-Hairpin Antimicrobial Peptides. *J. Cell. Biochem.* **2017**, *118*, 1764–1773. [CrossRef] [PubMed]

6.   Chen, J.; Xu, X.M.; Underhill, C.B.; Yang, S.; Wang, L.; Chen, Y.; Hong, S.; Creswell, K.; Zhang, L. Tachyplesin activates the classic complement pathway to kill tumor cells. *Cancer Res.* **2005**, *65*, 4614–4622. [CrossRef] [PubMed]

7.   Ding, H.; Jin, G.; Zhang, L.; Dai, J.; Dang, J.; Han, Y. Effects of tachyplesin I on human U251 glioma stem cells. *Mol. Med. Rep.* **2015**, *11*, 2953–2958. [CrossRef] [PubMed]

8.   Kuzmin, D.V.; Emel'yanova, A.A.; Kalashnikova, M.B.; Panteleev, P.V.; Ovchinnikova, T.V. In Vitro Study of Antitumor Effect of Antimicrobial Peptide Tachyplesin I in Combination with Cisplatin. *Bull. Exp. Biol. Med.* **2018**, *165*, 220–224. [CrossRef]

9.   Li, Q.F.; Ou Yang, G.L.; Li, C.Y.; Hong, S.G. Effects of tachyplesin on the morphology and ultrastructure of human gastric carcinoma cell line BGC-823. *World J. Gastroenterol.* **2000**, *6*, 676–680. [CrossRef]

10.   Li, X.; Dai, J.; Tang, Y.; Li, L.; Jin, G. Quantitative Proteomic Profiling of Tachyplesin I Targets in U251 Gliomaspheres. *Mar. Drugs* **2017**, *15*, 20. [CrossRef]

11.   Ouyang, G.L.; Li, Q.F.; Peng, X.X.; Liu, Q.R.; Hong, S.G. Effects of tachyplesin on proliferation and differentiation of human hepatocellular carcinoma SMMC-7721 cells. *World J. Gastroenterol.* **2002**, *8*, 1053–1058. [CrossRef]

12.   Paredes-Gamero, E.J.; Martins, M.N.; Cappabianco, F.A.; Ide, J.S.; Miranda, A. Characterization of dual effects induced by antimicrobial peptides: regulated cell death or membrane disruption. *Biochim. Biophys. Acta* **2012**, *1820*, 1062–1072. [CrossRef]

13.   Rothan, H.A.; Ambikabothy, J.; Ramasamy, T.S.; Rashid, N.N.; Yusof, R. A Preliminary Study in Search of Potential Peptide Candidates for a Combinational Therapy with Cancer Chemotherapy Drug. *Int. J. Pept. Res. Ther.* **2017**, *25*, 115–122. [CrossRef]

14.   Shi, S.L.; Wang, Y.Y.; Liang, Y.; Li, Q.F. Effects of tachyplesin and n-sodium butyrate on proliferation and gene expression of human gastric adenocarcinoma cell line BGC-823. *World J. Gastroenterol.* **2006**, *12*, 1694–1698. [CrossRef]

15.   Zhang, H.T.; Wu, J.; Zhang, H.F.; Zhu, Q.F. Efflux of potassium ion is an important reason of HL-60 cells apoptosis induced by tachyplesin. *Acta Pharmacol. Sin.* **2006**, *27*, 1367–1374. [CrossRef]

16.   Muta, T.; Fujimoto, T.; Nakajima, H.; Iwanaga, S. Tachyplesins Isolated from Hemocytes of Southeast Asian Horseshoe Crabs (Carcinoscorpius rotundicauda and Tachypleus gigas): Identification of a New Tachyplesin, Tachyplesin III, and a Processing Intermediate of Its Precursor1. *J. Biochem.* **1990**, *108*, 261–266. [CrossRef]

17.   Zasloff, M. Antimicrobial peptides of multicellular organisms. *Nature* **2002**, *415*, 389–395. [CrossRef]

18.   Teixeira, V.; Feio, M.J.; Bastos, M. Role of lipids in the interaction of antimicrobial peptides with membranes. *Prog. Lipid Res.* **2012**, *51*, 149–177. [CrossRef]

19.   Yeaman, M.R.; Yount, N.Y. Mechanisms of Antimicrobial Peptide Action and Resistance. *Pharmacol. Rev.* **2003**, *55*, 27–55. [CrossRef]

20.   Utsugi, T.; Schroit, A.J.; Connor, J.; Bucana, C.D.; Fidler, I.J. Elevated expression of phosphatidylserine in the outer membrane leaflet of human tumor cells and recognition by activated human blood monocytes. *Cancer Res.* **1991**, *51*, 3062–3066.

21.   Ran, S.; Downes, A.; Thorpe, P.E. Increased Exposure of Anionic Phospholipids on the Surface of Tumor Blood Vessels. *Cancer Res.* **2002**, *62*, 6132.

22.   Zwaal, R.F.; Comfurius, P.; Bevers, E.M. Surface exposure of phosphatidylserine in pathological cells. *Cell. Mol. Life Sci.* **2005**, *62*, 971–988. [CrossRef]

23.   Riedl, S.; Rinner, B.; Asslaber, M.; Schaider, H.; Walzer, S.; Novak, A.; Lohner, K.; Zweytick, D. In search of a novel target - phosphatidylserine exposed by non-apoptotic tumor cells and metastases of malignancies with poor treatment efficacy. *Biochim. Biophys. Acta* **2011**, *1808*, 2638–2645. [CrossRef]

24. Iwasaki, T.; Ishibashi, J.; Tanaka, H.; Sato, M.; Asaoka, A.; Taylor, D.; Yamakawa, M. Selective cancer cell cytotoxicity of enantiomeric 9-mer peptides derived from beetle defensins depends on negatively charged phosphatidylserine on the cell surface. *Peptides* **2009**, *30*, 660–668. [CrossRef]

25. Leite Natália, B.; Aufderhorst-Roberts, A.; Palma Mario, S.; Connell Simon, D.; Neto João, R.; Beales Paul, A. PE and PS Lipids Synergistically Enhance Membrane Poration by a Peptide with Anticancer Properties. *Biophys. J.* **2015**, *109*, 936–947. [CrossRef]

26. Matsuzaki, K.; Sugishita, K.; Fujii, N.; Miyajima, K. Molecular Basis for Membrane Selectivity of an Antimicrobial Peptide, Magainin 2. *Biochemistry* **1995**, *34*, 3423–3429. [CrossRef]

27. Hoskin, D.W.; Ramamoorthy, A. Studies on anticancer activities of antimicrobial peptides. *Biochim. Biophys. Acta* **2008**, *1778*, 357–375. [CrossRef]

28. Harris, F.; Dennison, S.R.; Singh, J.; Phoenix, D.A. On the selectivity and efficacy of defense peptides with respect to cancer cells. *Med. Res. Rev.* **2013**, *33*, 190–234. [CrossRef]

29. Gaspar, D.; Veiga, A.S.; Castanho, M.A. From antimicrobial to anticancer peptides. A review. *Front. Microbiol.* **2013**, *4*, 294. [CrossRef]

30. Matsuzaki, K. Why and how are peptide–lipid interactions utilized for self-defense? Magainins and tachyplesins as archetypes. *Biochim. Biophys. Acta* **1999**, *1462*, 1–10. [CrossRef]

31. Matsuzaki, K.; Fukui, M.; Fujii, N.; Miyajima, K. Interactions of an antimicrobial peptide, tachyplesin I, with lipid membranes. *Biochim. Biophys. Acta* **1991**, *1070*, 259–264. [CrossRef]

32. Iwanaga, S.; Muta, T.; Shigenaga, T.; Seki, N.; Kawano, K.; Katsu, T.; Kawabata, S. Structure-function relationships of tachyplesins and their analogues. *Ciba Found. Symp.* **1994**, *186*, 160–174.

33. Imura, Y.; Nishida, M.; Ogawa, Y.; Takakura, Y.; Matsuzaki, K. Action mechanism of tachyplesin I and effects of PEGylation. *Biochim. Biophys. Acta* **2007**, *1768*, 1160–1169. [CrossRef]

34. Matsuzaki, K.; Yoneyama, S.; Fujii, N.; Miyajima, K.; Yamada, K.; Kirino, Y.; Anzai, K. Membrane permeabilization mechanisms of a cyclic antimicrobial peptide, tachyplesin I, and its linear analog. *Biochemistry* **1997**, *36*, 9799–9806. [CrossRef]

35. Kuzmin, D.V.; Emelianova, A.A.; Kalashnikova, M.B.; Panteleev, P.V.; Balandin, S.V.; Serebrovskaya, E.O.; Belogurova-Ovchinnikova, O.Y.; Ovchinnikova, T.V. Comparative in vitro study on cytotoxicity of recombinant beta-hairpin peptides. *Chem. Biol. Drug Des.* **2017**, *91*, 294–303. [CrossRef]

36. Laederach, A.; Andreotti, A.H.; Fulton, D.B. Solution and micelle-bound structures of tachyplesin I and its active aromatic linear derivatives. *Biochemistry* **2002**, *41*, 12359–12368. [CrossRef]

37. Kushibiki, T.; Kamiya, M.; Aizawa, T.; Kumaki, Y.; Kikukawa, T.; Mizuguchi, M.; Demura, M.; Kawabata, S.; Kawano, K. Interaction between tachyplesin I, an antimicrobial peptide derived from horseshoe crab, and lipopolysaccharide. *Biochim. Biophys. Acta* **2014**, *1844*, 527–534. [CrossRef]

38. Kawano, K.; Yoneya, T.; Miyata, T.; Yoshikawa, K.; Tokunaga, F.; Terada, Y.; Iwanaga, S. Antimicrobial peptide, tachyplesin I, isolated from hemocytes of the horseshoe crab (Tachypleus tridentatus). NMR determination of the beta-sheet structure. *J. Biol. Chem.* **1990**, *265*, 15365–15367.

39. Mielke, S.P.; Krishnan, V.V. Characterization of protein secondary structure from NMR chemical shifts. *Prog. Nucl. Magn. Reson. Spectrosc.* **2009**, *54*, 141–165. [CrossRef]

40. Hutchinson, E.G.; Thornton, J.M. PROMOTIF–a program to identify and analyze structural motifs in proteins. *Protein Sci.* **1996**, *5*, 212–220. [CrossRef]

41. Wishart, D.S.; Bigam, C.G.; Yao, J.; Abildgaard, F.; Dyson, H.J.; Oldfield, E.; Markley, J.L.; Sykes, B.D. 1H, 13C and 15N chemical shift referencing in biomolecular NMR. *J. Biomol. NMR* **1995**, *6*, 135–140. [CrossRef]

42. Chan, L.Y.; Zhang, V.M.; Huang, Y.H.; Waters, N.C.; Bansal, P.S.; Craik, D.J.; Daly, N.L. Cyclization of the antimicrobial peptide gomesin with native chemical ligation: influences on stability and bioactivity. *Chembiochem.* **2013**, *14*, 617–624. [CrossRef]

43. Troeira Henriques, S.; Lawrence, N.; Chaousis, S.; Ravipati, A.S.; Cheneval, O.; Benfield, A.H.; Elliott, A.G.; Kavanagh, A.M.; Cooper, M.A.; Chan, L.Y.; et al. Redesigned Spider Peptide with Improved Antimicrobial and Anticancer Properties. *ACS Chem. Biol.* **2017**, *12*, 2324–2334. [CrossRef]

44. Kondejewski, L.H.; Farmer, S.W.; Wishart, D.S.; Kay, C.M.; Hancock, R.E.; Hodges, R.S. Modulation of structure and antibacterial and hemolytic activity by ring size in cyclic gramicidin S analogs. *J. Biol. Chem.* **1996**, *271*, 25261–25268. [CrossRef]

45. Tam, J.P.; Lu, Y.A.; Yang, J.L. Marked increase in membranolytic selectivity of novel cyclic tachyplesins constrained with an antiparallel two-beta strand cystine knot framework. *Biochem. Biophys. Res. Commun.* **2000**, *267*, 783–790. [CrossRef]

46. Edwards, I.A.; Elliott, A.G.; Kavanagh, A.M.; Blaskovich, M.A.T.; Cooper, M.A. Structure-Activity and -Toxicity Relationships of the Antimicrobial Peptide Tachyplesin-1. *ACS Infect. Dis.* **2017**, *3*, 917–926. [CrossRef]

47. Marggraf, M.B.; Panteleev, P.V.; Emelianova, A.A.; Sorokin, M.I.; Bolosov, I.A.; Buzdin, A.A.; Kuzmin, D.V.; Ovchinnikova, T.V. Cytotoxic Potential of the Novel Horseshoe Crab Peptide Polyphemusin III. *Mar. Drugs* **2018**, *16*, 466. [CrossRef]

48. Hilchie, A.L.; Hoskin, D.W.; Power Coombs, M.R. Anticancer Activities of Natural and Synthetic Peptides. In *Antimicrobial Peptides: Basics for Clinical Application*; Matsuzaki, K., Ed.; Springer Singapore: Singapore, 2019; pp. 131–147.

49. Kuzmin, D.V.; Emelianova, A.A.; Kalashnikova, M.B.; Panteleev, P.V.; Ovchinnikova, T.V. Effect of N- and C-Terminal Modifications on Cytotoxic Properties of Antimicrobial Peptide Tachyplesin I. *Bull. Exp. Biol. Med.* **2017**, *162*, 754–757. [CrossRef]

50. Rivera, M.; Bertasso, A.; McCaffrey, C.; Georgopapadakou, N.H. Porins and lipopolysaccharide of Escherichia coli ATCC 25922 and isogenic rough mutants. *FEMS Microbiol. Lett.* **1993**, *108*, 183–187. [CrossRef]

51. Treangen, T.J.; Maybank, R.A.; Enke, S.; Friss, M.B.; Diviak, L.F.; Karaolis, D.K.; Koren, S.; Ondov, B.; Phillippy, A.M.; Bergman, N.H.; et al. Complete Genome Sequence of the Quality Control Strain Staphylococcus aureus subsp. aureus ATCC 25923. *Genome Announc.* **2014**, *2*, e01110-14. [CrossRef]

52. Clark, D. Novel antibiotic hypersensitive mutants of Escherichia coli genetic mapping and chemical characterization. *FEMS Microbiol. Lett.* **1984**, *21*, 189–195. [CrossRef]

53. Pinto, S.N.; Dias, S.A.; Cruz, A.F.; Mil-Homens, D.; Fernandes, F.; Valle, J.; Andreu, D.; Prieto, M.; Castanho, M.; Coutinho, A.; et al. The mechanism of action of pepR, a viral-derived peptide, against Staphylococcus aureus biofilms. *J. Antimicrob. Chemother.* **2019**, dkz223. [CrossRef]

54. Watson, H. Biological membranes. *Essays Biochem.* **2015**, *59*, 43–69. [CrossRef]

55. Ingolfsson, H.I.; Melo, M.N.; van Eerden, F.J.; Arnarez, C.; Lopez, C.A.; Wassenaar, T.A.; Periole, X.; de Vries, A.H.; Tieleman, D.P.; Marrink, S.J. Lipid organization of the plasma membrane. *J. Am. Chem. Soc.* **2014**, *136*, 14554–14559. [CrossRef]

56. Van Meer, G.; Voelker, D.R.; Feigenson, G.W. Membrane lipids: where they are and how they behave. *Nat. Rev. Mol. Cell Biol.* **2008**, *9*, 112–124. [CrossRef]

57. Eftink, M.R. Fluorescence Quenching Reactions. In *Biophysical and Biochemical Aspects of Fluorescence Spectroscopy*; Dewey, T.G., Ed.; Springer US: Boston, MA, USA, 1991; pp. 1–41.

58. Ghisaidoobe, A.B.; Chung, S.J. Intrinsic tryptophan fluorescence in the detection and analysis of proteins: a focus on Forster resonance energy transfer techniques. *Int. J. Mol. Sci.* **2014**, *15*, 22518–22538. [CrossRef]

59. Santos, N.C.; Prieto, M.; Castanho, M.A.R.B. Quantifying molecular partition into model systems of biomembranes: an emphasis on optical spectroscopic methods. *Biochim. Biophys. Acta* **2003**, *1612*, 123–135. [CrossRef]

60. Fernandes, M.X.; García de la Torre, J.; Castanho, M.A.R.B. Joint determination by Brownian dynamics and fluorescence quenching of the in-depth location profile of biomolecules in membranes. *Anal. Biochem.* **2002**, *307*, 1–12. [CrossRef]

61. Lakowicz, J.R. *Topics in Fluorescence Spectroscopy Principles*; Springer US: Boston, MA, USA, 2002.

62. Chattopadhyay, A.; London, E. Parallax method for direct measurement of membrane penetration depth utilizing fluorescence quenching by spin-labeled phospholipids. *Biochemistry* **1987**, *26*, 39–45. [CrossRef]

63. Kyrychenko, A.; Ladokhin, A.S. Molecular Dynamics Simulations of Depth Distribution of Spin-Labeled Phospholipids within Lipid Bilayer. *J. Phys. Chem. B* **2013**, *117*, 5875–5885. [CrossRef]

64. Batoni, G.; Maisetta, G.; Esin, S. Antimicrobial peptides and their interaction with biofilms of medically relevant bacteria. *Biochim. Biophys. Acta* **2016**, *1858*, 1044–1060. [CrossRef]

65. Grassi, L.; Maisetta, G.; Esin, S.; Batoni, G. Combination Strategies to Enhance the Efficacy of Antimicrobial Peptides against Bacterial Biofilms. *Front. Microbiol.* **2017**, *8*, 2409. [CrossRef]

66. Minardi, D.; Ghiselli, R.; Cirioni, O.; Giacometti, A.; Kamysz, W.; Orlando, F.; Silvestri, C.; Parri, G.; Kamysz, E.; Scalise, G.; et al. The antimicrobial peptide Tachyplesin III coated alone and in combination with intraperitoneal piperacillin-tazobactam prevents ureteral stent Pseudomonas infection in a rat subcutaneous pouch model. *Peptides* **2007**, *28*, 2293–2298. [CrossRef]

67. Yu, K.; Lo, J.C.; Yan, M.; Yang, X.; Brooks, D.E.; Hancock, R.E.; Lange, D.; Kizhakkedathu, J.N. Anti-adhesive antimicrobial peptide coating prevents catheter associated infection in a mouse urinary infection model. *Biomaterials* **2017**, *116*, 69–81. [CrossRef]

68. Oishi, O.; Yamashita, S.; Nishimoto, E.; Lee, S.; Sugihara, G.; Ohno, M. Conformations and orientations of aromatic amino acid residues of tachyplesin I in phospholipid membranes. *Biochemistry* **1997**, *36*, 4352–4359. [CrossRef]

69. Doherty, T.; Waring, A.J.; Hong, M. Dynamic structure of disulfide-removed linear analogs of tachyplesin-I in the lipid bilayer from solid-state NMR. *Biochemistry* **2008**, *47*, 1105–1116. [CrossRef]

70. Torcato, I.M.; Huang, Y.H.; Franquelim, H.G.; Gaspar, D.; Craik, D.J.; Castanho, M.A.; Troeira Henriques, S. Design and characterization of novel antimicrobial peptides, R-BP100 and RW-BP100, with activity against Gram-negative and Gram-positive bacteria. *Biochim. Biophys. Acta* **2013**, *1828*, 944–955. [CrossRef]

71. Cheneval, O.; Schroeder, C.I.; Durek, T.; Walsh, P.; Huang, Y.H.; Liras, S.; Price, D.A.; Craik, D.J. Fmoc-based synthesis of disulfide-rich cyclic peptides. *J. Org. Chem.* **2014**, *79*, 5538–5544. [CrossRef]

72. Güntert, P. Automated NMR Structure Calculation with CYANA. In *Protein NMR Techniques*; Downing, A.K., Ed.; Humana Press: Totowa, NJ, USA, 2004; pp. 353–378.

73. Shen, Y.; Bax, A. Protein backbone and sidechain torsion angles predicted from NMR chemical shifts using artificial neural networks. *J. Biomol. NMR* **2013**, *56*, 227–241. [CrossRef]

74. Brunger, A.T. Version 1.2 of the Crystallography and NMR system. *Nat. Protoc.* **2007**, *2*, 2728. [CrossRef]

75. Chen, V.B.; Arendall, W.B., 3rd; Headd, J.J.; Keedy, D.A.; Immormino, R.M.; Kapral, G.J.; Murray, L.W.; Richardson, J.S.; Richardson, D.C. MolProbity: all-atom structure validation for macromolecular crystallography. *Acta Crystallogr. D Biol. Crystallogr.* **2010**, *66*, 12–21. [CrossRef]

76. Huang, Y.H.; Colgrave, M.L.; Clark, R.J.; Kotze, A.C.; Craik, D.J. Lysine-scanning mutagenesis reveals an amendable face of the cyclotide kalata B1 for the optimization of nematocidal activity. *J. Biol. Chem.* **2010**, *285*, 10797–10805. [CrossRef]

77. Huang, Y.; Liu, Y.; Zheng, C.; Shen, C. Investigation of Cross-Contamination and Misidentification of 278 Widely Used Tumor Cell Lines. *PLoS ONE* **2017**, *12*, e0170384. [CrossRef]

78. Riss, T.L.; Moravec, R.A.; Niles, A.L.; Duellman, S.; Benink, H.A.; Worzella, T.J.; Minor, L. Cell Viability Assays. In *Assay Guidance Manual*; Sittampalam, G.S., Coussens, N.P., Brimacombe, K., Grossman, A., Arkin, M., Auld, D., Austin, C., Baell, J., Bejcek, B., Chung, T.D.Y., et al., Eds.; Eli Lilly & Company and the National Center for Advancing Translational Sciences: Bethesda, MD, USA, 2016.

79. Mayer, L.D.; Hope, M.J.; Cullis, P.R. Vesicles of variable sizes produced by a rapid extrusion procedure. *BBA-Biomembranes* **1986**, *858*, 161–168. [CrossRef]

80. Henriques, S.T.; Pattenden, L.K.; Aguilar, M.-I.; Castanho, M.A.R.B. The Toxicity of Prion Protein Fragment PrP(106–126) is Not Mediated by Membrane Permeabilization as Shown by a M112W Substitution. *Biochemistry* **2009**, *48*, 4198–4208. [CrossRef]

81. Henriques, S.T.; Castanho, M.A.R.B. Environmental factors that enhance the action of the cell penetrating peptide pep-1: A spectroscopic study using lipidic vesicles. *Biochim. Biophys. Acta* **2005**, *1669*, 75–86. [CrossRef]

82. Caputo, G.A.; London, E. Using a novel dual fluorescence quenching assay for measurement of tryptophan depth within lipid bilayers to determine hydrophobic alpha-helix locations within membranes. *Biochemistry* **2003**, *42*, 3265–3274. [CrossRef]

83. Henriques, S.T.; Huang, Y.H.; Castanho, M.A.; Bagatolli, L.A.; Sonza, S.; Tachedjian, G.; Daly, N.L.; Craik, D.J. Phosphatidylethanolamine binding is a conserved feature of cyclotide-membrane interactions. *J. Biol. Chem.* **2012**, *287*, 33629–33643. [CrossRef]

84. Henriques, S.T.; Huang, Y.-H.; Rosengren, K.J.; Franquelim, H.G.; Carvalho, F.A.; Johnson, A.; Sonza, S.; Tachedjian, G.; Castanho, M.A.R.B.; Daly, N.L.; et al. Decoding the membrane activity of the cyclotide kalata B1: the importance of phosphatidylethanolamine phospholipids and lipid organization on hemolytic and anti- HIV activities. *J. Biol. Chem.* **2011**, *286*, 24231. [CrossRef]

85. Huang, Y.-H.; Colgrave, M.L.; Daly, N.L.; Keleshian, A.; Martinac, B.; Craik, D.J. Biological Activity of the Prototypic Cyclotide Kalata B1 Is Modulated by the Formation of Multimeric Pores. *J. Biol. Chem.* **2009**, *284*, 20699–20707. [CrossRef]
86. Stewart, J.C. Colorimetric determination of phospholipids with ammonium ferrothiocyanate. *Anal. Biochem.* **1980**, *104*, 10–14. [CrossRef]

International Journal of
Molecular Sciences

Article

Pisum sativum Defensin 1 Eradicates Mouse Metastatic Lung Nodules from B16F10 Melanoma Cells

Virginia Sara Grancieri do Amaral [1,2], Stephanie Alexia Cristina Silva Santos [1], Paula Cavalcante de Andrade [1,2], Jenifer Nowatzki [1,2], Nilton Silva Júnior [1], Luciano Neves de Medeiros [1], Lycia Brito Gitirana [3], Pedro Geraldo Pascutti [1], Vitor H. Almeida [2], Robson Q. Monteiro [2] and Eleonora Kurtenbach [1,*]

[1] Instituto de Biofísica Carlos Chagas Filho, Universidade Federal do Rio de Janeiro, Rio de Janeiro, RJ 21941-902, Brasil; vi_farma@biof.ufrj.br (V.S.G.d.A.); stephaniealexia@biof.ufrj.br (S.A.C.S.S.); capaulinha1@gmail.com (P.C.d.A.); jeninowatzki@gmail.com (J.N.); nlju1@biof.ufrj.br (N.S.J.); lnmedeiros@gmail.com (L.N.d.M.); pascutti@biof.ufrj.br (P.G.P.)
[2] Instituto de Bioquímica Médica Leopoldo de Meis, Universidade Federal do Rio de Janeiro, Rio de Janeiro, RJ 21941-902, Brasil; vhluna@bioqmed.ufrj.br (V.H.A.); robsonqm@bioqmed.ufrj.br (R.Q.M.)
[3] Instituto de Ciências Biomédicas, Universidade Federal do Rio de Janeiro, Rio de Janeiro, RJ 21941-902, Brasil; lyciabg@histo.ufrj.br
* Correspondence: kurten@biof.ufrj.br

Received: 21 January 2020; Accepted: 1 April 2020; Published: 11 April 2020

Abstract: *Psd*1 is a pea plant defensin which can be actively expressed in *Pichia pastoris* and shows broad antifungal activity. This activity is dependent on fungal membrane glucosylceramide (GlcCer), which is also important for its internalization, nuclear localization, and endoreduplication. Certain cancer cells present a lipid metabolism imbalance resulting in the overexpression of GlcCer in their membrane. In this work, in vitroassays using B16F10 cells showed that labeled fluorescein isothiocyanate FITC-*Psd*1 internalized into live cultured cells and targeted the nucleus, which underwent fragmentation, exhibiting approximately 60% of cells in the sub-G0/G1 stage. This phenomenon was dependent on GlcCer, and the participation of cyclin-F was suggested. In a murine lung metastatic melanoma model, intravenous injection of *Psd*1 together with B16F10 cells drastically reduced the number of nodules at concentrations above 0.5 mg/kg. Additionally, the administration of 1 mg/kg *Psd*1 decreased the number of lung inflammatory cells to near zero without weight loss, unlike animals that received melanoma cells only. It is worth noting that 1 mg/kg *Psd*1 alone did not provoke inflammation in lung tissue or weight or vital signal losses over 21 days, inferring no whole animal cytotoxicity. These results suggest that *Psd*1 could be a promising prototype for human lung anti-metastatic melanoma therapy.

Keywords: metastasis model of B16F10 melanoma; *Pisum sativum* defensin 1 (*Psd*1); anti-metastatic activity; glucosylceramide (GlcCer); cyclin F

1. Introduction

Plant defensins (PDs) belong to the superfamily of cationic rich antimicrobial peptides (AMPs) [1] and are produced by plants as part of their innate immunity [2]. To date, over 100 known plant defensin primary sequences have been described [3]. They display low primary sequence homology, apart from the cysteine residues that are common, two glycine residues (positions 12 and 33), and aromatic residues (positions 10 and 41) related to *Psd*1 [4,5]. Nevertheless, the tertiary structures of PDs show a common cysteine stabilized αβ-fold (CSαβ-fold), characterized by one α-helix and three antiparallel β-sheets [1,6,7]. Because of their amphipathic characteristics, their ability to kill microorganisms can involve nonspecific electrostatic and hydrophobic interactions with positive plasmatic membranes [8,9]. Additionally, membrane-specific targets for some plant defensins have been described, such as phosphatidic acid, phosphoinositides, mannosylinositolphosphoryl-containing sphingolipids, and glycosphingolipids (GSLs) [5,8,10–14].

Pisum sativum defensin 1 (*Psd*1) is a 46-amino acid residue plant defensin isolated from pea seeds that presents well-documented antimicrobial activity against several fungal species [4,7,15–17]. The mechanism of action proposed for *Psd*1 antifungal activity includes its interaction with specific cell wall/membrane lipid targets, such as C8-desaturated and C9-methylated glucosylceramide (GlcCer), a fungal exclusive GSL, and ergosterol [8,11,16]. NMR spectroscopy analysis has demonstrated that *Psd*1 activity is intimately linked to its structure, with glycine 12 anchored in the first loop (residues 7 to 17) and histidine 36 anchored in turn 3, which are important amino acid residues for interaction with the fungal plasmatic membrane [15]. Nevertheless, *Psd*1 does not show good affinity to cholesterol-enriched lipid bilayers, such as those found in mammalian cell membranes, which suggests its high human therapeutic potential [5,16].

Previous results from our group using a yeast two-hybrid system revealed that cyclin F from *Neurospora crassa* could be an intracellular partner for *Psd*1 [18]. At that time, we also described that *Psd*1 was internalized in *Fusarium solani* planktonic cells, directing their cell cycle impairment and causing fungal endoreduplication. Furthermore, we showed that the entrance of fluorescein isothiocyanate FITC-labeled *Psd*1 in *Candida albicans* cells was dependent on GlcCer synthesis [8], a dependence also shown for its full antifungal activity against *Aspergillus nidulans* [11].

In mammals, cyclin F is expressed during S phase and peaks during the G2 phase of the cell cycle, which is considered an emerging factor in genome maintenance [19,20]. Cyclin F is also known as F-box only protein 1 (FBXO1) with an F-box domain required for binding to Skp1. Skp1 recruits Cul1 (and RBX1 with Cul1), forming the SCF ubiquitin ligase machinery that recruits the E2 ligase for ubiquitylation of target substrates. It utilizes a hydrophobic patch within its cyclin box domain, also known as the WD repeat domain, to bind the CY motif (RxL), also known as cyclin binding domain, in the substrates following their ubiquitylation and degradation as a ribonuclease. Various cyclin F substrates have been identified in the last decade, such as ribonuclease RRM2 [21], in order to ensure genome stability and efficient DNA repair and synthesis [19]. Recently, Clijsters and colleagues showed that the three activators of the E2F family of transcription factors, E2F1, E2F2, and E2F3A, key regulators of the G1/S cell transitions, interact with the cyclin box of cyclin F, resulting in their degradation and impairment in cell fitness [22]. The carboxy-terminal region of cyclin F is the regulatory module that controls its nuclear and centrosome localization as well as its abundance during the cell cycle and following genotoxic stress.

More recently, new functions have been reported for AMPs, including chemotactic, immunomodulatory, oncolytic, and mitogenic activities, among others [9,23,24]. Indeed, some host defense peptides that selectively target cancer cell membrane components have excellent tumor tissue penetration and thus can reach the sites of both the primary tumor and distant metastases [25]. However, to date, only a few plant defensins have been reported to exhibit cytotoxic activity towards cancer cells in vitro [26].

It is well known that cancer cells suffer lipid metabolic reprogramming [27] that can lead to plasmatic membranes enriched with negatively charged phospholipid phosphatidylserine (PS), as previously reported [28] in melanoma cells when compared to non-neoplastic cells. Additionally, primary cultures and metastases in addition to other cancer types expose PS [29], in contrast to the normally neutral outer leaflet of the plasma membrane.

More specifically, cancer cells suffer dysregulation of sphingolipid metabolism, and increased expression of glucosylceramide synthase and the accumulation of glucosylceramide (GlcCer) in multidrug-resistant tumor cells have been described [30–32]. GlcCer is a neutral sphingolipid composed of a sphingoid base (or LCB, long chain base), a fatty acid chain and a glucose residue. It is found in most fungi, except in *Saccharomyces cerevisiae* and *Candida glabrata* [33] and is conserved in higher eukaryotes, such as plants and mammals. They are essential for cellular structural integrity and regulating the fluidity of the lipid bilayer and are involved in cell proliferation [34,35], differentiation [27,36], and oncogenic transformation [37,38].

The properties described so far have classified *Psd*1 as a putative candidate for the development of a prototype for cancer therapy, representing a novel family of oncolytic agents that can discriminate between the neutral surfaces of non-cancerous cells and the negatively charged surfaces of cancer membranes, being cytotoxic towards a broad spectrum of malignant cells without impairing normal body physiological functions [26,39].

Thus, the main goal of this work was to test the cytotoxic effects of *Psd*1 against cancer cells in vitro and in vivo using a mouse B16F10 lung metastatic model. We were able to demonstrate that *Psd*1 decreased the viability of several cancer cells in vitro. Confocal images showed that *Psd*1 caused permeabilization and was internalized in live cells, localizing to the nucleus. The participation of the protein cyclin F as an intracellular partner was reinforced by surface plasmon resonance (SPR) analysis and molecular docking when it was possible to detect several points of contact between these proteins. For the first time, the eradication of mouse metastatic B16F10 cell lung nodules by the plant defensin *Psd*1 was successfully recorded.

2. Results

2.1. Psd1 Presents Selective Cytotoxic Effects against Tumor Cells In Vitro

To investigate the cytotoxic activity of *Psd*1 on tumor cell viability, MTT-based colorimetric assays were performed using murine skin melanoma (B16F10), human epidermoid carcinoma (A-431), and healthy (Beas-2B, HEK, R8, HSP, and CHO) cell lines (Figure 1). Cell viability was detected by the ability of viable cells to transform yellow tetrazolium salt into purple formazan crystals [40]. As shown in Figure 1A, *Psd*1 significantly inhibited the viability of both types of tumor cells in a dose-dependent manner after 24 h of treatment. A-431 cells had decreased growth in the presence of 12.5 μM peptide, reaching a maximum reduction of 20% with 50 μM peptide. B16F10 cells were reduced to approximately 50% and 60% when treated with 25 and 50 μM of the peptide, respectively. In contrast, when *Psd*1 was incubated with healthy human bronchial epithelial cells Beas-2B (Figure 1B), HEK-293, R8, HSP-2, and CHO (Figure 1C) cells, no alteration in cell viability was observed even after 72 h of exposure in the case of Beas-2B cells (Figure 1B).

The conserved Gly12 residue in defensins [1] is crucial for *Psd*1 antifungal activity [15]. *Psd*1 was more effective against B16F10 cells, and therefore this lineage was chosen to evaluate the importance of this residue in antitumor activity. For this, *Psd*1 Gly12Glu, with a glutamic acid at position 12 [15], was incubated with B16F10 tumor cells for 24 h using the same previous concentrations (Figure 1D). The mutant peptide was not able to interfere with B16F10 viability at any of the concentrations tested, which is different from the results observed with native *Psd*1 (Figure 1A).

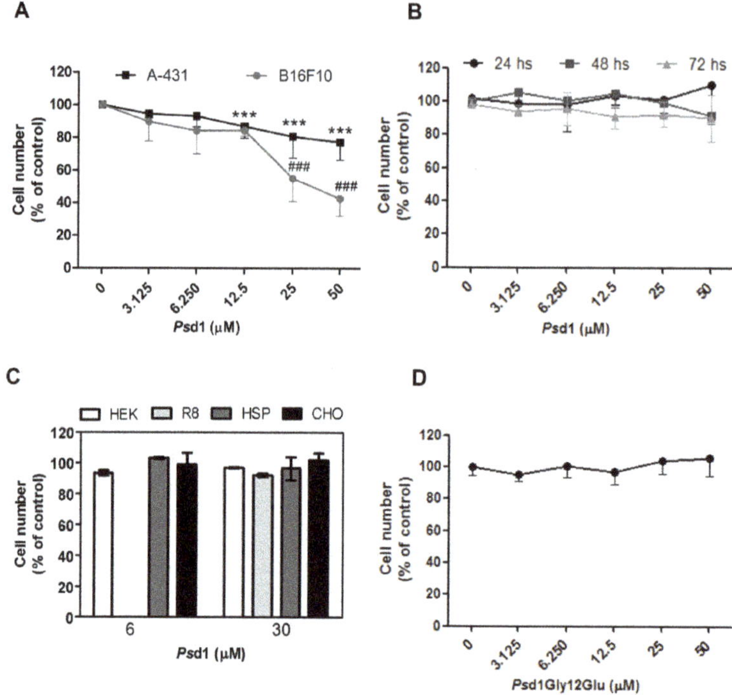

Figure 1. *Psd*1 has cytotoxic activity against tumor cells but not against healthy cells in vitro by the MTT assay. Cells lines (5×10^4 cells/well) were treated with different concentrations of *Psd*1 (3.125, 6.25, 12.5, 25, and 50 μM). Data were normalized against the untreated control, which was arbitrarily assigned as 100% cell viability. (**A**) A-431 and B16F10 cells were treated with *Psd*1 for 24 h. Both tumor cell lines had a significant reduction in viability in a dose-dependent manner. (**B**) Healthy Beas-2B cells were treated with *Psd*1 for 24, 48, and 72 h. No significant difference was observed at any time tested. (**C**) Cell viability of different mammalian health lineages by (lactate dehydrogenase) LDH release after incubation with *Psd*1 for 3 h. Values are the mean ± SEM of two experiments performed in triplicate. *Psd*1 at concentration of 6 μM was not tested against R8 cells. (**D**) B16F10 cells were incubated with *Psd*1 Gly12Glu for 24 h. The mutant did not show cytotoxic activity against this lineage. (**A,B,D**) Values are the mean ± SEM of two independent experiments performed in triplicate. Statistical significance was determined using one-way ANOVA with Dunnett's multiple comparison test compared to cells without treatment. *** $p < 0.001$ for A431 cells in the presence of 12.5 μM, 25 μM, or 50 μM *Psd*1 and ### $p < 0.001$ for B16F10 cells in the presence of 25 μM or 50 μM *Psd*1.

2.2. Insights into the Action Mechanism of Psd1 Antitumor Activity

2.2.1. Psd1 Permeabilizes the Plasma Membrane and Induces Apoptosis in B16F10 Cells

Confocal fluorescence microscopy was performed to detect whether *Psd*1 promotes B16F10 cell death by membrane permeability and apoptosis induction (Figure 2). For this, a nuclear dye impermeable to the plasma membrane, SYTOXGreen (SG), was used to monitor membrane integrity changes caused by *Psd*1. Additionally, mitochondrial viability was monitored with MitoTrackerCMRos dye accumulation, which is dependent upon membrane potential. In the absence of *Psd*1 (Figure 2A, a–c), the existence of attached well-formed cells with projections and an intact whole plasma membrane was observed. Further, intense MitoTrackerCMRos red marker indicated that the mitochondrial membrane potential was normal [41] and that mitochondria were metabolically active (Figure 2A, a). As expected, no SYTOXGreen fluorescence was detected (Figure 2A, b).

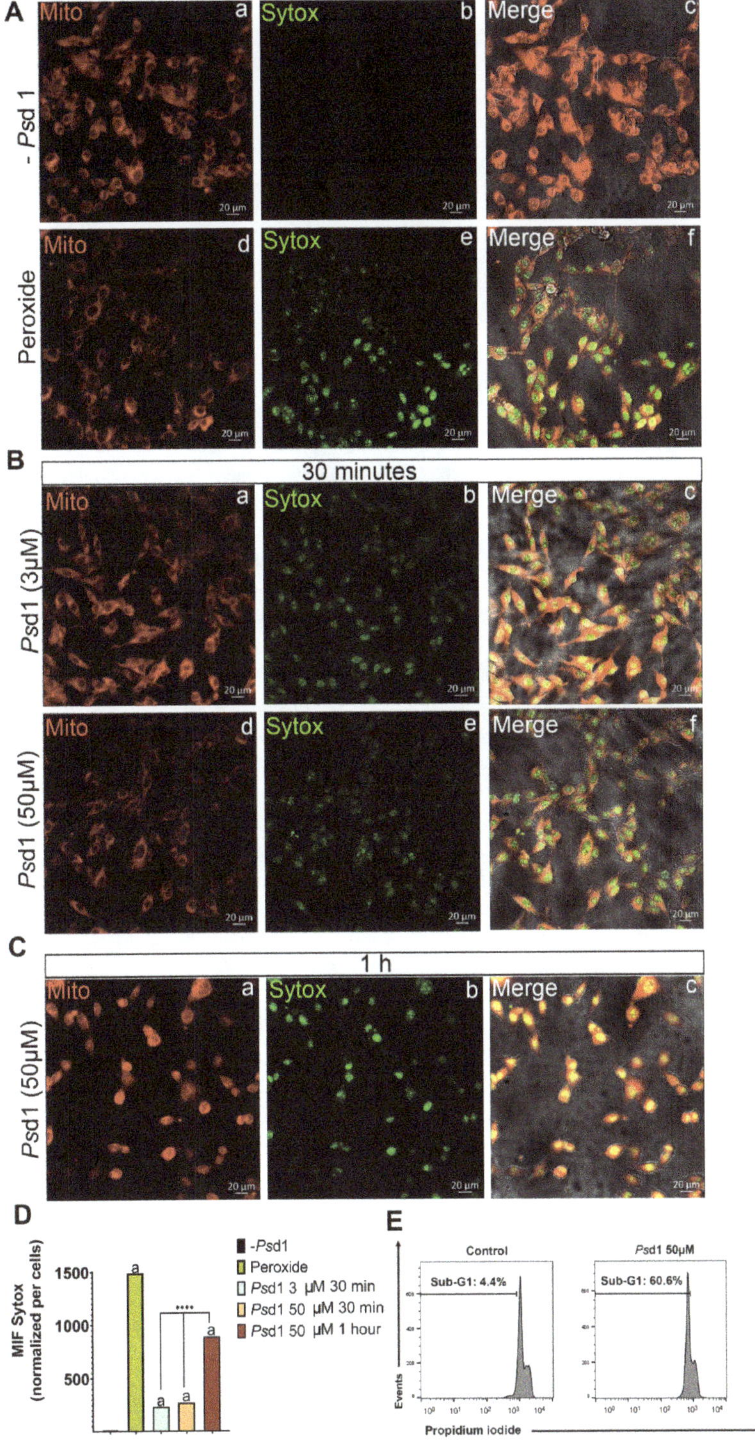

Figure 2. *Ps*d1 promotes B16F10 live cell membrane permeabilization as monitored by SYTOX Green fluorescence. (**A**) *Ps*d1 untreated and hydrogen peroxide-treated B16F10 cells. (**B**) A total of 1.5×10^4

B16F10 cells were incubated with 3 or 50 μM *Psd*1 for 30 min. (**C**) 1.5×10^4 B16F10 cells were incubated with 50 μM *Psd*1 for 1 h. All experiments were performed at 37 °C. Nucleus staining is shown in green (SYTOXGreen), and mitochondria are shown in red (MitoTracker Red CMXRos). Images are shown at 20× magnification. (**D**) Mean fluorescence intensities (MIF) of SITOXGreen fluorescence signals showed in (**A**, b,e), (**B**, b,e), and (**C**, b) per cell number are time-dependent. ****, $p < 0.0001$ treatment with 50 μM of *Psd*1 for 1 hour when compared with treatments with 3 and 50 μM of *Psd*1 for 30 min; letter a, $p < 0.0001$ treatments with peroxide and *Psd*1 when compared with the absence of *Psd*1 using one-way ANOVA and Bonferroni's multiple comparison test. (**E**) *Psd*1 induces apoptosis in B16F10 cells. B16F10 cells were treated with 50 μM *Psd*1 for 24 h and labeled with propidium iodide, and the DNA content was analyzed by flow cytometry for the sub-G0/G1 profile. At least two independent experiments were performed.

Contrarily, treatment with 3 μM and 50 μM *Psd*1 for 30 min caused the entrance of SYTOXGreen into the cell, resulting in intense green nuclear staining (Figure 2B, b/e). The green fluorescence signal increased significantly when cells were treated with 50 μM for 1 h (Figure 2C, b), as observed in the presence of hydrogen peroxide (Figure 2A, e), indicating that the promotion of membrane damage by *Psd*1 was time-dependent. When SYTOXGreen fluorescent signals were normalized by cell number it was possible to confirm a very significantly increase in treatment with 50 μM of *Psd*1 for 1 h when compared with treatments with 3 and 50 μM of *Psd*1 for 30 min. (****, $p < 0.0001$) (Figure 2D).

This effect was accompanied by a lower mitochondria red marker, mainly in the presence of 50 μM *Psd*1 for 30 min (Figure 2B, e). Surprisingly treatment with 50 μM *Psd*1 for 1 h, that displayed higher SYTOX permeability, showed a brighter red fluorescence signal, probably due to the superposition of SYTOXGreen and MitoTrackerCMRos dye signals inside the cells (Figure 2C, a). In addition, in this situation B16F10 cells started to lose their adhesion capacity, assuming a less spread shape (Figure 2C, c). Together, these results suggest that *Psd*1 altered the biophysical properties of plasmatic membrane of the tumor cells, which could be accompanied by intracellular death signaling events such as oxidative stress of mitochondria. However, more investigation about the mitochondrial effect caused by *Psd*1 treatment must be performed.

Also B16F10 cells treated with 50 μM *Psd*1 exhibited a high proportion of cells with sub-G0/G1 DNA content (~60%), as shown by flow cytometry (Figure 2D), which is indicative of cells undergoing DNA fragmentation. Internucleosomal DNA fragmentation is one of the hallmarks of apoptosis and is frequently used as a criterion for its detection [42]. Thus, these data corroborate the findings obtained in the MTT assay (Figure 1A), suggesting that the decrease in cell viability of *Psd*1-treated B16F10 cells is associated with the induction of apoptosis.

2.2.2. *Psd*1 Internalizes towards the Nucleus

The entrance of *Psd*1-FITC in the B16F10 live cells in real time was observed using spinning disk confocal fluorescence microscopy (Figure 3). Before the addition of *Psd*1 (time zero), B16F10 cells showed good plate adherence with nuclei well stained in blue (DAPI) and several mitochondria in red (MitoTracker Red CMXRos) (Figure 3A, a/c-d), indicating cell viability and normal mitochondrial transmembrane potential. No background concerning FITC signal was detected (Figure 3A, b). Then, 9 μM *Psd*1-FITC was added, and the same cell field was photographed for 2 h. After 30 min, it was possible to detect few and faint green fluorescence signals in the nuclei, suggesting the entrance of *Psd*1 in those cells (Figure 3A, g). The relative fluorescence of *Psd*1-FITC increased after 1 h (Figure 3A, l). *Psd*1-FITC fully accumulated in the nucleus after 2 h of treatment, where intense bright green fluorescent points were detected (Figure 3A, q). The nuclear location of *Psd*1 was confirmed by merging the FITC and DAPI signals detected in merge and orthogonal views, as shown in the right panels (Figure 3A, i, n, s (merge) and j, o, t (ortho)). Moreover, over time, it was possible to detect a decrease in the red fluorescence intensity at real time (Figure 3A, c, h, m and r).

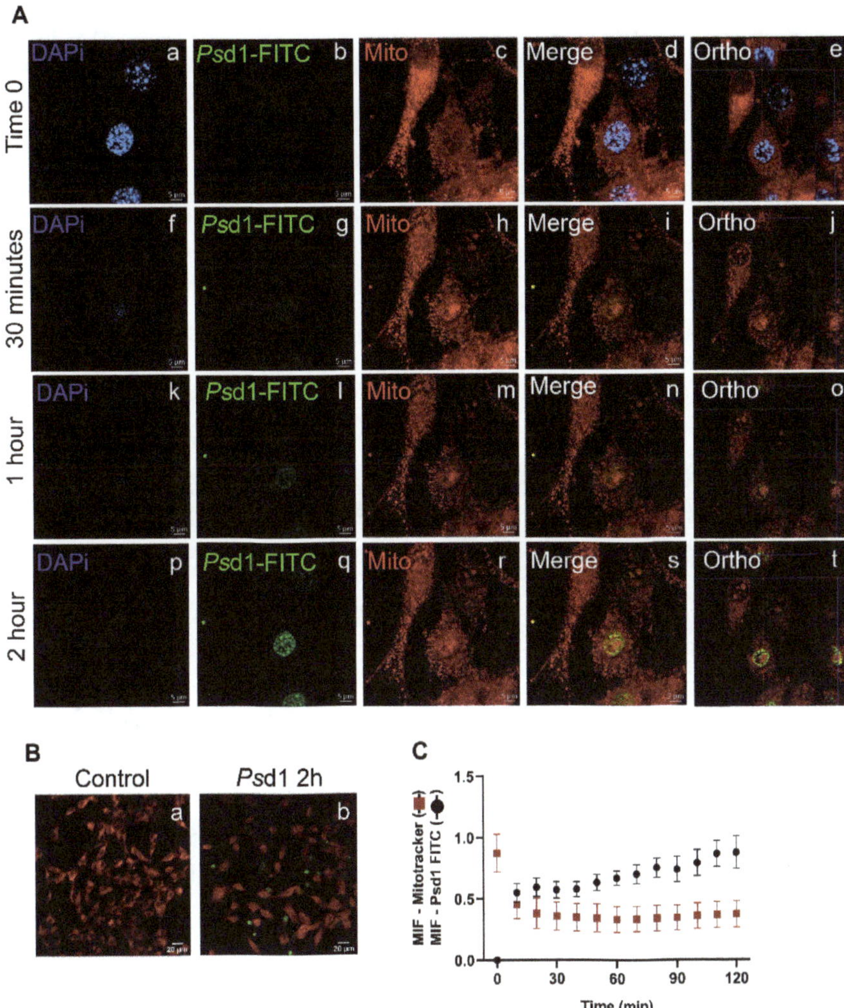

Figure 3. Fluorescein isothiocyanate (FITC)-*Psd*1 was localized in the B16F10 live cell nucleus. (**A**) B16F10 cells (1.5×10^4) were initially pre-stained with DAPI and MitoTracker Red CMXRos followed by the addition of 9 µM *Psd*1-FITC. Images were acquired in real time every 10 min for two hours at 100× magnification. (**B**) B16F10 cells in the absence of *Psd*1 (control time zero) and in the presence of 9 µM *Psd*1 for two hours (*Psd*1 2 h) at 10× magnification. (**C**) The mean fluorescence intensity (MIF) of MitoTracker and FITC-*Psd*1 labeled cells measured over time. Average values with standard deviation (SD) of two independent experiments are reported.

Pictures showing B16F10 cells marked with MitoTracker Red CMXRos and FITC-*Psd*1 for 2 h in a smaller order of magnitude were shown in Figure 3B. In this case it was possible to have an overview of the *Psd*1 internalization phenomena in a larger number of cells. Quantification of the regions of interest (ROI) relative to the mean fluorescence intensity (MFI) from the intensity of the pixels revealed that simultaneously, the red fluorescence decreased and the green fluorescence increased (Figure 3C). Together, these findings plus permeability results shown in Figure 2 strongly suggested that the *Psd*1 entrance caused loss of membrane barrier function, compromising the cell cycle and mitochondrial roles. However, it is not clear the order in which these events happened, that is, does the peptide need to interact with nuclear targets and then trigger apoptosis events, collapsing the mitochondria, or this occur just after membrane interaction? Additional experiments will be done to answer these questions.

2.2.3. *Psd*1 Entrance in B16F10 Cells is Glucosylceramide-Dependent

We evaluated whether glucosylceramide (GlcCer) could impact the entrance of *Psd*1 in B16F10 cells by reducing its amount on the plasmatic membrane using DL-threo-1-phenyl-2-palmitoylamino-3-morpholino-1-propanol (PPMP) (Figure 4). PPMP is a well-studied glucosylceramide synthase inhibitor [30,32,43–45] with several reports about its action in B16 melanoma cells [30,46]. The amount of GlcCer in the plasmatic membrane was determined using the red cholera toxin subunit B (CT-B) compound [47–49], which binds specifically to raft domains, followed by treatment with an anti-CT-B antibody labeled with Alexa Fluor 594. As seen in Figure 4c, cells treated with PPMP showed a significant reduction in the bright red signal (Figure 4c, +PPMP, −*Psd*1) relative to that observed in PPMP-free cells (Figure 4a, −PPMP, −*Psd*1).

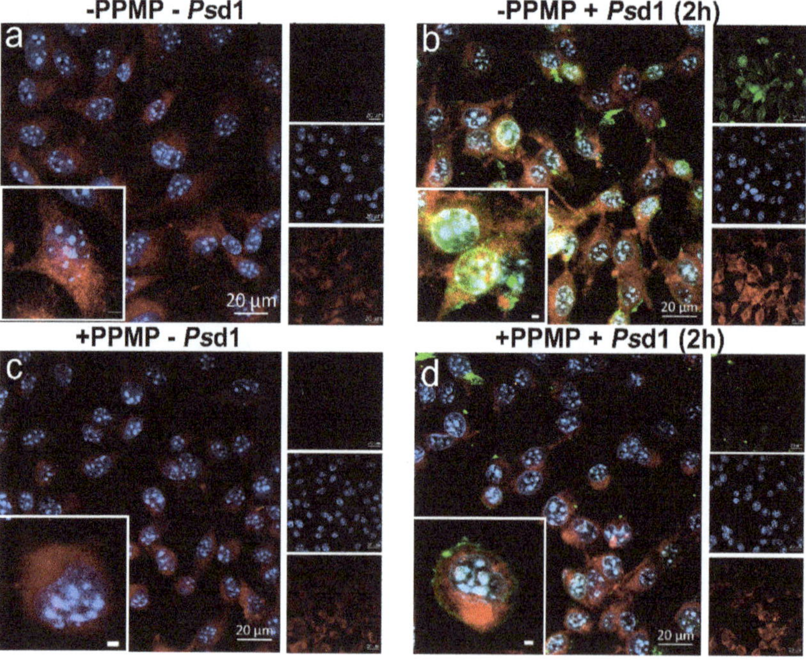

Figure 4. GlcCer reduction inhibits the entrance of FITC-*Psd*1 into B16F10 cancer cells. B16F10 cells (1.5 × 10⁴) were incubated with or without 20 µM PPMP for 1 h followed by treatment with or without 9 µM FITC-*Psd*1 for 2 h. Fixed cells were stained with Vybrant™ Alexa Fluor™ 594 and DAPI. (**a**) B16F10 cells without treatments (−PPMP, −*Psd*1); (**b**) B16F10 cells treated with *Psd*1 for 2 h (−PPMP, +*Psd*1 (2 h)); (**c**) B16F10 cells treated with PPMP (+PPMP, −*Psd*1); (**d**) B16F10 cells treated with PPMP plus *Psd*1 (+PPMP, +*Psd*1 (2 h)). Pictures were acquired at 40× magnification. Inset zoom images are at 100× magnification. At least two experiments were performed independently.

As GlcCer is the majority neutral GSL (87% of total) in B16F10 cells [50] and one of the major components of lipid rafts [51,52] we do believe that the diminished red fluorescent signal detected in the presence of PPMP could reflect the decrease of glucosylceramide in B16F10 cell membranes.

The image in Figure 4b (−PPMP, +*Psd*1 (2 h)) clearly showed that in the absence of PPMP, FITC-Psd1 was internalized into B16F10 cells being localized in the cells nuclei, confirmed by the superposition between green and blue dyes. Meanwhile when B16F10 cells were treated with PMPP plus 9 μM *Psd*1 for 2 h (Figure 4d, +PPMP, +*Psd*1 (2 h)), a different profile was observed. Inthis new situation the nuclear FITC-*Psd*1 green fluorescent signals were practically non-existent. *Psd*1 was kept outside the cell and retained in the membrane, as noted by the green signals around the outside of the cell (Figure 4d, +PPMP, +*Psd*1 (2 h), for details see the images insets with larger magnitude).

Together, the results using PPMP, a glucosylceramide synthase inhibitor, provided evidence that GlcCer actively participates in the *Psd*1 interaction with the B16F10 membrane, internalization, and further death signaling mechanisms.

2.3. Psd1–Cyclin F Interaction by Surface Plasmon Resonance (SPR)

Considering that cyclin F is an intracellular target for *Psd*1 in fungal cells [18], we evaluated the specificity of their interaction in real-time by SPR. Recombinant cyclin F was immobilized on the CM5 chip followed by a constant flow injection of *Psd*1. The obtained sensorgram revealed the cyclin F/*Psd*1 interaction as a function of *Psd*1 concentration, indicating a dose-dependent effect (Figure 5A). Increases in plasmon resonance signals (RU) due to protein–protein binding were observed 80 s after *Psd*1 injection. This response indicated a very fast association phase that reached a plateau until the injection ended. RU responses enhanced from zero to 2.5, 3.6, 11.5, 26, 64, and 119 RU when concentrations of 10, 20, 40, 80, 160, and 240 μM *Psd*1 were tested, showing a closed, linear rise pattern response. A full dissociation phase was observed when the injection of *Psd*1 ended, with all curves returning to baseline levels. The results of global fitting using the 1:1 Langmuir model allowed us to calculate an affinity constant (KD) of 1.5 mM, indicating that *Psd*1 interacts with cyclin F in this cell-free system.

In parallel, in silico molecular docking was used to predict the best binding mode of *Psd*1 to human cyclin F. This technique allows the generation of the most likely stable conformations and orientations, named poses, of *Psd*1 within the cyclin F binding sites. The results are displayed in Figure 5B. *Psd*1 fits into a small cavity of the cyclin F protein, with certain parts of its structure being in contact with the F-box and cyclin substrate recruitment domains (left image). Most of the *Psd*1 defensin residues found in the interaction interfaces belonged to its loop regions (15 residues out of 24 found). Likewise, most of the cyclin F residues detected in the interface belonged to the F-Box and cyclin domains (21 residues out of 30 found). The peptide–protein complex was formed by the cumulative contribution of hydrogen bonds and electrostatic, hydrophobic and van der Waals interactions, although its stability over time could not be evaluated (Figure 5B, right image, Table S1 and Video S1).

At the interfaces, a total of 15 hydrogen bonds between *Psd*1 and cyclin F were predicted. In the defensin structure, most of the participating residues belong to the loop regions (7 residues out of 11, namely, Leu6, Arg11, Gly12, Ala28, His29, Cys35, and Trp38), while in the cyclin F structure, a total of 11 residues took part in this type of interaction, with four residues belonging to the F-Box domain (Tyr147, Lys171, His175, and Tyr177) and another four residing in the WD-repeat domain (Lys470, Ile543, Glu545, and Arg546). Notably, both Arg11 of *Psd*1 and Tyr147 of cyclin F make three distinct hydrogen bonds, which is more than any other residue in the complex. In addition to hydrogen bonds, the protein–peptide interfaces exhibited several hydrophobic contacts. In the case of *Psd*1, all three engaging residues resided in loops (Leu6, Val13, and Phe15). With respect to cyclin F, four residues in total made meaningful contacts (Pro230, Pro233, Ile472, and Pro563). The inset zoom images in Figure 5B (right) represent the hydrophobic contacts between Leu6 from *Psd*1 with prolines in positions 230 and 233 belonging to F-box domain of cyclin F, and Val 13 and Phe 15 from *Psd*1, with Ile 472 and Pro 563 belonging to the cyclin domain of cyclin F. More details can be seen in Table S1 and Video S1.

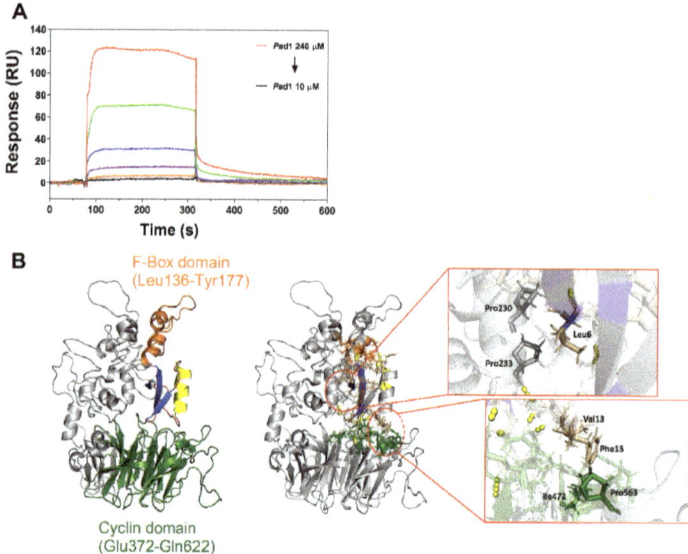

Figure 5. (**A**) The *Psd*1-cyclin F interaction was confirmed by surface Plasmon resonance SPR. Sensorgram curves demonstrating the association and dissociation phases of immobilized cyclin F on the CM5 chip surface with different concentrations of *Psd*1: 10 µM (black), 20 µM (orange), 40 µM (purple), 80 µM (blue), 160 µM (green) and 240 µM (red). Each experimental condition was performed at least twice. (**B**) The most favorable pose for the *Psd*1-cyclin F interaction predicted by molecular modeling simulations. Left cartoon: Human cyclin F domains were designated in orange for the F-box domain and green for the cyclin domain (also known as the WD domain); *Psd*1 secondary structures are shown in blue for β-strands and yellow for α-helices attached by pale pink loops(Protein Database Bank (PDB accession number 1JKZ). Right cartoon: Amino acid residues involved in the hydrophobic interface clusters are highlighted by red balloons and shown in the inset zoom images (Leu 6 from *Psd*1 with Pro 230 and Pro 233 belonging to F-box domain of cyclin F, and Val 13 and Phe 15 from Psd1 with Ile 472 and Pro 563 belonging to the cyclin domain of cyclin F).

2.4. Psd1 Impaired the Establishment of B16F10 Tumor Metastasis In Vivo

A well-characterized mouse model of experimental lung metastasis by intravenous injection of B16F10 melanoma cells was used to evaluate the effect of *Psd*1 in vivo. Different concentrations of *Psd*1 were injected together with B16F10 cells into C57BL/6 mice via the tail vein. One group of animals received only 1 mg/kg *Psd*1 to evaluate the possible toxicity of the peptide. Twenty-one days after injection, the mice were sacrificed and assayed. The lung nodule quantification from all groups is shown in Figure 6A, and representative lung images are shown in Figure 6B. Metastasis induction was successfully achieved, showing an average of 43 nodules in the lungs of mice that received only B16F10 melanoma cells (Figure 6A,B, *n* = 20). A significant reduction in lung metastasis colonization after treatment with *Psd*1 was achieved in a dose-dependent manner beginning at the dose of 0.5 mg/kg. Treatment with 3.0 mg/kg *Psd*1 completely abolished tumor development (Figure 6A, 0.14 ± 0.14 nodules). The number of pulmonary nodules decreased by 75% and 88% when 0.5 mg/kg (10.2 ± 2.6 nodules) and 1.0 mg/kg (4.82 ± 1.88 nodules) *Psd*1 was used, respectively. Notably, animals that received either phosphate buffer solution (PBS) pH 7.4 or 1.0 mg/kg *Psd*1 alone did not present lung nodules during the experimental period (Figure 6A).

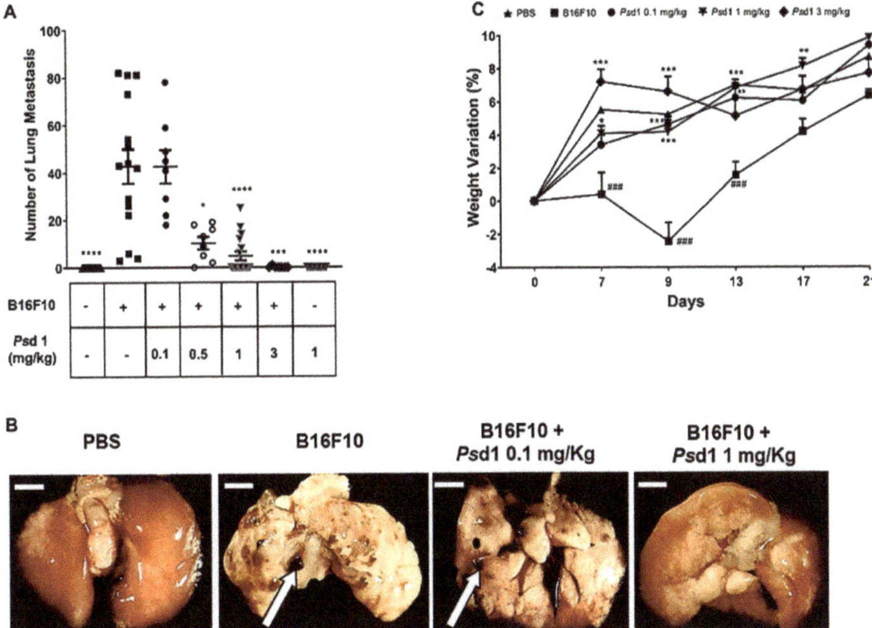

Figure 6. Protective effects of *Psd*1 on a model of experimental lung metastasis in vivo. B16F10 mouse melanoma tumors were established by intravenous injection of 2.5×10^5 B16F10 cells into C57BL/6 mice. (**A**) After 21 days, mice from each experimental group were sacrificed, and the number of visible lung metastatic nodules was quantified. Treatment of the animals inoculated with melanoma cells plus *Psd*1 caused a dose-dependent reduction in the number of lung metastases. Black lines represent the mean ± SEM of two independent experiments analyzed by one-way ANOVA and Bonferroni's multiple comparison test. **** $p < 0.0001$ B16F10 vs. phosphate buffer solution (PBS); **** $p < 0.0001$ B16F10 vs. *Psd*1 1 mg/kg; **** $p < 0.0001$ B16F10 vs. just *Psd*1 1 mg/kg (without B16F10 cells); *** $p < 0.001$ B16F10 vs. *Psd*1 3 mg/kg; * $p < 0.05$ B16F10 vs. *Psd*1 0.5 mg/kg. (**B**) Representative images of lung metastatic nodules from animals that received just PBS or B16F10 cells alone or together with 0.1 mg/kg or 1 mg/kg *Psd*1. Bar = 2 mm. (**C**) Animal weights were acquired every two days. Weight variation was calculated as (weight on that day (weight of the first day × 100 − 100)). Only the days with higher weight differences between animals injected with only B16F10 cells and those that received cells plus *Psd*1 or PBS alone are represented. All of the values represent the mean ± SEM of two independent experiments analyzed by two-way ANOVA. *** $p < 0.001$ B16F10 vs. PBS at 7, 9 and 13 weeks; *** $p < 0.001$ B16F10 vs. *Psd*1 0.1 mg/kg at 9 weeks; ** $p < 0.01$ B16F10 vs. *Psd*1 0.1 mg/kg at 13 weeks; *** $p < 0.001$ B16F10 vs. *Psd*1 1 mg/kg at 9 and 13 weeks; ** $p < 0.01$ B16F10 vs. *Psd*1 1 mg/kg at 17 weeks; * $p < 0.05$ B16F10 vs. *Psd*1 1 mg/kg at 7 weeks; *** $p < 0.001$ B16F10 vs. *Psd*1 3 mg/kg at 7 and 9 weeks.

The animal group that received just intravenous injection of PBS gained weight throughout the experimental period, as expected (Figure 6C). Meanwhile, animals from the B16F10 group, which were injected exclusively with B16F10 melanoma cells, showed a very large loss in body weight compared with the PBS group, being more pronounced on the ninth day. After this point, their weights partially recovered but remained lower than the other groups. Animals that received B16F10 cells plus 0.1, 1, and 3 mg/kg *Psd*1 showed a similar pattern as those injected with PBS; that is, these animals gained weight throughout the assay. The results obtained in the presence of 1 and 3 mg/kg *Psd*1 are consistent with less lung metastatic nodules occurring in these animals. However treatment with 0.1 mg/Kg was not effective in reducing the number of nodules but was effective in reducing their size when compared to animals that just received B16F10 cells (that not shown).We do believe that in this case these smaller nodules were not enough to trigger the pathways responsible to lose detectable weight.

It is important to note that mice treated with *Psd*1 alone did not present changes in important behavior signals, such as changes in locomotion, piloerection, diarrhea, or mortality as preconized by ANVISA through the Guide for Conducting Non-Clinical Drug Safety and Toxicology Studies Required for Drug Development/Safety and Efficacy Assessment Management (GESEF 2013 version 2) when compared to animals that received PBS solution alone.

To better characterize the inhibitory effect of *Psd*1 on the colonization of pulmonary metastasis in this model, animals from each group were euthanized, and their lungs were excised and stained with HE (Figure 7). Histological analysis revealed the presence of inflammatory cells around the blood vessels and bronchi of the lung and evidence of fibrotic lesions in animals injected with B16F10 cells only (Figure 7A, red head arrows and red arrows, respectively). This profile was almost absent in animals that received just PBS (−B16F10, −Psd1), indicating that accumulation of these cells, as well as fibrosis, was related to the presence of the tumor cells only. In accordance with the observed reduction in lung tumor nodule metastasis (Figure 5A), animals co-inoculated with B16F10 cells and 1 and 3 mg/kg *Psd*1 (+B16F10, +1 mg/kg *Psd*1, and +B16F10, +3 mg/kg *Psd*1) showed a great reduction in the accumulation of inflammatory cells (Figure 6A). The quantification of inflammatory foci (± SEM) per mm^2 was obtained in six fields, corresponding to 90% of the tissue as shown in Figure 7B. When compared with animals that received just B16F10, animals co-inoculated with B16F10 cells plus 1 and 3 mg/kg *Psd*1 (+B16F10, +1 mg/kg *Psd*1 and +B16F10, +3 mg/kg *Psd*1) showed a significant reduction in the accumulation of inflammatory cells (***, $p < 0.001$). We were also able to show that the injection of *Psd*1 alone (+ 1 mg/kg *Psd*1) provoked a very low inflammatory foci compared with animals that received B16F10 cells (*** $p < 0.001$), as observed in PBS control animals (−B16F10, −Psd1). This was an important result considering the potential mammalian use for *Psd*1. Together, these results indicated that *Psd*1 directly suppresses the lung metastasis of circulating B16F10 melanoma cells in vivo.

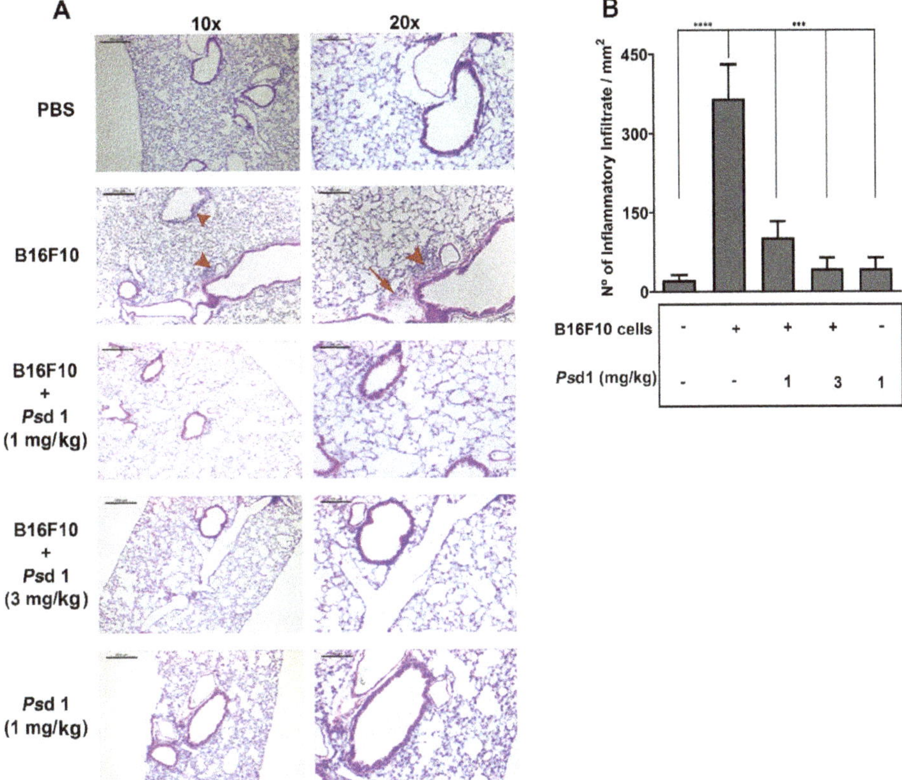

Figure 7. *Psd*1 inhibited the appearance of inflammatory cells in lung tissues. Two animals from each group of the in vivo experiment were anaesthetized and submitted to transcardial perfusion for further histological procedures. The lungs were removed and prepared for hematoxylin and eosin (HE) staining as described in the Materials and Methods. (**A**) Representative images from the lungs of one animal from each experimental group. Animals injected with just B16F10 cells showed the presence of infiltrating cells around blood vessels (red head arrows) and fibrotic lesions (red arrow) compared to animals from the other groups. The left and right panels indicate 10× and 20× magnifications, and the bars represent 200 and 100 μM, respectively. (**B**) Quantification of inflammatory infiltration (± SEM) was obtained in six fields, corresponding to 90% of the tissue. A significant reduction in inflammatory cells was observed in animals treated with *Psd*1. **** $p < 0.0001$ for B16F10 vs. PBS or *** $p < 0.001$ forB16F10 vs. *Psd*1 1 mg/kg or B16F10 vs. *Psd*1 3 mg/kg or *Psd*1 1 mg/kg alone. Values represent the mean ± SD of two independent experiments analyzed by one-way ANOVA and Bonferroni's multiple comparison test.

3. Discussion

Melanoma is a very aggressive metastatic cancer that results in a quick death. Treatments based on multiple combined therapies are not effective, and the ten-year survival rate of melanoma with distant metastasis that is lower than 10% [53].

In this work, we suggest that the pea defensin *Psd*1 could be a promising drug candidate for lung melanoma metastasis treatment using the syngeneic murine metastasis model of B16F10 melanoma cells.

In vitro data showed that the plant defensin *Psd*1 was able to inhibit both A-431 and B16F10 cancer cell growth in a concentration-dependent manner without promoting damage to healthy cell lineages.

No cell death was observed in Beas-2B human bronchial epithelial cells, where B16F10 melanoma metastasis can occur, or in the other immortalized healthy cells tested, such as HEK, R8, HSP, and CHO, suggesting that *Psd*1 could be safe for humans.

Few works have described the inhibitory effect of plant defensins on tumor cells in vitro. The group of T.B. Ng reported that some plant defensins from the Fabaceae family showed anticancer activity and repressed the growth of leukemia and breast cancer cells [54–56]. However, lunatusin isolated from lima beans (*Phaseolus lunatus* L.) showed cytotoxic effects towards normal cell types and tissues [55]. They also identified that plant defensins from *Phaseolus vulgaris* and *P.coccineus* had great potential to inhibit the multiplication of colon and breast cancer cell lines without exhibiting any cytotoxic effects on normal cell types [57–59]. Without any effect on immortalized bovine endothelial cells, the complete inhibition of HeLa cell viability was achieved by γ-thionin defensin from *Capsicum chinense* [60]. As far as we know, none of these defensins have been tested in animal models, and in general, the mechanism of their anticancer activity has been poorly elucidated.

To gain insights into the antitumor mechanism of *Psd*1, cell integrity studies showed that *Psd*1 defeats the barrier function of the plasma membrane of B16F10 cells, allowing SYTOXGreen input to bind to nuclear DNA. In addition, permeation assays revealed the presence of mitochondrial oxidative stress. In this case, MitoTracker fluorescence decreased throughout *Psd*1 treatment, indicating a reduction in the mitochondrial potential membrane [61]. After 30 min, FITC-*Psd*1 was detected in the nuclei of DAPI-stained B16F10 cancer cells, as seen by orthogonal confocal microscopy analysis.

The signaling order of the events provoked by *Psd*1 in B16F10 cells is not yet known, but some hypotheses may be formulated by these results. The intrinsic pathway of apoptosis is a well-established mechanism and can be activated by DNA damage via p53 protein activation or by metabolic stress. This latter can cause an increase in reactive oxygen species and a consequent loss in mitochondrial electrical potential with cytochrome C release and the activation of caspases [62]. Strong evidence has postulated that therapeutic agents that can induce ROS-mediated apoptosis in cancer cells are considered potential anticancer agents. Further studies must be carried out to determine the production level of ROS in this situation.

Some lipids and their metabolites are also involved in apoptosis, inflammation, angiogenesis, and cell proliferation signaling [63]. It is known that several cancer cell types suffer lipid metabolism reprogramming when compared to non-cancerous cells [27,36,64]. As an example, phosphatidylserine (PS), before being present in the membrane inner leaflet, is externalized [65], and cholesterol synthesis increases [27,66,67]. These changes result in a negative net charge on the cancer cell membrane, contrary to a neutral surface charge found on non-cancerous cell membranes [68–70].

Altered GSL metabolism leads to an upregulation of enzymes involved in this pathway, such as glucosylceramide synthase, which catalyses the transfer of a glucose residue from uridine 5′-diphospho-glucose (UDP-glucose) to the ceramide moiety [31,71,72]. The increase in GlcCer levels on cancer cell membranes has been associated with multidrug-resistant cancer cells [31,32,73] and has been proposed as a potential biomarker to evaluate the malignancy level of breast tumors [64].

An important dependence on glucosylceramide for *Psd*1 entrance into B16F10 cell membranes was reported in this work, as detected in cells treated with PPMP, an inhibitor of glucosylceramide synthase.

Previous work by our group showed that GlcCer and ergosterol are very important for the initial interaction of *Psd*1 with *C. albicans* [8] and *A. nidulans* [11] membranes. Mutants lacking glucosylceramide synthase are partially resistant to *Psd*1 antifungal activity and permeabilization [8,11,15]. *Psd*1 chemical shifts and dynamic property alterations were detected by NMR structural analysis in the presence of vesicles composed of phosphatidylcholine and GlcCer (POPC:GlcCer 90:10) [15]. The most sensitive regions in the peptide are the Gly12 and His36–Trp38 residues belonging to loop 1 and turn 3. As seen in the present work, the conserved Gly12 residue in the plant defensin family was also important for the in vitro death of B16F10 cells.

In 2007, Lobo and co-workers showed by yeast double hybrid and pull down assays that *Psd*1 interacted with cyclin F from *N. crassa* [18]. At this time, we proposed that this phenomenon could be in part responsible for the occurrence of the *N. crassa* endoreduplication observed.

In the present work, we were able to confirm the interaction of these two proteins by surface plasmon resonance and in silico molecular docking simulations. *Psd*1 fit well in a cavity formed between the F-box and cyclin domains of human cyclin F. Several hydrogen bonds and hydrophobic contacts were important for the maintenance of this complex. In fact, it has been proposed that disulfide bonds and polar contacts are the main forces responsible for defensin CSαβ folding stability, enabling the surface exposure of hydrophobic residues [74]. Recently, we showed that this was also valid for *Psd*2, a pea defensin that shares 42% identity and high 3D structural homology with *Psd*1 [6]. All hydrophobic residues are exposed on the surface, except for Leu6. They are clustered on the surface formed by two loops, between β1 and the α-helix and between β-sheets 2 and 3. We believe that these contacts between *Psd*1 and cyclin F can disturb the interaction of cyclin F with their endogenous substrates, some of which are related to cancer diseases [75].

Cyclin F expression is enhanced in the final G2 phase, where it controls genome integrity, cell proliferation, fitness, and transcription. It is a cyclin-dependent kinase CDK-independent cyclin that contains an F-box domain and is a member of the modules of SKP1-Cullin 1-F-box protein, SCF, and E3 ligase complexes [76,77]. Cyclin F controls the ubiquitination and subsequent proteasomal degradation [20] of several target substrates, such as CP110 [78], Nusap1 [79], RRM2 [19], Cdc6 [80], DNA exonuclease Exo1 [81], SLBP [82], and, as shown very recently, the three activators E2F1, E2F2, and E2F3A of the E2F family of transcription factors [22]. In the latter case, E2F1, E2F2, and E2F3A interact with the cyclin box of cyclin F via their conserved N-terminal cyclin binding motifs.

Together, our findings suggest that the interaction of *Psd*1 with the Fbox and the CY substrate recruitment domains of cyclin F could potentially inhibit the formation of the SCF ubiquitin–ligase complex, restricting the degradation of several substrates as E2F transcription factors. Failure to degrade E2F1, E2F2, or E2F3A in the late S and G2 phases maintains E2F activity. This, in turn, results in an imbalance of the transcriptional landscape in the G2 and M phases and in unscheduled DNA synthesis in the next cell cycle, which is accompanied by DNA replication stress and DNA damage. This is in line with the evidence that DNA replication stress can be caused by overexpression of oncoproteins [83,84] and that the control of the RB-E2F pathway is important for safeguarding genomic stability.

The capacity of *Psd*1 to inhibit B16F10 cell growth was also tested in an in vivo model of metastatic melanoma. We reported for the first time anti-metastatic activity of a plant defensin. *Psd*1 completely inhibited the formation of B16F10 lung metastasis nodules in mice at concentrations above 0.5 mg/kg when administered intravenously. Moreover, this effect was accompanied by the absence of weight loss in animals that received B16F10 cells plus *Psd*1, similar to the weight pattern observed for animals that received PBS. Important weight loss was observed in animals that received only B16F10 cells, which is consistent with metastatic lung nodule implementation.

It is well known that TNF-α can be produced by alveolar macrophages. It can promote the formation and proliferation of tumors through multiple signaling pathways, and it can also promote the formation of tumor neovascularization by promoting the stable expression of interleukin 8 IL-8. IL-8 can promote the proliferation of vascular endothelial cells and can activate G proteins so that the vascular endothelial cells undergo retraction and increase the cell gap, providing conditions for tumor cell metastasis and infection [85]. The activity of *Psd*1 was confirmed by a histological investigation that demonstrated massive infiltration into the surrounding lung parenchyma in mice that received only B16F10 cells. Infiltrating cells were not observed in animals that just received PBS and were statistically diminished when *Psd*1 was present, confirming its therapeutic effect.

Antimicrobial peptides are only beginning to encroach into the oncological sphere, and therefore efficacy data are relatively limited [86]. In addition to its dual effect in cancer cells, the human cathelicidin LL37 is in the ongoing phase I oncology trial NCT02225366. It is being administered intratumorally in patients with documented metastatic melanoma with at least three cutaneous lesions measuring with stage IIIB, IIIC, or IV or nodal lesions. Its optimal biological therapeutic dose against metastatic melanoma is being determined in this setting. To date, no results have been published.

The safety data already obtained for AMPs in infectious disease trials substantiate the notion that AMPs, such as the pea defensin *Psd*1, could also be well tolerated in cancer patients.

4. Materials and Methods

4.1. Expression and Purification of Psd1 and the Site-Directed Mutant Psd1 Gly12Glu

*Psd*1 (PDB accession number 1JKZ) and the Gly12 mutant were expressed and purified as previously described [11], except that a HiPrep 26/60 Sephacryl S-100 HR column (GE Healthcare, Amersham, UK) was used in the first purification step. The purified fractions were collected, dried, and solubilized in milliQ water, and the peptide concentration was estimated using Lowry's method [87]. The corrected amino acid primary sequences of *Psd*1 and *Psd*1 Gly12Glu were confirmed by LC/MS–MS analysis after peptide trypsin digestion with coverage of approximately 90%. In all cases, peptide fragments not directly related to the protein sequences of interest were not detected. LC/MS/MS analysis was performed at the CEMBIO facility (Centro de Espectrometria de Massas de Biomoléculas) at the Biophysics Federal Institute Carlos Chagas Filho at the Federal University of Rio de Janeiro.

4.2. Cell Lines and Culture Conditions

Murine skin melanoma (B16F10), human epidermoid carcinoma (A-431), HEK 293 (human embryonic kidney), R8 (rat lymphocyte), HSP (human hepatocyte), and CHO (Chinese hamster ovary) cell lines were maintained with Dulbecco's Modified Eagle's Medium - high glucose The bronchial human respiratory epithelial (Beas-2B) cell line was cultured in Roswell Park Memorial Institute Medium RPMI 1640. All media were supplemented with 10% fetal calf serum (FCS) and a solution of 100 U/mL penicillin, 100 µg/mL streptomycin, and 0.25 µg/mL Fungizone® antimycotic. All cell culture reagents (unless indicated) were purchased from Thermo Fisher Scientific Inc, Maryland, USA). The cells were routinely maintained in a humidified 5% CO_2 air incubator at 37 °C and sub-cultured every 3–4 days.

4.3. Cell Viability Assays

A-431, B16F10 and Beas-2B cells were seeded at a density of 1.5×10^4 cells/well in a 96-well plate. After overnight incubation at 37 °C, the cells were treated in octuplicate with *Psd*1 or *Psd*1 Gly12Glu (0, 3.12, 6.25, 12.5, 25, or 50 µM) for an additional 24, 48, or 72 h, as indicated in each figure. HSP-2 was incubated with 6 and 30 µM *Psd*1 for 3 h. HEK-293 for was incubated for 3 h at the same time. The cells were then washed twice with PBS (pH 7.4) and incubated with 5 mg/mL MTT (3-(4,5-dimethylthiazol-2-yl)-2,5-diphenyltetrazolium bromide) solution for 3 h. The cells were subsequently washed, and the formazan crystals formed were solubilized in DMSO. The optical density of each well was measured at a wavelength of 490 nm with an ELISA plate reader (UVM340-ASYS, Biochrom, Cambridge, UK). The effect of *Psd*1 on cell growth was assessed as the percent of cell viability calculated by the absorbance of the cells in culture media (100% viability) and those treated with 0.1% Triton X-100 (0% viability) × 100. R8 (rat lymphocyte) and CHO (hamster ovary) cell lines were incubated with 6 and 30 µM *Psd*1 for 3 h. Cell viability was detected by lactate dehydrogenase (LDH) release into cell culture medium through a colorimetric assay. Values are the mean ± SEM of two experiments performed in triplicate. The percent of cell viability was calculated by the absorbance of the cells in culture media (100% viability). Two independent experiments were performed in triplicate, and the values are expressed as ± standard error of the mean (SEM) after statistical analysis using one-way ANOVA with Dunnett's multiple comparison test. *** $p < 0.0001$.

4.4. Confocal Microscopy Scanning

Images were obtained in the X, Y, and Z planes of the image, giving the image a 3-dimensional depth (Z-stack) at 100× magnification with a Zeiss Cell Observer Yokogawa Spinning Disk confocal microscope (Cell Observed SD, Carl-Zeiss, Oberkochen, Germany) located in the Microscopia Óptica de Luz Gustavo de Oliveira Castro (Plamol) platform in Universidade Federal do Rio de Janeiro. To measure the mean fluorescence intensity (MFI), Zen Lite Blue (Carl-Zeiss, Oberkochen, Germany) software was used. On average, thirty cells per field were analyzed in fourteen fields by experiments. Then, the region of interest (ROI), specifically the nucleus and mitochondria, was tagged on the cells photographed for fluorescence analysis, which is given by the software from the intensity of the pixels. Subsequently, the statistical test was performed using GraphPad Prism version 7.4, and the values presented are averages with the corresponding standard deviations (SDs).

4.5. SYTOXGreen (SG) Uptake Assay

B16F10 cells were trypsinized and counted, and 1.5×10^4 cells/well in DMEM containing 0.1% (*w/v*) bovine serum albumin BSA were plated in a cell-view glass bottom culture dish with four compartments (Thermo Fisher Scientific Inc, Maryland, USA) and then incubated at 37 °C overnight to adhere. Post-treatment, the cells were treated with 3 or 50 µM *Psd*1 for 30 or 60 min and maintained at the same temperature. Cells treated with only water or 0.0001% H_2O_2 for 30 min were used as the negative and positive controls, respectively. Washes with saline were carried out between the additions of each probe followed by incubation at room temperature. The cells were incubated with 1 µL of SG, 1.5 µL of MitoTracker Red CMXRos, and 2 µL of DAPI for 10, 20, and 10 min, respectively. The fluorescent images were acquired by confocal microscopy.

4.6. Psd1 Localization Fluorescence Assays

B16F10 cells (1.5×10^4 cells/well) in DMEM containing 0.1% (*w/v*) BSA were plated in the glass-bottom of 4-well plates and then incubated at 37 °C overnight to adhere. For analysis of colocalization, MitoTracker Red CMXRos and DAPI were used to mark the mitochondria and nuclei, respectively, of live B16F10 cells in saline solution at 37 °C. The internalization of FITC-conjugated *Psd*1 was monitored in real time for 2 h by confocal microscopy. When tested by MTT assay (see Section 4.5), 25 µM of FITC-*Psd*1 decreased cell viability of B16F10 about 30%, a compatible value with the unlabeled peptide. Fluorescence quantification of the ROI is described in the Section 4.4 on confocal microscopy.

4.7. Glucosylceramide Depletion Studies on Psd1 B16F10 Cell Entrance

B16F10 cells (1.5×10^4) in cover glass were previously incubated with 20 µM DL-threo-1-phenyl-2-palmitoylamino-3-morpholino-1-propanol (PPMP, Sigma-Aldrich Brasil, São Paulo, Brazil) inhibitor for 60 min followed by treatment with 9 µM *Psd*1 for 30 min or 2 h. The cells were fixed on glass slides with 4% paraformaldehyde plus 4% sucrose for 10 min and then stained with a Vybrant Alexa Fluor 594 Lipid Raft Labeling Kit (Thermo Fisher Scientific Inc, Maryland, USA). Briefly, 1.25 µg/mL cholera toxin subunit B (CT-B) and 160× diluted anti-CT-B were incubated with cells for 10 min each at 4 °C. Three PBS wash steps were performed between each stage. PPMP-treated cell slides were mounted with ProLong Gold antifade reagent with DAPI (Life Technologies, Carlsbad, CA, USA).

4.8. Flow Cytometry-Based Apoptosis Detection

B16F10 cells (4×10^5 cells) were treated with 50 µM *Psd*1 for 24 h. After treatment, the cells were trypsinized, centrifuged, and washed twice with phosphate-buffered saline (PBS). The cells were then stained with Nicoletti buffer (0.1% sodium citrate, 0.1% NP-40, 200 µg/mL RNase, and 50 µg/mL propidium iodide). Doublets and debris were identified and excluded. Analysis of the DNA content was performed by collecting 20,000 events using a BD FACSCalibur flow cytometer (BD Biosciences, San Jose, CA, USA). Cells with fragmented DNA (sub-G0/G1 peak) were considered apoptotic cells.

4.9. Surface Plasmon Resonance Studies

Surface plasmon resonance (SPR) assays were run on a Biacore X (GE Healthcare Life Sciences, Amersham, UK) apparatus in real time using a CM5 sensor chip at 25 °C. Briefly, CM5 chip activation was performed by injection of 100 µL of a 1:1 mixture of amine coupling kit (750 mg 1-ethyl-3-(3-dimethylaminopropyl) carbodiimide hydrochloride (EDC), 115 mg *N*-hydroxysuccinimide (NHS), 2 × 10.5 mL 1.0 M ethanolamine-HCl pH 8.5) with a continuous flow of 2 µL/min. Then, 100 µL of 10 µg/µL recombinant cyclin F-GST diluted in 10 mM sodium acetate pH 4.0 buffer was immobilized by amine coupling onto the carboxylate dextran layer of CM5 with the same flow. This was injected with 80 µL of 1 M ethanolamine pH 8.0 to block the remaining binding free sites of the protein. To evaluate the interaction of *Ps*d1 with the cyclin F protein, increasing concentrations of the peptide (10–240 µM) in 100 µL of running buffer (10 mM HEPES, 150 mM NaCl, pH 7.4) were injected using a flow rate of 15 µL/min for 4 min. After the injection ended, the dissociation phase was measured in not less than 300 s. The *Ps*d1-GST-cyclin F interaction curves were subtracted from the respective curves obtained for *Ps*d1-GST to discount possible artifact interactions with GST alone. The sensorgrams obtained for each peptide–cyclin F interaction were processed by curve fitting with numerical integration analysis using BIA evaluation software 3.0.1. All analyses were run in at least duplicate.

4.10. Psd1–Cyclin F Molecular Docking

The ClusPro server (https://cluspro.org) was used for a blind docking simulation between *Ps*d1 and cyclin F [88]. The *Ps*d1 structure was obtained from the RCSB PDB database (code 1JKZ), while the structure of cyclin F from *Neurospora crassa* was built using the homology modeling program MODELLER. The human F-box/WD-repeat protein (code 1P22) was used as a template model, with its sequence showing 35% identity and 52% similarity with the sequence of *N. crassa* cyclin F. Only the interval between the residues Leu136 and Lys683, consisting of 548 residues out of 1010, was successfully built and used in the following computational studies. The F-box (Leu136 to Tyr177) and WD (Glu372 to Gln622) cyclin F domains were included. Every other residue outside of this interval was excluded in the final structure due to a lack of homologous templates. After these preliminary steps, both molecules (in PDB format) were sent as inputs to ClusPro selected for the advanced option "Others mode", an algorithm that usually yields better results for protein complexes not classified as enzyme–substrate/inhibitor or antibody–antigen complexes. After docking, the two most populated structural clusters were assessed, with the selection of one central structure of each for further analysis. Post-docking analyses used the PyMOL program for generation of the molecular images, determination of the interaction interface, and detection and quantification of the intermolecular hydrogen bonds. The PSAIA (Protein Structure and Interaction Analyzer) program was used for the identification and counting of ionic, hydrophobic and van der Waals contacts. Both the interface and the intermolecular forces were defined by geometric criteria.

4.11. Experimental Animals

Eight-week-old C57BL/6 female mice (CEMIB, Campinas–Brasil) weighing approximately 20 g were individually marked and separated according to their experimental groups. These animals were kept in cages 20 cm × 35 cm × 15 cm (width × length × height) with free access to water and food under day and night cycles that lasted twelve hours with a controlled temperature (20–25 °C) in our own facilities. This study was approved by the Ethics Committee on the Use of Animals of Health Science Centre of the Federal University of Rio de Janeiro, Brazil (CEUA/CCS/UFRJ, CONCEA registered number 01200.001568/2013.87, approved protocol IBCCF 163 at 28 August 2012). All animals received humane care in compliance with the "Principles of Laboratory Animal Care" formulated by the National Society for Medical Research and the "Guide for the Care and Use of Laboratory Animals" prepared by the National Academy of Sciences, USA, and the National Council for Controlling Animal Experimentation, Ministry of Science, Technology and Innovation (CONCEA/MCTI), Brazil.

4.12. Experimental B16F10 Melanoma Metastasis Assays

Mice were injected via the lateral tail vein with 100 μL of PBS only (137 mM NaCl, 2.7 mM KCl, 10 mM Na_2HPO_4, 2 mM KH_2PO_4, pH 7.4) ($n = 17$) or 100 μL of B16F10 cells (2.5×10^5 cells/animal suspended in DMEM) ($n = 15$). Other groups of animals were intravenously co-injected with a freshly prepared, in order to avoid B16F10 cells death, mix of B16F10 cells plus 0.1 mg/kg ($n = 8$), 0.5 mg/kg ($n = 8$), 1 mg/kg ($n = 17$), or 3 mg/kg *Psd*1 ($n = 7$).

Another group of animals received 1 mg/kg of *Psd*1 only ($n = 9$) to evaluate *Psd*1 toxicity. Twenty-one days later, the animals were anaesthetized and sacrificed, and the black lung melanoma nodules were counted and measured under a dissecting microscope (Zeiss AxioPhot fluorescence microscope (Carl-Zeiss, Oberkochen, Germany)).

4.13. Histological Analysis

After the above procedure, two animals from each group were anaesthetized and submitted to treatment with 4% paraformaldehyde in 0.1 M phosphate buffer (pH 7.4) by transcardial perfusion. The lung was removed and immersed in the same fixative solution. The lung specimens were sliced into 5 mm pieces, dehydrated through an ascending ethanol series (70, 95, and 100% ethanol for 30 min each) and then embedded in paraffin using standard procedures. Serial 5 μm-thick sections were prepared using a sliding microtome (Leica Microsystems, Wetzlar, Germany). Hematoxylin and eosin (HE) staining wereperformed to quantify possible foci of inflammatory infiltrates and fibrosis in experimental and control animals. The stained sections were observed, and digital images were taken with a Zeiss AxioPhot fluorescence microscope (Carl-Zeiss, Oberkochen, Germany). Specifically, two slices of each sample (six fields of each slice corresponding to approximately 90% of the slice) were used for quantitative analysis to obtain the mean value.

4.14. Data Analysis

The results are presented as the mean values ± SD and were interpreted using one-way ANOVA with Dunnett's or Bonferroni post-tests or by two-way ANOVA, according to each experiment and as indicated in each figure legend. Differences were statistically significant when the p value was less than 0.05.

Supplementary Materials: Supplementary materials can be found at http://www.mdpi.com/1422-0067/21/8/2662/s1.

Author Contributions: E.K., V.S.G.d.A., P.C.d.A., and R.Q.M. provided conception and design of research; V.S.G.d.A., P.C.d.A., S.A.C.S.S., L.B.G., N.S.J., and V.H.A. performed experiments; E.K., P.C.d.A., L.N.d.M., V.S.G.d.A., V.H.M., and R.Q.M. analyzed data; V.S.G.d.A., P.C.d.A., L.B.G., L.N.d.M., and E.K. interpreted results of experiments; V.S.G.d.A., S.A.C.S.S., J.N., P.C.d.A., V.H.A., N.S.J., and E.K. prepared the figures; V.S.G.d.A., J.N., and E.K. drafted the manuscript; V.S.G.A., V.H.A., R.Q.M., and E.K. edited and revised the manuscript; V.S.G.d.A., V.H.A., R.Q.M., and E.K. approved the final version of manuscript, E.K., R.Q.M., P.G.P., and L.B.G. provided funding acquisition. All authors have read and agreed to the published version of the manuscript.

Funding: This research was funded by Conselho Nacional de Desenvolvimento Científico e Tecnológico (CNPq, Brasil) grants numbers 429583/2016-8 (E.K.) and 309946/2018-2 (R.Q.M.), and Fundação de Amparo a Pesquisa do Rio de Janeiro (FAPERJ, Brasil) grant number E-26/203.059/2017 (E.K.) and E-26/202.871/2018 (R.Q.M.).

Acknowledgments: The authors acknowledge Aline Sol Valle for technical support in *Pichia pastoris* recombinant expression and purification of *Psd*1, Eliana Barreto Bergman and Rodrigo Rollin Pinheiro for kind donation of PPMP and Vybrant Alexa Fluor 594 lipid raft kit, and Dra. Isabella Guimarães for technical support in flow cytometry.

Conflicts of Interest: The authors declare no conflicts of interest.

References

1. Cools, T.L.; Struyfs, C.; Cammue, B.P.; Thevissen, K. Antifungal plant defensins: Increased insight in their mode of action as a basis for their use to combat fungal infections. *Future Microbiol.* **2017**, *12*, 441–454. [CrossRef] [PubMed]
2. Maroti, G.; Kereszt, A.; Kondorosi, E.; Mergaert, P. Natural roles of antimicrobial peptides in microbes, plants and animals. *Res. Microbiol.* **2011**, *162*, 363–374. [CrossRef] [PubMed]
3. Tam, J.P.; Wang, S.; Wong, K.H.; Tan, W.L. Antimicrobial Peptides from Plants. *Pharmaceuticals* **2015**, *8*, 711–757. [CrossRef] [PubMed]
4. Almeida, M.S.; Cabral, K.M.; Zingali, R.B.; Kurtenbach, E. Characterization of two novel defense peptides from pea (*Pisum sativum*) seeds. *Arch. Biochem. Biophys.* **2000**, *378*, 278–286. [CrossRef] [PubMed]
5. Parisi, K.; Shafee, T.M.A.; Quimbar, P.; van der Weerden, N.L.; Bleackley, M.R.; Anderson, M.A. The evolution, function and mechanisms of action for plant defensins. *Semin. Cell Dev. Biol.* **2019**, *88*, 107–118. [CrossRef] [PubMed]
6. Pinheiro-Aguiar, R.; do Amaral, V.S.G.; Pereira, I.B.; Kurtenbach, E.; Almeida, F.C.L. Nuclear magnetic resonance solution structure of *Pisum sativum* defensin 2 provides evidence for the presence of hydrophobic surface-clusters. *Proteins* **2019**, *88*, 242–246. [CrossRef] [PubMed]
7. Almeida, M.S.; Cabral, K.M.; Kurtenbach, E.; Almeida, F.C.; Valente, A.P. Solution structure of *Pisum sativum* defensin 1 by high resolution NMR: Plant defensins, identical backbone with different mechanisms of action. *J. Mol. Biol.* **2002**, *315*, 749–757. [CrossRef]
8. Medeiros, L.N.d.; Domitrovic, T.; Andrade, P.C.d.; Faria, J.; Barreto-Bergter, E.; Weissmüller, G.; Kurtenbach, E. *Psd*1 binding affinity toward fungal membrane components as assessed by SPR: The role of glucosylceramide in fungal recognition and entry. *Pept. Sci.* **2014**, *102*, 456–464. [CrossRef]
9. Pushpanathan, M.; Gunasekaran, P.; Rajendhran, J. Antimicrobial peptides: Versatile biological properties. *Int. J. Pept.* **2013**, *2013*, 675391. [CrossRef]
10. Amaral, V.S.G.; Fernandes, C.M.; Felício, M.R.; Valle, A.S.; Quintana, P.G.; Almeida, C.C.; Barreto-Bergter, E.; Gonçalves, S.; Santos, N.C.; Kurtenbach, E. *Psd*2 pea defensin shows a preference for mimetic membrane rafts enriched with glucosylceramide and ergosterol. *Biochim. Et Biophys. Acta (Bba)-Biomembr.* **2019**, *1861*, 713–728. [CrossRef]
11. Fernandes, C.; de Castro, P.; Singh, A.; Fonseca, F.; Pereira, M.; Vila, T.; Atella, G.; Rozental, S.; Savoldi, M.; Del Poeta, M. Functional characterization of the *Aspergillus nidulans* glucosylceramide pathway reveals that LCB Δ8-desaturation and C9-methylation are relevant to filamentous growth, lipid raft localization and *Psd*1 defensin activity. *Mol. Microbiol.* **2016**, *102*, 488–505. [CrossRef]
12. Ramamoorthy, V.; Cahoon, E.B.; Li, J.; Thokala, M.; Minto, R.E.; Shah, D.M. Glucosylceramide synthase is essential for alfalfa defensin-mediated growth inhibition but not for pathogenicity of *Fusarium graminearum*. *Mol. Microbiol.* **2007**, *66*, 771–786. [CrossRef] [PubMed]
13. Thevissen, K.; Warnecke, D.C.; Francois, I.E.; Leipelt, M.; Heinz, E.; Ott, C.; Zähringer, U.; Thomma, B.P.; Ferket, K.K.; Cammue, B.P. Defensins from insects and plants interact with fungal glucosylceramides. *J. Biol. Chem.* **2004**, *279*, 3900–3905. [CrossRef] [PubMed]
14. Thevissen, K.; Cammue, B.P.; Lemaire, K.; Winderickx, J.; Dickson, R.C.; Lester, R.L.; Ferket, K.K.; Van Even, F.; Parret, A.H.; Broekaert, W.F. A gene encoding a sphingolipid biosynthesis enzyme determines the sensitivity of *Saccharomyces cerevisiae* to an antifungal plant defensin from dahlia (*Dahlia merckii*). *Proc. Natl. Acad. Sci. USA* **2000**, *97*, 9531–9536. [CrossRef] [PubMed]
15. Medeiros, L.N.; Angeli, R.; Sarzedas, C.G.; Barreto-Bergter, E.; Valente, A.P.; Kurtenbach, E.; Almeida, F.C. Backbone dynamics of the antifungal *Psd*1 pea defensin and its correlation with membrane interaction by NMR spectroscopy. *Biochim. Biophys. Acta (BBA)-Biomembr.* **2010**, *1798*, 105–113. [CrossRef]
16. Gonçalves, S.; Teixeira, A.; Abade, J.; de Medeiros, L.N.; Kurtenbach, E.; Santos, N.C. Evaluation of the membrane lipid selectivity of the pea defensin *Psd*1. *Biochim. Biophys. Acta (BBA)-Biomembr.* **2012**, *1818*, 1420–1426. [CrossRef]
17. Goncalves, S.; Silva, P.M.; Felicio, M.R.; de Medeiros, L.N.; Kurtenbach, E.; Santos, N.C. *Psd*1 Effects on *Candida albicans* Planktonic Cells and Biofilms. *Front. Cell. Infect. Microbiol.* **2017**, *7*, 249. [CrossRef]

18. Lobo, D.S.; Pereira, I.B.; Fragel-Madeira, L.; Medeiros, L.N.; Cabral, L.M.; Faria, J.; Bellio, M.; Campos, R.C.; Linden, R.; Kurtenbach, E. Antifungal *Pisum sativum* defensin 1 interacts with *Neurospora crassa* cyclin F related to the cell cycle. *Biochemistry* **2007**, *46*, 987–996. [CrossRef]

19. D'Angiolella, V.; Donato, V.; Forrester, F.M.; Jeong, Y.-T.; Pellacani, C.; Kudo, Y.; Saraf, A.; Florens, L.; Washburn, M.P.; Pagano, M. Cyclin F-mediated degradation of ribonucleotide reductase M2 controls genome integrity and DNA repair. *Cell* **2012**, *149*, 1023–1034. [CrossRef]

20. D'Angiolella, V.; Esencay, M.; Pagano, M. A cyclin without cyclin-dependent kinases: Cyclin F controls genome stability through ubiquitin-mediated proteolysis. *Trends Cell Biol.* **2013**, *23*, 135–140. [CrossRef]

21. Cardozo, T.; Pagano, M. The SCF ubiquitin ligase: Insights into a molecular machine. *Nat. Rev. Mol. Cell Biol.* **2004**, *5*, 739–751. [CrossRef]

22. Clijsters, L.; Hoencamp, C.; Calis, J.J.; Marzio, A.; Handgraaf, S.M.; Cuitino, M.C.; Rosenberg, B.R.; Leone, G.; Pagano, M. Cyclin F controls cell-cycle transcriptional outputs by directing the degradation of the three activator E2Fs. *Mol. Cell* **2019**, *74*, 1264–1277.e7. [CrossRef]

23. Guaní-Guerra, E.; Santos-Mendoza, T.; Lugo-Reyes, S.O.; Terán, L.M. Antimicrobial peptides: General overview and clinical implications in human health and disease. *Clin. Immunol.* **2010**, *135*, 1–11. [CrossRef]

24. Kruse, T.; Kristensen, H.H. Using antimicrobial host defense peptides as anti-infective and immunomodulatory agents. *Expert Rev. Anti Infect. Ther.* **2008**, *6*, 887–895. [CrossRef]

25. Riedl, S.; Zweytick, D.; Lohner, K. Membrane-active host defense peptides-challenges and perspectives for the development of novel anticancer drugs. *Chem. Phys. Lipids* **2011**, *164*, 766–781. [CrossRef] [PubMed]

26. Gaspar, D.; Veiga, A.S.; Castanho, M.A. From antimicrobial to anticancer peptides. A review. *Front. Microbiol.* **2013**, *4*, 294. [CrossRef] [PubMed]

27. Beloribi-Djefaflia, S.; Vasseur, S.; Guillaumond, F. Lipid metabolic reprogramming in cancer cells. *Oncogenesis* **2016**, *5*, e189. [CrossRef] [PubMed]

28. Utsugi, T.; Schroit, A.J.; Connor, J.; Bucana, C.D.; Fidler, I.J. Elevated expression of phosphatidylserine in the outer membrane leaflet of human tumor cells and recognition by activated human blood monocytes. *Cancer Res.* **1991**, *51*, 3062–3066.

29. Riedl, S.; Rinner, B.; Asslaber, M.; Schaider, H.; Walzer, S.; Novak, A.; Lohner, K.; Zweytick, D. In search of a novel target - phosphatidylserine exposed by non-apoptotic tumor cells and metastases of malignancies with poor treatment efficacy. *Biochim. Biophys. Acta* **2011**, *1808*, 2638–2645. [CrossRef]

30. Lavie, Y.; Cao, H.; Bursten, S.L.; Giuliano, A.E.; Cabot, M.C. Accumulation of glucosylceramides in multidrug-resistant cancer cells. *J. Biol. Chem.* **1996**, *271*, 19530–19536. [CrossRef]

31. Liu, Y.-Y.; Han, T.-Y.; Giuliano, A.E.; Cabot, M.C. Expression of glucosylceramide synthase, converting ceramide to glucosylceramide, confers adriamycin resistance in human breast cancer cells. *J. Biol. Chem.* **1999**, *274*, 1140–1146. [CrossRef]

32. Morjani, H.; Aouali, N.; Belhoussine, R.; Veldman, R.J.; Levade, T.; Manfait, M. Elevation of glucosylceramide in multidrug-resistant cancer cells and accumulation in cytoplasmic droplets. *Int. J. Cancer* **2001**, *94*, 157–165. [CrossRef]

33. Saito, K.; Takakuwa, N.; Ohnishi, M.; Oda, Y. Presence of glucosylceramide in yeast and its relation to alkali tolerance of yeast. *Appl. Microbiol. Biotechnol.* **2006**, *71*, 515–521. [CrossRef]

34. Borradaile, N.M.; Han, X.; Harp, J.D.; Gale, S.E.; Ory, D.S.; Schaffer, J.E. Disruption of endoplasmic reticulum structure and integrity in lipotoxic cell death. *J. Lipid Res.* **2006**, *47*, 2726–2737. [CrossRef]

35. Li, J.-F.; Zheng, S.-J.; Wang, L.-L.; Liu, S.; Ren, F.; Chen, Y.; Bai, L.; Liu, M.; Duan, Z.P. Glucosylceramide synthase regulates the proliferation and apoptosis of liver cells in vitro by Bcl-2/Bax pathway. *Mol. Med. Rep.* **2017**, *16*, 7355–7360. [CrossRef] [PubMed]

36. De Oliveira, A.R.; da Silva, I.D.C.G.; Lo Turco, E.G.; Júnior, H.A.M.; Chauffaille, M.d.L.L.F. Initial analysis of lipid metabolomic profile reveals differential expression features in myeloid malignancies. *J. Cancer Ther.* **2015**, *6*, 1262–1272. [CrossRef]

37. Jeon, J.H.; Kim, S.K.; Kim, H.J.; Chang, J.; Ahn, C.M.; Chang, Y.S. Lipid raft modulation inhibits NSCLC cell migration through delocalization of the focal adhesion complex. *Lung Cancer* **2010**, *69*, 165–171. [CrossRef] [PubMed]

38. Wang, R.; Bi, J.; Ampah, K.K.; Zhang, C.; Li, Z.; Jiao, Y.; Wang, X.; Ba, X.; Zeng, X. Lipid raft regulates the initial spreading of melanoma A375 cells by modulating β1 integrin clustering. *Int. J. Biochem. Cell B* **2013**, *45*, 1679–1689. [CrossRef] [PubMed]

39. Al-Benna, S.; Shai, Y.; Jacobsen, F.; Steinstraesser, L. Oncolytic activities of host defense peptides. *Int. J. Mol. Sci.* **2011**, *12*, 8027–8051. [CrossRef] [PubMed]

40. Mosmann, T. Rapid colorimetric assay for cellular growth and survival: Application to proliferation and cytotoxicity assays. *J. Immunol. Methods* **1983**, *65*, 55–63. [CrossRef]

41. Sorvina, A.; Bader, C.A.; Darby, J.R.; Lock, M.C.; Soo, J.Y.; Johnson, I.R.; Caporale, C.; Voelcker, N.H.; Stagni, S.; Massi, M. Mitochondrial imaging in live or fixed tissues using a luminescent iridium complex. *Sci. Rep.* **2018**, *8*, 1–8. [CrossRef]

42. Guimaraes, C.A.; Linden, R. Programmed cell deaths. Apoptosis and alternative deathstyles. *Eur. J. Biochem.* **2004**, *271*, 1638–1650. [CrossRef]

43. Ablan, S.; Rawat, S.S.; Blumenthal, R.; Puri, A. Entry of influenza virus into a glycosphingolipid-deficient mouse skin fibroblast cell line. *Arch. Virol.* **2001**, *146*, 2227–2238. [CrossRef]

44. Chan, S.Y.; Hilchie, A.L.; Brown, M.G.; Anderson, R.; Hoskin, D.W. Apoptosis induced by intracellular ceramide accumulation in MDA-MB-435 breast carcinoma cells is dependent on the generation of reactive oxygen species. *Exp. Mol. Pathol.* **2007**, *82*, 1–11. [CrossRef]

45. Lee, H.S.; Park, C.B.; Kim, J.M.; Jang, S.A.; Park, I.Y.; Kim, M.S.; Cho, J.H.; Kim, S.C. Mechanism of anticancer activity of buforin IIb, a histone H2A-derived peptide. *Cancer Lett.* **2008**, *271*, 47–55. [CrossRef] [PubMed]

46. Inokuchi, J.; Usuki, S.; Jimbo, M. Stimulation of glycosphingolipid biosynthesis by L-threo-1-phenyl-2-decanoylamino-1-propanol and its homologs in B16 melanoma cells. *J. Biochem.* **1995**, *117*, 766–773. [CrossRef] [PubMed]

47. Coelho, V.d.M.; Nguyen, D.; Giri, B.; Bunbury, A.; Schaffer, E.; Taub, D.D. Quantitative differences in lipid raft components between murine CD4+ and CD8+ T cells. *BMC Immunol.* **2004**, *5*, 2.

48. Nichols, B.J.; Kenworthy, A.K.; Polishchuk, R.S.; Lodge, R.; Roberts, T.H.; Hirschberg, K.; Phair, R.D.; Lippincott-Schwartz, J. Rapid cycling of lipid raft markers between the cell surface and Golgi complex. *J. Cell Biol.* **2001**, *153*, 529–542. [CrossRef] [PubMed]

49. Merritt, E.A.; Sarfaty, S.; Akker, F.V.D.; L'hoir, C.; Martial, J.A.; Hol, W.G. Crystal structure of cholera toxin B-pentamer bound to receptor GM1 pentasaccharide. *Protein Sci.* **1994**, *3*, 166–175. [CrossRef] [PubMed]

50. Calorini, L.; Fallani, A.; Tombaccini, D.; Mugnai, G.; Ruggieri, S. Lipid composition of cultured B16 melanoma cell variants with different lung-colonizing potential. *Lipids* **1987**, *22*, 651–656. [CrossRef]

51. Diaz-Rohrer, B.B.; Levental, K.R.; Simons, K.; Levental, I. Membrane raft association is a determinant of plasma membrane localization. *Proc. Natl. Acad. Sci. USA* **2014**, *111*, 8500–8505. [CrossRef]

52. Diaz-Rohrer, B.; Levental, K.R.; Levental, I. Rafting through traffic: Membrane domains in cellular logistics. *Biochim. Biophys. Acta* **2014**, *1838*, 3003–3013. [CrossRef]

53. Bhatia, S.; Tykodi, S.S.; Thompson, J.A. Treatment of metastatic melanoma: An overview. *Oncology* **2009**, *23*, 488–496.

54. Wong, J.H.; Ng, T.B. Sesquin, a potent defensin-like antimicrobial peptide from ground beans with inhibitory activities toward tumor cells and HIV-1 reverse transcriptase. *Peptides* **2005**, *26*, 1120–1126. [CrossRef]

55. Wong, J.H.; Ng, T.B. Lunatusin, a trypsin-stable antimicrobial peptide from lima beans (*Phaseolus lunatus* L.). *Peptides* **2005**, *26*, 2086–2092. [CrossRef]

56. Wong, J.H.; Ng, T. Limenin, a defensin-like peptide with multiple exploitable activities from shelf beans. *J. Pept. Sci.* **2006**, *12*, 341–346. [CrossRef] [PubMed]

57. Lin, P.; Wong, J.H.; Ng, T.B. A defensin with highly potent antipathogenic activities from the seeds of purple pole bean. *Biosci. Rep.* **2009**, *30*, 101–109. [CrossRef] [PubMed]

58. Ngai, P.H.; Ng, T.B. Phaseococcin, an antifungal protein with antiproliferative and anti-HIV-1 reverse transcriptase activities from small scarlet runner beans. *Biochem. Cell Biol.* **2005**, *83*, 212–220. [CrossRef] [PubMed]

59. Ngai, P.H.; Ng, T. Coccinin, an antifungal peptide with antiproliferative and HIV-1 reverse transcriptase inhibitory activities from large scarlet runner beans. *Peptides* **2004**, *25*, 2063–2068. [CrossRef] [PubMed]

60. Anaya-Lopez, J.L.; Lopez-Meza, J.E.; Baizabal-Aguirre, V.M.; Cano-Camacho, H.; Ochoa-Zarzosa, A. Fungicidal and cytotoxic activity of a Capsicum chinense defensin expressed by endothelial cells. *Biotechnol. Lett.* **2006**, *28*, 1101–1108. [CrossRef] [PubMed]

61. Pendergrass, W.; Wolf, N.; Poot, M. Efficacy of MitoTracker Green™ and CMXrosamine to measure changes in mitochondrial membrane potentials in living cells and tissues. *Cytom. A* **2004**, *61*, 162–169. [CrossRef] [PubMed]

62. Lemasters, J.J. Charpter 1—Molecular Mechanisms of Cell Death. In *Molecular Pathology*; Academic Press: Cambridge, MA, USA, 2018; pp. 1–24. [CrossRef]

63. Liu, X.; Cheng, J.C.; Turner, L.S.; Elojeimy, S.; Beckham, T.H.; Bielawska, A.; Keane, T.E.; Hannun, Y.A.; Norris, J.S. Acid ceramidase upregulation in prostate cancer: Role in tumor development and implications for therapy. *Expert Opin. Ther. Targets* **2009**, *13*, 1449–1458. [CrossRef]

64. Yang, L.; Cui, X.; Zhang, N.; Li, M.; Bai, Y.; Han, X.; Shi, Y.; Liu, H. Comprehensive lipid profiling of plasma in patients with benign breast tumor and breast cancer reveals novel biomarkers. *Anal. Bioanal. Chem.* **2015**, *407*, 5065–5077. [CrossRef]

65. Birge, R.; Boeltz, S.; Kumar, S.; Carlson, J.; Wanderley, J.; Calianese, D.; Barcinski, M.; Brekken, R.; Huang, X.; Hutchins, J. Phosphatidylserine is a global immunosuppressive signal in efferocytosis, infectious disease, and cancer. *Cell Death Differ.* **2016**, *23*, 962–978. [CrossRef]

66. Li, A.; Yao, L.; Fang, Y.; Yang, K.; Jiang, W.; Huang, W.; Cai, Z. Specifically blocking the fatty acid synthesis to inhibit the malignant phenotype of bladder cancer. *Int. J. Biol. Sci.* **2019**, *15*, 1610–1617. [CrossRef]

67. Menendez, J.A.; Lupu, R. Fatty acid synthase and the lipogenic phenotype in cancer pathogenesis. *Nat. Rev. Cancer* **2007**, *7*, 763–777. [CrossRef] [PubMed]

68. Qi, H.; Priyadarsini, S.; Nicholas, S.E.; Sarker-Nag, A.; Allegood, J.; Chalfant, C.E.; Mandal, N.A.; Karamichos, D. Analysis of sphingolipids in human corneal fibroblasts from normal and keratoconus patients. *J. Lipid Res.* **2017**, *58*, 636–648. [CrossRef] [PubMed]

69. Van Meer, G.; Voelker, D.R.; Feigenson, G.W. Membrane lipids: Where they are and how they behave. *Nat. Rev. Mol. Cell Biol.* **2008**, *9*, 112–124. [CrossRef] [PubMed]

70. Van Meer, G.; de Kroon, A.I. Lipid map of the mammalian cell. *J. Cell Sci.* **2011**, *124*, 5–8. [CrossRef] [PubMed]

71. Ryland, L.K.; Fox, T.E.; Liu, X.; Loughran, T.P.; Kester, M. Dysregulation of sphingolipid metabolism in cancer. *Cancer Biol. Ther.* **2011**, *11*, 138–149. [CrossRef] [PubMed]

72. Ternes, P.; Wobbe, T.; Schwarz, M.; Albrecht, S.; Feussner, K.; Riezman, I.; Cregg, J.M.; Heinz, E.; Riezman, H.; Feussner, I.; et al. Two pathways of sphingolipid biosynthesis are separated in the yeast *Pichia pastoris*. *J. Biol. Chem.* **2011**, *286*, 11401–11414. [CrossRef]

73. Huang, H.; Tong, T.-T.; Yau, L.-F.; Chen, C.-Y.; Mi, J.-N.; Wang, J.-R.; Jiang, Z.-H. LC-MS based sphingolipidomic study on A549 human lung adenocarcinoma cell line and its taxol-resistant strain. *BMC Cancer* **2018**, *18*, 799. [CrossRef]

74. Machado, L.; De Paula, V.S.; Pustovalova, Y.; Bezsonova, I.; Valente, A.P.; Korzhnev, D.M.; Almeida, F.C.L. Conformational dynamics of a cysteine-stabilized plant defensin reveals an evolutionary mechanism to expose hydrophobic residues. *Biochemistry* **2018**, *57*, 5797–5806. [CrossRef]

75. Choudhury, R.; Bonacci, T.; Wang, X.; Truong, A.; Arceci, A.; Zhang, Y.; Mills, C.A.; Kernan, J.L.; Liu, P.; Emanuele, M.J. The E3 ubiquitin ligase SCF (cyclin F) transmits AKT signaling to the cell-cycle machinery. *Cell Rep.* **2017**, *20*, 3212–3222. [CrossRef]

76. Bai, C.; Richman, R.; Elledge, S.J. Human cyclin F. *EMBO J.* **1994**, *13*, 6087–6098. [CrossRef]

77. Zheng, N.; Zhou, Q.; Wang, Z.; Wei, W. Recent advances in SCF ubiquitin ligase complex: Clinical implications. *BBA-Rev. Cancer* **2016**, *1866*, 12–22. [CrossRef]

78. D'Angiolella, V.; Donato, V.; Vijayakumar, S.; Saraf, A.; Florens, L.; Washburn, M.P.; Dynlacht, B.; Pagano, M. SCF Cyclin F controls centrosome homeostasis and mitotic fidelity through CP110 degradation. *Nature* **2010**, *466*, 138–142. [CrossRef] [PubMed]

79. Emanuele, M.J.; Elia, A.E.; Xu, Q.; Thoma, C.R.; Izhar, L.; Leng, Y.; Guo, A.; Chen, Y.-N.; Rush, J.; Hsu, P.W.-C. Global identification of modular cullin-RING ligase substrates. *Cell* **2011**, *147*, 459–474. [CrossRef] [PubMed]

80. Walter, D.; Hoffmann, S.; Komseli, E.-S.; Rappsilber, J.; Gorgoulis, V.; Sorensen, C.S. SCF Cyclin F-dependent degradation of CDC6 suppresses DNA re-replication. *Nat. Commun.* **2016**, *7*, 10530. [CrossRef] [PubMed]

81. Elia, A.E.; Boardman, A.P.; Wang, D.C.; Huttlin, E.L.; Everley, R.A.; Dephoure, N.; Zhou, C.; Koren, I.; Gygi, S.P.; Elledge, S.J. Quantitative proteomic atlas of ubiquitination and acetylation in the DNA damage response. *Mol. Cell* **2015**, *59*, 867–881. [CrossRef]

82. Dankert, J.F.; Rona, G.; Clijsters, L.; Geter, P.; Skaar, J.R.; Bermudez-Hernandez, K.; Sassani, E.; Fenyö, D.; Ueberheide, B.; Schneider, R. Cyclin F-mediated degradation of SLBP limits H2A. X accumulation and apoptosis upon genotoxic stress in G2. *Mol. Cell* **2016**, *64*, 507–519. [CrossRef]

83. Jones, K.B.; Su, L.; Jin, H.; Lenz, C.; Randall, R.L.; Underhill, T.M.; Nielsen, T.O.; Sharma, S.; Capecchi, M.R. SS18-SSX2 and the mitochondrial apoptosis pathway in mouse and human synovial sarcomas. *Oncogene* **2013**, *32*, 2365–2375.e5. [CrossRef]

84. Macheret, M.; Halazonetis, T.D. Intragenic origins due to short G1 phases underlie oncogene-induced DNA replication stress. *Nature* **2018**, *555*, 112. [CrossRef]

85. Tchirkov, A.; Khalil, T.; Chautard, E.; Mokhtari, K.; Veronese, L.; Irthum, B.; Vago, P.; Kémény, J.; Verrelle, P. Interleukin-6 gene amplification and shortened survival in glioblastoma patients. *Br. J. Cancer* **2007**, *96*, 474–476. [CrossRef]

86. Roudi, R.; Syn, N.L.; Roudbary, M. Antimicrobial peptides as biologic and immunotherapeutic agents against cancer: A comprehensive overview. *Front. Immunol.* **2017**, *8*, 1320. [CrossRef]

87. Lowry, O.H.; Rosebrough, N.J.; Farr, A.L.; Randall, R.J. Protein measurement with the Folin phenol reagent. *J. Biol. Chem.* **1951**, *193*, 265–275.

88. Kozakov, D.; Hall, D.R.; Xia, B.; Porter, K.A.; Padhorny, D.; Yueh, C.; Beglov, D.; Vajda, S. The ClusPro web server for protein–protein docking. *Nat. Protoc.* **2017**, *12*, 255–278. [CrossRef]

International Journal of
Molecular Sciences

Article

A Synthetic Cell-Penetrating Heparin-Binding Peptide Derived from BMP4 with Anti-Inflammatory and Chondrogenic Functions for the Treatment of Arthritis

Da Hyeon Choi [1], Dongwoo Lee [2], Beom Soo Jo [2], Kwang-Sook Park [3], Kyeong Eun Lee [1], Ju Kwang Choi [1], Yoon Jeong Park [2,3], Jue-Yeon Lee [2,*] and Yoon Shin Park [1,*]

[1] School of Biological Sciences, College of Natural Sciences, Chungbuk National University, Cheongju 28644, Korea; dahyun6682@hanmail.net (D.H.C.); dlruddms1223@naver.com (K.E.L.); jukwang@chungbuk.ac.kr (J.K.C.)

[2] Central Research Institute, Nano Intelligent Biomedical Engineering Corporation (NIBEC), School of Dentistry, Seoul National University, Seoul 03080, Korea; ldw410@nibec.co.kr (D.L.); beomsoojo@nibec.co.kr (B.S.J.); parkyj@snu.ac.kr (Y.J.P.)

[3] Department of Dental Regenerative Bioengineering and Dental Research Institute, School of Dentistry, Seoul National University, Seoul 03080, Korea; soooook7424@snu.ac.kr

* Correspondence: yeon0417@nibec.co.kr (J.-Y.L.); pys@cbnu.ac.kr (Y.S.P.); Tel.: +82-2-765-1976 (J.-Y.L.); +82-43-261-2303 (Y.S.P.)

Received: 27 May 2020; Accepted: 12 June 2020; Published: 15 June 2020

Abstract: We report dual therapeutic effects of a synthetic heparin-binding peptide (HBP) corresponding to residues 15–24 of the heparin binding site in BMP4 in a collagen-induced rheumatic arthritis model (CIA) for the first time. The cell penetrating capacity of HBP led to improved cartilage recovery and anti-inflammatory effects via down-regulation of the iNOS-IFNγ-IL6 signaling pathway in inflamed RAW264.7 cells. Both arthritis and paw swelling scores were significantly improved following HBP injection into CIA model mice. Anti-rheumatic effects were accelerated upon combined treatment with Enbrel® and HBP. Serum IFNγ and IL6 concentrations were markedly reduced following intraperitoneal HBP injection in CIA mice. The anti-rheumatic effects of HBP in mice were similar to those of Enbrel®. Furthermore, the combination of Enbrel® and HBP induced similar anti-rheumatic and anti-inflammatory effects as Enbrel®. We further investigated the effect of HBP on damaged chondrocytes in CIA mice. Regenerative capacity of HBP was confirmed based on increased expression of chondrocyte biomarker genes, including aggrecan, collagen type II and TNFα, in adult human knee chondrocytes. These findings collectively support the utility of our cell-permeable bifunctional HBP with anti-inflammatory and chondrogenic properties as a potential source of therapeutic agents for degenerative inflammatory diseases.

Keywords: anti-inflammatory peptide; cell permeable peptide; heparin-binding peptide; collagen-induced arthritis; inducible nitric oxide; interferon gamma; interleukin-6; Enbrel

---

1. Introduction

Rheumatic arthritis (RA) is one of the most common autoimmune rheumatic diseases (AIRD) caused by dysregulation of tolerance to self-antigens, leading to chronic systemic inflammatory disorders involving the musculoskeletal system [1]. According to Korean National Health Insurance (NHI) Claims Database reports, the steadily increasing prevalence of RA poses a significant economic burden [2].

The causes and mechanisms of action of RA have been relatively unexplored to date. The primary treatment goal is control of joint inflammation and reduction of joint damage, the major determinant of functional prognosis in RA [3]. The most common therapeutic agents for RA are disease-modifying anti-rheumatic drugs (DMARD), traditionally comprising synthetic chemicals and small molecules that act as anti-inflammatory agents [4,5].

TNFα is a key cytokine in the inflammatory process in RA [6]. Five TNFα blockers have been approved by US-FDA for RA therapy, specifically, adalimumab, certolizumab pegol, etanercept, infliximab, and golimumab. In randomized controlled clinical trials, these drugs are reported to be effective in recovery of inflammation-related clinical signs in RA patients with response failure to synthetic DMARDs [7]. Multiple studies have demonstrated remarkable therapeutic benefits of early co-treatment with TNFα blockers and methotrexate [7]. Other FDA-approved drugs for treatment of moderate to severe RA include abatacept, rituximab, and tocilizumab. However, these drugs commonly induce severe side-effects, such as osteoporosis, liver function failure, and even lymphoma, due to the complexity of the inflammatory network [8,9]. Various attempts have been made to counteract the side-effects of these therapeutic agents [10]. Treatment with synthetic disease-modifying anti-rheumatic drugs (DMARD), including methotrexate, sulfasalazine, and leflunomide, represents an important paradigm shift that can lead to remarkable improvement of clinical symptoms and delayed joint damage. Despite the effectiveness of these medications, a significant number of RA patients continue to experience clinical symptoms of inflammation and progressive joint destruction, clearly suggesting that therapeutic blockade with any one cytokine is not necessarily sufficient for relief of RA [11,12].

To improve therapeutic efficacy and/or prolong drug persistence, various tissue-engineered materials have been developed. For instance, an injectable electrostatically interacting drug depot [13] and click-crosslinked hyaluronic acid depot [14] have been developed as a supplementary or combination therapeutic agent for RA with minocycline and methotrexate, respectively. Direct intra- articular injection resulted in increased duration of therapy and enhanced relief of symptoms by inducing sustained release of the ionic drug. Additionally, novel biomimetic scaffolds or drug carriers have been suggested as modes for delivering nanosized hydroxyapatite (HA) to promote damaged tissue recovery [15] through osteoblast adhesion [16] and bone adjuvant activities [17]. Limitations of tissue-engineered agents or scaffolds are mostly attributed to uncontrolled proinflammatory cytokines and inflammation-related signaling during RA pathogenesis.

Biological compounds, such as neutralizing antibodies against interleukin-17A (IL17A; secukinumab and ixekizumab) and janus kinase (JAK) inhibitors (tofacitinib, baricitinib, GLPG0634, and VX-509) [18,19], have been characterized as beneficial agents with significant therapeutic effects. In particular, IL6 neutralization is a recently developed therapeutic strategy for controlling RA. Neutralization of IL6 plays a critical role in both the initiation and perpetuation of immunologic dysfunction and inflammatory responses in RA via modulation of the IL6/IL6 receptor/gp130 pathway [20]. A novel role of IL6/IL6R/gp130 signaling in autoimmune arthritis has been reported. An apparent strong link exists between enhanced Toll-Like Receptor (TLR) expression as a consequence of RA development and production of IL6 accompanied by upregulation of MMP genes [21–23]. However, protein-based biological agents are expensive and associated with safety concerns, such as development of opportunistic and viral infections [24–26]. Biological agents, such as neutralizing antibodies that are mainly composed of protein polymers, can induce severe infections, drug resistance, and/or autoimmune reactions unless completely degraded in the body [27,28].

To overcome these limitations, the use of peptide drugs as substitutes for common RA therapeutics has been widely investigated in recent years [18,19,29]. The advantages of peptide agents include high selectivity, high potency, low immunogenicity, low toxicity, improved efficiency and safety, low tissue accumulation, easy combination with biomaterial surfaces via chemical modifications, and low cost [19,30]. Safe and effective novel peptide therapeutic agents for RA treatment are urgently required.

Here, we focused on the anti-inflammatory effects of a synthetic heparin-binding peptide (HBP) composed of 10 amino acids derived from the heparin binding site of bone morphogenetic protein 4 (BMP4) [31]. The HBP sequence examined in this study, RKKNPNCRRH, corresponds to residues 15–24 of the heparin binding site in BMP4, and is similar to the HBP "consensus sequence" BBBXTXXBBB (whereby X, B, and T represent hydropathic residue, basic residue, and turn, respectively) [31]. We previously identified an anti-angiogenic role of HBP in endothelial cells (EC) and a breast cancer xenograft model [32]. The collective results suggest that HBP may serve as potential antitumor agents by regulating the tumor microenvironment through interactions with several cytokines [32]. Human heparin binding domain is the most recently discovered member of the host defense peptide family involved in adaptive immunity and anti-inflammatory activity [33,34]. The domain itself is unable to penetrate immune cells for chemical production due to its high amino acid content. However, selected key peptide sequences of this domain with cell penetrating activity have successfully achieved inhibition of intracellular inflammatory signaling in cells.

The main objective of this study was to investigate the potential anti-inflammatory and chondrocyte regenerative effects of a cell permeable HBP, RKKNPNCRRH, derived from the heparin binding site of BMP4 in lipopolysaccharide (LPS)-treated macrophages and human articular chondrocytes in vitro and a CIA-induced RA mouse model in vivo.

2. Results

2.1. Characterization of HBP

The primary and secondary structures of the synthetic HBP derived from the human heparin binding domain are presented in Figure 1A,B. The HBP sequence was composed of 10 amino acids (RKKNPNCRRH, Figure 1A). The proline acts as a turning site and induces flexibility to the structure, which facilitates penetration of cells or tissues. The molecular mass of HBP was determined as 1307.7087, which is the optimal size for ease of cell penetration (Supplementary Figure S1A). Additionally, the positive surface charge of HBP (net charge, +5) enhanced cell penetration capacity. Secondary structures are presented in ball-and-stick and cartoon styles (Figure 1B).

A Primary structure

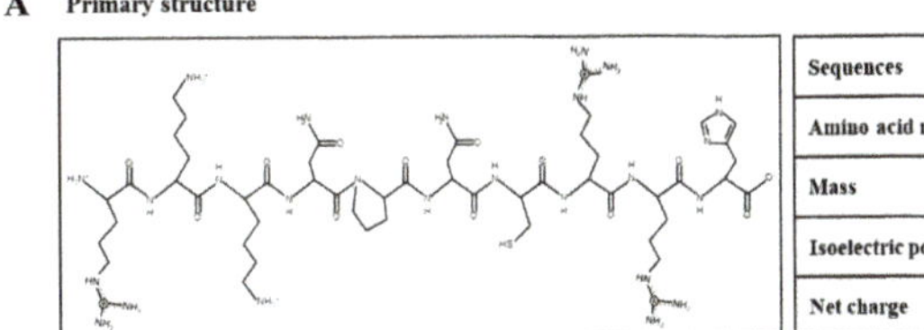

Sequences	RKKNPNCRRH
Amino acid number	10
Mass	1307.7087
Isoelectric point	12.21
Net charge	+5

B Secondary structure

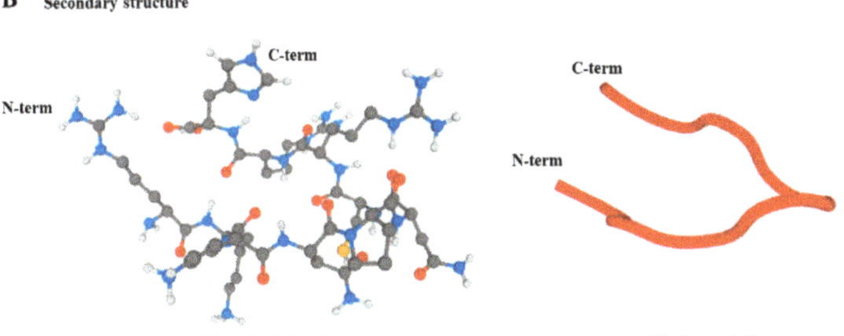

Figure 1. Structural illustration of heparin-binding peptide (HBP). (**A**) Primary structure of synthetic HBP composed of 10 amino acids and (**B**) secondary structure, presented as ball-and-stick and cartoon styles (Black; carbon, Blue; nitrogen, Red; Oxygen, Yellow; Sulfur; White; Hydrogen).

2.2. Cellular Uptake Activity of HBP

The cell penetration capacity of HBPs was examined by treatment of RAW264.7 cells with a range of HBP concentrations for 10 min, followed by confocal microscopy (Figure 2A). As shown in Figure 2A, rhodamine-labeled HBP penetrated the cytoplasm and even nucleus of RAW264.7 cells. The penetration capacity was increased in a concentration-dependent manner, with the highest infiltration observed at 100 μg/mL HBP. Our results support the utility of the newly synthesized HBP from BMP4 as a delivery agent for large therapeutic molecules, such as DNA, RNA, antibodies, and proteins, with minor safety concerns owing to its derivation from human protein. We further examined cell penetration capacity under different temperatures (4 and 37 °C) and incubation times (10 min, 1 h, and 4 h) (Figure 2B). HBP penetration of RAW264.7 cells at 4 °C was observed at time-points of 10 min (pink line) and 1 h (orange line), but not 4 h (yellow line), which may be attributed to decreased cell viability due to long-term exposure to low temperature. Penetration was improved in cells incubated at high temperature for a longer period (37 °C for 4 h) (Figure 2B) and evident after only 10 min of incubation. Penetration capacity was therefore slightly increased with temperature in a time-dependent manner (Figure 2B). Cell uptake of HBP occurs via both direct penetration and endocytosis.

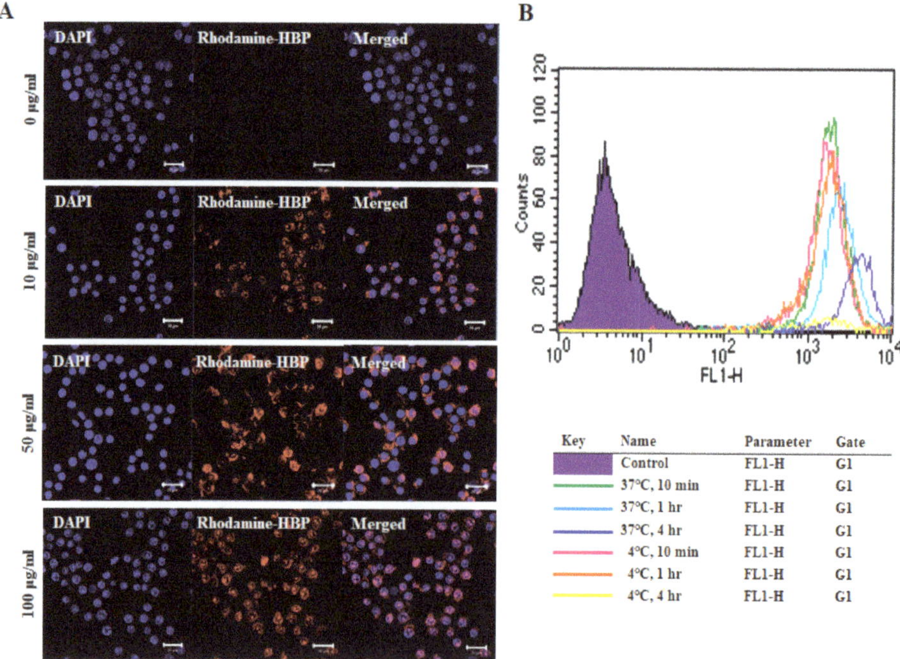

Figure 2. Images of intracellular HBP uptake in RAW264.7 cells. (**A**) Rhodamine-labeled HBP was detected in RAW264.7 cells via confocal microscopy. Intracellular HBP localization was increased in a dose-dependent manner (Scale bar = 20 μm, *n* = 3). (**B**) Fluorescence-activated cell sorting (FACS) analysis of differences in cellular uptake of HBP under different culture conditions, including temperature (4 °C, 37 °C) and incubation time (10 min, 1 h, and 4 h).

2.3. Anti-Inflammatory Effects of HBP on LPS-Treated RAW264.7 Cells

Upon treatment of RAW264.7 cells with LPS (1 μg/mL) for 24 h, their unique bubble-like shape altered to a fibroblast-like morphology, indicative of stimulation of the inflammatory response (Figure 3A, pre-HBP treatment). Treatment of LPS-stimulated cells with HBP (100 μg/mL) for 1 h led to recovery of the unique morphology of RAW264.7 cells (Figure 3A, post-HBP treatment) (Figure 3B,C).

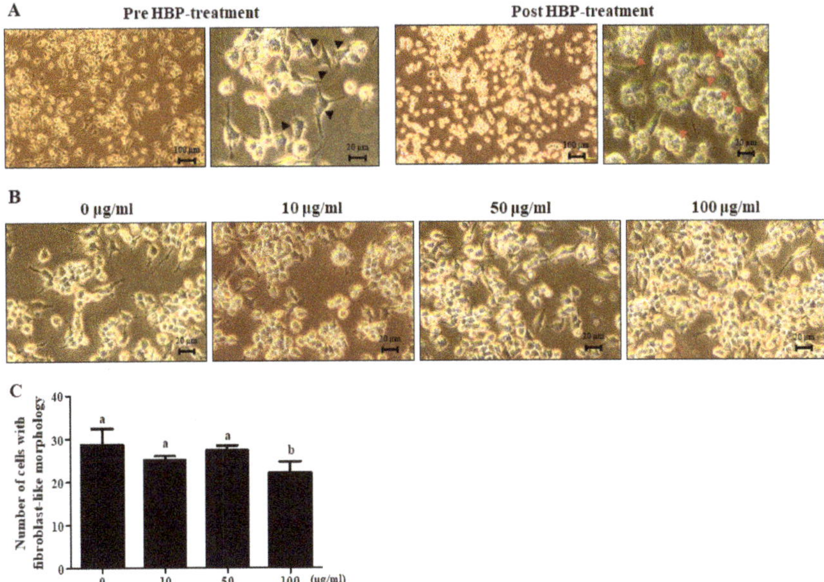

Figure 3. Light microscope view of morphological changes of lipopolysaccharide (LPS)-stimulated RAW264.7 and HBP treatment. (**A**) Cell morphology was examined before (left) and after HBP treatment (right) ($n = 3$). Black arrowheads signify LPS-stimulated inflammation of RAW264.7 cells. Red arrowheads represent RAW264.7 cells recovery following HBP treatment (magnification: ×40). (**B**) Morphology of LPS-stimulated RAW264.7 cells showing recovery following HBP treatment in a dose-dependent manner (magnification: ×200). (**C**) Bar graph indicating the number of cells showing fibroblast-like morphology.

2.4. Effects of HBP on Proteins Related to the Inflammation Pathway

To further confirm the anti-inflammatory activity of HBP, LPS-stimulated RAW264.7 cells were treated with varying concentrations of peptide (0, 10, 50, and 100 µg/mL) for 24 h, and changes in levels of inflammation-related proteins, including iNOS (Figure 4A,B), COX2 (Figure 4A,C), IFNγ (Figure 4A,D), and IL6 (Figure 4A,E), examined in cell lysates. Compared to the non-treated group (NT), iNOS, COX2, IFNγ, and IL6 protein levels presented as image densities were significantly increased following LPS treatment (62.4-, 28.6-, 1.2-, and 4.6-fold increase compared to the NT group, respectively). Inflammatory protein expression was significantly suppressed by HBP in a dose-dependent manner. At the highest HBP concentration (100 µg/mL), iNOS, COX2, IFNγ, and IL6 protein levels were significantly decreased to 0.41-, 0.74-, 0.30-, and 0.13-fold, respectively, relative the value of 1.0 at 0 µg/mL HBP. IL6 expression was the most significantly suppressed by HBP.

2.5. Chondrocyte Recovery Effect of HBP in Human Articular Chondrocytes

We first evaluated the effect of HBP on NHAC cells without LPS stimulation to clarify the chondrogenic potential of the peptide alone (Figure 5A–C). To determine whether our newly synthesized HBP could affect recovery of chondrocytes, aggrecan (AGG; Figure 5D), collagen type II (COLII; Figure 5E), and TNFα (Figure 5F) gene expression changes were evaluated in LPS-stimulated chondrocytes after 5 days of HBP treatment (Figure 5). Quantitative RT-PCR analyses revealed that LPS stimulation suppressed AGG and COLII and enhanced TNFα expression. HBP treatment induced significant recovery of AGG and COLII expression (0, 10, 50 and 100 µg/mL), but had a slight and not significant effect on TNFα expression. In view of the HBP-mediated recovery of damaged chondrocytes, we suggest that the peptide improves chondrocyte-specific characteristics through effects on AGG, COLII, and TNFα, even under inflammatory conditions.

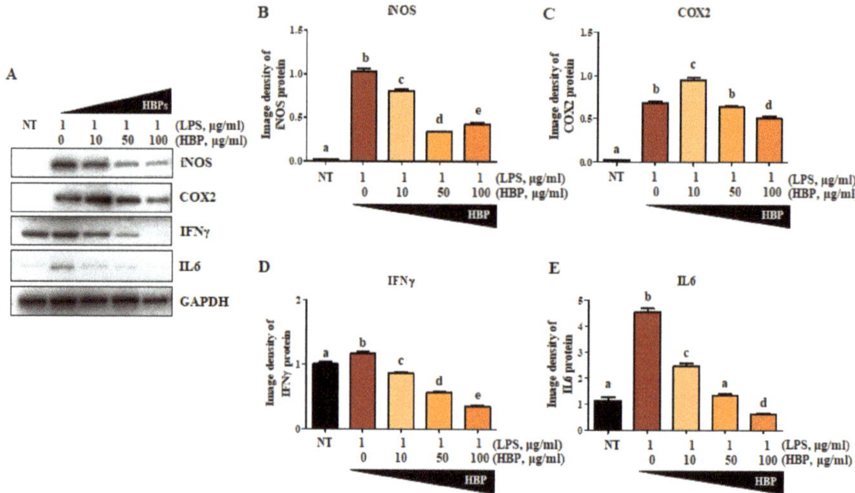

Figure 4. Effects of HBP on inflammatory marker protein expression in LPS-stimulated RAW264.7 macrophages. (**A**) Western blot analysis of inflammatory markers, including iNOS, COX2, IFNγ, and IL6, in LPS-stimulated RAW264.7 cells. (**B–E**) Band intensity of each protein (iNOS (**B**), COX2 (**C**), IFNγ (**D**), and IL6 (**E**)) presented as a bar graph normalized to the intensity of the corresponding GAPDH band (*n* = 3). Different alphabets (a, b, c, d, and e) in each Figure indicate significant differences among experimental groups (*p* < 0.05).

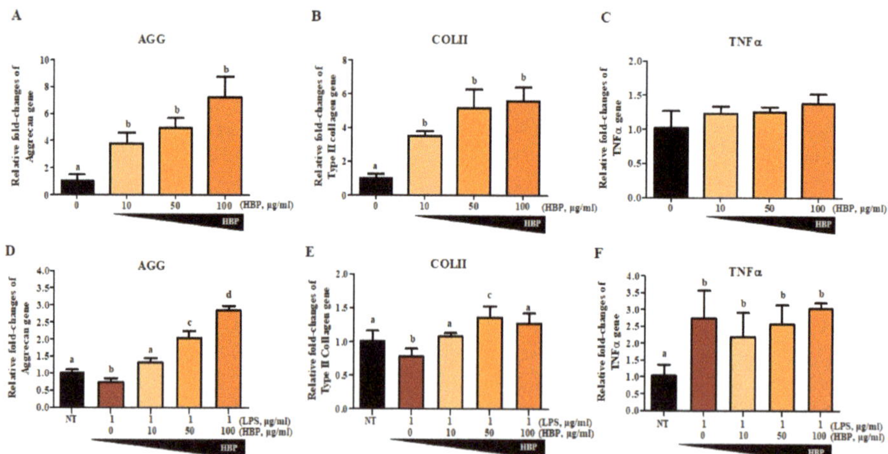

Figure 5. Gene expressions related to chondrocyte potentials with HBP treatment of human cartilage cells. The HBP itself increased (**A**) Aggrecan (AGG), (**B**) Collagen Type II (COLII), and (**C**) TNFα mRNA expressions in NHAC cells in a dose dependent manner (*p* < 0.05, *n* = 3). The LPS-stimulated were treated with various concentrations of HBP, followed by examination of cartilage regeneration-related gene expression. Expression changes in (**D**) AGG, (**E**) COLII and (**F**) TNFα were analyzed via quantitative PCR (*p* < 0.05, *n* = 3). Different alphabets (a, b, c, and d) in each Figure indicate significant differences among experimental groups (*p* < 0.05).

2.6. Antiarthritic Effects of HBP on CIA Mice

2.6.1. Hind Paw Swelling, Arthritis Score, and Histological Recovery of CIA Mice Injected with HBP

Compared with the normal control group (NT), the CIA control group (PBS) showed significant hind paw swelling. An experimental scheme depicting HBP activity in the CIA mouse model is presented in Figure 6A. Enbrel®, an agent widely used to treat rheumatoid arthritis in the clinic, was selected as a positive control [35]. Intraperitoneal injection of HBP and Enbrel® was performed into the lower right quadrant of the abdomen of mice for comparison of therapeutic efficacy.

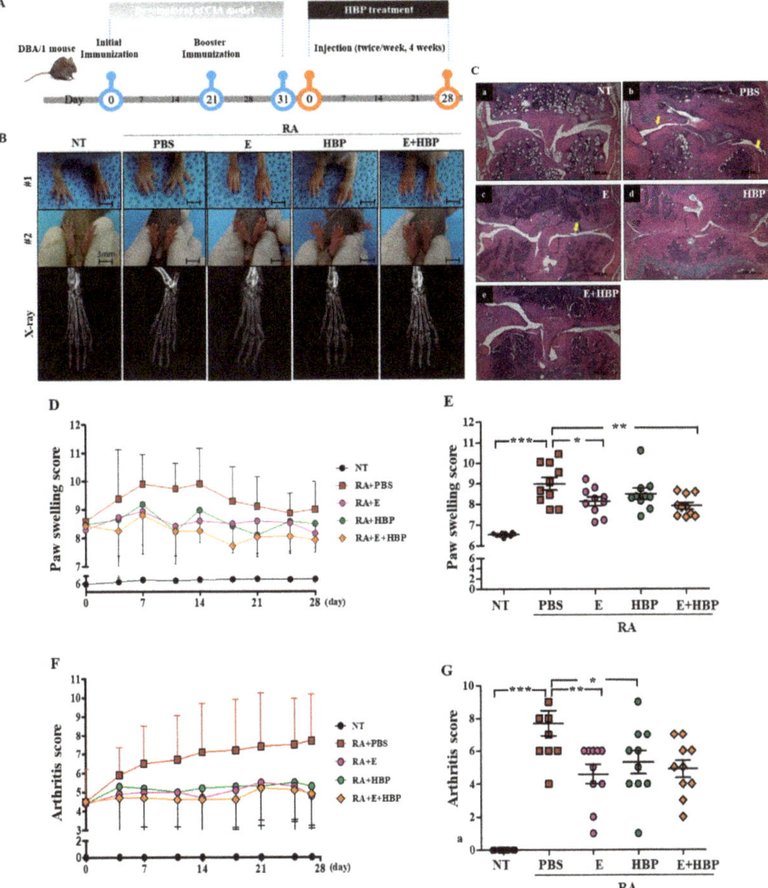

Figure 6. Development of a collagen-induced arthritis (CIA) mouse model and its application in examining the anti-rheumatic activity of HBP. (**A**) Schematic illustration of the induction of rheumatoid arthritis (RA) with collagen injection and HBP treatment. (**B**) Foot and x-ray images of all experimental groups are presented. (**C**) H&E staining of all experimental groups (SP; synovial proliferation, Yellow arrow; Pannus formation). (**D,E**) Changes in paw swelling scores over the entire experimental period. Results on day 28 from all experimental groups are presented in the bar graph (*n* = 10 per group). (**F,G**) Changes in arthritis scores throughout the entire experimental period (**F**). Results on day 28 from all experimental groups are presented in the bar graph (**G**). Statistical significance between groups was analyzed using the Student's *t*-test. The level of significance is represented as * $p < 0.05$, ** $p < 0.01$, or *** $p < 0.001$.

CIA mice displayed significantly swollen hind paws, which was confirmed with visual evaluation (Figure 6B, #1 and #2 panel). The anti-rheumatic effect of HBP in CIA mice was histologically evaluated (Figure 6C). The CIA-PBS group developed chronic inflammation in the synovial tissue, with synovial proliferation (sp) and pannus formation (yellow arrow, cartilage and bone destruction, and bone erosion) (Figure 6C-b), compared to the NT group (Figure 6C-a). The Enbrel®-treated CIA group displayed decreased inflammation to a lower extent than the HBP-treated CIA group (Figure 6C-c). Inflammation was markedly reduced in the HBP-treated CIA group (Figure 6C-d). Synovial cell hyperplasia, bone erosion, and pannus formation were additionally attenuated in groups treated with HBP. Recovery of cartilage morphology in the HBP-treated group was similar to that in the NT and Enbrel® treatment groups (Figure 6C-a,e).

Hind paw swelling score was notably increased in CIA mice (8.99 ± 0.99, $p < 0.05$), compared to the NT group (6.54 ± 0.09), at the end of the experimental period (day 28) (Figure 6D). Relative to the CIA-PBS group, the Enbrel®-treated CIA group showed significantly reduced hind paw swelling from days 11 to 28 ($p < 0.05$). The HBP-treated CIA group also presented markedly decreased hind paw swelling on day 11, which was maintained until day 28 ($p < 0.05$). No significant differences between the Enbrel® and HBP treatment groups were evident throughout the experimental period (Figure 6E), indicative of similar anti-rheumatic effects.

Consistent with data on hind paw swelling, the histological arthritis score in CIA mice was markedly lower following HBP treatment, compared to that in the CIA-PBS group (3.3 ± 0.67, $p < 0.05$; Figure 6F, Enbrel® group; 1.6 ± 0.2). In the HBP and Enbrel® co-injected CIA group, histological arthritis score (2.3 ± 0.3) was significantly decreased relative to that of the CIA-PBS group (3.3 ± 0.67). HBP treatment clearly suppressed inflammatory cell infiltration and reduced the arthritis score in CIA mice (Figure 6G).

The combined effects of HBP and Enbrel® were examined in CIA mice. The combination of peptide and Enbrel® exerted a marginally greater therapeutic effect on paw swelling than each single agent alone (Figure 6D,E). Arthritis score was improved with HBP as well as Enbrel® injection for 28 days (Figure 6F,G). In general, the degree of recovery of arthritis was further improved upon co-injection with the peptide and Enbrel® both agents.

2.6.2. HBP Induces Suppression of Serum Inflammatory Cytokine Levels

Serum concentrations of IFNγ and IL6 in CIA mice were examined to ascertain the therapeutic effect of HBP (Figure 7). IFNγ (37.28 ± 11.86 pg/mL) and IL6 (242.77 ± 77.48 pg/mL) levels were significantly increased in CIA mice (RA-PBS group) on day 28, compared to the NT group (8.86 ± 2.29 pg/mL and 98.09 ± 4.32 of IFNγ and IL6 concentrations, respectively; $p < 0.001$).

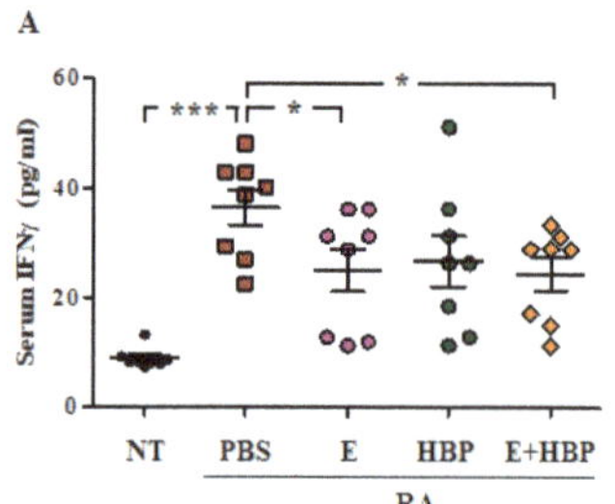
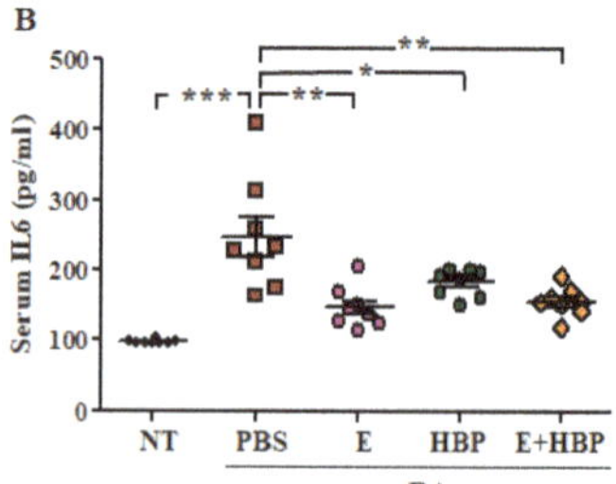

Figure 7. Comparison of serum cytokine levels among all experimental groups. Changes in serum IFNγ (**A**) and IL6 (**B**) concentrations in Enbrel® and HBP-treated RA mice ($n = 10$ per group). Statistical significance between groups was analyzed using the Student's *t*-test. The level of significance is represented as * $p < 0.05$, ** $p < 0.01$, or *** $p < 0.001$.

Serum concentrations of IFNγ (24.88 ±11.82 pg/mL) and IL6 (158.02 ± 55.09 pg/mL) in Enbrel®-treated CIA mice were significantly decreased, compared to CIA-PBS mice. The values obtained from HBP-treated CIA mice were similar to those of the Enbrel® group. Serum levels of IFNγ and IL6 were reduced to 27.79 ± 15.63 pg/mL and 198.18 ± 51.63 pg/mL, respectively, following HBP injection (Figure 7A,B). Co-treatment with HBP and Enbrel® additionally induced a marked reduction in serum IFNγ and IL6 levels to 23.61 ± 8.98 pg/mL and 161.27 ± 47.96 pg/mL, respectively.

2.6.3. Safranin O-Fast Green Staining and Immunohistochemical Analysis of IL6 in Cartilage of CIA Mice

To further validate the chondrogenic potential of HBP, we performed Safranin O-fast green staining and analyzed chondrocyte recovery using the Image J program (Figure 8A,C). The red-stained glycosaminoglycan (GAG) area was significantly damaged in CIA-induced RA mice, compared with the RA-PBS group (1.0 ± 0.01 vs. 0.21 ± 0.01, NT vs. RA-PBS; $p < 0.05$). Treatment with Enbrel® promoted recovery of damaged chondrocytes (0.79 ± 0.02 vs. 0.21 ± 0.01, Enbrel® vs. RA-PBS; $p < 0.05$). Notably, while the HBP-only treatment group showed significant recovery of chondrocyte regeneration potential relative to the PBS group, co-treatment with Enbrel® was required for optimal results. Our findings imply that HBP is more effective as a co-therapeutic complementing the activity of Enbrel® than as a single treatment agent.

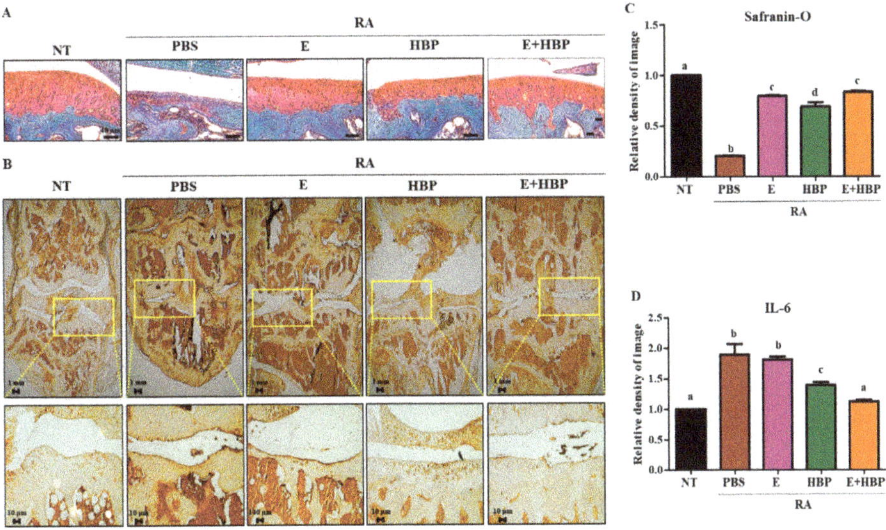

Figure 8. Effects of HBP on cartilage recovery determined via immunohistochemical analysis of IL6 expression and chondrogenic potential via Safranin-O staining. (**A**) Safranin-O-stained cartilage of RA mice. (**B**) IL6 protein in cartilage of RA mice was stained with avidin-biotin. (**C**) Quantification of the degree of Safranin-O and (**D**) IL6 staining of cartilage using Image J software, presented as bar graphs ($n = 10$ per group). Different alphabets (a, b, c, and d) in each Figure indicate significant differences among experimental groups ($p < 0.05$).

Immunohistochemical analysis was performed to evaluate HBP-induced changes in IL6 protein expression in CIA mice. Limb joints of male CIA mice were obtained at the end of the experimental period. IL6 expression was significantly higher in CIA mice (RA-PBS, 1.88-fold increase), compared to the NT group (Figure 8B,D).

Enbrel®-treated CIA mice showed a slight reduction in IL6 expression, compared with the RA-PBS group (1.88 ± 0.12 vs. 1.80 ± 0.03, Enbrel® vs. RA-PBS; $p > 0.05$). However, HBP treatment induced

significant IL6 reduction relative to the RA-PBS treatment group (1.39 ± 0.03 vs. 1.80 ± 0.03, Enbrel®
vs. RA-PBS; $p < 0.05$). Importantly, the most significant decrease in IL6 expression was observed in the
co-injection group (E+HBP) (1.12 ± 0.02 vs. 1.80 ± 0.03, E+HBP vs. RA-PBS; $p < 0.05$).

3. Discussion

Previously, we synthesized a heparin binding domain peptide (HBP) derived from human
heparin binding domain of BMP4 and evaluated its biological significance in tumor targeting
and anti-angiogenesis [32]. BMP4 is a multifunctional growth factor that mainly promotes bone
formation [36]. This growth factor is expected to be highly efficacious in clinical treatment of
various diseases owing to involvement in multiple physiological processes, such as regulation of
angiogenesis [32], muscle development, human embryo development, and mineralization of bone [31].
In lieu of BMP4 protein, a peptide composed of 10 amino acids, RKKNPNCRRH, corresponding to
residues 15–24 of the heparin binding site of BMP-4 was generated and evaluated for therapeutic
application in various diseases.

We initially examined the efficacy of the BMP-4-derived HBP against RA and whether the peptide
could reduce the side-effects of Enbrel®, with a view to assessing its utility as a candidate or alternative
agent for RA therapy. Additionally, the effects of HBP on chondrocyte recovery were investigated in
an RA animal model.

To trigger an anti-inflammatory effect, peptides need to be efficiently delivered into cells. The amino
acid sequence of HBP exhibits similarities with cell-penetrating peptides containing poly-arginine [37].
Cell penetration properties are attributed to eight positive amino acids [38]. The cell penetration
activity of the newly developed HBP was evaluated by our group using human breast cancer
cells in a preliminary study [32] and RAW264.7 cells and artificial skin tissue in the current study
(Figure 2A). The infiltration capacity of HBP into inflamed cells was demonstrated for the first time
using an LPS-stimulated in vitro model (Figure 5A). As shown in Figure 2A, rhodamine-tagged HBP
was successfully delivered into the cytoplasm and even nucleus of inflammatory RAW264.7 cells.
HBP penetration into cells may not be required owing to its heparan sulfate proteoglycan (HSPG)
binding affinity. Interactions of inflammatory mediators, such as chemokines, with HSPG are known
to modulate the inflammation process. HSPGs play a key role in inflammation and their mimetics
effectively act as anti-inflammatory agents through modulating the interactions between mediators
and HSPGs [39]. Therefore, the anti-inflammatory activity of HBP may be mediated through binding to
HSPGs. In addition to HSPG-binding affinity, HBP displayed cell uptake ability, even at 4 °C, indicative
of direct cell penetration (Figure 2B). Intracellular distribution patterns of rhodamine-labeled HBP,
observed via confocal microscopy, supported its successful delivery into cells (Figure 2A). Besides
bound HSPGs, HBP delivered into cells may affect intracellular signaling pathways, such as NF-kB,
involving production of proinflammatory cytokines, including TNF-α and IL6. Further studies are
required to establish whether HBP exerts anti-inflammatory effects through binding to HSPGs or
inhibition of specific intracellular targets.

Following delivery into inflammatory cells, HBP induced a decrease in expression of inducible
nitric oxide synthase (iNOS), cyclooxygenase 2 (COX2), and interferon gamma (IFNγ) proteins
triggered by the intracellular inflammatory response (Figure 4). Nitric oxide (NO) is closely associated
with inflammation, angiogenesis, and tissue destruction in the RA model. Interplay of pro- and
anti-inflammatory cytokines promotes iNOS production in the affected tissues of RA patients [40].
The iNOS enzyme is responsible for localized overproduction of NO in synovial joints affected by RA.
Poulami et al. [41] suggested that the determination of the cytokine signaling network underlying
regulation of iNOS is essential to understand the pathophysiology of RA progression. Microarray
data analysis by the group revealed upregulation of the gene network belonging to interferon gamma
(IFNγ) and interleukin 6 (IL6) pathways in the RA synovium. Conversely, genes contributing to the
anti-inflammatory transforming growth factor-beta (TGFβ) signaling pathway were downregulated [42].
In the current study, iNOS and COX2 protein levels were significantly decreased at HBP concentrations

of 50 μg/mL to 100 μg/mL (about two-fold), compared to control cells, with limited differences between the 50 and 100 μg/mL treatment groups. Complete recovery from acute inflammation induced by LPS to basal levels within a short time-frame (24 h in the current experiment) may be difficult to achieve. Although HBP successfully exerted an anti-inflammatory effect on LPS-treated RAW264.7 cells, as evident from decreased iNOS and COX2 protein expression, the initial significant increases in iNOS and COX2 induced by LPS were extremely high and could not be reduced to basal levels by HBP. Expression of IFNγ, a known representative downstream molecule of iNOS signaling in pre- and pro-inflammatory pathways, was markedly decreased upon treatment with 100 μg/mL HBP. Accordingly, the optimum concentration of HBP for controlling inflammatory reaction was determined as 100 μg/mL, and this concentration was also confirmed by an MTT assay (Supplementary Figure S1B). In this respect, inhibition of the iNOS-IFNγ signaling pathway and IL6 may be contributory factors to the anti-inflammatory mechanism of HBP.

Next, we examined the therapeutic effect of HBP in vivo via intraperitoneal injection twice per week for 4 weeks into collagen-induced induced arthritic (CIA) DBA/1 mice, a common autoimmune disease model widely employed to study rheumatoid arthritis (RA) [43].

Enbrel®, a tumor necrosis factor (TNF) blocker, is an FDA-approved RA therapeutic agent indicated for reducing signs and symptoms, inducing major clinical responses, inhibiting progression of structural damage, and improving physical function in patients with RA [44]. Serious side-effects of Enbrel® include susceptibility to new infections, induction of hepatitis B, nervous system problems, such as multiple sclerosis, seizures or inflammation of nerves of the eyes, blood problems, heart failure, and allergic or autoimmune reactions. Long-term use of Enbrel® raises further safety concerns, such as lymphoma and other malignancies. Owing to its wide range of side-effects, limited use over short periods of time is recommended, highlighting the need to develop new, effective therapies to replace or supplement Enbrel® [44]. Here, we compared the therapeutic effects of HBP and Enbrel® in the CIA model. The Enbrel® dose was obtained from a previous report by our group, where CIA mice received 100 μg Enbrel® intraperitoneally (5 mg/kg) twice per week for 4 weeks [45].

Severe RA is associated with deformation of joints, along with cartilage and bone destruction (Figure 6), which is the most representative symptom [46–48]. In the current study, bone structure and articular chondrocytes at the joint in our mouse model were recovered upon HBP treatment (Figure 6). CIA mice treated with Enbrel® showed significant improvements in arthritis and paw swelling scores (Figure 6D–G). Moreover, damaged chondrocytes in CIA mice detected via H&E staining were significantly reduced following Enbrel® treatment (Figure 6C).

Arthritis and hind paw swelling scores were not significantly different between Enbrel® and HBP-treated groups throughout the experimental period, implying that the HBP exerts similar therapeutic effects as Enbrel® (Figure 6).

To further elucidate the combined effects of HBP and Enbrel, mice were simultaneously co-injected with both agents. Co-injection induced a significant reduction in arthritis as well as hind paw swelling scores. In terms of inflammation, HBP was considered to exert a more significant effect when combined with Enbrel®. Levels of serum inflammatory cytokines, including IFNγ and IL6, were also markedly increased in CIA mice and notably suppressed with Enbrel®. Unlike arthritis and paw swelling scores, serum cytokine concentrations showed limited response to co-injection of HBP and Enbrel®. The patterns of inflammatory cytokine concentrations in serum are reported to differ from arthritis and paw swelling scores in CIA mice, which may be attributed to the potential involvement of other factors in improving pro-inflammatory IFNγ or IL6 levels in blood. Further research is essential to clarify the specific pre- and pro-inflammatory mechanisms. Co-treatment of CIA mice with HBP and Enbrel® exerted a marginally greater therapeutic effect on paw swelling relative to each single agent. While the results appear favorable, they may not be sufficiently beneficial to justify combination therapy in humans. Further research is required to establish whether Enbrel® enhances the therapeutic efficacy of HBP in the clinic.

We observed no apparent changes in organ weights (including liver, spleen, and kidney) in the HBP-treated group, compared with the normal control group (Supplementary Figure S2). Moreover, no animal deaths were recorded during the experimental period, implying no cytotoxic and organ-specific toxic effects of HBP in vivo.

In our experiments, HBP exerted similar anti-inflammatory effects to Enbrel®, even when used alone. The anti-inflammatory effect of the peptide appeared further enhanced upon co-treatment with Enbrel®. Additionally, injection of HBP led to recovery of the phenotype of inflammation-damaged chondrocytes. HBP-mediated chondrocyte recovery was confirmed based on reduced IL6 expression in cartilage tissue of treated mouse groups, which was accelerated with co-injection of HBP and Enbrel® (Figure 8). Decreased expression of chondrocyte markers, including AGG, COLII, and TNFα, in LPS-stimulated human chondrocytes was reversed upon HBP treatment (Figure 2). Chondrocytes damaged via inflammation in the CIA animal model were also regenerated in the presence of HBP (Figure 6C).

Our results collectively suggest that abnormal intracellular changes of joint chondrocytes in the CIA mouse model are effectively attenuated by HBP and/or Enbrel. In this regard, it may be advantageous to combine the peptide with Enbrel® to optimize treatment of inflammation and recovery of joint chondrocyte structures in chronic arthritis.

4. Materials and Methods

4.1. Peptide Preparation

Heparin-binding peptides were generated using a peptide synthesizer (Prelude, Protein Technologies Inc., Tucson, AZ, USA). The C-terminal amide form was produced using standard 9-fluorenylmethoxycarbonyl (Fmoc) chemistry. Rink amide-MBHA resin (GL Biochem, Shanghai, China) was pre-swollen in DMF (50 mg/mL) and Fmoc-protecting groups of resin and amino acids removed using 30% piperidine in DMF, 10 eq. DIPEA, 5 eq. HBTU, and 5 eq. Fmoc-protected amino acid, calculated according to resin loading. Cleavage and side-chain deprotection of the peptide resin was conducted for 4 h using a cleavage cocktail (TFA/water/thioanisole/phenol/ethanedithiol (8.5/0.5/0.5/0.5,0.25)). Solutions containing cleaved peptide were precipitated by the addition of chilled ether. Peptides were purified using preparative reverse-phase high-performance liquid chromatography (RP-HPLC; Waters, Milford, MA, USA) with a Vydac C18 column and 50 min gradient from 90% to 10% water/acetonitrile containing 0.1% trifluoroacetic acid (TFA). Peptide purity was determined as >98% via HPLC (Shimadzu, Kyoto, Japan) and liquid chromatography-mass spectrometry (LC-MS, Shimadzu, Kyoto, Japan). To establish the peptide translocation pathways, rhodamine was manually conjugated to the N-terminus during synthesis [49,50]. The purity and efficiency of rhodamine labeling was assayed via HPLC by monitoring absorbance at 230 nm and fluorescence at 440 nm. Fluorescence-labeled peptides were purified using HPLC (purity >95%), lyophilized, and stored at −20 °C in the dark until use.

4.2. Cell Culture

Human articular chondrocytes (NHAC, adult human knee chondrocytes) were purchased from Lonza (Walkersville, MD, USA) and used until passage 6. Cells were seeded onto 40 mm cell culture dishes (TPP, Trasadingen, Switzerland) at a concentration of 1×10^4 cells/dish. Chondrocyte basal medium (Lonza) was supplemented with SingleQuots chondrocyte growth medium BulletKit® (Lonza) composed of FBS, fibroblast growth factor 2 (FGF-2), gentamicin sulfate/amphotericin-B, insulin, insulin-like growth factor 1 (IGF-1), and transferrin. NHAC cells were sub-cultured two–three times to < 85% confluency and the medium changed twice a week.

Murine RAW264.7 macrophages were obtained from American Type Culture Collection (ATCC; Manassas, VA) and maintained in DMEM (Gibco-BRL, Grand Island, NY, USA) supplemented with

10% FBS (Gibco-BRL) and 1% antibiotic-antimycotic solution (Gibco-BRL) at 37 °C in a humidified atmosphere in 5% CO_2.

4.3. MTT Assay

Cell-viability assay was analyzed by using an MTT assay. RAW264.7 cells were seeded in 96-well plates (1×10^4 cells/well). After treatment with various concentrations of HBP (0, 0.01, 0.1, 1, 10, 100, and 200 µg/mL) for 24 h, the MTT solution was added to the cells and incubated for 4 h. After incubation, medium was removed and dimethyl sulfoxide (500 µL/well) was added to dissolve the formazan precipitates. Extracted formazan was transferred to a 96-well plate, and measured their absorbance at 570 nm using a microplate reader (BioTek Instruments, VT, USA). The cell viability assay was obtained from three independent experiments.

4.4. RNA Isolation and Quantitative RT-PCR

NHAC cells were plated onto 100 mm culture dishes and incubated at 37 °C under 5% CO_2. At 80–90% confluence, cells were incubated for 2 h in DMEM with 0.5% serum for starvation, followed by treatment with LPS (1 µg/mL; St. Louis, MO, USA) for 24 h and various concentrations of HBP for 1 h.

The medium was replaced after 24 h with fresh medium containing HBP with or without LPS for a total culture period of 3 days. Total RNA from human articular chondrocytes was extracted using a TRIzol reagent according to the manufacturer's instructions (Life Technologies, Darmstadt, Germany). Isolated RNA was treated with DNase I (Thermo Scientific, Schwerte, Germany) to remove possible genomic DNA contamination and used for cDNA synthesis with the aid of Superscript III Transcriptase (Life Technologies) and random hexamer primers (Thermo Scientific) at 50 °C for 1 h. Each sample was run in triplicate. Quantitative RT-PCR was conducted using POWER SYBR Green qPCR Master Mix (Life Technologies) and 0.2 µM primer (primer sequences are listed in Supplementary Table S1) on the StepOne Plus Real-Time PCR System (Applied Biosystems, Forster City, CA, USA) under the following cycle conditions: primary denaturation at 95 °C for 5 min, 40 cycles of 30 s at 95 °C, 40 s at 60 °C, and 30 s at 72 °C, followed by fluorescence measurements.

4.5. In Vitro Cellular Internalization

RAW264.7 cells (1×10^4 cells) were seeded on glass slides (Life Technologies, CA, USA). Following 24 h incubation to allow cell attachment, the culture medium was removed and 50 µM rhodamine-labeled HBP added along with fresh complete medium, followed by incubation for 10 min at 37 °C in a 5% CO_2 atmosphere. After incubation, cells were washed with PBS and incubated for 30 min at room temperature with 1 µg/mL 4′,6-diamidino-2-phenylindole (DAPI) to visualize nuclei. Cells were washed with PBS and imaged using an Olympus FV-300 confocal laser scanning microscope operated with FLUOVIEW software (Olympus, Tokyo, Japan).

4.6. Western Blot Analysis

RAW264.7 cells were plated on 10 mm diameter culture dishes (1×10^5 cells/dish). At 80–90% confluence, cells were incubated for 2 h in DMEM with 0.5% FBS for starvation [51]. Next, cells were treated with LPS (1 µg/mL) for 30 min or various concentrations of HBP for 1 h. Following cell lysis in RIPA containing protease and phosphatase inhibitors (Sigma, St. Louis, MO, USA) for 30 min on ice, total protein concentrations were determined with a Pierce BCA protein assay kit (Thermo Fisher Scientific, Waltham, MA, USA). Equal amounts of protein (30 µg) were boiled in 5× electrophoresis sample buffer (0.25 M Tris-HCl, pH 6.8 15% SDS, 50% glycerol, 25% β-mercaptoethanol (ME), 0.01% bromophenol blue) for 5 min, separated via sodium dodecyl sulfate-polyacrylamide gel electrophoresis (SDS-PAGE) and electrotransferred onto nitrocellulose membranes. Non-specific protein binding was blocked via incubation with 3% BSA in T-TBS for 1 h. Membranes were washed with T-TBS and incubated with primary antibodies in T-TBS containing 3% BSA for 4 h at 4 °C, including anti-iNOS, COX2, IFNγ, and IL6 antibodies (Santa Cruz, CA, USA). After three further washes, membranes were incubated with a secondary antibody (horseradish peroxidase (HRP)-conjugated goat anti-rabbit IgG, diluted 1:2000 in 3% BSA) for 60 min. Blots were visualized with chemiluminescence reagents (Thermo Scientific, Schwerte, Germany). The relative optical densities of protein bands were quantified using Image J software (National Institutes of Health, Bethesda, MD, USA).

4.7. Induction of CIA in Mice and Peptide Treatment

Male DBA/1 (7 weeks old, 21–24 g) were bred at Harlan Co., Ltd. (Indianapolis, IN, USA) and supplied by Orientbio Inc. (Seungnam, Korea). All animal experiments were approved by the Institutional Animal Care and Use Committee of Chungbuk National University (IACUC ID: CBNUA-1095-17-02). Arthritis was induced via collagen inoculation (CIA) as described previously [43,52]. Mice were intradermally immunized with 500 µg collagen type II (COLII, Sigma) dissolved in 0.5 mL 0.1 M acetic acid and emulsified in 0.5 mL Freund's incomplete adjuvant (Sigma) at 4 °C. Booster injections containing 250 µg COLII similarly dissolved and emulsified with Freund's incomplete adjuvant (1:1) were administered intradermally into the tail base 21 days after primary immunization, following which hind paw volumes, survival, and body weights were monitored.

Mice were divided into five experimental groups as follows: normal control group (NT, *n* = 10) not immunized with collagen, CIA control group (RA-PBS, *n* = 10), HBP-treated CIA group (RA-HBP, 30 mg/kg, *n* = 10), Enbrel® (Pfizer, Seoul, Korea)-treated CIA group (RA-E, 10 mg/kg, *n* = 10), and Enbrel® and HBP co-injection group (RA-E+HBP, *n* = 10). At 28 days after primary immunization, HBP (30 mg/kg) was subcutaneously injected into the neck skin of CIA mice twice a week for 4 weeks. Enbrel® was injected intraperitoneally three times a week for 4 weeks. The CIA control group was administered PBS. At the end of the experimental period, liver, spleen, and kidney weights were measured.

4.8. Assessment of Clinical Signs of Inflammation

Mice were monitored and absence of abnormal lesions confirmed for 28 days before the assessment of rheumatoid arthritis via visual evaluation of the limb joints. On the same schedule as HBP injection (days 0, 4, 7, 11, 14, 18, 21, and 25), clinical signs of inflammation were visually evaluated in the same manner and classified according to the clinical scoring system (arthritis score) as follows [53]: 0 (normal), 1 (slight swelling and/or erythema), 2 (pronounced edematous swelling), and 3 (Ankylosis). The average sum of the scores of the four limb joints was assessed through a blind test carried out by two trained investigators.

After visual evaluation, paw swelling was measured and thickness of the center of the sole of the limb measured using an electrical caliper (CD-15CPX, Kawasaki, Japan). The individual in charge of the measurement reduced the error by one person to the maximal extent and summed the measured values of edema at the same position in all four limbs.

4.9. Histological Examinations

On day 28, mice were killed for histological analysis. Both hind paws and ankles were harvested from each mouse and fixed overnight in 10% buffered formalin, decalcified in 30% citrate-buffered formic acid for 2 weeks at 4 °C, dehydrated in a graded series of methanol and xylene, and embedded in paraffin. Thin sections (5 µm) were stained with hematoxylin and eosin (HE) and histopathologic scoring performed under a light microscope by a blinded observer. The degree of inflammation around articular cartilage was analyzed according to a previously reported method and scored as follows: 0, mild; step 1, moderate; step 2, severe; step 3 [53].

4.10. Serum Cytokine Levels

At the end of the experimental period, serum samples were isolated from whole blood in each group following sacrifice via centrifugation at 4 °C for 30 min and stored at −80 °C until use. IFNγ and

IL6 levels in serum were measured via ELISA (R&D Systems, Minneapolis, MN, USA) according to the manufacturer's instructions.

4.11. Statistical Analysis

All data are presented as mean values ± standard deviation. Graphs were generated using GraphPad Prism 5 software (GraphPad Software Inc., La Jolla, CA, USA). Statistical significance was analyzed with one-way analysis of variance (ANOVA), followed by the Tukey's post hoc multiple comparison test. Significant differences among experimental groups are indicated with different letters. In this study, we conducted three independent sets of in vitro experiments and used 10 mice per experimental group in vivo. All differences were considered significant at p values < 0.05.

5. Conclusions

Experiments from the current study demonstrated that a cell-penetrating HBP derived from the human heparin binding domain of BMP-4 could effectively suppress inflammation via regulation of iNOS-IFNγ-IL6 signaling in murine macrophages and human chondrocytes. Furthermore, the HBP reduced several arthritis symptoms, including hind paw swelling, chondrocyte inflammation, pro-inflammatory cytokine production, and increased arthritis score in CIA mice, and further protected against chondrocyte damage and bone structure at the joint by enhancing cartilage recovery through induction of IL6 expression. The data collectively support the utility of HBP as effective therapeutic agents or supplementary treatments for RA.

Supplementary Materials: Supplementary Materials can be found at http://www.mdpi.com/1422-0067/21/12/4251/s1.

Author Contributions: Conceptualization and draft supervision of project, Y.J.P., J.-Y.L., and Y.S.P.; Biological experiments and preparation of original draft, D.H.C.; Writing a manuscript, J.-Y.L., Y.J.P., and Y.S.P.; Synthesis of peptide, B.S.J. and D.L.; Immunohistochemical analysis. K.-S.P.; Statistical analysis, J.K.C. and K.E.L. All authors have read and agreed to the published version of the manuscript.

Funding: This study was supported by the Basic Science Research Program through the National Research Foundation of Korea (NRF) funded by the Ministry of Science, ICT & Future Planning (No. 2017R1A2B4002611, 2017M3A9B3063635, 2019M3A9H1032376), and partly by the Technological innovation R&D program of SMBA (No. P0010145).

Conflicts of Interest: The authors declare no conflict of interest. The funders had no role in the design of the study; in the collection, analyses, or interpretation of data; in the writing of the manuscript, or in the decision to publish the results.

Abbreviations

BSA	bovine serum albumin
CD	circular dichroism
CIA	collagen-induced arthritis
COLII	Type II collagen
COX2	cyclooxygenase
DMARD	disease-modifying anti rheumatic drug
DMEM	Dulbecco's modified Eagle's medium
EDAC	N-ethyl-N0-(3-dimethylaminopropyl)-carbodiimide hydrochloride
EDTA	fetal bovine serum
FBS	ethylenediaminetetracetate
HBP	10-mer synthetic anti-inflammatory heparin binding peptides
HE	hematoxylin and eosin
HBSS	Hank's balanced salt solution
HRP	horse radish peroxidase
IFN	interferon
IL	interleukin
LPS	lipopolysaccharide

NHAC	Normal Human Articular Chondrocyte
NHS	N-hydroxysuccinimide
PBS	phosphate-buffered saline
RA	rheumatoid arthritis
SAHA	suberoylanilide hydroxamic acid
SDS-PAGE	sodium dodecyl sulfate-polyacrylamide gel electrophoresis
TFA	trifluoroacetic acid
TFE	trifluoroethanol
TNF-α	tumor necrosis factor-alpha
T-TBS	Tween-Tris-buffered saline

References

1. Kim, H.K.; Cho, S.K.; Kim, J.W.; Jung, S.Y.; Jang, E.J.; Bae, S.C.; Yoo, D.H.; Sung, Y.K. An Increased Disease Burden of Autoimmune Inflammatory Rheumatic Diseases in Korea. *Semin. Arthritis Rheum.* **2019**. [CrossRef]
2. Korean Statistical Information Service (KOSIS). Available online: http://kosis.kr/eng/statisticsList/statisticsListIndex.do?menuId=M0101vwcd=MTETITLE&parmTabId=M_01_01&statId=1962001&themaId=#SelectStatsBoxDiv (accessed on 13 March 2020).
3. Furst, D.E.; Emery, P. Rheumatoid Arthritis Pathophysiology: Update on Emerging Cytokine and Cytokine-Associated Cell Targets. *Rheumatology (Oxford)* **2014**, *9*, 1560–1569. [CrossRef]
4. Yoo, D.H.; Hrycaj, P.; Miranda, P.; Ramiterre, E.; Piotrowski, M.; Shevchuk, S.; Kovalenko, V.; Prodanovic, N.; Abello-Banfi, M.; Gutierrez-Urena, S.; et al. A Randomised, Double-Blind, Parallel-Group Study to Demonstrate Equivalence in Efficacy and Safety of Ct-P13 Compared with Innovator Infliximab When Coadministered with Methotrexate in Patients with Active Rheumatoid Arthritis: The Planetra Study. *Ann. Rheum. Dis.* **2013**, *10*, 1613–1620. [CrossRef] [PubMed]
5. Lee, C.K.; Lee, E.Y.; Chung, S.M.; Mun, S.H.; Yoo, B.; Moon, H.B. Effects of Disease-Modifying Antirheumatic Drugs and Antiinflammatory Cytokines on Human Osteoclastogenesis through Interaction with Receptor Activator of Nuclear Factor Kappab, Osteoprotegerin, and Receptor Activator of Nuclear Factor Kappab Ligand. *Arthritis Rheumatol.* **2004**, *12*, 3831–3843. [CrossRef] [PubMed]
6. Choy, E.H.; Panayi, G.S. Cytokine Pathways and Joint Inflammation in Rheumatoid Arthritis. *N. Engl. J. Med.* **2001**, *12*, 907–916. [CrossRef] [PubMed]
7. Agarwal, S.K. Biologic Agents in Rheumatoid Arthritis: An Update for Managed Care Professionals. *J. Manag. Care Pharm.* **2011**, *9*, S14–S18. [CrossRef] [PubMed]
8. Burmester, G.R.; Pope, J.E. Novel Treatment Strategies in Rheumatoid Arthritis. *Lancet* **2017**, *10086*, 2338–2348. [CrossRef]
9. Sepriano, A.; Kerschbaumer, A.; Smolen, J.S.; van der Heijde, D.; Dougados, M.; van Vollenhoven, R.; McInnes, I.B.; Bijlsma, J.W.; Burmester, G.R.; de Wit, M.; et al. Safety of Synthetic and Biological Dmards: A Systematic Literature Review Informing the 2019 Update of the Eular Recommendations for the Management of Rheumatoid Arthritis. *Ann. Rheum. Dis.* **2020**, *79*. [CrossRef] [PubMed]
10. Silvagni, E.; Di Battista, M.; Bonifacio, A.F.; Zucchi, D.; Governato, G.; Scire, C.A. One Year in Review 2019: Novelties in the Treatment of Rheumatoid Arthritis. *Clin. Exp. Rheumatol.* **2019**, *4*, 519–534.
11. Blanchard, F.; Chipoy, C. Histone Deacetylase Inhibitors: New Drugs for the Treatment of Inflammatory Diseases? *Drug Discov. Today* **2005**, *3*, 197–204. [CrossRef]
12. Minucci, S.; Pelicci, P.G. Histone Deacetylase Inhibitors and the Promise of Epigenetic (and More) Treatments for Cancer. *Nat. Rev. Cancer* **2006**, *1*, 38–51. [CrossRef] [PubMed]
13. Park, J.H.; Park, S.H.; Lee, H.Y.; Lee, J.W.; Lee, B.K.; Lee, B.Y.; Kim, J.H.; Kim, M.S. An Injectable, Electrostatically Interacting Drug Depot for the Treatment of Rheumatoid Arthritis. *Biomaterials* **2018**, *154*, 86–98. [CrossRef] [PubMed]
14. Seo, J.; Park, S.H.; Kim, M.J.; Ju, H.J.; Yin, X.Y.; Min, B.H.; Kim, M.S. Injectable Click-Crosslinked Hyaluronic Acid Depot to Prolong Therapeutic Activity in Articular Joints Affected by Rheumatoid Arthritis. *ACS Appl. Mater. Interfaces* **2019**, *28*, 24984–24998. [CrossRef] [PubMed]
15. Ain, Q.; Zeeshan, M.; Khan, S.; Ali, H. Biomimetic Hydroxyapatite as Potential Polymeric Nanocarrier for the Treatment of Rheumatoid Arthritis. *J. Biomed. Mater. Res. A* **2019**, *12*, 2595–2600. [CrossRef]

16. Chen, L.; McCrate, J.M.; Lee, J.C.; Li, H. The Role of Surface Charge on the Uptake and Biocompatibility of Hydroxyapatite Nanoparticles with Osteoblast Cells. *Nanotechnology* **2011**, *10*, 105708. [CrossRef]

17. Shi, Z.; Huang, X.; Cai, Y.; Tang, R.; Yang, D. Size Effect of Hydroxyapatite Nanoparticles on Proliferation and Apoptosis of Osteoblast-Like Cells. *Acta Biomater.* **2009**, *1*, 338–345. [CrossRef]

18. Kaspar, A.A.; Reichert, J.M. Future Directions for Peptide Therapeutics Development. *Drug Discov. Today* **2013**, *18*, 807–817. [CrossRef]

19. Craik, D.J.; Fairlie, D.P.; Liras, S.; Price, D. The Future of Peptide-Based Drugs. *Chem. Biol. Drug Des.* **2013**, *1*, 136–147. [CrossRef]

20. Malemud, C.J. Recent Advances in Neutralizing the Il-6 Pathway in Arthritis. *Open Access Rheumatol.* **2009**, *1*, 133–150. [CrossRef]

21. Lee, J.H.; Cho, M.L.; Kim, J.I.; Moon, Y.M.; Oh, H.J.; Kim, G.T.; Ryu, S.; Baek, S.H.; Lee, S.H.; Kim, H.Y.; et al. Interleukin 17 (Il-17) Increases the Expression of Toll-Like Receptor-2, 4, and 9 by Increasing Il-1beta and Il-6 Production in Autoimmune Arthritis. *J. Rheumatol.* **2009**, *36*, 684–692. [CrossRef]

22. Ospelt, C.; Brentano, F.; Rengel, Y.; Stanczyk, J.; Kolling, C.; Tak, P.P.; Gay, R.E.; Gay, S.; Kyburz, D. Overexpression of Toll-Like Receptors 3 and 4 in Synovial Tissue from Patients with Early Rheumatoid Arthritis: Toll-Like Receptor Expression in Early and Longstanding Arthritis. *Arthritis Rheumatol.* **2008**, *58*, 3684–3692. [CrossRef] [PubMed]

23. Palmer, C.D.; Mutch, B.E.; Workman, S.; McDaid, J.P.; Horwood, N.J.; Foxwell, B.M. Bmx Tyrosine Kinase Regulates Tlr4-Induced Il-6 Production in Human Macrophages Independently of P38 Mapk and NF kappa B Activity. *Blood* **2008**, *111*, 1781–1788. [CrossRef] [PubMed]

24. Smolen, J.S.; Schoels, M.M.; Nishimoto, N.; Breedveld, F.C.; Burmester, G.R.; Dougados, M.; Emery, P.; Ferraccioli, G.; Gabay, C.; Gibofsky, A.; et al. Consensus Statement on Blocking the Effects of Interleukin-6 and in Particular by Interleukin-6 Receptor Inhibition in Rheumatoid Arthritis and Other Inflammatory Conditions. *Ann. Rheum. Dis* **2013**, *72*, 482–492. [CrossRef]

25. Kremer, J.M.; Bloom, B.J.; Breedveld, F.C.; Coombs, J.H.; Fletcher, M.P.; Gruben, D.; Krishnaswami, S.; Burgos-Vargas, R.; Wilkinson, B.A.; Zerbini, C.A.; et al. The Safety and Efficacy of a Jak Inhibitor in Patients with Active Rheumatoid Arthritis: Results of a Double-Blind, Placebo-Controlled Phase Iia Trial of Three Dosage Levels of Cp-690,550 Versus Placebo. *Arthritis Rheumatol.* **2009**, *60*, 1895–1905. [CrossRef] [PubMed]

26. Winthrop, K.L. The Emerging Safety Profile of Jak Inhibitors in Rheumatic Disease. *Nat. Rev. Rheumatol.* **2017**, *13*, 234–243. [CrossRef]

27. Atzeni, F.; Gianturco, L.; Talotta, R.; Varisco, V.; Ditto, M.C.; Turiel, M.; Sarzi-Puttini, P. Investigating the Potential Side Effects of Anti-Tnf Therapy for Rheumatoid Arthritis: Cause for Concern? *Immunotherapy* **2015**, *7*, 353–361. [CrossRef]

28. Keiserman, M.; Codreanu, C.; Handa, R.; Xibille-Friedmann, D.; Mysler, E.; Briceno, F.; Akar, S. The Effect of Antidrug Antibodies on the Sustainable Efficacy of Biologic Therapies in Rheumatoid Arthritis: Practical Consequences. *Expert Rev. Clin. Immunol.* **2014**, *10*, 1049–1057. [CrossRef]

29. Fosgerau, K.; Hoffmann, T. Peptide Therapeutics: Current Status and Future Directions. *Drug Discov. Today* **2015**, *20*, 122–128. [CrossRef]

30. Karaman, O.; Kelebek, S.; Demirci, E.A.; Ibis, F.; Ulu, M.; Ercan, U.K. Synergistic Effect of Cold Plasma Treatment and Rgd Peptide Coating on Cell Proliferation over Titanium Surfaces. *Tissue Eng. Regen. Med.* **2018**, *15*, 13–24. [CrossRef]

31. Choi, Y.J.; Lee, J.Y.; Park, J.H.; Park, J.B.; Suh, J.S.; Choi, Y.S.; Lee, S.J.; Chung, C.P.; Park, Y.J. The Identification of a Heparin Binding Domain Peptide from Bone Morphogenetic Protein-4 and Its Role on Osteogenesis. *Biomaterials* **2010**, *31*, 7226–7238. [CrossRef]

32. Choi, S.H.; Lee, J.Y.; Suh, J.S.; Park, Y.S.; Chung, C.P.; Park, Y.J. Dual-Function Synthetic Peptide Derived from Bmp4 for Highly Efficient Tumor Targeting and Antiangiogenesis. *Int. J. Nanomed.* **2016**, *11*, 4643–4656.

33. Ganz, T. Defensins: Antimicrobial Peptides of Innate Immunity. *Nat. Rev. Immunol.* **2003**, *3*, 710–720. [CrossRef] [PubMed]

34. Schibli, D.J.; Hunter, H.N.; Aseyev, V.; Starner, T.D.; Wiencek, J.M.; Mc, P.B.; Tack, B.F.; Vogel, H.J. The Solution Structures of the Human β Defensins Lead to a Better Understanding of the Potent Bactericidal Activity of Hbd3 against Staphylococcus Aureus. *J. Biol. Chem.* **2002**, *277*, 8279–8289. [CrossRef] [PubMed]

35. Joosten, L.A.; Helsen, M.M.; Loo, F.A.; Berg, W.B. Anticytokine Treatment of Established Type Ii Collagen- Induced Arthritis in Dba/1 Mice. A Comparative Study Using Anti-Tnfα, Anti-Il-1α/Beta, and Il-1ra. *Arthritis Rheumatol.* **1996**, *39*, 797–809. [CrossRef] [PubMed]

36. Chen, D.; Zhao, M.; Mundy, G.R. Bone Morphogenetic Proteins. *Growth Factors* **2004**, *22*, 233–241. [CrossRef] [PubMed]

37. Lee, J.Y.; Suh, J.S.; Kim, J.M.; Kim, J.H.; Park, H.J.; Park, Y.J.; Chung, C.P. Identification of a Cell-Penetrating Peptide Domain from Human Beta-Defensin 3 and Characterization of Its Anti-Inflammatory Activity. *Int. J. Nanomed.* **2015**, *10*, 5423–5434.

38. Patel, S.G.; Sayers, E.J.; He, L.; Narayan, R.; Williams, T.L.; Mills, E.M.; Allemann, R.K.; Luk, L.Y.P.; Jones, A.T.; Tsai, Y.H. Cell-Penetrating Peptide Sequence and Modification Dependent Uptake and Subcellular Distribution of Green Florescent Protein in Different Cell Lines. *Sci. Rep.* **2019**, *9*, 6298. [CrossRef]

39. Farrugia, B.L.; Lord, M.S.; Melrose, J.; Whitelock, J.M. The Role of Heparan Sulfate in Inflammation, and the Development of Biomimetics as Anti-Inflammatory Strategies. *J. Histochem. Cytochem.* **2018**, *66*, 321–336. [CrossRef]

40. Alunno, A.; Carubbi, F.; Giacomelli, R.; Gerli, R. Cytokines in the Pathogenesis of Rheumatoid Arthritis: New Players and Therapeutic Targets. *BMC Rheumatol.* **2017**, *13*, 3.

41. Dey, P.; Panga, V.; Raghunathan, S. A Cytokine Signalling Network for the Regulation of Inducible Nitric Oxide Synthase Expression in Rheumatoid Arthritis. *PLoS ONE* **2016**, *11*, e0161306. [CrossRef]

42. Tong, S.; Zhang, C.; Liu, J. Platelet-Rich Plasma Exhibits Beneficial Effects for Rheumatoid Arthritis Mice by Suppressing Inflammatory Factors. *Mol. Med. Rep.* **2017**, *16*, 4082–4088. [CrossRef] [PubMed]

43. Pietrosimone, K.M.; Jin, M.; Poston, B.; Liu, P. Collagen-Induced Arthritis: A Model for Murine Autoimmune Arthritis. *Bio-Protocol* **2015**, *20*, e1626. [CrossRef] [PubMed]

44. Haraoui, B.; Bykerk, V. Etanercept in the Treatment of Rheumatoid Arthritis. *Ther. Clin. Risk Manag.* **2007**, *3*, 99–105. [CrossRef] [PubMed]

45. Yi, H.; Kim, J.; Jung, H.; Rim, Y.A.; Kim, Y.; Jung, S.M.; Park, S.-H.; Ju, J.H. Induced Production of Anti-Etanercept Antibody in Collagen-Induced Arthritis. *Mol. Med. Rep.* **2014**, *9*, 2301–2308. [CrossRef] [PubMed]

46. Goldring, S.R. Pathogenesis of Bone Erosions in Rheumatoid Arthritis. *Curr. Opin. Rheumatol.* **2002**, *14*, 406–410. [CrossRef] [PubMed]

47. Van der Kraan, P.M. The Interaction between Joint Inflammation and Cartilage Repair. *Tissue Eng. Regen. Med.* **2019**, *16*, 327–334. [CrossRef]

48. Irawan, V.; Sung, T.C.; Higuchi, A.; Ikoma, T. Collagen Scaffolds in Cartilage Tissue Engineering and Relevant Approaches for Future Development. *Tissue Eng. Regen. Med.* **2018**, *15*, 673–697. [CrossRef]

49. Lee, J.Y.; Seo, Y.N.; Park, H.J.; Park, Y.J.; Chung, C.P. The Cell-Penetrating Peptide Domain from Human Heparin-Binding Epidermal Growth Factor-Like Growth Factor (Hb-Egf) Has Anti-Inflammatory Activity In Vitro and In Vivo. *Biochem. Biophys. Res. Commun.* **2012**, *419*, 597–604. [CrossRef]

50. Yang, Y.; Xia, M.; Zhang, S.; Zhang, X. Cell-Penetrating Peptide-Modified Quantum Dots as a Ratiometric Nanobiosensor for the Simultaneous Sensing and Imaging of Lysosomes and Extracellular Ph. *Chem. Commun. (Cambridge)* **2019**, *56*, 145–148. [CrossRef] [PubMed]

51. Yoon, J.Y.; Kim, D.W.; Ahn, J.H.; Choi, E.J.; Kim, Y.H.; Jeun, M.; Kim, E.J. Propofol Suppresses Lps-Induced Inflammation in Amnion Cells Via Inhibition of Nf-Kappab Activation. *Tissue Eng. Regen. Med.* **2019**, *16*, 301–309. [CrossRef] [PubMed]

52. Nishikawa, M.; Myoui, A.; Tomita, T.; Takahi, K.; Nampei, A.; Yoshikawa, H. Prevention of the Onset and Progression of Collagen-Induced Arthritis in Rats by the Potent P38 Mitogen-Activated Protein Kinase Inhibitor Fr167653. *Arthritis Rheumatol.* **2003**, *48*, 2670–2681. [CrossRef] [PubMed]

53. Mossiat, C.; Laroche, D.; Prati, C.; Pozzo, T.; Demougeot, C.; Marie, C. Association between Arthritis Score at the Onset of the Disease and Long-Term Locomotor Outcome in Adjuvant-Induced Arthritis in Rats. *Arthritis Res. Ther.* **2015**, *17*, 184. [CrossRef] [PubMed]

International Journal of
Molecular Sciences

Article

Modification of Luffa Sponge for Enrichment of Phosphopeptides

Lili Dai, Zhe Sun and Ping Zhou *

Key Laboratory of Analytical Chemistry for Biology and Medicine (Ministry of Education), College of Chemistry and Molecular Sciences, Wuhan University, Wuhan 430072, China; lilydai@whu.edu.cn (L.D.); 2014282030161@whu.edu.cn (Z.S.)

* Correspondence: zbping@whu.edu.cn

Received: 8 November 2019; Accepted: 18 December 2019; Published: 22 December 2019

Abstract: The enrichment technique is crucial to the comprehensive analysis of protein phosphorylation. In this work, a facile, green and efficient synthetic method was set up for quaternization of luffa sponge. The resultant luffa sponge showed strong anion-exchange characteristics and a high adsorption ability for phosphate ions. Along with the unique physical properties, e.g., tenacity and porous texture, quaternized luffa sponge was demonstrated to be a well-suited solid-phase extraction (SPE) material. The quaternized luffa sponge-based SPE method was simple, cost-effective and convenient in operation, and was successfully applied to the capture of phosphopeptides from protein digests. The enrichment approach exhibited exceptionally high selectivity, sensitivity and strong anti-interference ability. Four phosphopeptides were still detected by using the digest mixture of β-casein and bovine serum albumin with a molar ratio of 1:100. 21 phosphopeptides were identified from the tryptic digest of non-fat milk.

Keywords: luffa sponge; phosphopeptide; mass spectrometry; Matrix-assisted laser desorption ionization; solid-phase extraction

1. Introduction

Protein phosphorylation is a reversible posttranslational modification regulated by phosphatases and kinases. Abnormal protein phosphorylation events are often correlated with diseases [1,2]. Research of phosphopeptides provides valuable information to elucidate the biological regulatory mechanisms [3]. About 30% of cellular proteins would be phosphorylated during the physiological processes [4]. However, because of the dynamic and reversibility of phosphorylation, only 1–2% of the entire amount of protein is estimated to be phosphorylated at a specific moment [5]. In addition, the abundance of phosphopeptides received after digestion is much lower [6]. Matrix-assisted laser desorption ionization time-of-flight mass spectrometry (MALDI-TOF MS) is considered to be a powerful tool for the detection of phosphopeptides, however, of major concern is the poor ionization efficiency of phosphopeptides, caused by the low occupancy ratio of phosphopeptides and signal suppression of non-phosphopeptides. Therefore, prefractionation and the selective enrichment strategy of phosphopeptides is crucial for a comprehensive phosphoproteomics analysis [7].

Various methods have been extensively studied in recent decades, including chemical derivatization [8], immunoprecipitation [9], affinity chromatography with immobilized metal ions (IMAC) such as Fe^{3+} [10] and Ti^{4+} [11], affinity chromatography with metal oxides (MOAC) such as TiO_2 [12] and ZrO_2 [13], strong cation exchange (SCX) chromatography [14], strong anion exchange (SAX) chromatography [15], and hydrophilic interaction chromatography [16]. The effectiveness of a combined strategy, e.g., the use of SAX or SCX together with an affinity enrichment method (MOAC or IMAX), has also been demonstrated [17]. Owing to the existence of negatively charged phosphate groups (pKa = 1–2), the SAX materials would prefer to catch the phosphopeptides and screen the non-phosphopeptides via the charge difference between phosphopeptides and non-phosphopeptides [18]. SAX can not only enrich phosphopeptides but also fractionate phosphopeptides [19]. Amine-based materials are expected to provide a means for enriching phosphorylated proteins/peptides by SAX [20]. The electrostatic attraction and hydrogen bonding between amine and phosphate groups provide a foundation for selective enrichment [21]. Up to now, some amine-based materials have been used for selective enrichment of phosphopeptides, such as amino-functionalized materials [22], guanidyl-functionalized materials [23], polyethylenimine-functionalized materials [24], arginine-functionalized materials [25] and quaternary ammonium-functionalized materials [26,27]. Among them, quaternary ammonium materials, bearing permanent positive charges, provide the possibility of optimizing the best conditions for specific enrichment of phosphopeptides over a broad pH range.

Natural polymers have received increasing attention because they possess many virtues such as renewability, non-toxicity, inexpensiveness, biocompatibility and biodegradability. Luffa sponge, a natural macromolecular material, is obtained from the ripened dried fruit of *Luffa cylindrical*. Luffa sponge has a fibro-vascular reticulated structure with high porosity (79%–93%) and simultaneously low density (0.02–0.04 g/cm^3) [28]. It is mainly composed of cellulose, hemicellulose and lignin, and may also contain a small amount of pectin, protein and trace elements such as calcium, magnesium, phosphorus, potassium and so on [29]. Luffa sponge exhibits excellent mechanical properties and a tough texture, which is acid and alkali resistant [30]. Owing to the abundance of active groups on the surface of luffa sponge, a variety of chemical modifications may be carried out to increase its functionality and the scope of its use [31,32]. Luffa sponge and its modified products have been used as sorbents for heavy metals [33], dyes [34] or phenols [35], biomatrixes for cell immobilization, and structural supports for biosorption in diverse biotechnological applications [36,37].

Luffa sponge has great potential to be used as a solid phase extraction material due to its physical and chemical properties. In this work, the strategy for quaternization of luffa sponge was investigated, and a new solid phase extraction approach based on quaternized luffa sponge was developed for selective enrichment of phosphopeptides. The performance of the enrichment approach was assessed.

2. Results and Discussion

2.1. Preparation and Characterization

The fiber of luffa sponge was principally made up of cellulose, hemicellulose and lignin. Small organic molecules were removed by methanol extraction. Alkali treatment had a beneficial effect on further modification. The general scheme for the preparation of quaternized luffa sponge A (QA) and quaternized luffa sponge B (QB) is presented in Scheme S1. The synthetic route to QA included three main steps. Firstly, epichlorohydrin (ECH) reacted with the hydroxyl groups of luffa cellulose to form epoxy cellulose ether. Then, the epoxide ring was opened by reacting with ethylenediamine. In the third step, triethylamine was connected to ethylenediamine moiety that had been grafted on the polymer chain to obtain QA. For the synthesis of QB, 3-chloro-2-hydroxypropyltrimethylammonium chloride (CHPTAC) was chosen as the etherification agent. Under alkaline conditions, CHPTAC was cyclized to epoxy propyl quaternary ammonium salt, and the latter reacted with cellulose through a nucleophilic ring opening reaction to generate cationic cellulose. To ensure QA and QB were carrying more quaternary amine groups, the reaction conditions such as feeding ratio, reaction time and temperature were optimized and has been explained in the experimental section. The obtained materials were characterized by scanning electron microscope (SEM) and X-ray photoelectron spectroscopy (XPS).

The differences in the morphology of untreated luffa sponge, QA and QB are displayed in Figure 1. The untreated luffa sponge material was composed of interwoven fibers finer than 100 nm. After modification, the microcosmic reticular structure on the surface of QA disappeared, implying that the fibers on the surface were partly impaired during the chemical modification process. QB had a smooth surface and was uniform in fiber microstructure.

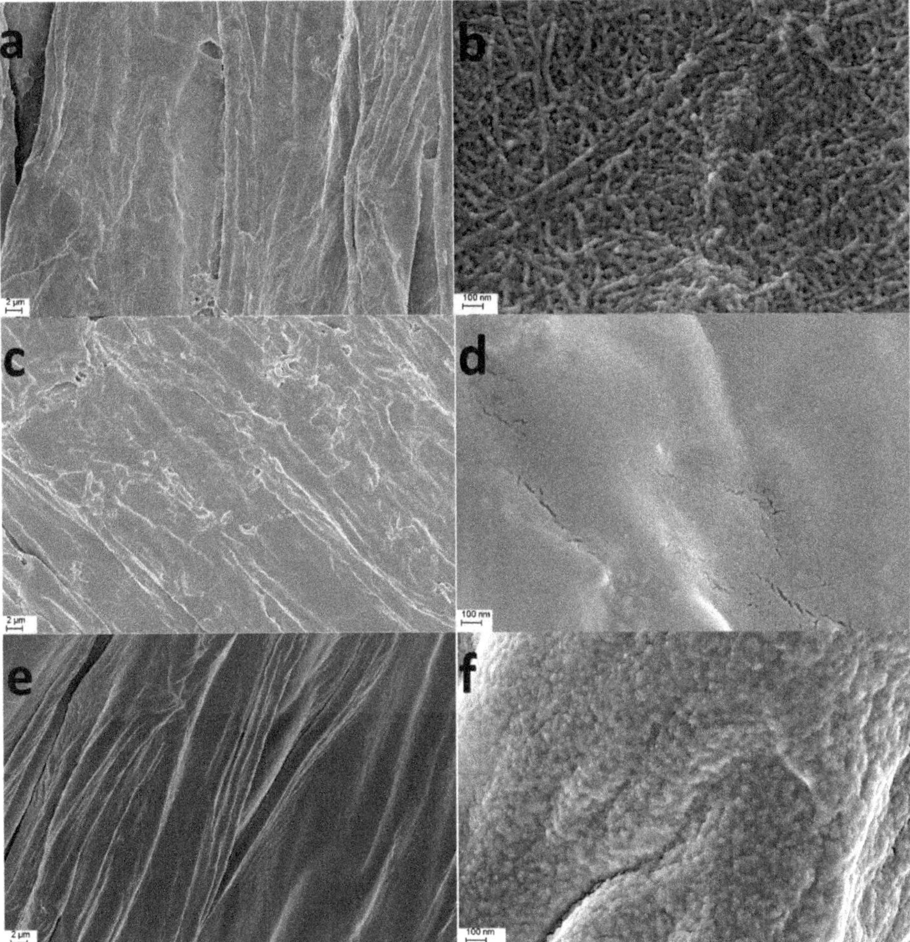

Figure 1. SEM images of (**a**,**b**) untreated luffa sponge, (**c**,**d**) modified luffa sponge QA, and (**e**,**f**) modified luffa sponge QB.

The atomic percentages (C 1s, N 1s and O 1s) of untreated luffa sponge, QA and QB revealed by XPS are listed in Supplementary Table S1. The relative contents of N element on the surfaces of untreated luffa sponge, QA and QB were 0.67%, 5.60% and 2.56%, respectively, indicating that the ammonium groups were introduced to QA and QB. As shown in Figure 2, in comparison with untreated luffa sponge, both QA and QB displayed an additional peak at 402.5 eV in N 1s XPS high resolution spectra. This peak was attributable to the N element in quaternary ammonium groups, indicating that both synthesis routes were successful for the quaternization of luffa sponge.

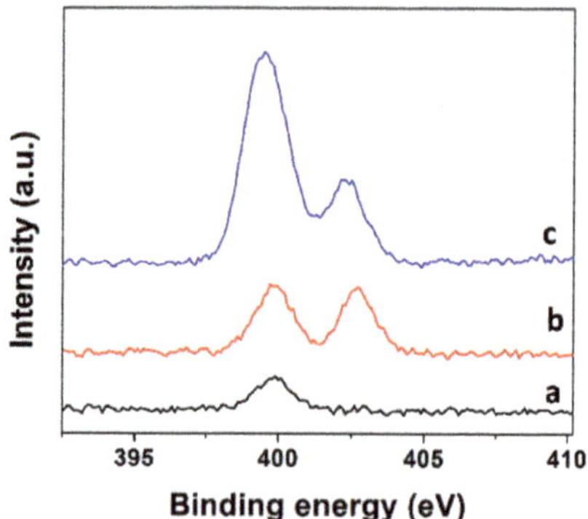

Figure 2. XPS spectra of N 1s of (**a**) untreated luffa sponge, (**b**) modified luffa sponge QB, and (**c**) modified luffa sponge QA.

2.2. Adsorption of Anions by Modified Luffa Sponge

The adsorption performances of luffa sponge materials for inorganic anions were investigated. As shown in Figure 3, the untreated luffa sponge had almost no effect on adsorption of anions, however, both modified luffa sponge materials exhibited a significant effect, especially on sulfate and phosphate ions. 77.1% of phosphate ions were removed from the solution of mixed anions by QA, and 88.9% of phosphate ions were removed by QB. Although the total content of nitrogen element on the surface of QA revealed by XPS was higher than that of QB, the adsorption performance of QB was better, which was attributable to more quaternary ammonium groups on QB. Once the incubation temperature was set at 40 °C, 98.3% of phosphate ions were removed by QB.

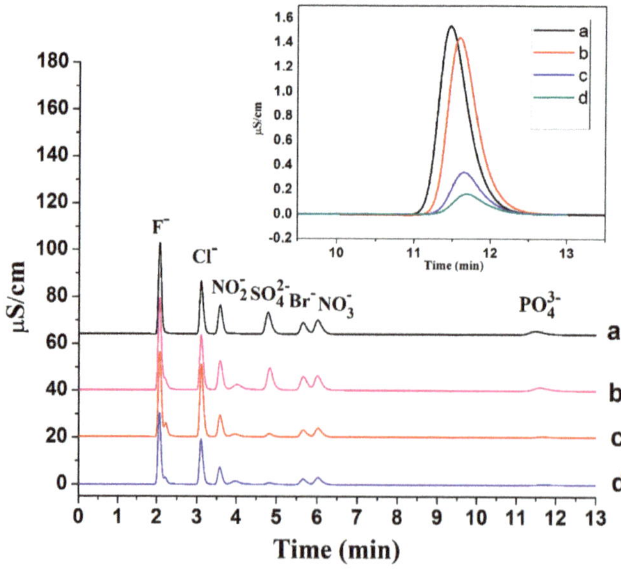

Figure 3. Ion exchange chromatograms of anions (**a**) before and after adsorption by (**b**) untreated luffa sponge, (**c**) QA, and (**d**) QB.

2.3. Enrichment of Phosphopeptides by QB

Inspired by the aforementioned experimental results, we selected QB as a potential solid phase material for phosphopeptide enrichment. The schematic diagram of the procedure for solid phase extraction of phosphopeptides is shown in Figure 4. The enrichment mechanism was based on anion exchange. Phosphorylated peptides, carrying one or more phosphate groups, are more electronegative than other peptides in the same environment. When a proper loading solution was used, non-phosphopeptides would be exchanged by the anions in the solution while phosphopeptides could be retained on the solid phase. The pH of the loading solution was set at 4.0 to reduce the adsorption of non-phosphopeptides caused by electrostatic interaction. A small amount of acetonitrile was added to reduce the adsorption of non-phosphopeptides caused by hydrophobic interaction. Especially, owing to the characteristics of texture and fibro-vascular reticulated structure, the luffa sponge had open and free space for the exchange of matter, and was able to tolerate remarkable stresses and resume its original shape when brought again to rest. The extraction cartridge was assembled by packing QB within the barrel of a syringe between a sieve plate and the piston. The extraction was very convenient in operation. Since luffa sponge is almost an inexhaustible resource and the modification method for QB is facile and efficient, quaternized luffa sponge is low-cost.

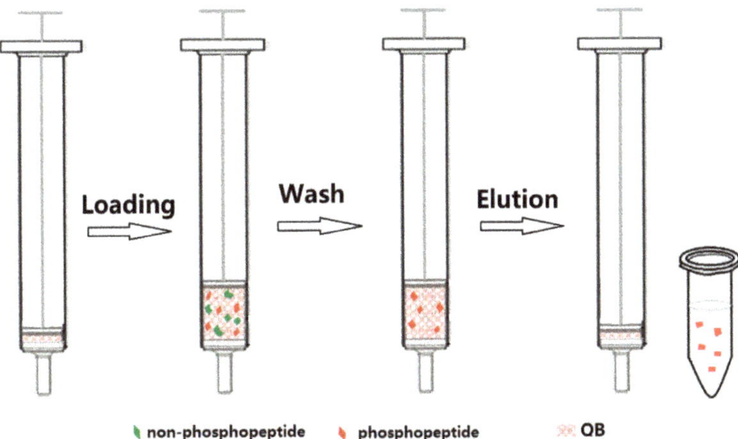

Figure 4. Schematic diagram of the procedure for enrichment of phosphopeptides with QB. Loading solution: 30 mM acetic acid-ammonium acetate in 20% acetonitrile, pH 4.0; washing solution: deionized water; elution solution: 5% trifluoroacetic acid.

The performance of QB for phosphopeptide enrichment was investigated by using tryptic digests of two typical phosphorylated proteins, namely α-casein and β-casein. When the tryptic digest of α-casein (3 pmol) or β-casein (3 pmol) was directly analyzed by MALDI-TOF MS, only weak signals of one or two phosphopeptides could be detected. After enrichment, the signals from phosphopeptides dominated the spectra. Taking α-casein digest as an example, the relative peak area of the phosphopeptides versus the total area of the peptide-ions in the mass spectrum was 69.2%. In contrast, the ratio $I_{phos}/(I_{phos} + I_{non})$ was 5.9% before enrichment. Eighteen phosphopeptides from α-casein digest (Figure 5b) and nine from β-casein digest (Figure 5d) were detected. The number of phosphopeptides enriched from α-casein was improved in comparison with the previously reported results enriched by commercial TiO_2 [38,39], in which thirteen phosphopeptides were identified. The information about the phosphopeptides involved in this paper is displayed in Supplementary Table S2.

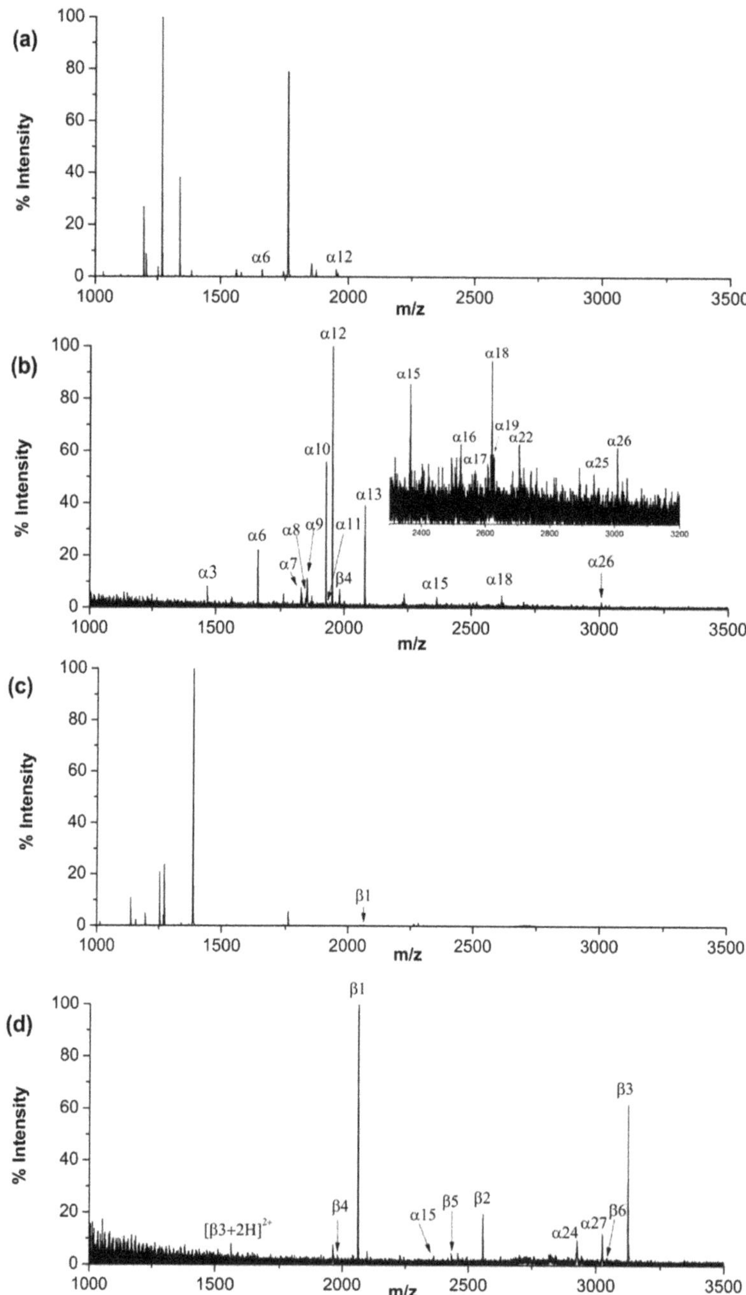

Figure 5. MALDI mass spectra of tryptic digests of α-casein (3 pmol) by direct analysis (**a**) or after enrichment using QB (**b**), and β-casein (3 pmol) by direct analysis (**c**) or after enrichment using QB (**d**).

To assess the selectivity of this enrichment approach to phosphopeptides, a non-phosphorylated protein, BSA, was chosen as a reference to examine the anti-interference ability. Enzymatic digest mixtures of β-casein and BSA with a molar ratio of 1:10 and 1:100 were tested. The MALDI-TOF MS spectra for the digest mixtures prior to enrichment are displayed in Supplementary Figure S1a,c. The majority of observed signals corresponded to non-phosphopeptides. After enrichment, peaks of phosphopeptides dominated in the mass spectra (Supplementary Figure S1b,d). When the molar ratio of β-casein to BSA was 1:100, four phosphopeptides could still be detected. Two among them were from α-casein due to protein impurities in β-casein. The results demonstrated the excellent selectivity of this method toward phosphopeptides in spite of the presence of a large amount of non-phosphopeptides.

To investigate the sensitivity of this approach, less amounts (300 fmol and 30 fmol) of β-casein digests were tested. As shown in Supplementary Figure S2, three phosphopeptides could still be detected even when the amount of β-casein was as low as 30 fmol.

2.4. Enrichment of Phosphopeptides from Non-Fat Milk

The application potential of QB to real samples was examined by selective enrichment of phosphopeptides from non-fat milk. Prior to enrichment, most peaks in the mass spectrum indeed corresponded to non-phosphopeptides in the tryptic digest (Figure 6a). After solid phase extraction using QB, three phosphopeptides from β-casein and eighteen phosphopeptides from α-casein were detected (Figure 6b). When the digest was enriched by TiO_2 nanoparticles, only thirteen phosphopeptides in total were observed (Figure 6c). The number of phosphopeptides (21) captured from non-fat milk digest by QB also exceeded those captured by some other materials, such as mesoporous TiO_2 nanoparticles (12) [40], Zr^{4+}-immobilized magnetic covalent organic frameworks (14) [41], Ti(IV) and Nb(V) modified magnetic microspheres (19) [42], chitosan and polyethylenimine coated magnetic particles (17) [43] and magnetic guanidyl-functionalized metal–organic framework nanospheres (19) [44], and was comparable to that of titanium dioxide/ions on magnetic microspheres (23) [45]. The results demonstrated that the QB-based method exhibited good selectivity toward phosphopeptides in complex samples.

3. Materials and Methods

3-Chloro-2-hydroxypropyltrimethylammonium chloride aqueous solution (69 wt%) was purchased from Dibo Chemical Reagent (Shanghai, China) and used as etherifying reagents without further purification. Epichlorohydrin was purchased from Lingfeng Chemical Reagent (Shanghai, China). N, N-dimethylformamide (DMF), ethylenediamine, triethylamine, sodium hydroxide, methanol, ethanol, acetonitrile (ACN), isopropanol, urea, acetic acid, ammonium acetate and sodium bicarbonate were analytical grade and purchased from Sinopharm Chemical Reagent Co., Ltd. (Shanghai, China). The standard solutions of phosphate, nitrate, sulfate, nitrite, fluoride ion, chloride ion and bromine ion were purchased from the National Center for Analysis and Testing of Nonferrous Metals and Electronic Materials (Beijing, China). Iodoacetamide (IAA), 1,4-dithiothreitiol (DTT), trifluoroacetic acid (TFA) and 2,5-dihydroxybenzoic acid (DHB) were purchased from Aladdin Chemical Reagent Co., Ltd. (Shanghai, China). Bovine β-casein, bovine α-casein, trypsin (from porcine pancreas, TPCK-treated) and bovine serum albumin (BSA) were purchased from Sigma–Aldrich (St. Louis, MO, USA). The luffa sponge was purchased from Henan Ledu Pharmacy (Henan, China). Ultra-high molecular weight polyethylene sieve plate (UHMW-PE, diameter of 4.9 mm, thickness of 1.6 mm, and pore size of 20 μm) was purchased from Haohai Linfeng Technology Co., Ltd. (Wuhan, China). The syringe (1 mL) was purchased from Jinta Medical Equipment Co., Ltd. (Shanghai, China). Non-fat milk was purchased from a local supermarket.

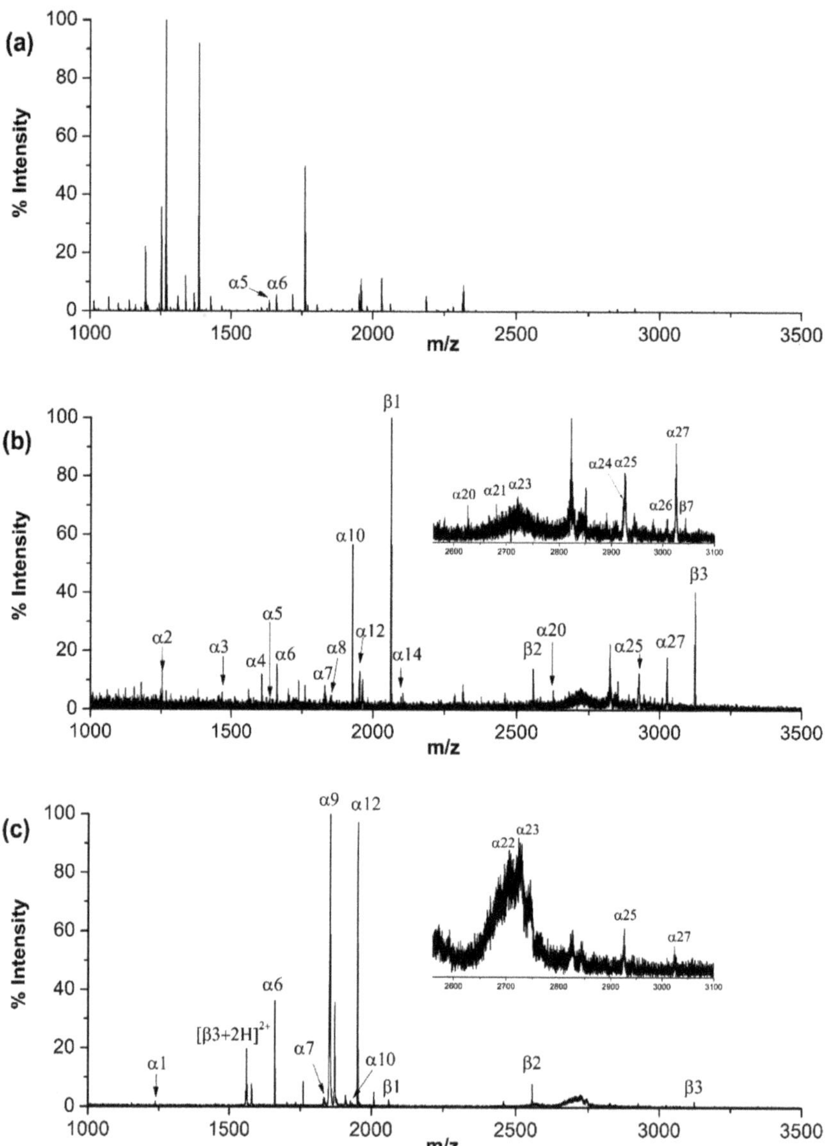

Figure 6. MALDI mass spectra of (**a**) tryptic digest of non-fat milk without any enrichment, (**b**) with QB enrichment, and (**c**) with TiO$_2$ enrichment.

3.1. Pretreatment of Luffa Sponge

The raw material of luffa sponge was peeled to remove the seeds and cut to small pieces. 10 g of luffa sponge was suspended in 300 mL of methanol in a 500 mL round-bottomed flask fitted with a condenser tube. The luffa sponge suspension was refluxed for 24 h at 60 °C. When dry, the luffa sponge was soaked in 5% (*w/w*) sodium hydroxide aqueous solution at room temperature for 24 h. The luffa sponge was then washed with deionized water, dried and sealed.

3.2. Synthesis of Quaternized Luffa Sponge

The first synthesis route was similar to that of Zhang et al. [46] whereby 1 g of pretreated luffa sponge was suspended in 40 mL of DMF and 20 mL of ECH in a 250-mL round-bottomed flask with a constant-temperature water bath at 85 °C and magnetic stirring. After 1 h, 2.6 mL of ethylenediamine was added dropwise into the flask. One hour later, 3.5 mL of triethylamine was added dropwise into the flask. Over a period of 60 min, the product was washed sequentially with deionized water and ethanol. When dry, the material was sealed for later use. The as-prepared material was coded as QA.

Quaternized luffa sponge was prepared by another route: 2 g of pretreated luffa sponge was soaked in 100 mL of sodium hydroxide aqueous solution (30%, *w/w*) overnight. After being washed with deionized water and dried, the luffa sponge was added to 80 mL of isopropanol in a flask with a constant-temperature water bath at 45 °C and magnetic stirring. Then 5 mL of sodium hydroxide aqueous solution (30%, *w/w*) was added and 7 g of CHPTAC aqueous solution was added dropwise into the flask. Over a period of 3 h, the product was washed sequentially with deionized water and ethanol. The material was dried and coded as QB.

3.3. Characterization of Quaternized Luffa Sponge Materials

The morphology of the materials was observed by a Sigma field emission scanning electron microscope (Zeiss, Germany). X-ray photoelectron spectroscopy measurement was performed on an ESCALAB 250Xi electron spectrometer (Thermo Scientific, Waltham, MA, USA) using a radiation source of Al Kα radiation with the energy of 1486.6 eV.

3.4. Adsorption of Anions

A mixture of anions containing F^-, Cl^-, NO_2^-, Br^-, SO_4^{2-}, NO_3^- and PO_4^{3-} with the concentration of 10 mg/L each was made by diluting anion standard solutions with deionized water. 15 mg of untreated luffa sponge, QA or QB, was added to 1.5 mL of the solution of mixed anions. After being incubated at room temperature for 1 h, the suspension was centrifugated and the supernatant was collected for further ion chromatography analysis.

3.5. Ion Chromatography

An ICS 2500 ion chromatography system (Dionex, CA, USA) consisting of a GP50 gradient pump, AERS 500 suppressor, RFC-30 eluent generator and ED50 electrochemical detector was used in this study. IonPac^{TM} AS 11-HC (4 × 250 mm) and AG 11-HC guard (4 × 50 mm) columns (Dionex, CA, USA) packed with anion-exchange resin were used as the separation columns. The analysis was performed at 35 °C with the flow rate set at 1.5 mL/min in isocratic mode. The injection volume was 25 µL, and 36 mM potassium hydroxide was employed as mobile phase.

3.6. Peptide Sample Pretreatment

Bovine α-casein or bovine β-casein was originally made into stock solution at a concentration of 1 mg/mL with ammonium bicarbonate solution (50 mM, pH 8.0). Proteins were digested with trypsin by using an enzyme to substrate ratio of 1:50 (*w/w*), and the digestion was performed at 37 °C for 24 h.

BSA (1 mg) was dissolved in 100 µL of denaturing solution (8 M urea in 50 mM ammonium bicarbonate solution, pH 8.0). Then BSA was reduced by 10 mM DTT (final concentration) at 56 °C for 30 min. The reduced cysteine residues were alkylated with 20 mM IAA (final concentration) in the dark at room temperature for 30 min. The reduced and alkylated protein sample was diluted with 300 µL of 50 mM ammonium bicarbonate solution and then digested with trypsin at an enzyme to substrate ratio of 1:50 (*w/w*) by incubating at 37 °C for 24 h.

100 µL of non-fat milk was lyophilized to dryness and then denatured by adding 100 µL of denaturing solution (8 M urea in 50 mM ammonium bicarbonate solution, pH 8.0) at 56 °C for 30 min.

The proteins were then reduced, alkylated, and digested the same way as BSA. All the tryptic digests were stored at −20 °C for future use.

3.7. Phosphopeptide Enrichment Procedure

The enrichment procedure involved three steps: load, wash and elution. One sieve plate was loaded into the bottom of a syringe barrel to construct a simple solid phase extraction device. Five milligrams of QB was put in between the sieve plate and the syringe piston. Ten microlitres of peptide mixture was added to 500 µL of loading solution (30 mM acetic acid-ammonium acetate buffer containing 20% ACN, pH 4.0). After being drawn in the syringe and incubated with QB for 1 h, the sample solution was pushed out. The QB material was then washed with loading solution and water in succession. The trapped peptides were eluted with 50 µL of 5% TFA.

The phosphopeptide enrichment by TiO_2 was performed according to the literature [47]. TiO_2 nanoparticles were prepared as previously described [48]. Before enrichment, 2 mg of TiO_2 nanoparticles were dispersed in 500 µL of loading buffer (1 M glycolic acid, 5% TFA, 80% acetonitrile). Ten microlitres of protein digests were mixed with TiO_2 suspension, and incubated for 1 h at room temperature, followed by centrifugation at 12,000g for 5 min. The supernatant was discarded and the remaining materials were washed sequentially with loading buffer and washing buffer (80% acetonitrile, 1% TFA). The phosphopeptides were eluted with 100 µL of NH_4OH, pH 11 and the eluate was acidified with 10 µL of 100% formic acid. The acidified phosphopeptides were desalted on a Sep-Pak-C18 microcolumn. The purified phosphopeptides were eluted from the microcolumn directly onto the MALDI-target using DHB matrix (20 mg/mL DHB in 70% acetonitrile, 1% phosphoric acid and 0.1% TFA).

3.8. Mass Spectrometry Analysis

Peptides were analyzed using a 5800 MALDI-TOF/TOF mass spectrometry (AB SCIEX, Framingham, MA, USA) in positive ion mode. Matrix solution (mixture of 25 mg/mL DHB in 50% (*v/v*) ACN, 1% (*v/v*) phosphoric acid [49]) was mixed with equal volume of the eluate, from which 0.6 µL of solution was loaded onto the MALDI target. MS data were acquired in reflection mode. Four hundred shots were accumulated for each MS spectrum, and the data were processed by using Data Explore (AB SCIEX, Framingham, MA, USA).

4. Conclusions

In summary, we developed a feasible method for synthesizing quaternized luffa sponge and a new approach for the enrichment of phosphopeptides. Quaternized luffa sponge was demonstrated to be a well-suited solid phase extraction material. The quaternized luffa sponge-based extraction approach is simple, cost-effective, and convenient in operation. The enrichment approach exhibited exceptionally high selectivity and sensitivity toward phosphopeptides, and was successfully applied to the analysis of phosphorylated proteins in a complex sample. Due to the unique structural properties it bears, quaternized luffa sponge should be an attractive solid phase extraction material for a wide range of applications.

Supplementary Materials: Supplementary Materials can be found at http://www.mdpi.com/1422-0067/21/1/101/s1.

Author Contributions: P.Z. conceived and supervised the study; L.D. and Z.S. performed the experiments and analyzed the data. All authors have read and agreed to the published version of the manuscript.

Funding: This work was funded by grants from the Natural Science Foundation of Hubei Province of China (no. 214CFB179).

Conflicts of Interest: The authors declare that they have no conflicts of interest.

Abbreviations

MALDI	Matrix-assisted laser desorption ionization
TOF	Time-of-flight
MS	Mass spectrometry
SAX	Strong anion exchange
SPE	Solid-phase extraction
CHPTAC	3-Chloro-2-hydroxypropyltrimethylammonium chloride
ECH	Epichlorohydrin
IAA	Iodoacetamide
DTT	1,4-dithiothreitiol
BSA	Bovine serum albumin

References

1. Hunter, T. Signaling–2000 and Beyond. *Cell* **2000**, *100*, 113–127. [CrossRef]
2. Ruprecht, B.; Lemeer, S. Proteomic analysis of phosphorylation in cancer. *Expert. Rev. Proteom.* **2014**, *11*, 259–267. [CrossRef] [PubMed]
3. Huang, J.F.; Wang, F.J.; Ye, M.L.; Zou, H.F. Enrichment and separation techniques for large-scale proteomics analysis of the protein post-translational modifications. *J. Chromatogr. A* **2014**, *1372*, 1–17. [CrossRef] [PubMed]
4. Ficarro, S.; Mccleland, M.L.; Stukenberg, P.T.; Burke, D.J.; Ross, M.M.; Shabanowitz, J.; Hunt, D.F.; White, F.M. Phosphoproteome analysis by mass spectrometry and its application to saccharomyces cerevisiae. *Nat. Biotechnol.* **2002**, *20*, 301–305. [CrossRef]
5. Ubersax, J.A.; Ferrell, J.E. Mechanisms of specificity in protein phosphorylation. *Nat. Rev. Mol. Cell Biol.* **2007**, *8*, 530–541. [CrossRef]
6. Joerg, R.; Albert, S. State-of-the-art in phosphoproteomics. *Proteomics* **2005**, *5*, 4052–4061.
7. Ren, L.L.; Li, C.Y.; Shao, W.L.; Lin, W.R.; He, F.C.; Jiang, Y. TiO$_2$ with tandem fractionation (TAFT): An approach for rapid, deep, reproducible and high-throughput phosphoproteome analysis. *J. Proteome Res.* **2018**, *17*, 710–721. [CrossRef]
8. Mclachlin, D.T.; Chait, B.T. Improved beta-elimination-based affinity purification strategy for enrichment of phosphopeptides. *Anal. Chem.* **2003**, *75*, 6826–6836. [CrossRef]
9. Rush., J.; Moritz, A.; Lee, K.A.; Guo, A.; Goss, V.L.; Spek, E.J.; Zhang, H.; Zha, X.M.; Polakiewicz, R.D.; Comb, M.J. Immunoaffinity profiling of tyrosine phosphorylation in cancer cells. *Nat. Biotechnol.* **2004**, *23*, 94–101. [CrossRef]
10. Stensballe, A.; Andersen, S.; Jensen, O.N. Characterization of phosphoproteins from electrophoretic gels by nanoscale Fe(III) affinity chromatography with off-line mass spectrometry analysis. *Proteomics* **2001**, *1*, 207–222. [CrossRef]
11. Zhou, H.J.; Ye, M.L.; Dong, J.; Han, G.H.; Jiang, X.N.; Wu, R.N.; Zou, H.F. Specific phosphopeptide enrichment with immobilized titanium ion affinity chromatography adsorbent for phosphoproteome analysis. *J. Proteome Res.* **2008**, *7*, 3957–3967. [CrossRef] [PubMed]
12. Larsen, M.R.; Thingholm, T.E.; Jensen, O.N.; Roepstorff, P.; Jorgensen, T.J.D. Highly selective enrichment of phosphorylated peptides from peptide mixtures using titanium dioxide microcolumns. *Mol. Cell. Proteom.* **2005**, *4*, 873–886. [CrossRef] [PubMed]
13. Kweon, H.K.; Hakansson, K. Selective zirconium dioxide-based enrichment of phosphorylated peptides for mass spectrometric analysis. *Anal. Chem.* **2006**, *78*, 1743–1749. [CrossRef] [PubMed]
14. Hennrich, M.L.; van den Toorn, H.W.P.; Groenewold, V.; Heck, A.J.; Mohammed, S. Ultra acidic strong cation exchange enabling the efficient enrichment of basic phosphopeptides. *Anal. Chem.* **2012**, *84*, 1804–1808. [CrossRef] [PubMed]
15. Motoyama, A.; Xu, T.; Ruse, C.I.; Wohlschlegel, J.A.; Yates, J.R. Anion and cation mixed-bed ion exchange for enhanced multidimensional separations of peptides and phosphopeptides. *Anal. Chem.* **2007**, *79*, 3623–3634. [CrossRef] [PubMed]

16. Mcnulty, D.E.; Annan, R.S. Hydrophilic interaction chromatography reduces the complexity of the phosphoproteome and improves global phosphopeptide isolation and detection. *Mol. Cell. Proteom.* **2008**, *7*, 971–980. [CrossRef] [PubMed]

17. Dehghani, A.; Gödderz, M.; Winter, D. Tip-based fractionation of batch-enriched phosphopeptides facilitates easy and robust phosphoproteome analysis. *J. Proteome Res.* **2018**, *17*, 46–54. [CrossRef]

18. Li, X.S.; Yuan, B.F.; Feng, Y.Q. Recent advances in phosphopeptide enrichment: Strategies and techniques. *Trac-Trends Anal. Chem.* **2016**, *78*, 70–83. [CrossRef]

19. Han, G.H.; Ye, M.L.; Zhou, H.J.; Jiang, X.N.; Feng, S.; Jiang, X.G.; Tian, R.J.; Wan, D.F.; Zou, H.F.; Gu, J.R. Large-scale phosphoproteome analysis of human liver tissue by enrichment and fractionation of phosphopeptides with strong anion exchange chromatography. *Proteomics* **2008**, *8*, 1346–1361. [CrossRef]

20. Wang, Z.G.; Lv, N.; Bi, W.Z.; Zhang, J.L.; Ni, J.Z. Development of the affinity materials for phosphorylated proteins/peptides enrichment in phosphoproteomics analysis. *Acs Appl. Mater. Interfaces* **2015**, *7*, 8377–8392. [CrossRef]

21. Hargrove, A.E.; Nieto, S.; Zhang, T.; Sessler, J.L.; Anslyn, E.V. Artificial receptors for the recognition of phosphorylated molecules. *Chem. Rev.* **2011**, *111*, 6603–6782. [CrossRef] [PubMed]

22. Zhang, Y.; Wang, H.J.; Lu, H.J. Sequential selective enrichment of phosphopeptides and glycopeptides using amine-functionalized magnetic nanoparticles. *Mol. Biosyst.* **2013**, *9*, 492–500. [CrossRef] [PubMed]

23. Xiong, Z.C.; Chen, Y.J.; Zhang, L.Y.; Ren, J.; Zhang, Q.Q.; Ye, M.L.; Zhang, W.B.; Zou, H.F. Facile synthesis of guanidyl-functionalized magnetic polymer microspheres for tunable and specific capture of global phosphopeptides or only multiphosphopeptides. *Acs Appl. Mater. Interfaces* **2014**, *6*, 22743–22750. [CrossRef] [PubMed]

24. Chen, C.T.; Wang, L.Y.; Ho, Y.P. Use of polyethylenimine-modified magnetic nanoparticles for highly specific enrichment of phosphopeptides for mass spectrometric analysis. *Anal. Bioanal. Chem.* **2011**, *399*, 2795–2806. [CrossRef] [PubMed]

25. Chang, C.K.; Wu, C.C.; Wang, Y.S.; Chang, H.C. Selective extraction and enrichment of multiphosphorylated peptides using polyarginine-coated diamond nanoparticles. *Anal. Chem.* **2008**, *80*, 3791–3797. [CrossRef]

26. Dai, L.L.; Jin, S.X.; Fan, M.Y.; Zhou, P. Preparation of quaternized cellulose/chitosan microspheres for selective enrichment of phosphopeptides. *Anal. Bioanal. Chem.* **2017**, *409*, 3309–3317. [CrossRef]

27. Dong, M.M.; Wu, M.H.; Wang, F.J.; Qin, H.Q.; Han, G.H.; Dong, J.; Wu, R.A.; Ye, M.L.; Liu, Z.; Zou, H.F. Coupling strong anion-exchange monolithic capillary with MALDI-TOF MS for sensitive detection of phosphopeptides in protein digest. *Anal. Chem.* **2010**, *8*, 2907–2915. [CrossRef]

28. Saeed, A.; Iqbal, M. Loofa (Luffa cylindrica) sponge: Review of development of the biomatrix as a tool for biotechnological applications. *Biotechnol. Prog.* **2013**, *29*, 573–600. [CrossRef]

29. Rowell, R.M.; Han, J.S.; Rowell, J.S. Characterization and factors effecting fiber properties. In *Natural Polymers and Agrofibers Based Composites*; Frollini, E., Leao, A.L., Mattoso, L.H.C., Eds.; Embrapa Agricultural Instrumentation: San Carlos, Brazil, 2000; pp. 115–134.

30. Chen, Q.; Shi, Q.; Gorb, S.N.; Li, Z.Y. A multiscale study on the structural and mechanical properties of the luffa sponge from Luffa cylindrica plant. *J. Biomech.* **2014**, *47*, 1332–1339. [CrossRef]

31. Roy, D.; Semsarilar, M.; Guthrie, J.T.; Perrier, S. Cheminform abstract: Cellulose modification by polymer grafting: A review. *Chem. Soc. Rev.* **2009**, *38*, 2046–2064. [CrossRef]

32. Figueiredo, P.C.; Lintinen, K.; Hirvonen, J.T.; Kostiainen, M.A.; Santos, H.I.A. Properties and chemical modifications of lignin: Towards lignin-based nanomaterials for biomedical applications. *Prog. Mater. Sci.* **2018**, *93*, 233–269. [CrossRef]

33. Akhtar, N.; Iqbal, J.; Iqbal, M. Microalgal-luffa sponge immobilized disc: A new efficient biosorbent for the removal of Ni(II) from aqueous solution. *Lett. Appl. Microbiol.* **2013**, *37*, 149–153. [CrossRef] [PubMed]

34. Gupta, V.K.; Pathania, D.; Agarwal, S.; Sharma, S. Amputation of congo red dye from waste water using microwave induced grafted Luffa cylindrica cellulosic fiber. *Carbohydr. Polym.* **2014**, *111*, 556–566. [CrossRef] [PubMed]

35. Ye, C.L.; Hu, N.; Wang, Z.K. Experimental investigation of Luffa cylindrica as a natural sorbent material for the removal of a cationic surfactant. *J. Taiwan Inst. Chem. Eng.* **2013**, *44*, 74–80. [CrossRef]

36. Iqbal, M.; Zafar, S.I. The use of fibrous network of matured dried fruit of Luffa aegyptica as immobilizing agent. *Biotechnol. Tech.* **1993**, *7*, 15–18. [CrossRef]

37. Iqbal, M.; Zafar, S.I. Vegetable sponge as a matrix to immobilize micro-organisms: A trial study for hyphal fungi, yeast and bacteria. *Lett. Appl. Microbiol.* **1994**, *18*, 214–217. [CrossRef]

38. Lu, Z.; Duan, J.; He, L.; Hu, Y.; Yin, Y. Mesoporous TiO_2 Nanocrystal Clusters for Selective Enrichment of Phosphopeptides. *Anal. Chem.* **2010**, *82*, 7249–7258. [CrossRef]

39. Li, X.; Su, X.; Zhu, G.; Zhao, Y.; Yuan, B.; Guo, L.; Feng, Y. Titanium-containing magnetic mesoporous silica spheres: Effective enrichment of peptides and simultaneous separation of nonphosphopeptides and phosphopeptides. *J. Sep. Sci.* **2012**, *35*, 1506–1513. [CrossRef]

40. Li, K.; Zhao, S.; Yan, Y.; Zhang, D.; Peng, M.; Wang, Y.; Guo, G.; Wang, X. In-tube solid-phase microextraction capillary column packed with mesoporous TiO_2 nanoparticles for phosphopeptide analysis. *Electrophor.* **2019**, *40*, 2142–2148. [CrossRef]

41. Gao, C.; Bai, J.; He, Y.; Zheng, Q.; Ma, W.; Lei, Z.; Zhang, M.; Wu, J.; Fu, F.; Lin, Z. Covalent organic frameworks for selective enrichment of phosphopeptides. *Acs Appl. Mater. Interfaces* **2019**, *11*, 13735–13741. [CrossRef]

42. Jiang, J.; Sun, X.; She, X.; Li, J.; Li, Y.; Deng, C.; Duan, G. Magnetic microspheres modified with Ti(IV) and Nb(V) for enrichment of phosphopeptides. *Microchim. Acta* **2018**, *185*, 309–316. [CrossRef] [PubMed]

43. Hussain, D.; Musharraf, S.G.; Fatima, B.; Saeed, A.; Jabeen, F.; Ashiq, M.N.; Najam-ul-Haq, M. Magnetite nanoparticles coated with chitosan and polyethylenimine as anion exchanger for sorptive enrichment of phosphopeptides. *Microchim. Acta* **2019**, *186*, 852–861.

44. Luo, B.; Yang, M.; Jiang, P.; Lan, F.; Wu, Y. Multi-affinity sites of magnetic guanidyl-functionalized metal–organic framework nanospheres for efficient enrichment of global phosphopeptides. *Nanoscale* **2018**, *10*, 8391–8396. [CrossRef] [PubMed]

45. Wang, J.; Wang, Z.; Sun, N.; Deng, C. Immobilization of titanium dioxide/ions on magnetic microspheres for enhanced recognition and extraction of mono- and multi-phosphopeptides. *Microchim. Acta* **2019**, *186*, 236–244.

46. Zhang, M.; Zhang, T.; Song, S.; Wang, J.; Chen, Z.; Xia, S. Preparation and adsorption properties of a quaternary ammonium net anion-exchange fiber. *Chin. J. Env. Eng.* **2015**, *9*, 2144–2148.

47. Bahl, J.M.C.; Jensen, S.S.; Larsen, M.R.; Heegaard, N.H.H. Characterization of the human cerebrospinal fluid phosphoproteome by titanium dioxide affinity chromatography and mass Spectrometry. *Anal. Chem.* **2008**, *80*, 6308–6316. [CrossRef]

48. Jin, S.; Liu, L.; Zhou, P. Amorphous titania modified with boric acid for selective capture of glycoproteins. *Microchim. Acta* **2018**, *185*, 308–314. [CrossRef]

49. Kjellstrom, S.; Jensen, O.N. Phosphoric acid as a matrix additive for MALDI MS analysis of phosphopeptides and phosphoproteins. *Anal. Chem.* **2004**, *76*, 5109–5117. [CrossRef]

 International Journal of
Molecular Sciences

Article

Characterization and Identification of Natural Antimicrobial Peptides on Different Organisms

Chia-Ru Chung [1], Jhih-Hua Jhong [2], Zhuo Wang [2], Siyu Chen [3], Yu Wan [3], Jorng-Tzong Horng [1,4] and Tzong-Yi Lee [2,3,*]

[1] Department of Computer Science and Information Engineering, National Central University, Taoyuan 32001, Taiwan; jjrchris@g.ncu.edu.tw (C.-R.C.); horng@db.csie.ncu.edu.tw (J.-T.H.)
[2] Warshel Institute for Computational Biology, The Chinese University of Hong Kong, Shenzhen 518172, China; zhongzhihua@cuhk.edu.cn (J.-H.J.); wangzhuo@cuhk.edu.cn (Z.W.)
[3] School of Life and Health Sciences, The Chinese University of Hong Kong, Shenzhen 518172, China; 117010024@link.cuhk.edu.cn (S.C.); 117010252@link.cuhk.edu.cn (Y.W.)
[4] Department of Bioinformatics and Medical Engineering, Asia University, Taichung 41359, Taiwan
* Correspondence: leetzongyi@cuhk.edu.cn; Tel.: +86-755-8427-3211

Received: 18 December 2019; Accepted: 30 January 2020; Published: 2 February 2020

Abstract: Because of the rapid development of multidrug resistance, conventional antibiotics cannot kill pathogenic bacteria efficiently. New antibiotic treatments such as antimicrobial peptides (AMPs) can provide a possible solution to the antibiotic-resistance crisis. However, the identification of AMPs using experimental methods is expensive and time-consuming. Meanwhile, few studies use amino acid compositions (AACs) and physicochemical properties with different sequence lengths against different organisms to predict AMPs. Therefore, the major purpose of this study is to identify AMPs on seven categories of organisms, including amphibians, humans, fish, insects, plants, bacteria, and mammals. According to the one-rule attribute evaluation, the selected features were used to construct the predictive models based on the random forest algorithm. Compared to the accuracies of iAMP-2L (a web-server for identifying AMPs and their functional types), ADAM (a database of AMP), and MLAMP (a multi-label AMP classifier), the proposed method yielded higher than 92% in predicting AMPs on each category. Additionally, the sensitivities of the proposed models in the prediction of AMPs of seven organisms were higher than that of all other tools. Furthermore, several physicochemical properties (charge, hydrophobicity, polarity, polarizability, secondary structure, normalized van der Waals volume, and solvent accessibility) of AMPs were investigated according to their sequence lengths. As a result, the proposed method is a practical means to complement the existing tools in the characterization and identification of AMPs in different organisms.

Keywords: antimicrobial peptides; organisms; sequence analysis; machine learning; feature selection

1. Introduction

Antimicrobial peptides (AMPs), naturally encoded by genes and usually containing 12–100 amino acids, are the essential components of the innate immune system and can protect the host from viruses and various pathogenic bacteria [1,2]. They are produced by various organisms, including protozoa, bacteria, and animals, and can cause the cell death of microbes by disrupting either their cell membrane or intracellular functions [3]. In recent years, the prevalent use of antibiotics has resulted in the rapid growth of antibiotic-resistant microorganisms that often induce severe infection and pathogenesis. Since antibiotic resistance is a growing phenomenon in contemporary medicine, the low drug-resistance development of AMPs can provide a possible solution [4].

Several studies have been dedicated to the prediction of AMPs, such as AntiBP [5], AntiBP2 [6], CAMP [7], ClassAMP [8], AVPpred [9], AMPER [10], iAMP-2L [11], iAMPred [12], AmPEP [13], and EFC-FCBF [14]. Specifically, the AMP database, namely APD, has collected 123 human host-defense peptides, 220 AMPs from mammals, 1050 active peptides from amphibians, 116 AMPs from fish, 35 reptile peptides, 40 AMPs from birds, 509 AMPs from arthropods, 160 AMPs from chelicerata, 42 AMPs from molluscs, and 6 AMPs from protozoa [15]. PhytAMP currently contains 271 entries of plant AMPs [16]. Moreover, previous studies have shown that there is a difference in amino acid composition (AAC) among different organisms. Cysteine is a major residue in AMPs from plants, probably because of the advantage of disulfide-bonded and defensive-like molecules [17]. In addition to AACs, the physicochemical property, sequence order, and the pattern of terminal residues have also been adopted in AMP prediction [13]. Furthermore, the net charge, isoelectric point, composition, and tendency for secondary structure are related to the activities of AMPs, such as antibacterial, antifungal, and antiviral activities [6,12,18].

With the rapid development of high-throughput proteomic technologies in recent years, machine learning (ML) algorithms have been the primary techniques for building up sequence-based classifiers to distinguish between AMPs and non-AMPs [13]. Mishra and Wang used AACs, physicochemical, and structural features to predict AMPs with different activities based on support vector machine (SVM) [17]. Meher et al. proposed the concept of the adoption of physicochemical features as the features used in ML [12]. Bhadra et al. adopted seven physicochemical classes and three distribution features, identifying where the first residue of a given group is located, and where 25%, 50%, 75%, and 100% of occurrences are contained, to differentiate between AMPs and non-AMPs [13]. Specifically, they proposed the concept of using distribution patterns as features. Additionally, there are several online tools available for the prediction of AMPs. i-AMP2L is a two-level multilabel predictor based on pseudo amino acid composition (PseAAC) and the fuzzy K-nearest neighbor (FKNN) algorithm [11]. It can identify an uncharacterized peptide as AMP or non-AMP based on the amino acid composition and physicochemical properties of sequences [11]. ADAM is a database of AMPs and allows users to predict sequences using SVM and hidden Markov models with amino acid composition adopted as the features [19]. DBAASP is an AMP prediction tool developed from SVM and artificial neural network (ANN) that incorporates hydrophobicity, amphipathicity, location of the peptide in relation to membrane, charge density, propensities to disordered structure, and aggregation being the features [20]. MLAMP adopted ML, synthetic minority oversampling technique (SMOTE), AACs, and physicochemical properties to construct a two-level AMP predictor [21]. CAMPR3 is a database that collects sequences, structures, and family-specific signatures of experimentally validated prokaryotic and eukaryotic AMPs [2]. It also provides AMP prediction tools based on random forest (RF), SVM, ANN, and discriminant analysis (DA), which use AACs, secondary structural propensities, and physicochemical properties as features.

Although AMPs are considered as an alternative drug to conventional antibiotics and has become a model for the development of new drugs that can solve the problem of multidrug resistance, using experimental methods to identify AMPs is expensive and time-consuming. Additionally, few studies have used AACs and physicochemical properties with different sequence lengths against different organisms to predict AMPs. In other words, research devoted to investigating the correlations between AACs/physicochemical properties and different sequence lengths on different organisms is scarce. Therefore, the major purpose of this study is to identify AMPs on seven organisms, including amphibians, humans, fish, insects, plants, bacteria, and mammals. Note that AACs, amino acid pairs, and the physicochemical properties (charge, hydrophobicity, polarity, polarizability, secondary structure, normalized van der Waals volume, and solvent accessibility) of each class are the major features that will be considered. After constructing the AMP classifiers for seven organisms, feature selection methods will be adopted to obtain a better understanding of the sequential characteristics of AMPs with respect to the seven categories of organisms. In addition, we will investigate these features on positions of the sequence to explore their relations.

2. Results

2.1. Characterization of AMPs

2.1.1. Compositional Characteristics of AMPs

Figure 1A demonstrates the average AACs of AMPs and non-AMPs. Specifically, "L", "G", and "K" were abundant amino acids for AMPs, while "L", "A", and "G" were abundant amino acids for non-AMPs. Additionally, there was an obvious difference in the composition of "C" (cysteine) between AMPs and non-AMPs. Previous research has indicated that the reason should be due to the dominance from disulfide-bonded and defensing-like molecules [17]. Meanwhile, the composition of "K" (lysine) was different between AMPs and non-AMPs, since the AMP structural cores mainly had positive net charges [22]. The composition of "G" (glycine) of AMPs was higher than the one for non-AMPs. This observation is consistent with that of a previous study, which indicated that the glycine-rich proteins (GRPs) are a group of proteins that occurs in a wide variety of organisms [23].

Figure 1B shows the AACs of AMPs with respect to the seven categories of organisms. There were some obvious differences among these organisms. The AACs related to a hydrophobic property ("C", "L", "V", "I", "M", "F", and "W") were different among these organisms. Additionally, the composition of "L" (leucine) in Amphibia was much higher than that in the other organisms; the composition of "C" in plants was the highest among the seven categories of organisms; the composition of "K" and "R", which have positive charges, were higher than that of "E" and "D", which have negative charges, for each organism. Moreover, the composition of "R" in humans and mammals was higher than that in other organisms. Because of these differences, the AACs were the critical features that differentiated identification of AMPs on different organisms.

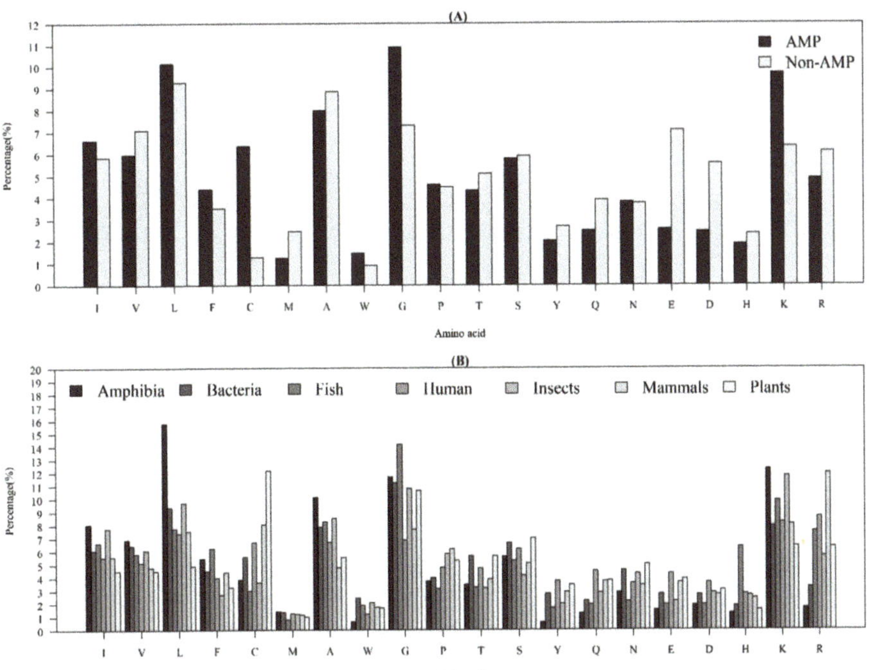

Figure 1. Average AACs of (**A**) AMPs and non-AMPs, and (**B**) AMPs with respect to the seven categories of organisms.

2.1.2. Investigation of Physicochemical Properties

Among the seven physicochemical properties we have collected, it was obvious that there was a significant difference between AMPs and non-AMPs. Figure 2 demonstrates the comparisons of three physicochemical properties between AMPs and non-AMPs. Hydrophobicity was obviously different between AMPs and non-AMPs for the polar class (Figure 2A). The result could be due to the hydrophobic interaction of the hydrophobic face with the lipidic moieties of membranes, which also drives peptide–cell binding [24]. The value of polarity between 4.9 and 6.2 in AMPs was higher than that in non-AMPs (Figure 2B). On the other hand, the value of polarity between 10.4 and 13 in AMPs was lower than that in non-AMPs. The activities of AMPs were found to decrease with an increase in polarity [25]. AMPs tend to be positively charged, which is consistent with previous research where the positive charges were influential in determining AMP activities (Figure 2C) [26]. Appendix A Figure A1 also demonstrates that the AMPs mainly had positive net charges. About half of the AMPs had net charges between +2 and +4, and less than 5% of the AMPs had negative net charges. In addition, the distribution of charges among non-AMPs was different from that of AMPs. Based on these differences in physicochemical properties between AMPs and non-AMPs, we considered these physicochemical features as the important features in the prediction of AMPs. The comparisons of polarizability, normalized van der Waals volume, secondary structure, and solvent accessibility are shown in Appendix A Figure A2. These observations can provide useful information for the construction of AMP classifiers for different classes of organisms and figure out the possible reasons for the high performance of the models.

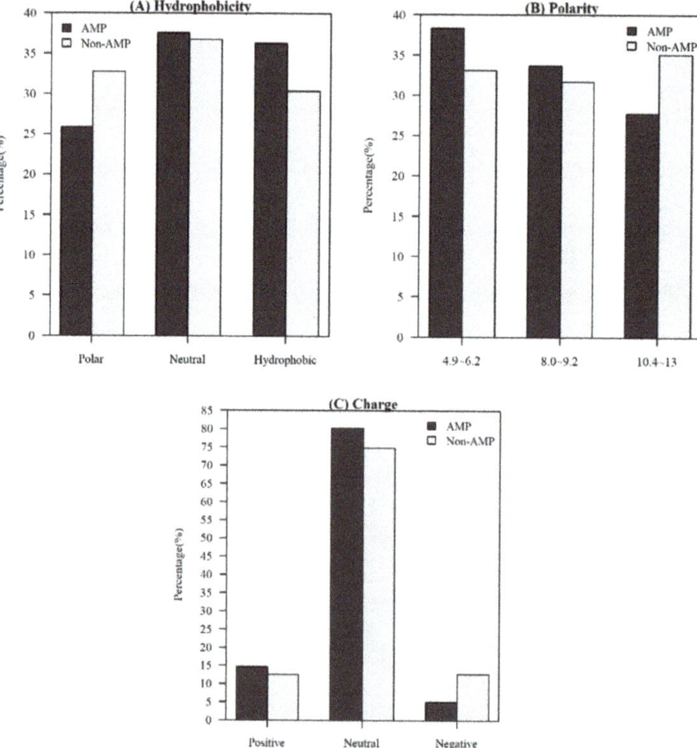

Figure 2. Comparisons of physicochemical properties between AMPs and non-AMPs for (**A**) hydrophobicity, (**B**) polarity, and (**C**) charge.

2.1.3. Physicochemical Properties with Respect to Different Sequence Lengths

In addition to observing physicochemical properties on AMPs and non-AMPs for different organisms, we also investigated them in different quantiles of sequence length. Figure 3A demonstrates that the majority of AMPs with positive charges were in the 90~100th percentile of sequence length. This is probably because charged amino acids at the tethered C-terminal increased the activity of the peptide. According to these distributions of AMP and non-AMPs, charge is an important feature to predict AMPs. In addition, Figure 3B illustrates the hydrophobicity in different percentiles of sequence length. The majority of AMPs with hydrophobicity were in the 90~100th percentile of sequence length. Previous research has indicated that a more hydrophobic and amphiphilic C-terminal obviously infiltrated into the hydrophobic part of the target cell membrane [27]. Moreover, many physicochemical properties vary among AMPs and different effects on AMP activities such as antibacterial, antifungal, and antiviral activities [22]. Differences can be found in the terminal residue profiles between AMP and non-AMPs. The remaining physicochemical properties also differed at different percentiles of sequence length. The comparisons of polarity, polarizability, normalized van der Waals volume, secondary structure, and solvent accessibility at different percentiles of sequence length are shown in Appendix A Figure A3. These observations can provide some indications on the investigation on the relations between the positions of the sequence and the physicochemical properties of AMPs and non-AMPs.

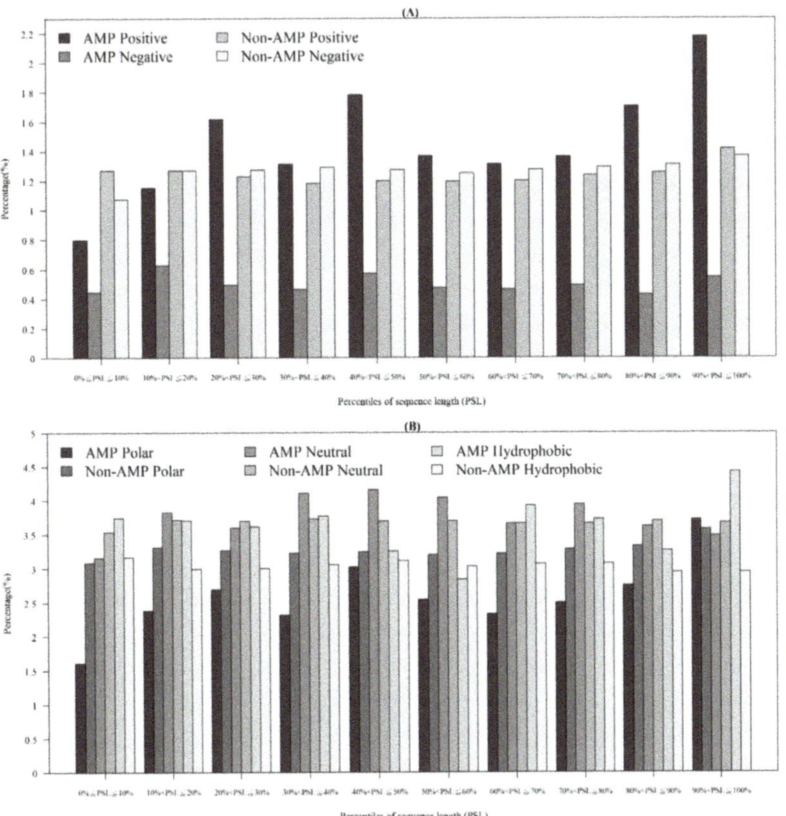

Figure 3. Comparisons of (**A**) charge on different positions of sequence between AMPs and non-AMPs, and (**B**) hydrophobicity at different positions of sequence between AMPs and non-AMPs.

2.1.4. Physicochemical Properties of AMPs with Respect to Different Categories of Organism

As shown in Table 1, the distribution of AMP sequence lengths among seven categories of organisms indicated that most of AMPs had 20–40 amino acids. Moreover, the number of AMPs with lengths over 100 for human and mammals were much higher than that of other organisms. Figure 4A shows that the AMPs from Amphibia tended to be hydrophobic compared with other organisms. Furthermore, Figure 4B investigates the hydrophobicity of different percentiles of sequence length for each organism. Most of the AMPs from Amphibia, bacteria, insects, and mammals had hydrophobicity in the 90–100th percentile of sequence length. In contrast, the AMPs from humans in the 10–20th and plants in the 30–40th percentiles of sequence length were hydrophobic. Appendix A Figure A4A shows that the percentage of positively charged AMPs was larger than that of the negatively charged AMPs for each category of organism. Appendix A Figure A4B indicates that the positively charged AMPs from Amphibia, insects, and mammals tended to be at larger percentiles of sequence length. Moreover, the distributions of charges in the AMPs from seven organisms are shown in Appendix A Figure A5. We found that the charge distribution was quite different among different organisms. The majority of AMPs from Amphibia had charges between +1 and +4. However, the AMPs from humans and mammals tended to have charges larger than +10 because of the sequence length. Specifically, the number of sequence lengths over 100 from humans and mammals were the largest ones among seven categories of organisms.

Table 1. Distribution of AMP sequence lengths among different organisms on training datasets.

Organisms	Number of Peptides with Length L						
	L ≤ 20	20 < L ≤ 40	40 < L ≤ 60	60 < L ≤ 80	80 < L ≤ 100	100 < L	Total
Amphibia	269	437	28	3	0	4	741
Bacteria	117	111	61	16	13	27	345
Fish	18	54	10	5	3	5	95
Human	11	53	13	26	7	76	186
Insects	67	94	32	12	7	8	220
Mammals	78	180	51	43	11	85	448
Plants	63	153	95	7	14	32	364

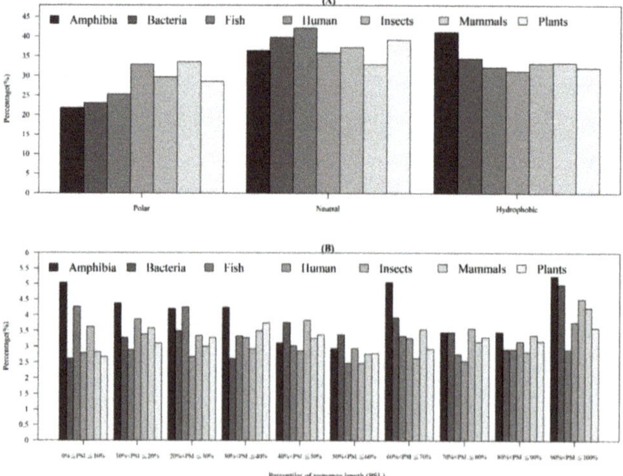

Figure 4. Comparisons of AMP hydrophobicity (**A**) in different categories of organisms and (**B**) at different positions of sequence (percentiles of sequence length) in each category of organism.

2.2. The Identification of Important Features

The order of importance was derived from the random forest algorithm and ranked the features for each category of organism. Appendix A Figure A6 shows that the patterns were accurate when the forward selection method was used to attain the approximate optimal results. These features were included in the prediction model one by one based on the rank order of feature selection. The performance would become better and better when more and more features were put into the prediction model. After a certain number of features were added, the performance curves converged, and further addition of the remaining features only affected the performance slightly. These features were thus selected and adopted in the prediction models, which helped us to reduce the size of the feature set. As shown in Appendix A Figure A6, the final feature sets of Amphibia, bacteria, fish, human, insects, mammals, and plants included the top 49, 65, 53, 64, 20, 77, and 65 features, respectively.

Appendix A Figure A7 demonstrates the details of the top 100 features for each organism after feature selection. These results indicated that the selected features differed among different organisms. As shown in Figure 5, the number of selected features in charge class for Amphibia was much higher than that of the other organisms that could also be found in Appendix A Figure A7A. Therefore, charge is important for the prediction of AMPs of Amphibia. Indeed, a previous study showed that the increase in charge could improve the antimicrobial activity of magainin peptides [28], which are a class of AMPs found in the African clawed frog. In addition, the number of selected features in the hydrophobicity class for bacteria was much higher than that of the other organisms, which could also be found in Appendix A Figure A7B, because the increase in peptide hydrophobicity caused an improvement in antimicrobial activity [29]. The number of selected features in the amino acid pair composition (AAPC) for humans was much higher than that of other organisms, which could also be found in Appendix A Figure A7C. Specifically, the AAPCs of "CC", "TC", "CR", "CY", and "CA" were ranked in the top 25. Plots of humans are also shown in AAPC heat map (Appendix A Figure A8A), where the color of the regions of "CC", "TC", "CR", "CY", and "CA" were darker than that of the other amino acid pairs, and these pairs were from human AMPs rather than non-AMPs. The AAPC heat map plots of other organisms are shown in Appendix A Figure A8. Moreover, "C" (cysteine) was the top-ranked feature in plants. Because of the benefit of disulfide-bonded and defensive-like molecules, "C" was the major amino acid residue in AMPs of plants.

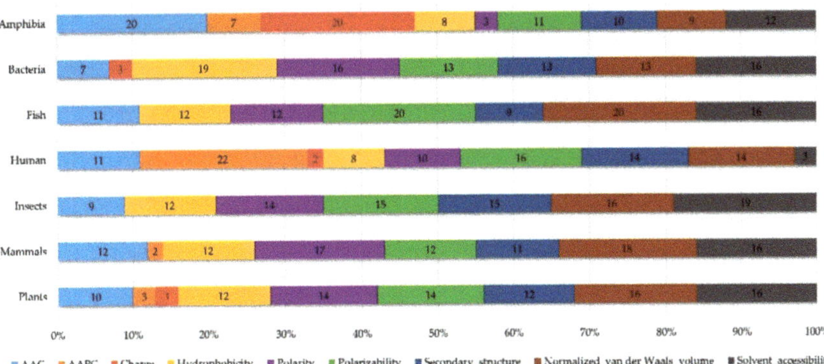

Figure 5. Distribution of features (top 100). Shows the performance of AAC and amino acid pair composition (AAPC), as well as physicochemical composition in different organisms.

2.3. Prediction Performance

The positive training datasets of Amphibians, bacteria, fish, humans, insects, mammals, and plants contained 741, 345, 95, 186, 220, 448, and 364 AMPs, respectively. Accordingly, the negative training dataset contained 1993, 6040, 1469, 6595, 1800, 6919, and 5432 non-AMPs, respectively. The performance of the four classifiers are given in Appendix A Table A1. According to the results, the prediction model can predict not only positive, but also negative data efficiently. Obviously, random forest (RF) was the best classifier for predicting AMPs in these seven categories of organisms. The accuracies of all the models were higher than 93%, and the sensitivities of all categories of organisms were higher than 94%. These results indicate that the used features and RF are efficient for predicting AMPs in each organism.

Furthermore, based on the performance in cross-validation, the RF model was selected to predict the independent test data. The positive test dataset in amphibians, bacteria, fish, humans, insects, mammals, and plants included 185, 86, 23, 46, 54, 111, and 90 AMPs, respectively. Accordingly, the negative test dataset contained 398, 1509, 367, 1648, 450, 1729, and 1358 non-AMPs, respectively. The prediction performance of the independent test is shown in Table 2. All the prediction accuracies of AMPs were above 94%, except that of humans, which was 92.23% but still high. Moreover, the MCCs for all the organisms were larger than 0.650.

Table 2. Performance of the models using data from different types of organisms in the independent test.

Organisms	Sensitivity	Specificity	Accuracy	Matthews Correlation Coefficient
Amphibia	100.00%	98.24%	98.80%	0.973
Bacteria	96.51%	96.36%	96.36%	0.746
Fish	100.00%	97.00%	97.18%	0.810
Human	97.83%	92.17%	92.33%	0.482
Insects	100.00%	97.56%	97.82%	0.900
Mammals	92.79%	94.56%	94.46%	0.673
Plants	97.78%	97.94%	97.93%	0.851

2.4. Comparison with Other AMP Prediction Tools

The performance of predicting the AMPs of different types of organisms was compared with that of other web tools: iAMPpred [12], iAMP-2L [11], ADAM [19], DBAASP [30], MLAMP [31], and CAMPR3 [2]. It should be noted that DBSSAP can only predict peptides with sequence lengths less than 100; therefore, peptides longer than that were removed from our test set to fulfill the requirement. The ROC curves of different models are shown in Figure 6. The comparisons of predicting AMPs for each organism compared with other tools were covered under the ROC curves obtained from our models.

The detailed performance of predicting AMPs in different categories of organisms with the proposed models and other tools are shown in Appendix A Table A2. The accuracies of iAMP-2L, ADAM, MLAMP, and our proposed models were higher than 92% for predicting AMPs from each organism. Additionally, our proposed models reached the highest accuracies when predicting AMPs from insects and plants. Although the accuracies of our proposed models in predicting AMPs in some organisms were not the best, the sensitivities of all our models were the highest. Therefore, the proposed models are efficient in predicting AMPs from different types of organisms.

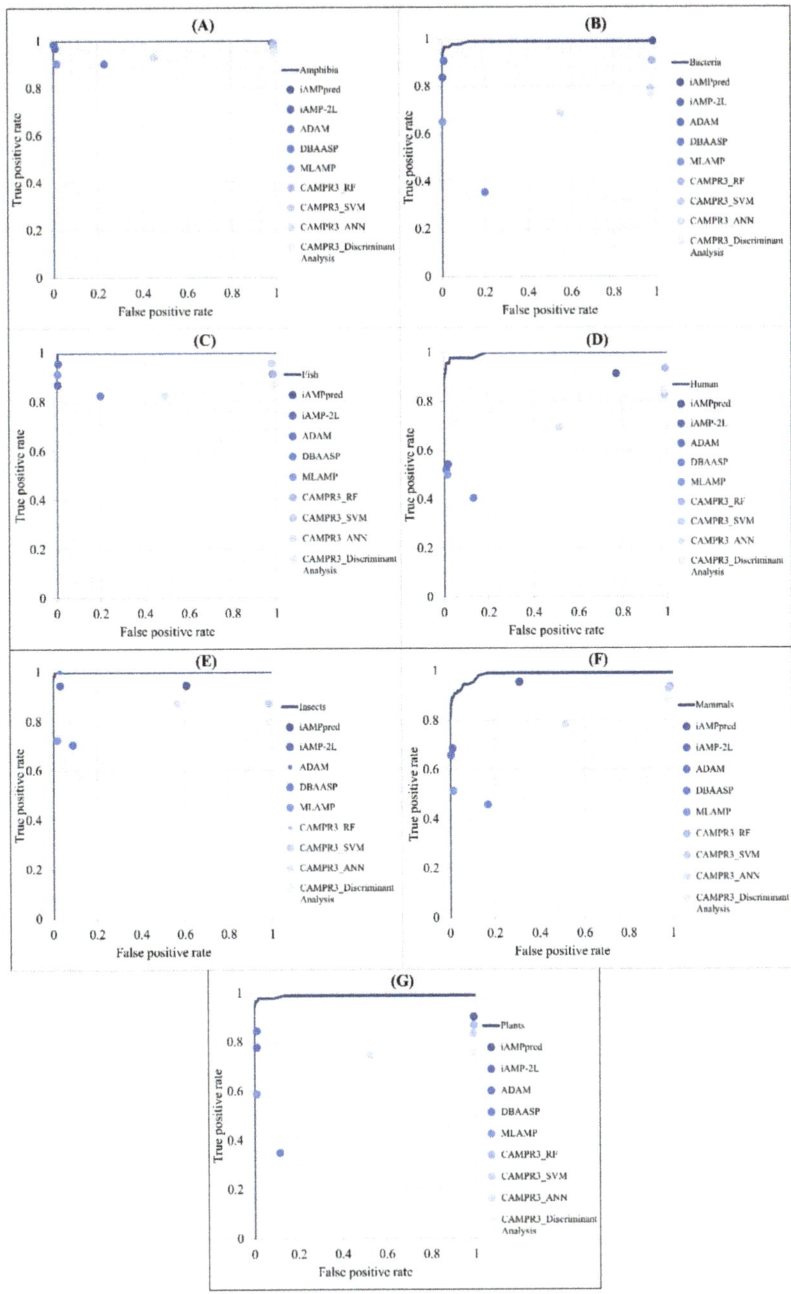

Figure 6. Comparison of ROC curves between our method and other prediction tools in the identification of AMPs on (**A**) Amphibians, (**B**) bacteria, (**C**) fish, (**D**) humans, (**E**) insects, (**F**) mammals, and (**G**) plants.

3. Discussion and Conclusions

Because of the rapid development of multidrug resistance, conventional treatment of antibiotics cannot kill pathogenic bacteria efficiently. Additionally, the identification of AMPs using experimental methods is expensive and time-consuming. Computational identification can efficiently and effectively discover candidate peptides as antimicrobial peptides for subsequent experimental assessment, which helps shorten the process of drug discovery [32,33]. In addition, because of the obvious differences in amino acid composition and physicochemical properties (charge, hydrophobicity, etc.) between AMPs and non-AMPs, and the difference in AMPs between different types of organisms, we believe that AMPs can be predicted effectively using these features. Additionally, AMPs from different types of organisms can be differentiated.

This study employed the one-rule attribute evaluation (OneR) method and forward-selection method, reducing the number of features from 630 to 49, 65, 53, 64, 20, 77, and 65, respectively, in amphibians, bacteria, fish, humans, insects, mammals, and plants. Then, four different classification algorithms were used to build predictive models. The performance of the models in five-fold cross-validation indicated that the feature sets were effective in the predictions. Accuracies and AUCs for all organisms were observed to be larger than 93%, which shows that the feature set and random forest method were efficient in predicting AMPs of different organisms. Moreover, we observed the feature sets of the seven types of organisms and found differences among organisms. For instance, electric charge was an important feature in the prediction of AMPs for Amphibia, because the charged residues in Amphibia were the most important features, which had a very high rank among all features of Amphibia. According to these differences in feature sets of the seven categories of organisms, we conclude that AMPs from different types of organisms can be differentiated well.

Furthermore, the performance of the models was compared with that of iAMPpred, iAMP-2L, ADAM, DBAASP, MLAMP, and CAMPR3 using the same testing dataset. The accuracies of iAMP-2L, ADAM, MLAMP, and proposed models were higher than 92% in predicting each organism. In addition, the sensitivity of the proposed models in predicting AMPs of seven organisms were the highest. As a result, the proposed models are believed to complement the existing tools in predicting AMPs and differentiate AMPs on different types of organisms. Last but not least, the proposed methods also lead a promising way to the design of new AMPs, which will enlighten the future of drug development. Accordingly, we believe that the proposed model in preclinical characterization of predicting AMPs will improve the long-term efficiency of AMP drug development.

4. Materials and Methods

4.1. Data Collection and Preprocessing

This study was divided into three parts as shown in Figure 7, data collection and preprocessing, feature investigation, and model training and evaluation. At first, positive datasets were collected from several databases. Then, AMPs were classified based on the types of organisms they came from. Negative datasets were downloaded from UniProt. After filtering conditions, all the non-AMPs were classified into seven types of organisms. Then, the sequence analysis tool, CD-HIT, was used to remove sequences that were 40% similar to positive dataset sequences in the negative dataset. The independent testing datasets of each organism were generated by drawing 20% of the data from the corresponding organism dataset. The AAC, amino acid pair composition (AAPC), and physicochemical properties in different sequence lengths of data were included in our feature sets. Then, the feature sets of each organism were analyzed by feature-selection methods to dig out the important features. With these selected features, prediction models were designed by four different kinds of algorithms. Finally, the predictive performances were compared after 5-fold cross validation and independent testing.

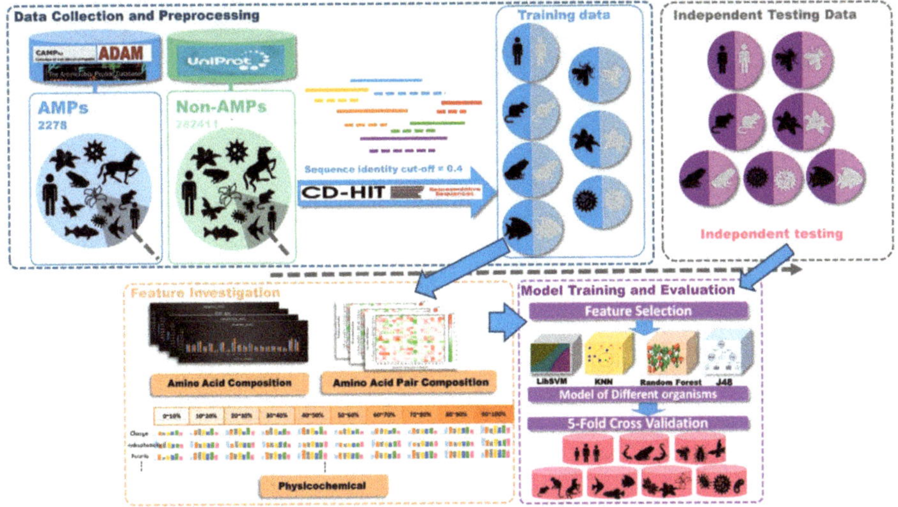

Figure 7. Conceptual framework. This study was divided into three parts: data collection and preprocessing, feature investigation, and model training and evaluation.

AMPs are common in nature and have been discovered in almost all forms of life, from single-celled bacteria to multicellular organisms such as animals and plants [17]. In this study, we collected the positive dataset by capturing naturally existing and experimentally validated AMP sequences from different organisms from several databases, CAMP [7], APD [15], ADAM [19], and DRAMP [21]. We collected all the AMPs and deleted the duplicated ones. Then, all the AMPs were classified into seven organisms, which contained 232, 926, 118, 274, 454, 431, and 559 from humans, amphibians, fish, insects, plants, bacteria, and mammals. We followed the data preparation procedure conducted in other studies to generate our negative dataset [11,34]. For the construction of negative data, we extracted protein sequences without the annotations of membrane, toxic, secretory, defensive, antibiotic, anticancer, antiviral, and antifungal properties from UniProt. Unique sequences were collected, which contained 11,275, 3656, 3005, 5225, 24,443, 281,434, and 33,483 non-AMPs from humans, amphibians, fish, insects, plants, bacteria, and mammals. In order to prevent the overestimation of predictive performance in this investigation, the CD-HIT program [35] was applied to remove similar sequences from the training dataset. It would be possible that some negative data were identical to some of the positive data in the training dataset, potentially causing "false positive" or "false negative" predictions. Consequently, CD-HIT was further applied by running CD-HTT-2D across positive and negative training datasets with 100% to 40% sequence identity to solve this problem. In this study, we reduced sequence redundancy of the negative dataset by removing the data with a 40% sequence similarity in all seven negative datasets. Then, for different types of organisms, we compared the sequence similarity between positive and negative datasets, and we removed sequences that were 40% similar to positive dataset sequences in the negative dataset. After filtering, our negative datasets had 8243, 1993, 1836, 2250, 6790, 7549, and 8648 non-AMPs from humans, amphibians, fish, insects, plants, bacteria, and mammals. The independent testing datasets of each organism were generated by separating 20% from the corresponding dataset. A summary of the positive and negative datasets is given in Table 3.

Table 3. Number of peptides in training and testing datasets among different organisms.

Organisms	Training Dataset		Testing Dataset	
	Positive	Negative	Positive	Negative
Amphibia	741	1595	185	398
Bacteria	345	6040	86	1509
Fish	95	1469	23	367
Human	186	6595	46	1648
Insects	220	1800	54	450
Mammals	448	6919	111	1729
Plants	364	5432	90	1358

4.2. Feature Constructions

AACs were obtained separately for each sequence, so were the ratios of all 20 amino acids. There are 20 amino acids, so this feature set had 20 dimensions. The following is an example of how to obtain AAC from a sequence "AIFIFIRWLLKLGHHGRAPP". First, we calculated the frequency of the 20 amino acid residues in this sequence. Then, the frequency of isoleucine (I) in this sequence was computed as (3(Number of I)/20(Sequence length)) = 0.15. Finally, the frequency of amino acid residues of this sequence will be calculated as AAC features.

AAPC is the ratio of the occurrences of the amino acids in pairs of two in each sequence. There are 20 amino acids, so this feature was 20 by 20 and equaled 400 dimensions. The same example was adopted to illustrate the determination of AAPC. First, we calculated the number of occurrences for 400 amino acid pairs in this sequence. Then, the frequency of "IF" pairs in the sequence was computed as (3(Number of IF)/19(Sequence length − 1)) = 0.105. Finally, the frequencies of 400 amino acid pairs of this sequence were taken as 400 AAPC features.

Previous studies have organized amino acids into several physicochemical property groups [13,17]. As shown in Appendix A Table A3, seven physicochemical properties were used in the grouping: (1) charge, (2) hydrophobicity, (3) polarity, (4) polarizability, (5) secondary structure, (6) normalized van der Waals volume, and (7) solvent accessibility. For each of these seven physicochemical properties, 20 amino acids were grouped into 3 classes. For example, for the charge property, the 3 classes were positive (K and R), neutral (A, N, C, Q, G, H, I, L, M, F, P, S, T, W, Y and V), and negative (D and E). For each 21 (= 7 × 3) classes, we generated 10 classes based on the percentiles of sequence length, such as 0~ 0, 10~20th, 20~30th, ... , and 90~100th percentiles of sequence length. The ratio of each amino acid of each physicochemical property class in each quantile class was calculated. We illustrated these computations with the sample sequence "AALKGCWTKSIPPKPCFGKR" according to the charge property and its three classes, positive, neutral, and negative. First, we split the sequence into 10 partitions, and then we calculated the ratio of the representative amino acids in each partition. The first partition (0–10th quantile) was the sequence "AA", which did not contain Class 1 and Class 3, but 2 of them were in the Class 2 charge. It means that the number of Class 2 sequences in the 0~10th percentile of sequence length was 2. Finally, the frequency of charge of Class 2 was computed as (2(0–10th percentile contained Class 2)/20(Sequence length)) = 0.1. After these calculations, we could obtain results at ten different positions, seven physicochemical properties of amino acids, three classes for each property, and final 210 (= 7 × 3 × 10) features in total for each sequence. Therefore, each sequence was transformed into 630 features (AAC (20) +AAPC (400) + physicochemical properties in different sequence length (210)).

4.3. Model Construction and Feature Selection Methods

In this study, OneR feature selection method was used to select features. This feature selection method can be found in Weka, which was the major analytic tool in this study [36]. OneR is a simple classification algorithm. As its name indicates, it generates a rule to predict the data. A contingency table was constructed for each predictor against the target, and then the best rule with the lowest total error, also named as "one rule", was selected.

RF is a classifier proposed by Breiman L., who published the ensemble of multiple classifiers based on random feature selection. The main idea about random forests is constructing a multitude of decision trees, and each tree is construct by random sampling of the training data. This machine learning method is considered as an appropriate classifier for processing a large-scale dataset, especially an imbalanced dataset. It corrects the habit of decision trees overfitting their training sets. This method was used in this study and generated by Weka. SVM is a supervised learning model based on associated learning algorithms using regression analysis to classify data [37]. The positive and negative training datasets were used for building a predictive model with the identified support vectors. In this study, a binary classification problem (AMP versus non-AMP) has been considered. The discriminatory ability of an SVM classifier is determined by a hyperplane in a high-dimensional space that can discriminate the AMPs from the non-AMPs. K-nearest neighbor models (KNN) is an instance-based algorithm used in classification. In a binary classification between positive and negative samples, every data point is a vector in a multidimensional feature space with a class label (AMPs or non-AMPs). Users can decide a value k, related to the scale of the subgroup, for prediction. A testing data point without a label was classified using k nearest training samples. In this study, many values of k tried to achieve the best performance. Decision tree (DT) is a tree-like model in which each internal node represents a "test" on an attribute, each branch represents the outcome of the test, and each leaf node represents a class label (positive or negative data) [38]. J48 is a classification model based on constructing a decision tree with the top-down process. The process starts from the test of the root node and follows the appropriate branch based on the test. A tree-like graph with a model of decisions was generated during the prediction. The outcome is the contents of the leaf node, and the conditions along the path is decided by a decision rule. Decision rules can be generated by constructing association rules and can denote temporal or causal relations.

4.4. Evaluation Matrics

The predictive models in this study based on machine learning methods have been trained and validated via five-fold cross-validation. The training dataset was divided into five non-overlapping subgroups with approximately equal sizes. In each round, four subgroups were used for training, and one for testing, and then the validation process was repeated five times. Then, the five validation results were combined to generate a single estimation. The performance of the trained models was estimated using sensitivity (S_n), specificity (S_p), accuracy (A_{cc}), and Matthews correlation coefficient (MCC). The definitions are given below.

$$S_n = \frac{TP}{TP + FN} \tag{1}$$

$$S_p = \frac{TN}{FP + TN} \tag{2}$$

$$A_{cc} = \frac{TP + TN}{TP + TN + FP + FN} \tag{3}$$

$$MCC = \frac{(TP \times TN) - (FP \times FN)}{\sqrt{(TP + FP) \times (TP + FN) \times (FP + TN) \times (TN + FN)}} \tag{4}$$

where TP, TN, FP, and FN represent the number of true positives, true negatives, false positives, and false negatives, respectively. In this study, to evaluate the performance of the ML models, a ranking list of features was generated by feature selection methods. After using the forward-selection method, the features that resulted in the best performance were used to design the models.

Author Contributions: C.-R.C. and J.-H.J. drafted the manuscript. C.-R.C., J.-H.J., Z.W., S.C., Y.W., and T.-Y.L. participated in the design of the study and performed the draft revision. J.-T.H. and T.-Y.L. conceived of the study and participated in its design and coordination. Z.W. and S.C. helped to revise the manuscript. All authors have read and agreed to the published version of the manuscript.

Funding: This research was funded by the Warshel Institute for Computational Biology, The Chinese University of Hong Kong, Shenzhen, China and the Ganghong Young Scholar Development Fund of Shenzhen Ganghong Group Co., Ltd.

Conflicts of Interest: The authors declare no conflict of interest.

Abbreviations

AMPs	Antimicrobial peptides
AACs	Amino acid compositions
ML	Machine learning
SVM	Support vector machine
PseAAC	Pseudo amino acid composition
FKNN	Fuzzy K-nearest neighbor
ANN	Artificial neural network
SMOTE	Synthetic minority oversampling technique
RF	Random forest
DA	Discriminant analysis
AAPC	Amino acid pair composition
OneR	One rule attribute evaluation
KNN	K-nearest neighbor models
Sn	Sensitivity
Sp	Specificity
Acc	Accuracy
MCC	Matthews correlation coefficient
TP	True positives
TN	True negatives
FP	False positives
FN	False negatives

Appendix A

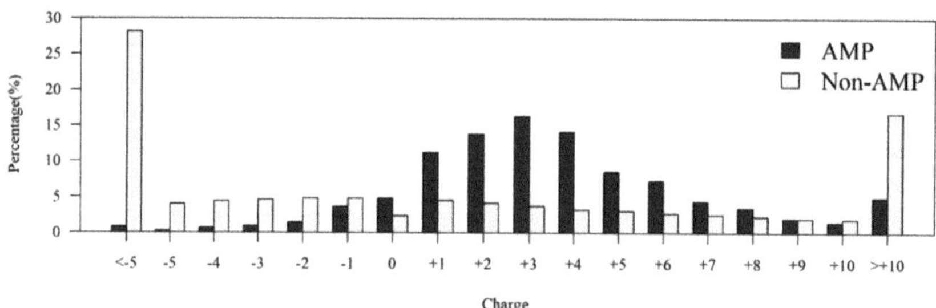

Figure A1. Comparisons of charge distributions between AMPs and non-AMPs.

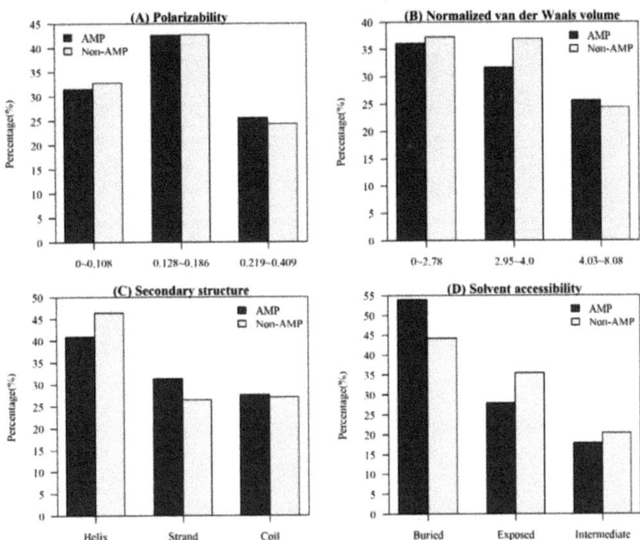

Figure A2. Comparisons of physicochemical properties between AMPs and non-AMPs for (**A**) polarizability, (**B**) normalized van der Waals volume, (**C**) secondary structure, and (**D**) solvent accessibility.

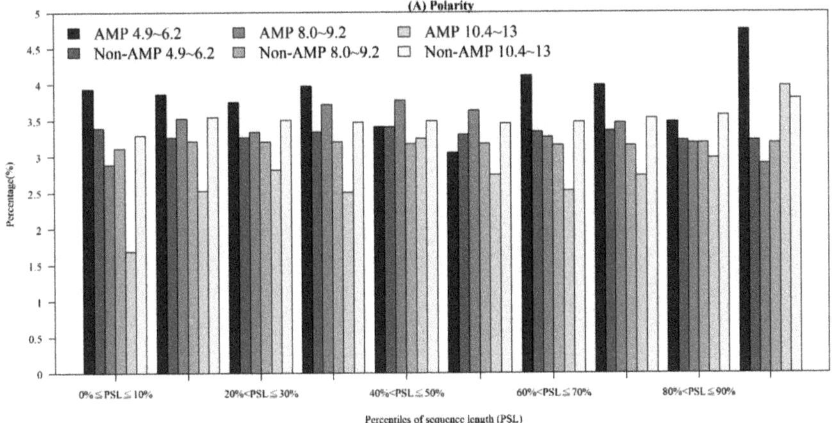

Figure A3. *Cont.*

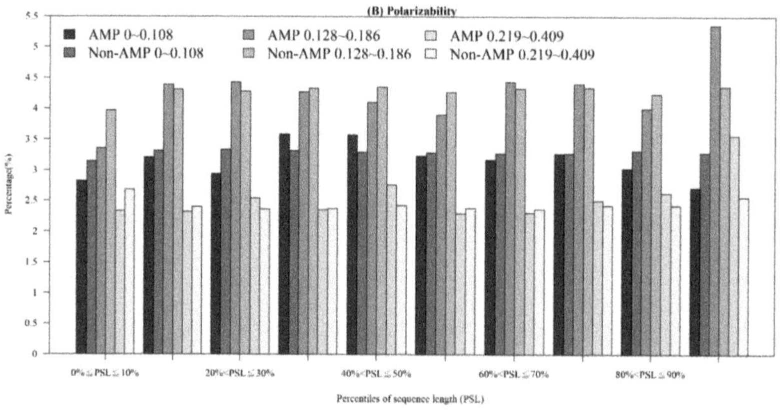

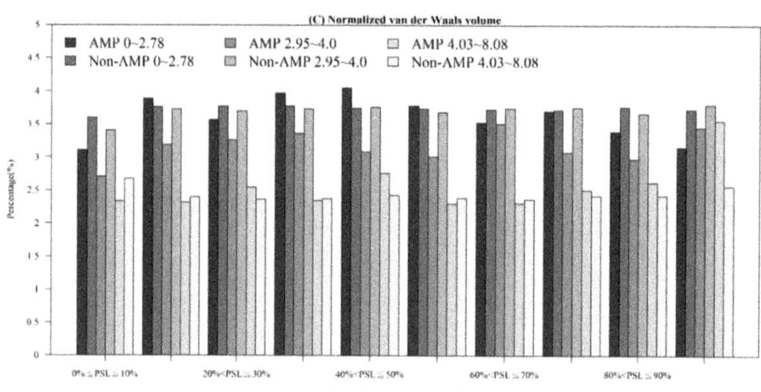

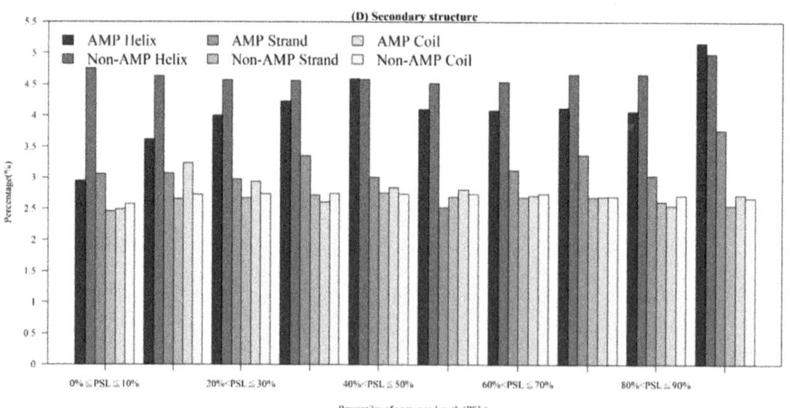

Figure A3. *Cont.*

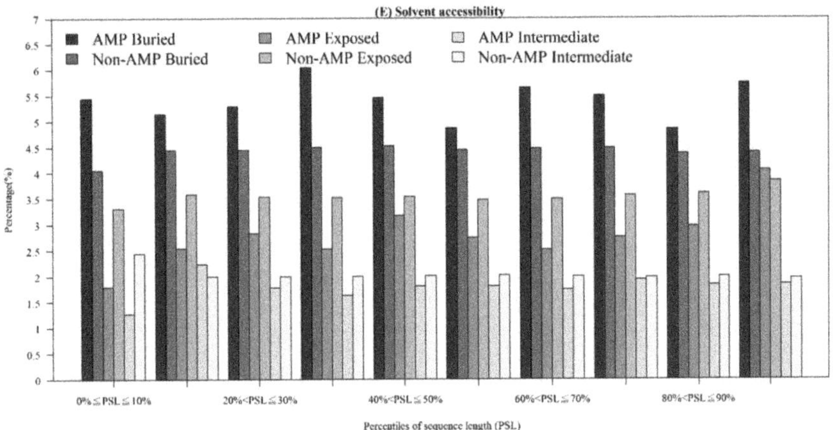

Figure A3. Comparisons of physicochemical properties between AMPs and non-AMPs at different positions (quantiles of sequence length) for (**A**) polarity, (**B**) polarizability, (**C**) normalized van der Waals volume, (**D**) secondary structure, and (**E**) solvent accessibility.

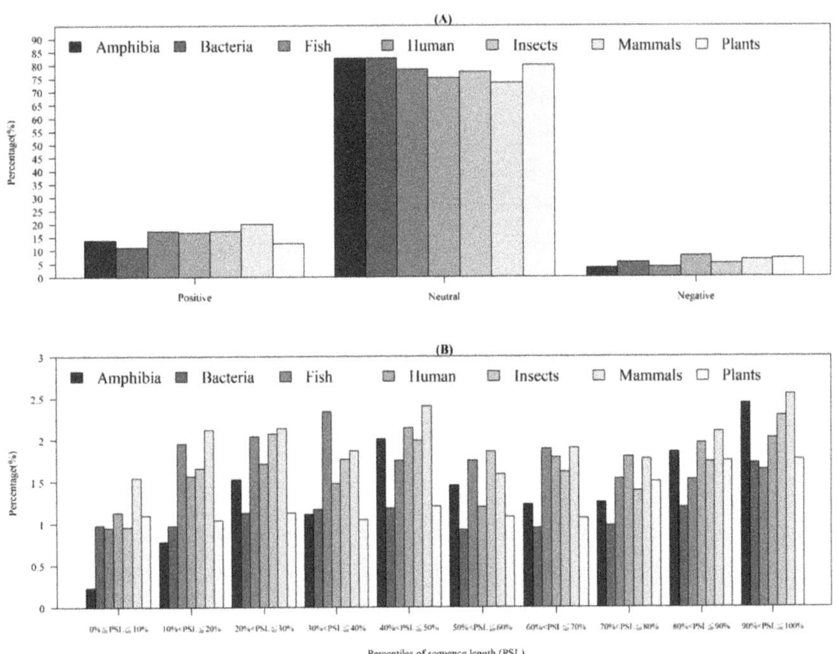

Figure A4. Comparisons of AMP charges (**A**) for different categories of organisms and (**B**) at different positions of sequence (percentiles of sequence length) in each category of organism.

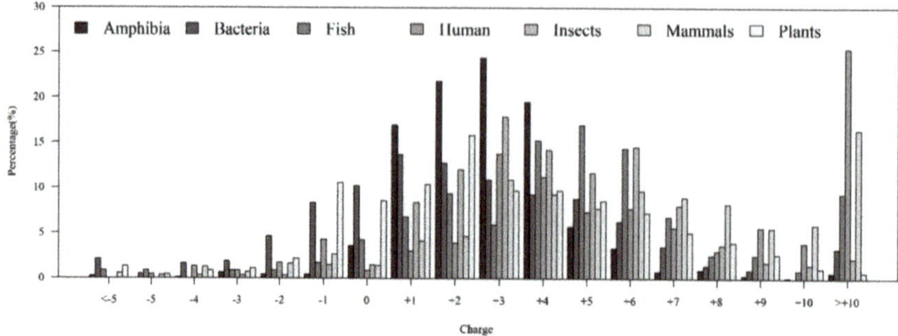

Figure A5. Charge distribution of AMPs from different organisms.

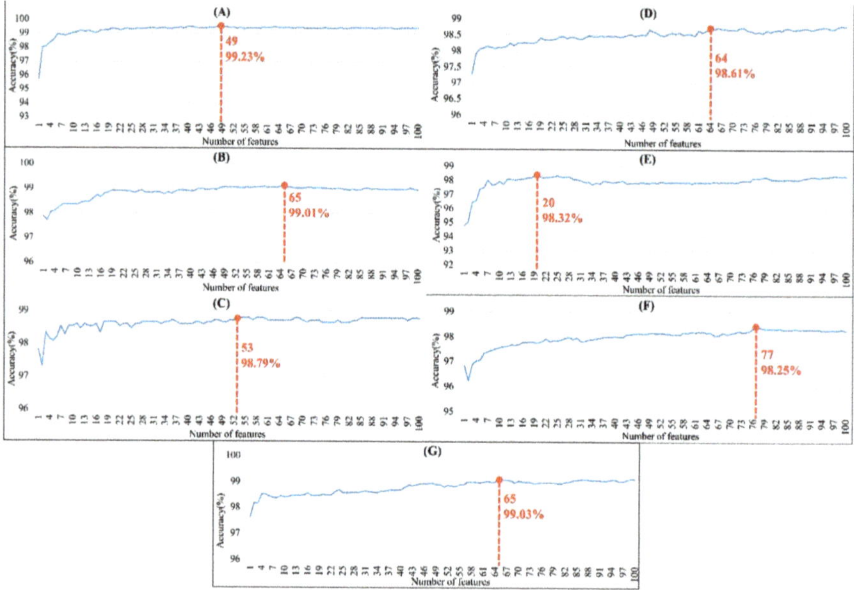

Figure A6. Performance with different numbers of features using forward selection method for (**A**) amphibians, (**B**) bacteria, (**C**) fish, (**D**) humans, (**E**) insects, (**F**) mammals, and (**G**) plants. Note that the red point means the number of features associated with the accuracy for the optimal model.

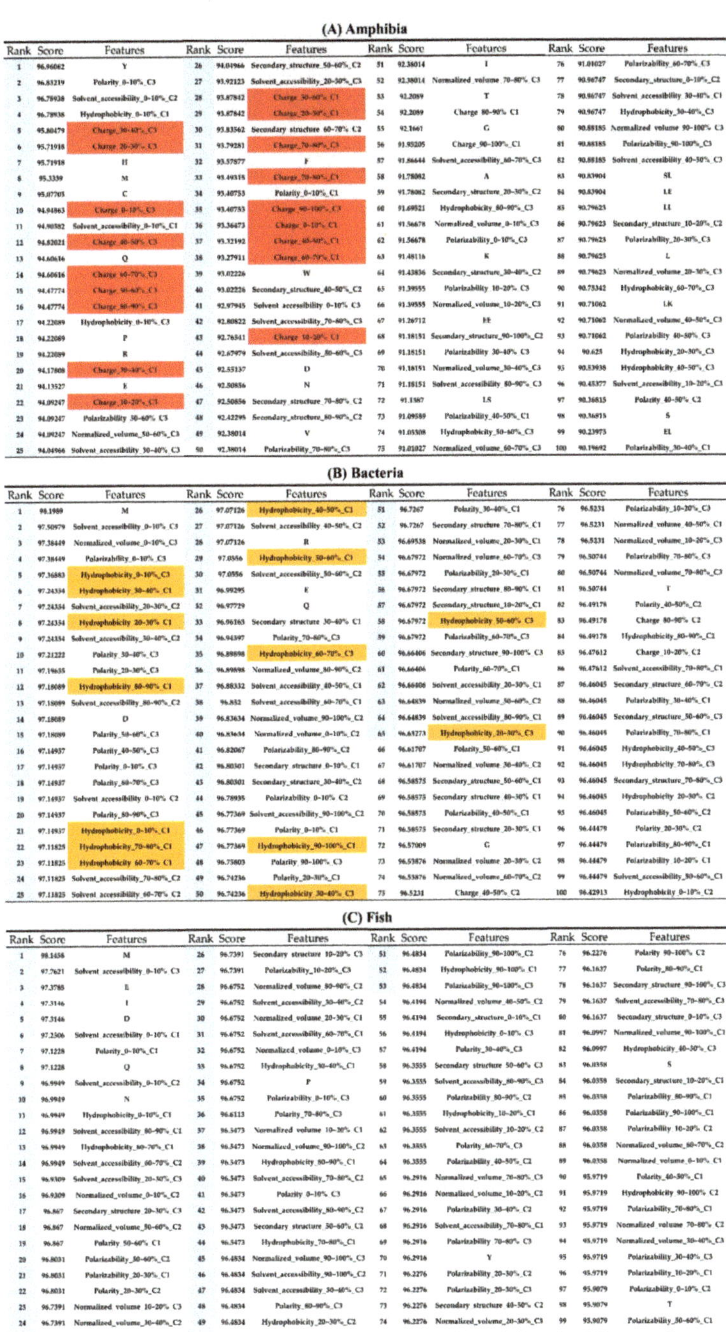

Figure A7. *Cont.*

(D) Human

Rank	Score	Features	Rank	Score	Features	Rank	Score	Features	Rank	Score	Features
1	98.30408	M	26	97.56673	RL	51	97.50774	Hydrophobicity_50-60%_C2	76	97.4635	Q
2	97.80268	Secondary_structure_10-20%_C2	27	97.56673	Polarizability_90-100%_C1	52	97.50774	Polarity_60-70%_C2	77	97.44875	Polarity_80-90%_C3
3	97.78794	Normalized_volume_10-20%_C1	28	97.56673	P	53	97.50774	YC	78	97.44875	GT
4	97.75844	Hydrophobicity_0-10%_C3	29	97.55198	Polarizability_30-40%_C2	54	97.50774	Solvent_accessibility_40-50%_C1	79	97.44875	Polarizability_50-60%_C2
5	97.7437	Secondary_structure_10-20%_C3	30	97.55198	Polarity_90-100%_C2	55	97.493	Polarizability_70-80%_C2	80	97.44875	Hydrophobicity_30-40%_C2
6	97.72895	CC	31	97.53724	Normalized_volume_80-90%_C1	56	97.493	Polarizability_60-70%_C1	81	97.44875	Secondary_structure_90-100%_C1
7	97.69945	Hydrophobicity_90-100%_C2	32	97.53724	Secondary_structure_50-60%_C1	57	97.493	Polarizability_80-90%_C1	82	97.44875	LY
8	97.69945	Normalized_volume_0-10%_C3	33	97.53724	YR	58	97.493	Polarizability_60-70%_C2	83	97.44875	TK
9	97.69945	Polarizability_0-10%_C3	34	97.53724	Secondary_structure_80-90%_C3	59	97.47825	A	84	97.44875	AC
10	97.69945	Polarity_0-10%_C1	35	97.53724	Polarity_40-50%_C2	60	97.47825	FC	85	97.43401	L
11	97.69945	TC	36	97.53724	Solvent_accessibility_0-10%_C1	61	97.47825	Secondary_structure_90-100%_C2	86	97.43401	T
12	97.68471	Polarizability_30-40%_C1	37	97.53724	Secondary_structure_60-70%_C2	62	97.47825	S	87	97.43401	Normalized_volume_10-20%_C2
13	97.66996	Normalized_volume_40-50%_C1	38	97.53724	Polarizability_40-50%_C1	63	97.47825	E	88	97.43401	Polarity_50-60%_C3
14	97.66996	Polarity_10-20%_C2	39	97.53724	Normalized_volume_60-70%_C1	64	97.47825	Secondary_structure_0-10%_C1	89	97.43401	Polarizability_40-50%_C2
15	97.66996	Hydrophobicity_10-20%_C2	40	97.52249	Polarizability_40-50%_C3	65	97.4635	LC	90	97.43401	Polarizability_10-20%_C2
16	97.64047	Hydrophobicity_40-50%_C2	41	97.52249	Normalized_volume_30-40%_C2	66	97.4635	Polarizability_10-20%_C3	91	97.43401	KC
17	97.64047	Polarizability_10-20%_C1	42	97.52249	CK	67	97.4635	Secondary_structure_0-10%_C3	92	97.43401	AV
18	97.62572	CR	43	97.52249	Normalized_volume_40-50%_C3	68	97.4635	Secondary_structure_50-60%_C2	93	97.41926	PI
19	97.62572	Secondary_structure_0-10%_C2	44	97.52249	V	69	97.4635	QG	94	97.41926	QR
20	97.62572	Charge_40-50%_C2	45	97.50774	Normalized_volume_30-40%_C1	70	97.4635	Secondary_structure_50-60%_C3	95	97.40451	Secondary_structure_10-20%_C1
21	97.62572	Normalized_volume_90-100%_C1	46	97.50774	C	71	97.4635	NC	96	97.40451	Polarity_80-90%_C2
22	97.59622	CY	47	97.50774	Normalized_volume_90-100%_C2	72	97.4635	Normalized_volume_50-60%_C1	97	97.40451	Polarizability_70-80%_C3
23	97.59622	Solvent_accessibility_0-10%_C3	48	97.50774	Secondary_structure_90-100%_C3	73	97.4635	Normalized_volume_10-20%_C3	98	97.40451	I
24	97.58148	Charge_40-50%_C1	49	97.50774	Normalized_volume_70-80%_C2	74	97.4635	Polarity_0-10%_C3	99	97.40451	FT
25	97.58148	CA	50	97.50774	Hydrophobicity_60-70%_C2	75	97.4635	Polarity_10-20%_C1	100	97.40451	Hydrophobicity_20-30%_C2

(E) Insects

Rank	Score	Features	Rank	Score	Features	Rank	Score	Features	Rank	Score	Features
1	96.0396	M	26	94.25743	Secondary_structure_60-70%_C2	51	93.81188	Polarity_50-60%_C1	76	93.56436	P
2	95.69307	Secondary_structure_0-10%_C2	27	94.25743	Normalized_volume_10-20%_C3	52	93.81188	Polarity_70-80%_C1	77	93.56436	Solvent_accessibility_90-100%_C1
3	95.29703	Normalized_volume_0-10%_C3	28	94.20792	Solvent_accessibility_70-80%_C1	53	93.76238	Normalized_volume_10-20%_C2	78	93.56436	Hydrophobicity_40-50%_C3
4	95.29703	Polarizability_0-10%_C3	29	94.20792	Secondary_structure_80-90%_C3	54	93.76238	Polarizability_80-90%_C1	79	93.56436	Solvent_accessibility_70-80%_C2
5	95.29703	Solvent_accessibility_0-10%_C3	30	94.20792	Normalized_volume_90-100%_C1	55	93.71287	Hydrophobicity_70-80%_C3	80	93.51485	D
6	95	Hydrophobicity_60-70%_C3	31	94.20792	E	56	93.71287	Hydrophobicity_80-90%_C3	81	93.46535	Normalized_volume_80-90%_C2
7	94.90099	Solvent_accessibility_20-30%_C3	32	94.20792	Secondary_structure_70-80%_C2	57	93.71287	Polarity_30-40%_C1	82	93.46535	Polarizability_40-50%_C3
8	94.80198	Polarizability_90-100%_C1	33	94.15842	Hydrophobicity_20-30%_C3	58	93.71287	Polarizability_70-80%_C1	83	93.46535	Normalized_volume_40-50%_C3
9	94.75248	Secondary_structure_90-100%_C3	34	94.15842	Normalized_volume_50-60%_C2	59	93.71287	Hydrophobicity_10-20%_C2	84	93.46535	Solvent_accessibility_50-60%_C1
10	94.75248	Solvent_accessibility_30-40%_C3	35	94.15842	C	60	93.66337	Hydrophobicity_50-60%_C1	85	93.46535	Polarity_0-10%_C2
11	94.65347	Solvent_accessibility_80-90%_C3	36	94.10891	Solvent_accessibility_80-90%_C1	61	93.66337	Polarity_40-50%_C1	86	93.46535	Normalized_volume_70-80%_C2
12	94.65347	Normalized_volume_60-70%_C2	37	94.10891	Normalized_volume_80-90%_C3	62	93.66337	Polarizability_30-40%_C3	87	93.41584	Polarity_60-70%_C2
13	94.65347	Secondary_structure_80-90%_C2	38	94.10891	Polarizability_80-90%_C3	63	93.66337	Normalized_volume_30-40%_C3	88	93.41584	Polarity_80-90%_C1
14	94.60396	Polarity_90-100%_C2	39	94.05941	Normalized_volume_50-60%_C3	64	93.66337	Secondary_structure_20-30%_C3	89	93.41584	Hydrophobicity_30-40%_C3
15	94.60396	Solvent_accessibility_90-100%_C3	40	94.05941	Secondary_structure_20-30%_C2	65	93.66337	Solvent_accessibility_50-60%_C2	90	93.36634	Polarizability_90-100%_C3
16	94.50495	Solvent_accessibility_50-60%_C3	41	94.05941	Polarizability_50-60%_C3	66	93.61386	Polarizability_20-30%_C3	91	93.36634	Polarity_50-60%_C3
17	94.50495	Polarity_60-70%_C1	42	94.0099	Y	67	93.61386	Normalized_volume_60-70%_C3	92	93.36634	Polarizability_60-70%_C1
18	94.50495	Secondary_structure_90-100%_C2	43	93.9604	Solvent_accessibility_60-70%_C1	68	93.61386	Secondary_structure_30-40%_C2	93	93.36634	Solvent_accessibility_40-50%_C1
19	94.45545	Solvent_accessibility_40-50%_C3	44	93.9604	Solvent_accessibility_10-20%_C3	69	93.61386	Polarizability_60-70%_C3	94	93.36634	Normalized_volume_90-100%_C3
20	94.40594	Hydrophobicity_90-100%_C3	45	93.91089	Secondary_structure_50-60%_C2	70	93.61386	Normalized_volume_20-30%_C3	95	93.31683	Normalized_volume_70-80%_C3
21	94.35644	Secondary_structure_40-50%_C2	46	93.91089	Hydrophobicity_50-60%_C3	71	93.56436	H	96	93.31683	Secondary_structure_70-80%_C3
22	94.30693	Secondary_structure_10-20%_C2	47	93.86139	Solvent_accessibility_60-70%_C2	72	93.56436	Hydrophobicity_70-80%_C1	97	93.31683	Polarizability_70-80%_C3
23	94.30693	Polarity_20-30%_C1	48	93.86139	Hydrophobicity_60-70%_C1	73	93.56436	F	98	93.26733	Solvent_accessibility_10-20%_C1
24	94.25743	Q	49	93.86139	Polarity_60-70%_C3	74	93.56436	Polarizability_20-30%_C1	99	93.26733	Polarity_10-20%_C2
25	94.25743	Polarizability_10-20%_C3	50	93.81188	Polarity_90-100%_C1	75	93.56436	Solvent_accessibility_20-30%_C1	100	93.26733	Secondary_structure_30-40%_C3

(F) Mammals

Rank	Score	Features	Rank	Score	Features	Rank	Score	Features	Rank	Score	Features
1	97.32482	M	26	95.79033	Solvent_accessibility_40-50%_C3	51	95.55948	Normalized_volume_50-60%_C2	76	95.4101	C
2	96.30636	Polarizability_90-100%_C1	27	95.77675	Polarizability_60-70%_C1	52	95.5459	Secondary_structure_90-100%_C2	77	95.4101	Q
3	96.29278	Normalized_volume_0-10%_C3	28	95.74959	A	53	95.5459	Normalized_volume_60-70%_C1	78	95.39652	CC
4	96.29278	Polarizability_0-10%_C3	29	95.72243	P	54	95.5459	Polarizability_50-60%_C1	79	95.39652	Polarizability_20-30%_C1
5	96.2113	Secondary_structure_90-100%_C3	30	95.70885	Polarity_60-70%_C2	55	95.53232	Hydrophobicity_60-70%_C2	80	95.38294	Hydrophobicity_90-100%_C3
6	96.18414	Secondary_structure_10-20%_C3	31	95.70885	Solvent_accessibility_60-70%_C3	56	95.51874	Normalized_volume_40-50%_C2	81	95.38294	Polarity_50-60%_C2
7	96.11624	Polarizability_10-20%_C1	32	95.69527	Polarity_40-50%_C2	57	95.50516	Polarity_80-90%_C2	82	95.38294	Polarity_0-10%_C1
8	96.11624	Polarizability_40-50%_C1	33	95.66812	Hydrophobicity_10-20%_C2	58	95.49158	Normalized_volume_50-60%_C1	83	95.38294	Normalized_volume_30-40%_C1
9	96.11624	Solvent_accessibility_0-10%_C3	34	95.66812	Polarizability_40-50%_C3	59	95.478	Solvent_accessibility_80-90%_C3	84	95.38294	Polarizability_70-80%_C1
10	96.10266	D	35	95.66812	Normalized_volume_40-50%_C3	60	95.45084	Polarity_0-10%_C3	85	95.35578	Hydrophobicity_60-70%_C1
11	96.04834	Normalized_volume_90-100%_C1	36	95.64096	Secondary_structure_10-20%_C2	61	95.45084	S	86	95.35578	Solvent_accessibility_60-70%_C2
12	96.03476	Polarizability_80-90%_C1	37	95.62738	Secondary_structure_50-60%_C3	62	95.45084	Polarity_70-80%_C3	87	95.35578	Polarity_70-80%_C1
13	96.02118	Hydrophobicity_90-100%_C2	38	95.62738	Normalized_volume_30-40%_C2	63	95.45084	Normalized_volume_0-10%_C2	88	95.34221	Hydrophobicity_20-30%_C2
14	95.98043	Polarity_90-100%_C2	39	95.62738	Solvent_accessibility_50-60%_C3	64	95.43726	Polarity_0-10%_C2	89	95.34221	Polarity_80-90%_C3
15	95.96687	Normalized_volume_90-100%_C2	40	95.62738	Secondary_structure_0-10%_C3	65	95.43726	Secondary_structure_70-80%_C2	90	95.34221	Secondary_structure_50-60%_C2
16	95.93971	Polarizability_30-40%_C1	41	95.62738	Normalized_volume_10-20%_C2	66	95.43726	V	91	95.34221	Secondary_structure_20-30%_C3
17	95.92613	Solvent_accessibility_10-20%_C3	42	95.6138	Secondary_structure_40-50%_C3	67	95.43726	Hydrophobicity_10-20%_C3	92	95.32863	Solvent_accessibility_90-100%_C1
18	95.92613	E	43	95.60022	Hydrophobicity_70-80%_C1	68	95.43726	R	93	95.32863	Polarity_90-100%_C1
19	95.89897	Normalized_volume_10-20%_C1	44	95.60022	Solvent_accessibility_70-80%_C2	69	95.43726	Hydrophobicity_50-60%_C2	94	95.31505	Polarity_30-40%_C2
20	95.88539	Polarity_10-20%_C2	45	95.58664	CR	70	95.42368	Solvent_accessibility_70-80%_C3	95	95.31505	Polarity_60-70%_C3
21	95.87181	Solvent_accessibility_30-40%_C3	46	95.58664	Solvent_accessibility_20-30%_C3	71	95.42368	Polarity_80-90%_C1	96	95.31505	Hydrophobicity_50-60%_C1
22	95.85823	Solvent_accessibility_90-100%_C3	47	95.57306	Normalized_volume_80-90%_C1	72	95.4101	Polarizability_50-60%_C3	97	95.31505	Hydrophobicity_30-40%_C1
23	95.84465	Normalized_volume_40-50%_C1	48	95.57306	Normalized_volume_20-30%_C2	73	95.4101	T	98	95.31505	Solvent_accessibility_30-40%_C1
24	95.81749	Secondary_structure_80-90%_C3	49	95.55948	Hydrophobicity_40-50%_C2	74	95.4101	Polarity_10-20%_C1	99	95.31505	Solvent_accessibility_50-60%_C2
25	95.79033	H	50	95.55948	Normalized_volume_70-80%_C2	75	95.4101	Normalized_volume_50-60%_C3	100	95.31505	Solvent_accessibility_30-40%_C2

Figure A7. *Cont.*

(G) Plants

Rank	Score	Features	Rank	Score	Features	Rank	Score	Features	Rank	Score	Features
1	97.9124	M	26	96.118	Polarizability_50-60%_C3	51	95.9455	L	76	95.7729	T
2	97.3602	C	27	96.118	Secondary_structure_30-40%_C1	52	95.9282	V	77	95.7729	Normalized_volume_30-60%_C2
3	96.8427	Normalized_volume_60-70%_C2	28	96.118	Normalized_volume_50-60%_C3	53	95.8937	Secondary_structure_80-90%_C3	78	95.7729	Solvent_accessibility_80-90%_C1
4	96.5839	Normalized_volume_0-10%_C3	29	96.1008	Hydrophobicity_20-30%_C1	54	95.8937	Hydrophobicity_50-60%_C3	79	95.7729	Hydrophobicity_30-40%_C3
5	96.5839	Polarizability_0-10%_C3	30	96.1008	Polarizability_40-50%_C2	55	95.8765	Polarizability_70-80%_C1	80	95.7557	SC
6	96.5666	D	31	96.1008	Solvent_accessibility_20-30%_C2	56	95.8765	Polarizability_80-90%_C3	81	95.7557	Hydrophobicity_80-90%_C2
7	96.4976	Polarizability_80-90%_C1	32	96.1008	Hydrophobicity_80-90%_C3	57	95.8765	E	82	95.7557	Charge_60-70%_C2
8	96.3941	Secondary_structure_40-50%_C2	33	96.0835	Normalized_volume_80-90%_C1	58	95.8765	Normalized_volume_40-50%_C2	83	95.7212	Polarizability_60-70%_C1
9	96.3396	Polarity_50-60%_C3	34	96.049	Secondary_structure_30-40%_C2	59	95.8765	Normalized_volume_80-90%_C3	84	95.7212	Hydrophobicity_90-100%_C1
10	96.3423	Solvent_accessibility_50-60%_C2	35	96.0317	Secondary_structure_80-90%_C2	60	95.8765	Polarizability_0-10%_C2	85	95.7212	VC
11	96.3425	Polarizability_90-100%_C1	36	96.0317	Hydrophobicity_40-50%_C1	61	95.8592	Secondary_structure_20-30%_C3	86	95.7212	Solvent_accessibility_90-100%_C2
12	96.3423	Secondary_structure_60-70%_C1	37	96.0317	Solvent_accessibility_40-50%_C2	62	95.8592	Solvent_accessibility_60-70%_C1	87	95.7212	Solvent_accessibility_10-20%_C3
13	96.3423	Hydrophobicity_50-60%_C1	38	96.0317	Polarity_80-90%_C2	63	95.8592	Polarity_50-60%_C1	88	95.7039	Solvent_accessibility_40-50%_C1
14	96.2908	Polarity_30-40%_C3	39	96.0145	Normalized_volume_70-80%_C2	64	95.8592	Polarity_20-30%_C1	89	95.6867	Solvent_accessibility_90-100%_C1
15	96.2733	Polarizability_10-20%_C3	40	96.0145	Normalized_volume_90-100%_C2	65	95.8592	Secondary_structure_40-50%_C3	90	95.6694	Polarity_40-50%_C2
16	96.2733	Normalized_volume_10-20%_C3	41	96.0145	H	66	95.842	Polarity_40-50%_C3	91	95.6694	Solvent_accessibility_40-50%_C3
17	96.256	Charge_40-50%_C2	42	95.9972	Normalized_volume_60-70%_C3	67	95.842	Polarity_80-90%_C1	92	95.6694	Normalized_volume_30-40%_C2
18	96.256	Hydrophobicity_30-40%_C1	43	95.9972	Polarizability_60-70%_C3	68	95.842	Charge_30-40%_C2	93	95.6522	Secondary_structure_50-60%_C2
19	96.256	Solvent_accessibility_30-40%_C2	44	95.98	Secondary_structure_20-30%_C2	69	95.8247	Solvent_accessibility_50-60%_C1	94	95.6522	Secondary_structure_90-100%_C2
20	96.2218	Polarity_90-100%_C2	45	95.9627	Normalized_volume_90-100%_C1	70	95.8247	Polarity_30-40%_C1	95	95.6349	Polarizability_30-40%_C2
21	96.2043	Solvent_accessibility_60-70%_C2	46	95.9627	Polarity_60-70%_C3	71	95.8075	Hydrophobicity_40-50%_C3	96	95.6177	Polarity_0-10%_C3
22	96.2043	Hydrophobicity_60-70%_C1	47	95.9627	CG	72	95.8075	Polarizability_40-50%_C3	97	95.6177	Hydrophobicity_90-100%_C3
23	96.187	Solvent_accessibility_0-10%_C3	48	95.9455	Polarity_20-30%_C3	73	95.8075	Normalized_volume_40-50%_C3	98	95.6177	I
24	96.1925	Polarity_40-50%_C1	49	95.9455	Solvent_accessibility_80-90%_C3	74	95.7902	Solvent_accessibility_50-60%_C3	99	95.6004	Normalized_volume_80-90%_C2
25	96.1385	Q	50	95.9455	Polarizability_30-40%_C1	75	95.7729	Secondary_structure_50-60%_C1	100	95.6004	Normalized_volume_20-30%_C2

Figure A7. Top 100 features for (**A**) Amphibians, (**B**) bacteria, (**C**) fish, (**D**) humans, (**E**) insects, (**F**) mammals, and (**G**) plants. The rank column with blue background color indicates that the feature was selected from the feature-selection method. The features marked red in (**A**) are related to charge property which is the majority member among the top 100 features for Amphibians. The features marked yellow in (**B**) are associated with the hydrophobicity which is the majority member among the top 100 features for bacteria. The features marked orange in (**D**) are related to AAPC which is the majority member among the top 100 features for human.

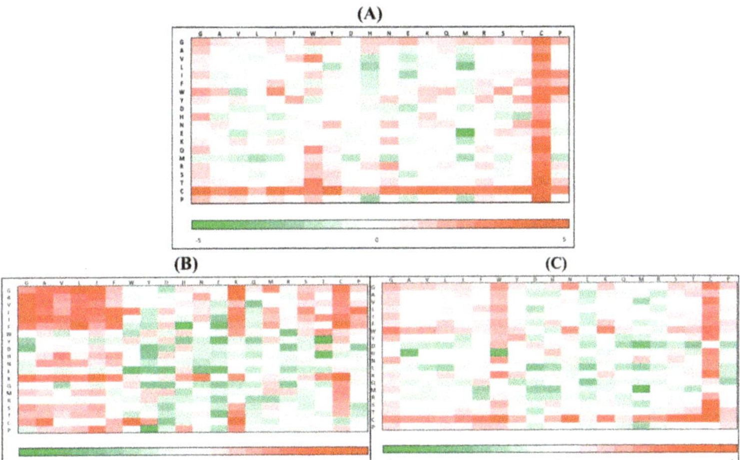

Figure A8. *Cont.*

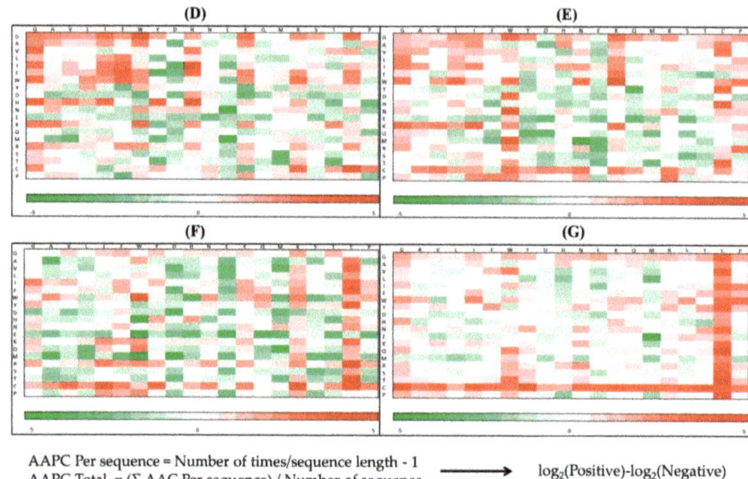

Figure A8. AAPC heatmaps for (**A**) human, (**B**) amphibians, (**C**) bacteria, (**D**) fish, (**E**) insects, (**F**) mammals, and (**G**) plants.

Table A1. Performance of training datasets for the AMPs derived from different organisms. The optimal models which contain best prediction performance are marked in blue background color. It would be noted that the optimal model was determined as the one with the minimum difference between sensitivity and specificity.

Organisms	Classifier	Sensitivity	Specificity	Accuracy	Matthews Correlation Coefficient
Amphibia	RF	99.19%	99.18%	99.19%	0.981
	DT	97.84%	98.81%	98.50%	0.965
	KNN	96.76%	99.81%	98.84%	0.973
	SVM	98.92%	98.93%	98.93%	0.975
Bacteria	RF	95.94%	96.18%	96.16%	0.735
	DT	86.67%	97.95%	97.34%	0.769
	KNN	73.62%	99.44%	98.04%	0.7959
	SVM	95.94%	95.94%	95.94%	0.725
Fish	RF	96.84%	96.87%	96.87%	0.789
	DT	73.68%	98.43%	96.93%	0.728
	KNN	68.42%	99.52%	97.63%	0.774
	SVM	82.11%	99.86%	98.79%	0.889
Human	RF	94.09%	93.07%	93.10%	0.489
	DT	74.19%	98.15%	97.49%	0.615
	KNN	68.28%	98.94%	98.10%	0.654
	SVM	88.17%	87.82%	87.83%	0.354
Insects	RF	96.36%	96.33%	96.34%	0.838
	DT	91.36%	97.56%	96.88%	0.849
	KNN	85.91%	98.28%	96.93%	0.842
	SVM	95.00%	95.11%	95.10%	0.793
Mammals	RF	94.42%	95.24%	95.19%	0.708
	DT	83.71%	92.60%	92.06%	0.560
	KNN	74.55%	98.92%	97.43%	0.767
	SVM	93.97%	93.97%	93.97%	0.662
Plants	RF	97.53%	97.39%	97.39%	0.822
	DT	88.74%	98.82%	98.19%	0.851
	KNN	80.49%	99.45%	98.26%	0.845
	SVM	96.70%	96.70%	96.70%	0.786

Note. RF = random forest; DT = decision tree; KNN = K-nearest neighbor; SVM = support vector machine.

Table A2. Comparisons of independent testing results between our method and other prediction tools in the identification of AMPs on different organisms.

Organisms	Classifier	Sensitivity	Specificity	Accuracy	Matthews Correlation Coefficient
Amphibia	Our method	100.00%	98.24%	98.80%	0.973
	iAMPpred	98.92%	1.51%	32.42%	0.017
	iAMP-2L	96.76%	98.99%	98.28%	0.960
	ADAM	98.38%	99.50%	99.14%	0.980
	DBAASP	90.22%	76.92%	89.34%	0.477
	MLAMP	90.27%	98.24%	95.71%	0.900
	CAMPR3_RF	98.92%	1.01%	32.08%	−0.004
	CAMPR3_SVM	97.30%	1.01%	31.56%	−0.064
	CAMPR3_ANN	92.97%	54.77%	66.90%	0.454
	CAMPR3_DA	95.14%	0.75%	30.70%	−0.135
Bacteria	Our method	96.51%	96.36%	96.36%	0.746
	iAMPpred	84.88%	1.99%	6.46%	−0.183
	iAMP-2L	83.72%	99.54%	98.68%	0.867
	ADAM	90.70%	98.87%	98.43%	0.855
	DBAASP	35.44%	80.00%	57.86%	0.173
	MLAMP	65.12%	99.47%	97.62%	0.743
	CAMPR3_RF	90.70%	1.99%	6.77%	−0.108
	CAMPR3_SVM	79.07%	2.72%	6.83%	−0.218
	CAMPR3_ANN	68.60%	45.00%	46.27%	0.062
	CAMPR3_DA	76.74%	2.78%	6.77%	−0.239
Fish	Our method	100.00%	97.00%	97.18%	0.810
	iAMPpred	91.30%	1.63%	6.92%	−0.117
	iAMP-2L	86.96%	99.46%	98.72%	0.882
	ADAM	95.65%	99.18%	98.97%	0.912
	DBAASP	82.61%	80.00%	81.58%	0.620
	MLAMP	91.30%	99.46%	98.97%	0.908
	CAMPR3_RF	91.30%	1.36%	6.67%	−0.130
	CAMPR3_SVM	95.65%	2.18%	7.69%	−0.034
	CAMPR3_ANN	82.61%	50.68%	52.56%	0.157
	CAMPR3_DA	86.96%	1.36%	6.41%	−0.194
Human	Our method	97.83%	92.17%	92.33%	0.482
	iAMPpred	91.30%	22.88%	24.73%	0.055
	iAMP-2L	54.35%	98.18%	96.99%	0.482
	ADAM	52.17%	98.91%	97.64%	0.534
	DBAASP	40.54%	86.84%	64.00%	0.310
	MLAMP	50.00%	98.36%	97.05%	0.464
	CAMPR3_RF	93.48%	0.85%	3.36%	−0.092
	CAMPR3_SVM	82.61%	1.09%	3.31%	−0.215
	CAMPR3_ANN	69.57%	48.67%	49.23%	0.059
	CAMPR3_DA	84.78%	1.46%	3.72%	−0.167
Insects	Our method	100.00%	97.56%	97.82%	0.900
	iAMPpred	94.44%	39.11%	45.04%	0.217
	iAMP-2L	94.44%	96.67%	96.43%	0.835
	ADAM	100.00%	96.67%	97.02%	0.870
	DBAASP	70.37%	90.91%	73.85%	0.469
	MLAMP	72.22%	98.00%	95.24%	0.740
	CAMPR3_RF	87.04%	1.33%	10.52%	−0.227
	CAMPR3_SVM	87.04%	1.33%	10.52%	−0.227
	CAMPR3_ANN	87.04%	43.33%	48.02%	0.192
	CAMPR3_DA	79.63%	1.56%	9.92%	−0.314
Mammals	Our method	92.79%	94.56%	94.46%	0.673
	iAMPpred	95.50%	68.94%	70.54%	0.322
	iAMP-2L	68.47%	98.73%	96.90%	0.712
	ADAM	65.77%	99.48%	97.45%	0.753
	DBAASP	45.88%	83.02%	60.14%	0.295
	MLAMP	51.35%	98.44%	95.60%	0.568
	CAMPR3_RF	93.69%	1.27%	6.85%	−0.096
	CAMPR3_SVM	92.79%	1.91%	7.39%	−0.085
	CAMPR3_ANN	78.38%	48.58%	50.38%	0.129
	CAMPR3_DA	88.29%	2.14%	7.34%	−0.140

Table A2. *Cont.*

Organisms	Classifier	Sensitivity	Specificity	Accuracy	Matthews Correlation Coefficient
Plants	Our method	97.78%	97.94%	97.93%	0.851
	iAMPpred	90.00%	0.81%	6.35%	−0.190
	iAMP-2L	77.78%	98.67%	97.38%	0.773
	ADAM	84.44%	98.67%	97.79%	0.815
	DBAASP	34.94%	88.46%	47.71%	0.219
	MLAMP	58.89%	98.82%	96.34%	0.654
	CAMPR3_RF	86.67%	0.59%	5.94%	−0.264
	CAMPR3_SVM	83.33%	0.88%	6.01%	−0.282
	CAMPR3_ANN	74.44%	47.57%	49.24%	0.107
	CAMPR3_DA	75.56%	1.10%	5.73%	−0.357

RF = random forest; DT = decision tree; KNN = K-nearest neighbor; SVM = support vector machine; ANN = artificial neural network; DA = discriminant analysis.

Table A3. Physicochemical properties and groupings of amino acids [13].

Physicochemical Properties	Group		
	Class 1	Class 2	Class 3
Charge	Positive K, R	Neutral A, N, C, Q, G, H, I, L, M, F, P, S, T, W, Y, V	Negative D, E
Hydrophobicity	Polar R, K, E, D, Q, N	Neutral G, A, S, T, P, H, Y	Hydrophobic C, L, V, I, M, F, W
Polarity	Polarity value 4.9~6.2 L, I, F, W, C, M, V, Y	Polarity value 8.0~9.2 P, A, T, G, S	Polarity value 10.4~13 H, Q, R, K, N, E, D
Polarizability	Polarizability value 0~0.108 G, A, S, D, T	Polarizability value 0.128~0.186 C, P, N, V, E, Q, I, L	Polarizability value 0.219~0.409 K, M, H, F, R, Y, W
Secondary Structure	Helix E, A, L, M, Q, K, R, H	Strand V, I, Y, C, W, F, T	Coil G, N, P, S, D
Normalized van der Waals volume	Volume range 0~2.78 G, A, S, T, P, D	Volume range 2.95~4.0 N, V, E, Q, I, L	Volume range 4.03~8.08 M, H, K, F, R, Y, W
Solvent accessibility	Buried A, L, F, C, G, I, V, W	Exposed R, K, Q, E, N, D	Intermediate M, P, S, T, H, Y

References

1. Huang, K.Y.; Chang, T.H.; Jhong, J.H.; Chi, Y.H.; Li, W.C.; Chan, C.L.; Robert Lai, K.; Lee, T.Y. Identification of natural antimicrobial peptides from bacteria through metagenomic and metatranscriptomic analysis of high-throughput transcriptome data of Taiwanese oolong teas. *BMC Syst. Biol.* **2017**, *11*, 131. [CrossRef] [PubMed]

2. Waghu, F.H.; Barai, R.S.; Gurung, P.; Idicula-Thomas, S. CAMPR3: A database on sequences, structures and signatures of antimicrobial peptides. *Nucleic Acids Res.* **2016**, *44*, D1094–D1097. [CrossRef] [PubMed]

3. Yeaman, M.R.; Yount, N.Y. Mechanisms of antimicrobial peptide action and resistance. *Pharmacol. Rev.* **2003**, *55*, 27–55. [CrossRef] [PubMed]

4. Gabere, M.N.; Noble, W.S. Empirical comparison of web-based antimicrobial peptide prediction tools. *Bioinformatics* **2017**, *33*, 1921–1929. [CrossRef] [PubMed]

5. Lata, S.; Sharma, B.K.; Raghava, G.P. Analysis and prediction of antibacterial peptides. *BMC Bioinform.* **2007**, *8*, 263. [CrossRef]

6. Lata, S.; Mishra, N.K.; Raghava, G.P. AntiBP2: Improved version of antibacterial peptide prediction. *BMC Bioinform.* **2010**, *11* (Suppl. 1), S19. [CrossRef]

7. Thomas, S.; Karnik, S.; Barai, R.S.; Jayaraman, V.K.; Idicula-Thomas, S. CAMP: A useful resource for research on antimicrobial peptides. *Nucleic Acids Res.* **2010**, *38*, D774–D780. [CrossRef]

8. Joseph, S.; Karnik, S.; Nilawe, P.; Jayaraman, V.K.; Idicula-Thomas, S. ClassAMP: A prediction tool for classification of antimicrobial peptides. *IEEE/ACM Trans. Comput. Biol. Bioinform.* **2012**, *9*, 1535–1538. [CrossRef]

9. Thakur, N.; Qureshi, A.; Kumar, M. AVPpred: Collection and prediction of highly effective antiviral peptides. *Nucleic Acids Res.* **2012**, *40*, W199–W204. [CrossRef]

10. Fjell, C.D.; Hancock, R.E.; Cherkasov, A. AMPer: A database and an automated discovery tool for antimicrobial peptides. *Bioinformatics* **2007**, *23*, 1148–1155. [CrossRef]

11. Xiao, X.; Wang, P.; Lin, W.Z.; Jia, J.H.; Chou, K.C. iAMP-2L: A two-level multi-label classifier for identifying antimicrobial peptides and their functional types. *Anal. Biochem.* **2013**, *436*, 168–177. [CrossRef] [PubMed]

12. Meher, P.K.; Sahu, T.K.; Saini, V.; Rao, A.R. Predicting antimicrobial peptides with improved accuracy by incorporating the compositional, physico-chemical and structural features into Chou's general PseAAC. *Sci. Rep.* **2017**, *7*, 42362. [CrossRef] [PubMed]

13. Bhadra, P.; Yan, J.; Li, J.; Fong, S.; Siu, S.W.I. AmPEP: Sequence-based prediction of antimicrobial peptides using distribution patterns of amino acid properties and random forest. *Sci. Rep.* **2018**, *8*, 1697. [CrossRef] [PubMed]

14. Veltri, D.; Kamath, U.; Shehu, A. Improving Recognition of Antimicrobial Peptides and Target Selectivity through Machine Learning and Genetic Programming. *IEEE/ACM Trans. Comput. Biol. Bioinform.* **2017**, *14*, 300–313. [CrossRef] [PubMed]

15. Wang, G.; Li, X.; Wang, Z. APD3: The antimicrobial peptide database as a tool for research and education. *Nucleic Acids Res.* **2016**, *44*, D1087–D1093. [CrossRef]

16. Hammami, R.; Ben Hamida, J.; Vergoten, G.; Fliss, I. PhytAMP: A database dedicated to antimicrobial plant peptides. *Nucleic Acids Res* **2009**, *37*, D963–D968. [CrossRef]

17. Mishra, B.; Wang, G. The Importance of Amino Acid Composition in Natural AMPs: An Evolutional, Structural, and Functional Perspective. *Front. Immunol.* **2012**, *3*, 221. [CrossRef]

18. Chung, C.R.; Kuo, T.R.; Wu, L.C.; Lee, T.Y.; Horng, J.T. Characterization and identification of antimicrobial peptides with different functional activities. *Brief Bioinform.* **2019**. [CrossRef]

19. Lee, H.T.; Lee, C.C.; Yang, J.R.; Lai, J.Z.; Chang, K.Y. A large-scale structural classification of antimicrobial peptides. *Biomed. Res. Int.* **2015**, *2015*, 475062. [CrossRef]

20. Vishnepolsky, B.; Pirtskhalava, M. Prediction of Linear Cationic Antimicrobial Peptides Based on Characteristics Responsible for Their Interaction with the Membranes. *J. Chem. Inf. Model.* **2014**, *54*, 1512–1523. [CrossRef]

21. Fan, L.; Sun, J.; Zhou, M.; Zhou, J.; Lao, X.; Zheng, H.; Xu, H. DRAMP: A comprehensive data repository of antimicrobial peptides. *Sci. Rep.* **2016**, *6*, 24482. [CrossRef] [PubMed]

22. Chang, K.Y.; Lin, T.P.; Shih, L.Y.; Wang, C.K. Analysis and prediction of the critical regions of antimicrobial peptides based on conditional random fields. *PLoS ONE* **2015**, *10*, e0119490. [CrossRef] [PubMed]

23. Tavares, L.S.; Rettore, J.V.; Freitas, R.M.; Porto, W.F.; Duque, A.P.; Singulani Jde, L.; Silva, O.N.; Detoni Mde, L.; Vasconcelos, E.G.; Dias, S.C.; et al. Antimicrobial activity of recombinant Pg-AMP1, a glycine-rich peptide from guava seeds. *Peptides* **2012**, *37*, 294–300. [CrossRef]

24. Matsuzaki, K. Control of cell selectivity of antimicrobial peptides. *Biochim. Biophys. Acta* **2009**, *1788*, 1687–1692. [CrossRef] [PubMed]

25. Tadeg, H.; Mohammed, E.; Asres, K.; Gebre-Mariam, T. Antimicrobial activities of some selected traditional Ethiopian medicinal plants used in the treatment of skin disorders. *J. Ethnopharmacol.* **2005**, *100*, 168–175. [CrossRef] [PubMed]

26. Hilpert, K.; Elliott, M.; Jenssen, H.; Kindrachuk, J.; Fjell, C.D.; Korner, J.; Winkler, D.F.; Weaver, L.L.; Henklein, P.; Ulrich, A.S.; et al. Screening and characterization of surface-tethered cationic peptides for antimicrobial activity. *Chem. Biol.* **2009**, *16*, 58–69. [CrossRef] [PubMed]

27. Johnsen, L.; Fimland, G.; Nissen-Meyer, J. The C-terminal domain of pediocin-like antimicrobial peptides (class IIa bacteriocins) is involved in specific recognition of the C-terminal part of cognate immunity proteins and in determining the antimicrobial spectrum. *J. Biol. Chem.* **2005**, *280*, 9243–9250. [CrossRef] [PubMed]

28. Dathe, M.; Nikolenko, H.; Meyer, J.; Beyermann, M.; Bienert, M. Optimization of the antimicrobial activity of magainin peptides by modification of charge. *FEBS Lett.* **2001**, *501*, 146–150. [CrossRef]

29. Chen, Y.; Guarnieri, M.T.; Vasil, A.I.; Vasil, M.L.; Mant, C.T.; Hodges, R.S. Role of peptide hydrophobicity in the mechanism of action of alpha-helical antimicrobial peptides. *Antimicrob. Agents Chemother.* **2007**, *51*, 1398–1406. [CrossRef]

30. Pirtskhalava, M.; Gabrielian, A.; Cruz, P.; Griggs, H.L.; Squires, R.B.; Hurt, D.E.; Grigolava, M.; Chubinidze, M.; Gogoladze, G.; Vishnepolsky, B.; et al. DBAASP v.2: An enhanced database of structure and antimicrobial/cytotoxic activity of natural and synthetic peptides. *Nucleic Acids Res.* **2016**, *44*, D1104–D1112. [CrossRef]

31. Lin, W.Z.; Xu, D. Imbalanced multi-label learning for identifying antimicrobial peptides and their functional types. *Bioinformatics* **2016**, *32*, 3745–3752. [CrossRef] [PubMed]

32. Torres, M.D.T.; de la Fuente-Nunez, C. Toward computer-made artificial antibiotics. *Curr. Opin. Microbiol.* **2019**, *51*, 30–38. [CrossRef] [PubMed]

33. Porto, W.F.; Irazazabal, L.; Alves, E.S.; Ribeiro, S.M.; Matos, C.O.; Pires, Á.S.; Fensterseifer, I.C.; Miranda, V.J.; Haney, E.F.; Humblot, V. In silico optimization of a guava antimicrobial peptide enables combinatorial exploration for peptide design. *Nature Commun.* **2018**, *9*, 1–12. [CrossRef] [PubMed]

34. Wang, P.; Hu, L.; Liu, G.; Jiang, N.; Chen, X.; Xu, J.; Zheng, W.; Li, L.; Tan, M.; Chen, Z.; et al. Prediction of antimicrobial peptides based on sequence alignment and feature selection methods. *PLoS ONE* **2011**, *6*, e18476. [CrossRef]

35. Huang, Y.; Niu, B.; Gao, Y.; Fu, L.; Li, W. CD-HIT Suite: A web server for clustering and comparing biological sequences. *Bioinformatics* **2010**, *26*, 680–682. [CrossRef]

36. Hall, M.; Frank, E.; Holmes, G.; Pfahringer, B.; Reutemann, P.; Witten, I.H. The WEKA data mining software: An update. ACM SIGKDD explorations newsletter. *ACM SIGKDD Explor. Newsl.* **2009**, *11*, 10–18. [CrossRef]

37. Vapnik, V.N. An overview of statistical learning theory. *IEEE Trans. Neural. Netw.* **1999**, *10*, 988–999. [CrossRef]

38. Salzberg, S. Locating protein coding regions in human DNA using a decision tree algorithm. *J. Comput. Biol.* **1995**, *2*, 473–485. [CrossRef]

International Journal of
Molecular Sciences

Article

Novel ACE Inhibitory Peptides Derived from Simulated Gastrointestinal Digestion in Vitro of Sesame (*Sesamum indicum* L.) Protein and Molecular Docking Study

Ruidan Wang, Xin Lu, Qiang Sun, Jinhong Gao, Lin Ma and Jinian Huang *

Research Center for Agricultural and Sideline Products Processing, Henan Academy of Agricultural Sciences, Zhengzhou 450002, China; wangrd199403@163.com (R.W.); xinlu1981@foxmail.com (X.L.); qiangsunxy@126.com (Q.S.); gaji.good@163.com (J.G.); malin199003@163.com (L.M.)
* Correspondence: jinianhuang71@gmail.com; Tel.: +86-371-6585-1580

Received: 28 December 2019; Accepted: 3 February 2020; Published: 5 February 2020

Abstract: The aim of this study was to isolate and identify angiotensin I-converting enzyme (ACE) inhibitory peptides from sesame protein through simulated gastrointestinal digestion in vitro, and to explore the underlying mechanisms by molecular docking. The sesame protein was enzymatically hydrolyzed by pepsin, trypsin, and α-chymotrypsin. The degree of hydrolysis (DH) and peptide yield increased with the increase of digest time. Moreover, ACE inhibitory activity was enhanced after digestion. The sesame protein digestive solution (SPDS) was purified by ultrafiltration through different molecular weight cut-off (MWCO) membranes and SPDS-VII (< 3 kDa) had the strongest ACE inhibition. SPDS-VII was further purified by NGC Quest™ 10 Plus Chromatography System and finally 11 peptides were identified by Nano UHPLC-ESI-MS/MS (nano ultra-high performance liquid chromatography-electrospray ionization mass spectrometry/mass spectrometry) from peak 4. The peptide GHIITVAR from 11S globulin displayed the strongest ACE inhibitory activity (IC_{50} = 3.60 ± 0.10 µM). Furthermore, the docking analysis revealed that the ACE inhibition of GHIITVAR was mainly attributed to forming very strong hydrogen bonds with the active sites of ACE. These results identify sesame protein as a rich source of ACE inhibitory peptides and further indicate that GHIITVAR has the potential for development of new functional foods.

Keywords: sesame protein; ACE inhibitory peptides; simulated gastrointestinal digestion; amino acid sequence; molecular docking

1. Introduction

Hypertension is one of the diseases with the highest mortality in the world, and it is the main pathogenesis factor of coronary heart disease, stroke, and heart and kidney failure [1–3]. According to epidemiological studies, more than one billion people suffer from hypertension nowadays [4]. Therefore, the prevention and treatment of hypertension has become a difficult task for the global medical community.

Relevant studies have shown that human blood pressure is regulated by many factors, among which renin-angiotensin system (RAS) and kallikrein kinin system (KKS) are the main ways to control the stability of blood pressure. ACE is a dipeptide carboxyl metalloproteinase which plays a key role in these two systems. In the RAS system, ACE can convert the inactive angiotensin I into angiotensin II, which has the function of constricting vascular smooth muscle, thus, increases the blood pressure. Meanwhile, ACE can inactivate the vasodilator bradykinin in the KKS system, leading to vasoconstriction, which causes an increase in the blood pressure [5]. Currently, many synthetic ACE inhibitors such as captopril, enalapril, alacepril, and lisinopril can effectively reduce high blood pressure. However, these drugs have severe adverse effects on patients such as persistent cough, taste distortion and skin rashes in long term usage [6,7]. Therefore, it is crucial to develop more effective bioactive agents or functional foods as substitutes of synthetic drugs in the prevention and treatment of hypertension.

Bioactive peptides are small fragments of food protein, mostly composed of 2–20 amino acid residues. They have the characteristics of easy absorption, low sensitivity and good solubility. Meanwhile, they have the function of nutrition and physiological regulation, which have obvious effects on the conditioning and treatment of modern chronic diseases and sub-health state. During processing, enzymatic hydrolysis, fermentation, or gastrointestinal digestion, foods can release bioactive peptides [8,9] which have been reported to exhibit different biological activities including antioxidant [10], antibacterial [11], inhibition activity on dipeptidyl peptidase IV [12], inhibition activity on ACE [13], and so on. So far, lots of ACE inhibitory peptides have been isolated from various food proteins such as milk protein [14], egg protein [15], and marine protein [16].

Sesame (*Sesamum indicum* L.) is one of the important oil seed crops worldwide and it is widely used in food, health care, and medical applications because of its high nutritional value [17,18]. Sesame seeds are mainly used to produce sesame oil due to high content of unsaturated fatty acids and lignans. In addition, sesame meal containing almost 50% proteins could be a valuable source of proteins for comprehensive use. Sesame protein has been reported to have ACE inhibitory peptides. Nakano et al. [19] have isolated six ACE inhibitory peptides from sesame protein hydrolyzed by thermolysin. However, there was little information about the ACE inhibitory peptides of sesame protein hydrolysate via simulated gastrointestinal digestion in vitro and molecular docking study. Here, the purpose of this study was to find the changing rules of ACE inhibitory peptides generated from sesame protein during simulated gastrointestinal digestion in vitro and to isolate and identify the sequence of new peptides. Moreover, the binding interaction of the screened ACE inhibitory peptide within the enzymatic active site was further studied through molecular docking simulation. Our results are expected to provide more evidence for the utility of sesame as a functional food for the treatment of hypertension.

2. Results

2.1. Changes of ACE Inhibitory Activity during Simulated Gastrointestinal Digestion

The degree of hydrolysis (DH) represents the extent of protein degradation, which has been widely used in hydrolysis efficiency assessments. As shown in Figure 1a, the DH of sesame protein showed an overall rising trend when simulating gastrointestinal digestion in vitro. In the stage of gastric digestion, sesame protein began to be hydrolyzed, and the DH ranged from 2.59% to 17.69% at 0–4 h. However, the DH increased suddenly when trypsin and α-chymotrypsin was added. In the stage of intestinal digestion, the DH increased slowly and tended to be stable due to the decrease of protein substrates and enzyme cutting sites in the digestive system. The DH eventually reached 36.70%. Figure 1b shows the changes of peptide yield at different time points during simulated gastrointestinal digestion. After adding pepsin, the peptide yield increased significantly ranging 95.46% with the increase of time. After being treated by trypsin and α-chymotrypsin, the peptide yield remained basically stable over a 6 h period. The changing rules of ACE inhibitory activity at different time points during simulated gastrointestinal digestion are shown in Figure 1c. Gastric digestive products of sesame protein exhibited weak ACE inhibitory activity without obvious upward or downward trend at 0–4 h, but intestinal digestive products had strong ACE inhibitory activity at 4–10 h and tended to be stable gradually. The ACE inhibitory activity reached 81.21% at 10 h. It was suggested that pepsin had less ability to hydrolyze sesame protein to produce polypeptide in simulated gastric digestion. There were more ACE inhibitory peptides generated from simulated intestinal digestion, which implied that trypsin and α-chymotrypsin offered the ability to achieve stronger ACE inhibitory peptides. It also would be supposed that the peptide sequences were buried deeper in the original protein structures from which they come or the sequences were embedded in the structured parts in which they were forced to be in the conformations that could not fit the active site of ACE. These results were also similar to that of other reports [20–23].

2.2. Separation of ACE Inhibitory Peptide from SPDS by Ultrafiltration

It is well known that molecular weights of peptide fragments are crucial for their biological activities. Ultrafiltration is a membrane separation technique that is of great use in the concentration, purification, fractionation, and clarification of solutes. It has high throughput of product, low operational cost, energy saving potential, and ease of equipment cleaning. In the separation of polypeptides, ultrafiltration can effectively remove insoluble substrates, large molecular weight proteins and peptides, and obtain bioactive peptides with smaller molecular weight [22,24]. The SPDS was subjected to fractionation by ultrafiltration through 100, 50, 30, 10, 5, and 3 kDa MWCO membranes to get seven fractions with >100, 50–100, 30–50, 10–30, 5–10, 3–5, and <3 kDa. The IC_{50} values inhibiting ACE activity were calculated from the results of a series of concentrations. As shown in Table 1, the IC_{50} values of these fractions were 35.143 ± 1.122, 15.066 ± 0.042, 9.146 ± 0.005, 6.108 ± 0.001, 5.106 ± 0.003, 4.583 ± 0.003, 2.720 ± 0.003 µg/mL, respectively. A lower IC_{50} value corresponds to a stronger inhibitory activity. SPDS-VII with molecular weight <3kDa exhibited the strongest ACE inhibitory activity with the lowest IC_{50} value among these fractions, indicating that lower molecular weight peptides or small molecules showed stronger ACE inhibition. This result was in agreement with previous studies reporting that most ACE inhibitory peptides derived from food protein were generally short sequences containing 2–20 amino acids, or the hydrolyzates were generally filtered through a 3 kDa membrane [25,26]. In addition, the high molecular weight fraction could not enter the ACE active site easily. Therefore, it was hard to change the catalytic activity of ACE compared with the low molecular weight fraction [27].

Table 1. Results of IC_{50} values of fractions from sesame protein digestive solution (SPDS) on ACE inhibitory activity.

Fraction	IC_{50} Values (µg/mL)
SPDS-I (>100 kDa)	35.143 ± 1.122 [a]
SPDS-II (50–100 kDa)	15.066 ± 0.042 [b]
SPDS-III (30–50 kDa)	9.146 ± 0.005 [c]
SPDS-IV (10–30 kDa)	6.108 ± 0.001 [cd]
SPDS-V (5–10 kDa)	5.106 ± 0.003 [ce]
SPDS-VI (3–5 kDa)	4.583 ± 0.003 [cf]
SPDS-VII (<3 kDa)	2.720 ± 0.003 [cg]

Values are expressed as the mean ± standard deviation ($n = 3$). Values with different letters are significantly different ($p < 0.05$).

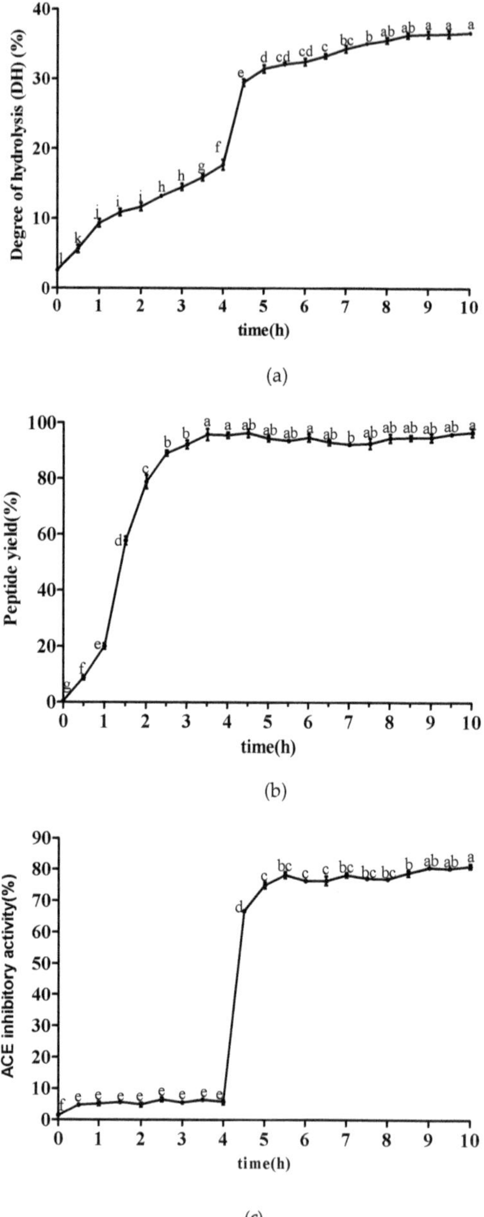

Figure 1. The changing rules of ACE inhibitory peptides at different time points during simulated gastrointestinal digestion. (**a**) Degree of hydrolysis changes at different time points during simulated gastrointestinal digestion. (**b**) The changes of peptide yield at different time points during simulated gastrointestinal digestion. (**c**) The changes of angiotensin I-converting enzyme (ACE) inhibitory activity at different time points during simulated gastrointestinal digestion. Data are expressed as the mean ± standard deviation (n = 3) and different letters marked are significantly different by one-way analysis of variation multiple test ($p < 0.05$).

2.3. Purification of ACE Inhibitory Peptide by NGC Quest™ 10 Plus Chromatography System

To enhance the ACE inhibition, SPDS-VII was further purified by NGC Quest™ 10 Plus Chromatography System. It is an effective way to purify proteins or peptides and collect them in the preparation phase. The chromatogram is shown in Figure 2. Five major peaks were eluted and collected for evaluating the ACE inhibition. As data represented in Table 2, peak 1 with the retention time from 5.11 to 6.52 min showed an IC_{50} value of (2.847 ± 0.045) µg/mL, peak 2 with the retention time from 6.52 to 7.12 min exhibited an IC_{50} value of (1.421 ± 0.035) µg/mL. The IC_{50} value of peak 3 with a retention time from 7.12 to 7.42 min was (1.838 ± 0.026) µg/mL, peak 4 with a retention time from 7.49 to 8.27 min was (0.558 ± 0.003) µg/mL and peak 5 with a retention time from 8.27 to 10.80 min was (0.757 ± 0.014) µg/mL. These results demonstrated that peak 4 possessed the most potent ACE inhibitory activity among the five peaks and thus was selected for further identification.

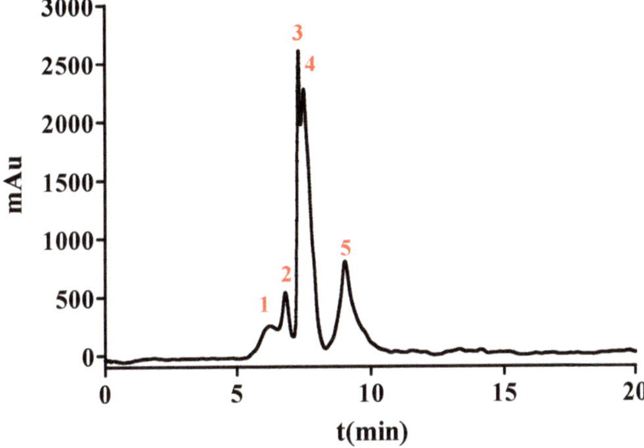

Figure 2. The chromatograms measured at 220 nm of SPD-VII by NGC Quest™ 10 Plus Chromatography System. The red numbers 1-5 represent peaks 1-5 respectively.

Table 2. The IC_{50} values inhibiting ACE activity of peaks 1-5 eluted from SPDS-VII.

Fraction	Retention Time (min)	IC_{50} Values (µg/mL)
peak 1	5.11–6.52	2.847 ± 0.045 [a]
peak 2	6.52–7.12	1.421 ± 0.035 [c]
peak 3	7.12–7.42	1.838 ± 0.026 [b]
peak 4	7.49–8.27	0.558 ± 0.003 [e]
peak 5	8.27–10.80	0.757 ± 0.014 [d]

Values are expressed as the mean ± standard deviation ($n = 3$). Values with different letters are significantly different ($p < 0.05$).

2.4. Characterization of ACE Inhibitory Peptide by Nano UHPLC-ESI-MS/MS from Peak 4

The peptides from peak 4 were identified by Nano UHPLC-ESI-MS/MS. The amino acid sequences from *Sesamum indicum* L. were found in Uniprot database (http://www.uniprot.org/). In this study, a total of 11 peptides were identified by analysis of the MS/MS spectrum (see Figure S1a–k). The sequences of the peptides were GHIITVAR, IGGIGTVPVGR, HIGNILSL, FMPGVPGPIQR, PNYHPSPR, AFPAGAAHW, HIITLGR, LAGNPAGR, MPGVPGPIQR, AGALGDSVTVTR, and INTLSGR, respectively. These peptides were chemically synthesized, and then their IC_{50} values inhibiting ACE activity were measured and the corresponding protein names are listed in Table 3. These peptides contained 7–12 amino acids with the molecular weight ranging from 754.8 to 1198.4

kDa; nevertheless, they had stronger ACE inhibitory activity in this study than the identified ACE inhibitory peptide YAHYSYA with the IC_{50} value of 0.6 mg/mL (686.58 μM) from sesame protein hydrolyzed by alcalase [20], suggesting that stronger ACE inhibitory activity was due to the action of endogenous enzymes in the body when simulating gastrointestinal digestion. Alternatively, sesame protein was hydrolyzed more thoroughly by multiple enzymes resulting in more cleavage sites via simulated gastrointestinal digestion in vitro.

Table 3. Peptides identified from peak 4 and corresponding physicochemical characteristics.

Peptide	Protein Name	Molecular Weight	IC_{50} Value (μM)
GHIITVAR	11S globulin isoform 4 (Q2XSW6_SESIN)	866.0	3.60 ± 0.10 [k]
IGGIGTVPVGR	Elongation factor 1-alpha-like	1025.2	6.97 ± 0.18 [j]
HIGNILSL	TBCC domain-containing protein 1	866.0	36.69 ± 0.33 [f]
FMPGVPGPIQR	Oil body-associated protein 1A	1198.4	11.08 ± 0.15 [i]
PNYHPSPR	11S globulin seed storage protein 2 precursor (Q9XHP0_SESIN)	967.0	18.98 ± 0.26 [h]
AFPAGAAHW	11S globulin isoform 4 (Q2XSW6_SESIN)	927.0	29.00 ± 0.20 [g]
HIITLGR	Protein NDH-DEPENDENT CYCLIC ELECTRON FLOW 5	808.9	74.65 ± 0.13 [d]
LAGNPAGR	11S globulin isoform 4 (Q2XSW6_SESIN)	754.8	148.41 ± 0.35 [b]
MPGVPGPIQR	Oil body-associated protein 1A	1051.2	54.79 ± 0.37 [e]
AGALGDSVTVTR	60S ribosomal protein L22-2-like	1146.2	68.49 ± 0.14 [c]
INTLSGR	11S globulin isoform 4 (Q2XSW6_SESIN)	759.8	149.63 ± 0.33 [a]

Values are expressed as the mean ± standard deviation (n = 3). Values with different letters are significantly different ($p < 0.05$).

According to previous literature reports, most of the proteins present in sesame seeds are storage proteins composed of globulins (67.3%), albumins (8.6%), prolamines (1.4%), and glutelins (6.9%), of which sesame 11S globulins includes 11S globulin seed storage protein 2 precursor, 11S globulin isoform 3, and 11S globulin isoform 4 [28,29]. As shown in Table 3, there were five peptides derived from sesame 11S globulins, of which four were from11S globulin isoform 4 and one was from 11S globulin seed storage protein 2 precursor. This result indicated that sesame 11S globulins, especially 11S globulin isoform 4 mainly contributed to the generation of ACE inhibitory peptides.

It is known that the ACE inhibition is strongly influenced by the type of amino acid composition of peptides. A series of studies have figured out that binding to ACE is influenced by hydrophobic amino acid residues (aromatic or branched chain) at three positions from the C-terminus of the peptide [30,31]. Some branched chain aliphatic amino acids, such as Ile and Val, are predominant in potent peptide inhibitors [32]. Moreover, the positively charged amino acids (Lys and Arg) have also been implied to increase the potency of ACE inhibitory peptides. Ryan et al. [33] concluded that hydrophilic peptides possess weak or no ACE inhibition. The hydrophobic nature of the N-terminus is another common feature of ACE inhibitory peptides. As a whole, the overall hydrophobicity of the peptide is important for ACE inhibitory activity. In the present study, a difference between MPGVPGPIQR (IC_{50} = 54.79 ± 0.37 μM) and FMPGVPGPIQR (IC_{50} = 11.08 ± 0.15 μM) in ACE inhibition further supported the above conclusion. Nevertheless, there were some exceptions, for example, the peptide EN and EVD from whey protein and the peptide CRQNTLGHNTQTSIAQ from *Stichopus horrens* do not have any reported specific amino acid at the C-terminal [34,35]. Thus, the relationship between the structure and function of ACE inhibitory peptides has not been fully established to date.

There were nine peptides (GHIITVAR, IGGIGTVPVGR, FMPGVPGPIQR, PNYHPSPR, HIITLGR, LAGNPAGR, MPGVPGPIQR, AGALGDSVTVTR, INTLSGR) containing Arg in this study (Table 3). Arg is a basic aliphatic amino acid and also a positively charged amino acid, which was most frequently observed at the C-terminal of the ACE inhibitory peptides according to a lot of studies, such as YR and IR from the hydrolysates of marine sponge [36], APER from trevally hydrolysate [37]. It was noteworthy that the peptide GHIITVAR exhibiting the lowest IC_{50} value of 3.60 ± 0.10 μM among the 11 peptides. The inhibition potential of GHIITVAR was still less than the synthetic ACE inhibitor Captopril (IC_{50} = 0.0071 μM). GHIITVAR contained a hydrophobic amino acid Arg at the C-terminus and two aliphatic amino acids (Val and Ala) at the antepenultimate and penultimate positions, respectively. Besides, the presence of two aliphatic amino acids Ile may contribute to ACE inhibitory activity. However, the IC_{50} value of the peptide HIITLGR was 74.65 ± 0.13 μM, which further confirmed the fact that the presence of hydrophobic amino acids, especially those with aliphatic chains such as Gly at the N-terminus might enhance the ACE inhibition.

In addition, the cleavage sites of pepsin were hydrophobic amino acids such as Phe, Trp, Leu, and Tyr. Trypsin preferentially cleaved at Arg and Lys and α-chymotrypsin preferentially cleaved at Trp, Tyr, and Phe. The cleavage site specificity of pepsin, trypsin, and α-chymotrypsin was found in bioinformatics tools (https://web.expasy.org/peptide_cutter/). From these results mentioned above, it could be concluded that the addition of trypsin and α-chymotrypsin significantly increase the exposure of active sites and thus increase ACE inhibitory activity. This was also consistent with the results in Figure 1c and explained why the inhibitory activity rose after pepsin digestion.

2.5. Molecular Docking Simulation between Peptide and ACE

Molecular docking was performed to estimate the binding affinities between peptide and ACE by using Surflex-Dock in Sybyl. In this study, lisinopril (5362119.pdb) was docked again with the prepared ACE. Taking GHIITVAR as an example, the C scores of the 20 conformations were different, but the Total scores gradually decreased (Figure S2). Therefore, the conformations of ranked No. 1 based on Total scores for each ligand were selected as the best docking conformation and the docking scores of 11 peptides and lisinopril were calculated. C score, as a synthetic scoring function was used for screening the binding affinity of ligand bound to ACE. As shown in Table 4, the Total score and C score of lisinopril were 11.24 and 4, respectively, which was in accordance with the results of the report basically [38]. Three peptides (AGALGDSVTVTR, HIITLGR, GHIITVAR) had the same highest C scores as lisinopril, whereas the C scores of the other peptides were below 4 (Table S1), suggesting that these peptides had stronger binding affinity with ACE. The computational visualization of the best docking conformation of the screened peptides and lisinopril by using Molecular Operating Environments (MOE; 2015.10 Chemical Computing Groups, Montreal, Quebec, Canada) is shown in Figure 3. Results revealed that these peptides were inserted into the binding site of ACE and formed hydrogen bonds with ACE residues.

Table 4. Surflex-Dock scores (kcal/mol) of peptides and lisinopril.

Peptide	Total_Score [1]	Crash [2]	Polar [3]	D_score [4]	PMF_Score [5]	G_Score [6]	Chem Score [7]	C Score [8]
AGALGDSVTVTR	13.66	−7.05	11.00	−395.98	−306.63	−707.54	−41.93	4
HIITLGR	12.35	−6.48	8.63	−288.74	−264.56	−519.33	−26.18	4
GHIITVAR	10.03	−9.60	6.79	−357.58	−308.79	−685.80	−40.87	4
Lisinopril	11.24	−2.48	7.27	−160.13	−179.32	−284.82	−23.95	4

[1] Total_score: represents the total surflex dock score expressed as ratio of concentrations of a compound in a mixture of two immiscible phases at equilibrium (−logKd). [2] Crash: stands for the capacity of penetration of a ligand into the active site of the protein. Crash scores close to 0 are favorable. Negative numbers indicate penetration. [3] Polar: describes the polar interaction of protein and the ligand. [4] D_score: stands for van der waals interaction between protein and the ligand. [5] PMF_score: (Potential of Mean Force, PMF) the free energies of interactions for protein-ligand atom pairs. [6] G_score: is based on hydrogen bonding, ligand-protein complex, and internal (ligand-ligand) energies. [7] Chem score: includes provisions for hydrogen bonding, lipophilic contact, and rotational entropy, metal-ligand interaction, along with an intercept term. [8] C score (the consensus score): gives a number of scoring functions of affinity of ligand bound to protein.

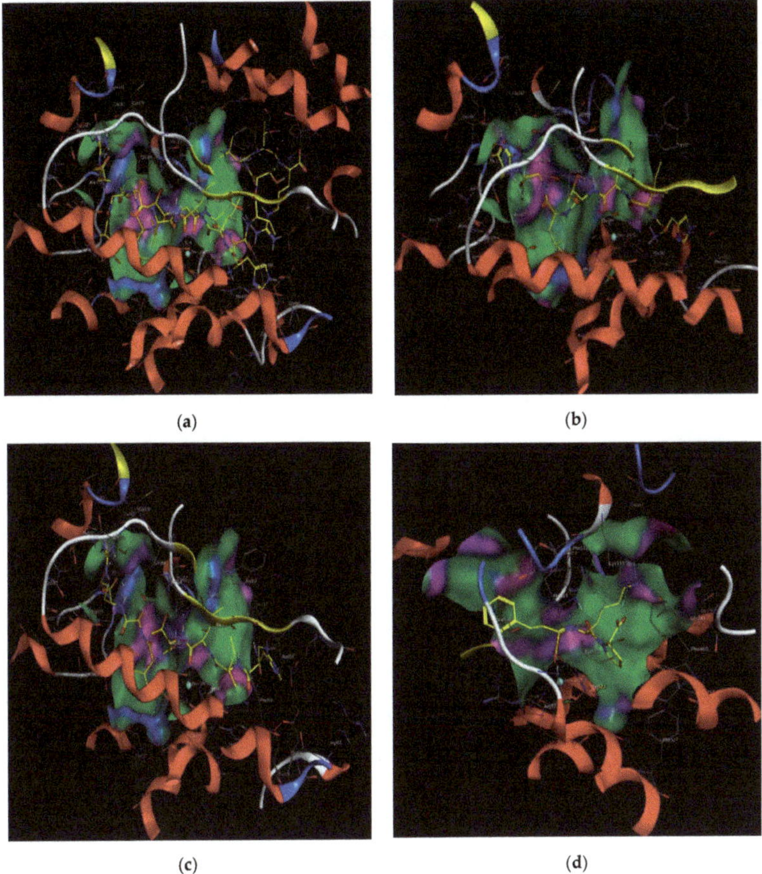

Figure 3. Computational visualization of the optimal docking conformation of AGALGDSVTVTR (**a**), HIITLGR (**b**), GHIITVAR (**c**), and lisinopril (**d**) with the active site of ACE. Carbon is in yellow, hydrogen is in grey, nitrogen is in dark blue, and oxygen is in red. The dotted lines represent hydrogen bonds.

Moreover, molecular modeling of the interaction between the peptides and ACE was further analyzed to explain the difference in the ACE inhibitory activities by using MOE. Figure 4 demonstrated that the results of the peptides' interaction with ACE and lisinopril was taken as a reference. According to previous reports, the active sites of ACE were divided three main pockets (S1, S2, and S1'). S1 pocket includes Ala354, Glu384, and Tyr523 residues and S2 pocket includes Gln281, His353, Lys511, His513, and Tyr520 residues, while S1'contains Glu162 residue. ACE had a zinc ion (Zn^{2+}) in its active site that coordinates with His383, His387, and Glu411 [39–41]. The hydrogen bonds play an important role in binding of the inhibitors to ACE potentially [41,42]. As shown in Figure 4a, lisinopril (5362119.pdb) was docked again with ACE and the interaction mode was the same as ACE-lisinopril complex (1O86.pdb), which indicated that the molecular docking procedure by using MOE was feasible. Lisinopril, as a synthetic antihypertensive drug, shared hydrogen bond interactions at Tyr523, Glu384, His353, Tyr520, Gln281, Lys511, Glu162, His383 and could bind the enzymatic active site through a metal ion interaction (Zn701). The peptide GHIITVAR displayed an excellent match with the active sites of ACE and formed hydrogen bonds with the S1 pocket (Ala354 and Tyr523), the S2 pocket (Gln281, His353, Tyr520, Lys511), and the S1' pocket (Glu162) after docking. It also bound to the active sites using a metal ion interaction (Zn701) and a hydrogen bond interaction with His383. In addition, the six residues surrounding the ACE active site, Ala356, Arg522, Glu123, Asp377, and Glu376, contributed significantly in stabilization of this peptide-ACE complex. Therefore, it was not hard to see that GHIITVAR may have a competitive inhibition through binding at active sites when compared to lisinopril. As shown in Figure 4c, the peptide AGALGDSVTVTR was only able to interact with Ala354, His353, Glu162, Glu411, and coordinate with Zn^{2+}. Similarly, the peptide HIITLGR displayed in Figure 4d established hydrogen bonds with Ala354, His353, Glu411, and coordinated with Zn^{2+}. These results revealed that GHIITVAR could effectively interact with the active sites of ACE and might explain why this peptide exhibited exceptional ACE inhibition.

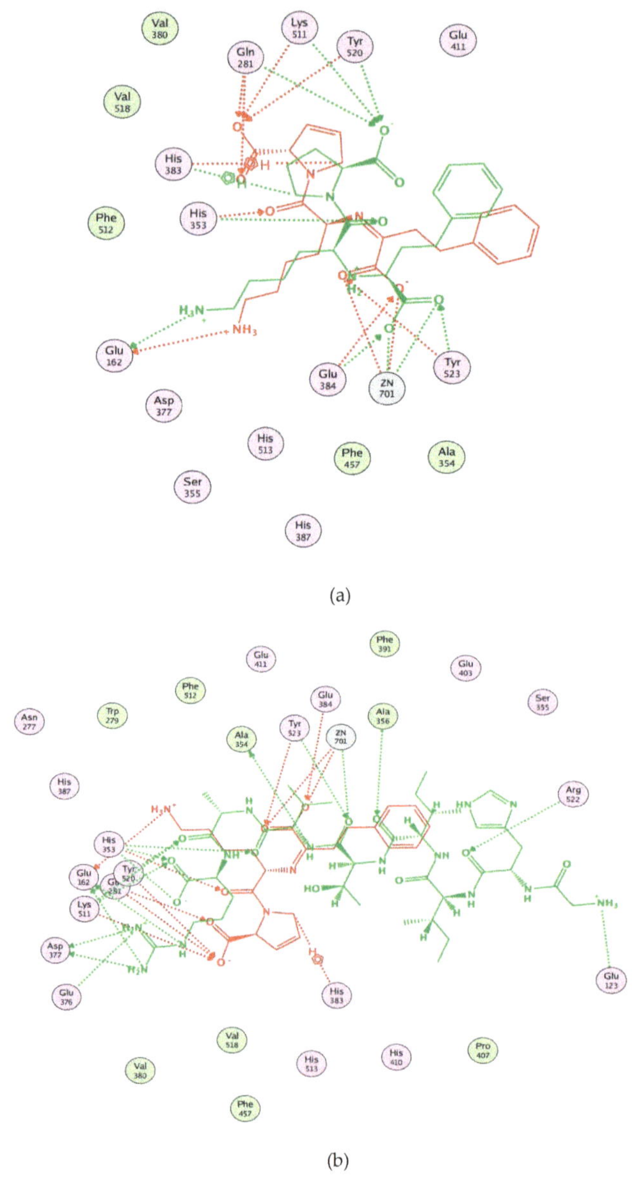

(a)

(b)

Figure 4. *Cont.*

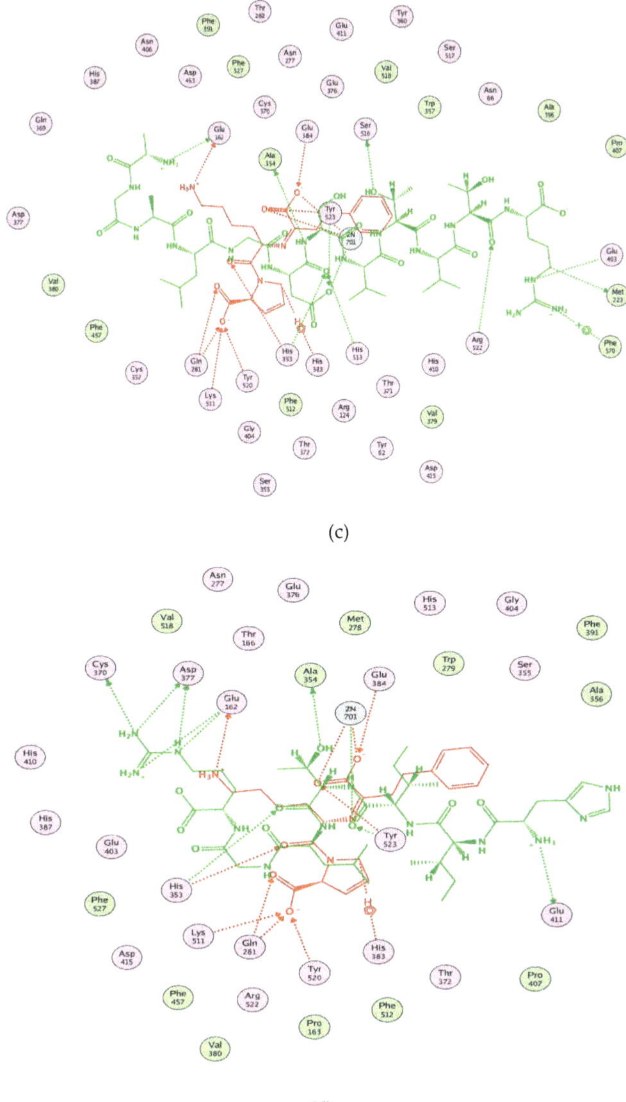

(c)

(d)

Figure 4. Molecular modeling of the interaction between inhibitors and ACE. (**a**) Molecular modeling of the interaction between lisinopril and ACE. (**b**) Molecular modeling of the interaction between GHIITVAR and ACE. (**c**) Molecular modeling of the interaction between AGALGDSVTVTR and ACE. (**d**) Molecular modeling of the interaction between HIITLGR and ACE. The ACE-lisinopril complex (1O86.pdb) was used as a template. The structure of lisinopril is in red and the peptides is in green. Hydrophobic, polar, and acidic residues of ACE are represented by green, violet, and red rings, respectively. Green and red arrows show hydrogen bonds from donor atom to acceptor.

3. Discussion

Typically, bioactive peptides are delivered via oral administration. Therefore, it is crucial to study their changes during gastrointestinal digestion. These peptides by enzymatic hydrolysis of exogenous proteases may be easily metabolized or further hydrolyzed in the digestive system, which may result in reduced or increased potency, or to a complete loss of activity [43]. In the present study, the digestion in vitro with pepsin, trypsin, and α-chymotrypsin enzymes was applied to simulate the gastrointestinal digestion procedure. After simulated gastrointestinal digestion, the action of endogenous digestive enzymes increased the ACE inhibitory activity of sesame protein as a result of the release of smaller peptide fragments and additive and synergistic biological effects [44]. Compared with in vivo digestion, simulated gastrointestinal digestion in vitro has the advantages of controllable conditions, simple operation, good reproducibility, and cost saving. This method has been widely used to study the digestion and formation of functional peptides of dietary proteins in recent years.

ACE inhibitory tripeptide generally requires a hydrophobic, a positively charged, and an aromatic amino acid residue at the N-terminal, middle, and C-terminal, respectively. Indeed, exact length of the C-terminal region to efficiently inhibit ACE activity remains unknown. Long-chain peptides could not always be speculated from that of their C-terminal tripeptide residues, implying the importance of other positions such as the fourth and so on [45]. García-Mora found that C-terminal heptapeptide was crucial for their antioxidant and ACE inhibitory activities [46].

Additionally, searching against the BIOPEP database (http://www.uwm.edu.pl/biochemia/index.php/pl/biopep) revealed that it is the first time the peptides with potent ACE inhibitory activity extracted from *Sesamum indicum* L were obtained. Therefore, sesame protein would be an attractive raw material generating ACE inhibitory peptides for controlling hypertension. Of course, further studies on the in vivo effects and constructing a three-dimensional quantitative structure–activity relationship (3D-QSAR) model still needs further investigation.

4. Materials and Methods

4.1. Materials and Reagents

White sesame (*Sesamum indicum* L.) seeds were purchased from China Oil and Foodstuffs Corporation (Beijing, China). Angiotensin Converting Enzyme (ACE) from rabbit lung, N-Hippuryl-His-Leu (HHL) hydrate power (≥98%, HPLC), trifluoroacetic acid (TFA), 2,4,6 trinitrobenzene sulfonic acid (TNBS), and O-phthaldialdehyde (OPA), were purchased from Sigma-Aldrich (St. Louis, MO, USA). Pepsin (3000-3500 NFU/g), trypsin (250.N.F.U/mg), and α-chymotrypsin (1200 u/mg) were obtained from Solarbio Science and Technology Co., Ltd. (Beijing, China). Nanofiltration and ultrafiltration membranes were purchased from Rising Sun Membrane Technology Co., Ltd. (Beijing, China). All other reagents used in this study were of analytical grade.

4.2. Extraction of Sesame Protein

The extraction of sesame protein was carried out pursuant to the method demonstrated by Saatchi et al. [47] with several modifications. Sesame seeds were ground in a crusher and were extracted with petroleum ether to remove the fat. The defatted sesame meal was dissolved in distilled water at a solid/liquid (S/L) ratio of 1 g/20 mL and the pH of the suspension was adjusted to 11.0 by using 5 mol/L NaOH. Then it was stirred in a blender for 1 h and centrifuged at 5000 r/min for 20 min at 4 °C. The supernatant was collected, and the residue was repeatedly extracted under the above conditions. Finally, the supernatant was combined and the protein in the supernatant was precipitated by adjusting the pH to 4.3 with 5 mol/L HCL. The precipitate was collected, freeze dried, and stored at −20 °C for further use.

4.3. Preparation of in Vitro Simulated Gastrointestinal Digestion of Sesame Protein

The simulated gastrointestinal digestion process was modified slightly according to the method used by Sangsawad et al. [48]. Briefly, a certain amount of sesame protein was added to distilled water and stirred evenly to form an 8% protein solution. The solution was heated in a water bath at 95 °C for 30 min to denature the protein completely after neutralizing the pH to 7.0. In the simulated gastric digestion phase, the suspension was digested by pepsin (0.4% of substrate, dry basis) at 37 °C for 4 h during which time the pH was adjusted to 2.0 with HCL. During the simulated intestinal digestion phase, trypsin (0.3% of substrate, dry basis) and α-chymotrypsin (0.1% of substrate, dry basis) was added and the reaction was carried out at 37 °C for 6 h, during which time NaOH was used to keep the pH of the digested solution at 7.6. Samples were collected every 30 min during digestion for determination. After 10 h, the digested solution was cooled to room temperature and centrifuged at 5000 r/min for 20 min. The supernatant was collected and nanofiltrated through a flat-membrane test cell C60F (Nitto Denko Co., Ltd., Osaka, Japan) to remove the salt [29]. Nanofiltration membrane NF1 (MWCO of 150 Da) was put into the test cell. The 100 mL of the supernatant and 100 mL distilled water was poured into the test cell and then the nanofiltration began at pressure of 3.0 MPa and stirring speed of 400 rpm. When the volume of permeate reached 100 mL, the first nanofiltration terminated. Nanofiltration was repeated thrice. The retentate was collected as the final SPDS and stored at 4 °C for further separation and purification.

4.4. Separation of ACE Inhibitory Peptide from SPDS by Ultrafiltration

The ACE inhibitory peptides from SPDS were separated by using MWCO of 100, 50, 30, 10, 5, and 3 kDa ultrafiltration membranes in a stirred cell C60F. The 250 mL of SPDS was poured into the test cell and then the ultrafiltration began at room temperature. The pressure was 2.5 MPa and the stirring speed was 400 rpm. When the permeate reached 200 mL, the first step ultrafiltration ended. Then the permeate was treated for the next step ultrafiltration under the above conditions. Each retentate was collected separately and seven fractions were generated, named SPDS-I (molecular weight over 100 kDa), SPDS-II (molecular weight between 50 and 100 kDa), SPDS-III (molecular weight between 30 and 50 kDa), SPDS-IV (molecular weight between 10 and 30 kDa), SPDS-V (molecular weight between 5 and 10 kDa), SPDS-VI (molecular weight between 3 and 5 kDa), and SPDS-VII (molecular weight below 3 kDa). These fractions were evaporated and lyophilized for subsequent ACE inhibitory study.

4.5. Purification of ACE Inhibitory Peptide by NGC Quest™ 10 Plus Chromatography System

The screened fraction with the strongest ACE inhibitory activity was purified by NGC Quest™ 10 Plus Chromatography System. First, 0.5 mL of the sample was filtered through a 0.22 μm membrane and injected into the system (Bio-Rad Laboratories, Hercules, CA, USA) with a Shim-pack GIS C18 preparative column (20 cm × 250 mm I.D., 10 μm, Shimadzu, Kyoto, Japan). The experiment was executed by using mobile phase A (ultrapure water containing 0.1% (*v/v*) TFA) and mobile phase B (methanol containing 0.1% (*v/v*) TFA). The sample was separated by a linear gradient elution from 55% to 60% of mobile phase A at a flow rate of 7.5 mL/min. The absorbance was monitored at 220 nm with a UV detector. Each chromatographic peak was pooled by repeated injection with corresponding test tube respectively through the fraction collector and lyophilized to determine the ACE inhibitory activity.

4.6. Identification of the Purified ACE Inhibitory Peptide by Nano UHPLC-ESI-MS/MS

The eluted peak composition with the strongest ACE inhibitory activity was firstly performed by Thermo Ultimate 3000 UHPLC (Thermo Fisher Scientific,). The sample was enriched and salt-removed in a trap column, and then was injected onto the C_{18} column (25 cm × 75 μm I.D., 3 μm). The mobile phase A was 2% acetonitrile, 0.1% formic acid, and the mobile phase B was 98% acetonitrile, 0.1% formic acid. The separation was carried out at a flow rate of 300 NL/min through the following gradient elution: 0–5 min, 5% mobile phase B; 5–45 min, mobile phase B linearly increased from 5% to 25%; 45–50 min, mobile phase B increased from 25% to 35%; 50–52 min, mobile phase B increased from 35% to 80%; 52–54 min, 80% mobile phase B; and 54–60 min, 5% mobile phase B.

Then mass analysis was performed on a Thermo Q-Exactive HF tandem mass spectrometer (Thermo Fisher Scientific, San Jose, CA, USA) after nano ESI ionization. The ion source voltage was set to 1.6 kV. The full scan range of mass spectrometry was 350–1600 m/z and the resolution was set to 60,000. The initial m/z of the secondary mass spectrometry was fixed at 100 and the resolution was set to 15,000. The ion fragmentation mode of high energy collisional dissociation (HCD) was set and the fragment ions were detected in Orbitrap, which was an ultrahigh field mass analyzer that could provide higher resolution and faster scanning speed. Databases including UniProt protein database, genomic annotation-based protein database, such as NCBI were selected for protein identification. The Mascot v2.3.02 software ((Matrix Sciences, London, UK) was used to process the MS/MS data for analyzing the peptide sequences.

4.7. Peptide Match and Peptide Synthesis

The protein names of the peptides were obtained from the Protein Information Resource (PIR) database (https://research.bioinformatics.udel.edu/peptidematch/index.jsp). Peptide match service is one of the most popular tools on the PIR website that plays an important role in locating occurrences of a specific peptide [49]. Meanwhile, in order to certify the ACE inhibitory activity of the purified peptides identified by mass spectroscopy, the selected peptides were chemically synthesized by GL Biochem Ltd. (Shanghai, China). The purity of the synthesized peptides reached 95% and was also verified by HPLC coupled with ESI-MS. These peptides were then dissolved in distilled water and assessed for ACE inhibitory activity.

4.8. Analysis of Degree of Hydrolysis

The degree of hydrolysis (DH) was determined according to the method of Adler-Nissen [50] with some modifications. Briefly, each sample at various points in time of simulated gastrointestinal digestion was appropriately diluted (125 µL) and then were mixed with phosphate buffer (1 mL, pH 8.2) and 1 mL of 0.1% TNBS reagent. The mixture was incubated at 50 °C for 60 min in the dark. At last, 2 mL of 0.1 mol/L HCL was used for the termination reaction and the samples were left at room temperature for 30 min. The standard curve was made with L-leucine and the absorbance of the reaction mixture was measured at 340 nm with a UV-visible spectrophotometer UV-3600Plus (Shimadzu Corporation, Kyoto, Japan). The degree of hydrolysis was determined from the following equation:

$$\text{DH (\%)} = h/h_{tot} \times 100 \tag{1}$$

where h means concentration of peptide bond hydrolyzed (mmol/g) and h_{tot} means total amount of the peptide bond (8 mmol/g) [51].

4.9. Determination of Protein and Peptide Content

The protein content before simulated gastrointestinal digestion was analyzed by Kjeldahl method as described by Marcó et al. [52]. The peptide content of each sample at various points in time of simulated gastrointestinal digestion was analyzed by using OPA spectrophotometric assay described by Agrawal et al. [53]. First, 40 mg of OPA was dissolved in 1 mL of methanol and mixed with 25 mL of 100 mmol/L sodium tetrahydroborate, 2.5 mL of 20% (*w/w*) sodium dodecyl sulfate, and 100 µL of β-mercaptoethanol. Distilled water was added to bring the total volume of the mixed solution up to 50 mL. Then the freshly prepared OPA solution was obtained. Subsequently, 100 µL of peptide solution was mixed with 2 mL of OPA solution. After the incubation for 2 min at room temperature, the absorbance of the reaction mixture was recorded spectrophotometrically at 340 nm by using a UV-visible spectrophotometer UV-3600Plus. Casein tryptone was used as a standard for quantification of peptide content. The peptide yield was determined by the ratio of peptide mass to protein mass.

4.10. Assay of the ACE Inhibitory Activity

The ACE inhibitory activity was measured by using cary eclipse fluorescence spectrophotometer (Agilent Technologies Australia Pty Ltd., Mulgrave, Australia) according to a modified method of Li et al. [54]. The sample was dissolved by distilled water to the appropriate concentration and the 96-well plate was used as the reaction vessel. Briefly, 30 µL of HHL substrate containing 0.3 mol/L

NaCl in 50 mM borate buffer (pH 8.3) and 15 μL of samples were mixed. Then the reaction was started by adding 30 μL of ACE (12.5 mU/mL) to the mixture at 37 °C for 30 min. The reaction was stopped by using 125 μL of 1.2 mol/L NaOH. Finally, 30 μL of methanol solution containing 2% OPA was added, mixed evenly, and then stood at room temperature for 20 min. The derivative reaction was terminated by adding 40 μL of 6 mol/L HCl. The reaction solution was diluted 10 times and added into the fluorescence cell. The fluorescence absorption intensity was measured within 30–90 min. The determination conditions were as follows: the excitation wavelength was 340 nm, the emission wavelength was 455 nm, and the slit was 5 nm wide. Furthermore, the half inhibitory concentration (IC_{50}) values, which meant the amount of inhibitor required to inactivate 50% of ACE activity under the experimental conditions were calculated from regression lines of a plot of % inhibition versus concentrations by using SPSS 13.0 software. All assays were performed in triplicate. ACE inhibitory activity was calculated by using the following equation:

$$\text{ACE inhibitory activity (\%)} = [1 - (a-c)/(b-d)] \times 100 \tag{2}$$

where a is the fluorescence absorption intensity in the presence of both ACE and inhibitor, b is the fluorescence absorption intensity in the presence of ACE but not inhibitor, c is the fluorescence absorption intensity in the presence of inhibitor but not ACE, d is the fluorescence absorption intensity in the absence of both ACE and inhibitor.

4.11. Molecular Docking Simulation between Peptide and ACE

The molecular docking was carried out by using Surflex-Dock module that had an empirical scoring function and a patented searching engine [55,56] in Sybyl-X 2.1.1 software (Tripos, Inc., St. Louis, MO, USA). The 3D structure of the selected peptide was constructed, of which the atomic charge was calculated in MMFF94. Furthermore, the structure was energy minimized by using the Powell conjugate gradient optimization algorithm with a convergence criterion of 0.0005 kcal/mol Å and the maximum iterations of 10,000. The crystal structure of human ACE-lisinopril complex (1O86.pdb) was derived from the Protein Data Bank (http://www.rcsb.org/pdb/home/home.do). To prepare the structure of ACE, all substrates, chloride ions, and water molecules were excluded in ACE model. Then the polar hydrogens were added to the ACE model. The number of poses per ligand was set to 20 to perform the molecular docking and the conformations for each peptide based on Total scores of Surflex-Dock were ranked [38]. The conformations of ranked No. 1 for each peptide were selected and the other four kinds of docking score, D_score, G_score, PMF_score, and Chem score were estimated together using the C score module [57]. The interaction model between the ACE residues and peptides was analyzed by Molecular Operating Environments (MOE; 2015.10 Chemical Computing Groups, Montreal, Quebec, Canada). The ACE–lisinopril complex (1O86.pdb) was used as a template. In the docking process, semiflexible docking mode was adopted to treat the receptor and ligand, which meant that the conformation of receptor was rigid and that of ligand was flexible.

4.12. Statistical Analysis

All analyses were performed in triplicate. Data was expressed as the mean ± standard deviation and performed by one-way analysis of variation by using SPSS 13.0 for Windows (Chicago, IL, USA) with Least Significant Difference (LSD) multiple range tests. The significant difference level was set at $p < 0.05$.

5. Conclusions

In conclusion, this study isolated and purified new ACE inhibitory peptides from sesame protein by using two stages of simulated gastrointestinal digestion in vitro. The DH, peptide yield, and ACE inhibition at different time points were determined during simulated gastrointestinal digestion, which demonstrated that the DH, peptide yield increased with the increase of digest time. The action of endogenous digestive enzymes increased the ACE inhibitory activity after simulated gastrointestinal digestion. SPDS was purified by ultrafiltration and SPDS-VII (<3 kDa) had strongest ACE inhibition. SPDS-VII was further purified by NGC Quest™ 10 Plus Chromatography System and finally 11 major peptides were identified by Nano UHPLC-ESI-MS/MS from peak 4. The peptide GHIITVAR derived from 11S globulin exhibited superior ACE inhibitory activity with the IC_{50} value of (3.60 ± 0.10) μM among the peptides. Molecular docking results suggested that the ACE inhibition of GHIITVAR was mainly attributed to forming very strong hydrogen bonds with the S1 pocket (Ala354 and Tyr523), the S2 pocket (Gln281, His353, Tyr520, Lys511) and the S1′ pocket (Glu162) after docking. It also bound to the active sites using a metal ion interaction (Zn701) and a hydrogen bond interaction with His383. Based on these results, sesame protein can be a promising ACE inhibitor in the prevention and treatment of hypertension. In addition, GHIITVAR can be used as a potential nutraceutical for development of functional foods.

Supplementary Materials: Supplementary materials can be found at http://www.mdpi.com/1422-0067/21/3/1059/s1.

Author Contributions: Conceptualization: R.W., J.H.; data curation: R.W., X.L.; investigation: J.G., L.M.; formal analysis: R.W.; software: R.W., X.L.; methodology: R.W.; resources: Q.S.; supervision: Q.S.; funding acquisition: J.H.; project administration: J.H.; validation: J.G., L.M.; writing—original draft preparation: R.W.; writing—review and editing: X.L., Q.S. All authors have read and agreed to the published version of the manuscript.

Funding: This research was funded by Leading Talents of Science and Technology Innovation in zhongyuan (No. 194200510014) and National technology system for special oilseed industry (Grant No. CARS-14).

Acknowledgments: We thank Guohui Song, Lixia Zhang, Cong Jia, and Songli Wei for their technical support during the course of this work.

Conflicts of Interest: The authors declare no conflict of interest.

Abbreviations

ACE	Angiotensin I-converting enzyme
Nano UHPLC-ESI-MS/MS	Nano ultra-high performance liquid chromatography-electrospray ionization mass spectrometry/mass spectrometry
SPD	Sesame protein digestive solution
DH	Degree of hydrolysis
MWCO	Molecular weight cut-off
RAS	Renin-angiotensin system
KKS	Kallikrein kinin system
MOE	Molecular Operating Environments

References

1. Vokonas, P.S.; Kannel, W.B.; Cupples, L.A. Epidemiology and risk of hypertension in the elderly: The Framingham-Study. *J. Hypertens Suppl.* **1988**, *6*, S3–S9.
2. Global Burden of Metabolic Risk Factors for Chronic Diseases Collaboration. Cardiovascular disease, chronic kidney disease, and diabetes mortality burden of cardiometabolic risk factors from 1980 to 2010: A comparative risk assessment. *Lancet Diabetes Endocrinol.* **2014**, *2*, 634–647. [CrossRef]
3. Lewington, S.; Clarke, R.; Qizilbash, N.; Peto, R.; Collins, R. Prospective Studies Collaboration. Age-specific relevance of usual blood pressure to vascular mortality: A meta-analysis of individual data for one million adults in 61 prospective studies. *Lancet* **2002**, *360*, 1903–1913. [PubMed]
4. Asoodeh, A.; Haghighi, L.; Chamani, J.; Ansari-Ogholbeyk, M.A.; Mojallal-Tabatabaei, Z.; Lagzian, M. Potential angiotensin I converting enzyme inhibitory peptides from gluten hydrolysate: Biochemical characterization and molecular docking study. *J. Cereal Sci.* **2014**, *60*, 92–98. [CrossRef]
5. López-Barrios, L.; Gutiérrez-Uribe, J.A.; Serna-Saldívar, S.O. Bioactive peptides and hydrolysates from pulses and their potential use as functional ingredients. *J. Food Sci.* **2014**, *79*, 273–283. [CrossRef] [PubMed]
6. Cheung, I.W.; Nakayama, S.; Hsu, M.N.; Samaranayaka, A.G.; Li-Chan, E.C. Angiotensin-I converting enzyme inhibitory activity of hydrolysates from oat (*Avena Sativa*) proteins by in silico and in vitro analyses. *J Agric Food Chem.* **2009**, *57*, 9234–9242. [CrossRef] [PubMed]

7. Gu, Y.; Wu, J. LC-MS/MS coupled with QSAR modeling in characterising of angiotensin I-converting enzyme inhibitory peptides from soybean proteins. *Food Chem.* **2013**, *141*, 2682–2690. [CrossRef]

8. Korhonen, H.; Pihlanto, A. Bioactive peptides: Production and functionality. *Int. Dairy J.* **2006**, *16*, 945–960. [CrossRef]

9. Moller, N.P.; Scholz-Ahrens, K.E.; Roos, N.; Schrezenmeir, J. Bioactive peptides and protein from foods: Indication for health effects. *Eur J. Nutr.* **2008**, *47*, 171–182. [CrossRef]

10. Sarbon, N.M.; Badii, F.; Howell, N.K. Purification and characterization of antioxidative peptides derived from chicken skin gelatin hydrolysate. *Food Hydrocoll.* **2018**, *85*, 311–320. [CrossRef]

11. Sun, Y.; Chang, R.; Li, Q.; Li, B. Isolation and characterization of an antibacterial peptide from protein hydrolysates of *Spirulina platensis. Eur Food Res Technol.* **2016**, *242*, 685–692. [CrossRef]

12. Nongonierma, A.B.; FitzGerald, R.J. Inhibition of dipeptidyl peptidase IV (DPP-IV) by proline containing casein-derived peptides. *J. Funct Foods.* **2013**, *5*, 1909–1917.

13. Paiva, L.; Lima, E.; Neto, A.I.; Baptista, J. Isolation and characterization of angiotensin I-converting enzyme (ACE) inhibitory peptides from *Ulva rigida* C. Agardh protein hydrolysate. *J. Funct Foods.* **2016**, *26*, 65–76. [CrossRef]

14. Rani, S.; Pooja, K.; Pal, G.K. Exploration of potential angiotensin converting enzyme inhibitory peptides generated from enzymatic hydrolysis of goat milk proteins. *Biocatal. Agric. Biotechnol.* **2017**, *11*, 83–88. [CrossRef]

15. Tanzadehpanah, H.; Asoodeh, A.; Saberi, M.R.; Chamani, J. Identification of a novel angiotensin-I converting enzyme inhibitory peptide from ostrich egg white and studying its interactions with the enzyme. *Innov. Food Sci. Emerg. Technol.* **2013**, *18*, 212–219. [CrossRef]

16. He, H.L.; Liu, D.; Ma, C.B. Review on the angiotensin-I-converting enzyme (ACE) inhibitor peptides from marine proteins. *Appl. Biochem. Biotechnol.* **2013**, *169*, 738–749. [CrossRef]

17. Anilakumar, K.R.; Pal, A.; Khanum, F.; Bawa, A.S. Nutritional, medicinal and industrial uses of sesame (*Sesamum indicum* L.) seeds—An overview. *Agric. Conspec. Sci.* **2010**, *75*, 159–168.

18. Al-Bachir, M. Some microbial, chemical and sensorial properties of gamma irradiated sesame (*Sesamum indicum* L.) seeds. *Food Chem.* **2016**, *197*, 191–197. [CrossRef]

19. Nakano, D.; Ogura, K.; Miyakoshi, M.; Ishii, F.; Kawanishi, H.; Kurumazuka, D.; Kwak, C.J.; Ikemura, K.; Takaoka, M.; Moriguchi, S.; et al. Antihypertensive effect of angiotensin I-converting enzyme inhibitory peptides from a sesame protein hydrolysate in spontaneously hypertensive rats. *Biosci. Biotechnol. Biochem.* **2006**, *70*, 1118–1126. [CrossRef]

20. Liu, W.P.; Cheng, G.Y.; Liu, H.M.; Kong, Y. Purification and identification of a novel angiotensin I-converting enzyme inhibitory peptide from sesame meal. *Int. J. Pept. Res. Ther.* **2015**, *21*, 433–442. [CrossRef]

21. Jiang, S.; Xia, D.; Zhang, D.; Chen, G.; Liu, Y. Analysis of protein profiles and peptides during in vitro gastrointestinal digestion of four Chinese dry-cured hams. *LWT-Food Sci. Technol.* **2020**, *120*, 108881. [CrossRef]

22. Fan, J.; He, J.T.; Zhuang, Y.L.; Sun, L.P. Purification and identification of antioxidant peptides from enzymatic hydrolysates of Tilapia (*Oreochromis niloticus*) frame protein. *Molecules* **2012**, *17*, 12836–12850. [CrossRef] [PubMed]

23. Mirzaei, M.; Mirdamadi, S.; Ehsani, M.R.; Aminlari, M. Production of antioxidant and ACE-inhibitory peptides from *Kluyveromyces marxianus* protein hydrolysates: Purification and molecular docking. *J. Food Drug Anal.* **2018**, *26*, 696–705. [CrossRef] [PubMed]

24. Zhu, Z.; Yuan, F.; Xu, Z.; Wang, W.; Di, X.; Barba, F.J.; Shen, W.; Koubaa, M. Stirring-assisted dead-end ultrafiltration for protein and polyphenol recovery from purple sweet potato juices: Filtration behavior investigation and HPLC-DAD-ESI-MS^2 profiling. *Sep. Purif. Technol.* **2016**, *169*, 25–32. [CrossRef]

25. Boschin, G.; Scigliuolo, G.M.; Resta, D.; Arnoldi, A. ACE-inhibitory activity of enzymatic protein hydrolysates from Lupin and other Legumes. *Food Chem.* **2014**, *145*, 34–40. [CrossRef]

26. Fu, W.; Chen, C.; Zeng, H.; Lin, J.; Zhang, Y.; Hu, J.; Zheng, B. Novel angiotensin-converting enzyme inhibitory peptides derived from Trichiurus lepturus myosin: Molecular docking and surface plasmon resonance study. *LWT-Food Sci. Technol.* **2019**, *110*, 54–63. [CrossRef]

27. Lin, K.; Zhang, L.; Han, X.; Cheng, D. Novel angiotensin I-converting enzyme inhibitory peptides from protease hydrolysates of Qula casein: Quantitative structure-activity relationship modeling and molecular docking study. *J. Funct. Foods.* **2017**, *32*, 266–277. [CrossRef]

28. Achouri, A.; Nail, V.; Boye, J.I. Sesame protein isolate: Fractionation; secondary structure and functional properties. *Food Res. Int.* **2012**, *46*, 360–369. [CrossRef]

29. Lu, X.; Zhang, L.; Sun, Q.; Song, G. Extraction; identification and structure-activity relationship of antioxidant peptides from sesame (*Sesamum indicum* L.) protein hydrolysate. *Food Res. Int.* **2019**, *116*, 707–716. [CrossRef]

30. Salampessy, J.; Reddy, N.; Phillips, M.; Kailasapathy, K. Isolation and characterization of nutraceutically potential ACE-Inhibitory peptides from leatherjacket (*Meuchenia* sp.) protein hydrolysates. *LWT-Food Sci. Technol.* **2017**, *80*, 430–436. [CrossRef]

31. Ahn, C.-B.; Jeon, Y.-J.; Kim, Y.-T.; Je, J.-Y. Angiotensin I converting enzyme (ACE) inhibitory peptides from salmon byproduct protein hydrolysate by alcalase hydrolysis. *Process Biochem.* **2012**, *47*, 2240–2245. [CrossRef]

32. Toopcham, T.; Mes, J.J.; Wichers, H.J.; Roytrakul, S.; Yongsawatdigul, J. Bioavailability of angiotensin I-converting enzyme (ACE) inhibitory peptides derived from *Virgibacillus halodenitrificans* SK1–3-7 proteinases hydrolyzed tilapia muscle proteins. *Food Chem.* **2017**, *220*, 190–197. [CrossRef] [PubMed]

33. Ryan, J.T.; Ross, R.P.; Bolton, D.; Fitzgerald, G.F.; Stanton, C. Bioactive peptides from muscle sources: Meat and fish. *Nutrients* **2011**, *3*, 765–791. [CrossRef] [PubMed]

34. Guo, Y.; Jiang, X.; Xiong, B.; Zhang, T.; Zeng, X.; Wu, Z.; Sun, Y.; Pan, D. Production and transepithelial transportation of angiotensin-I-converting enzyme (ACE)-inhibitory peptides from whey protein hydrolyzed by immobilized *Lactobacillus helveticus* proteinase. *J. Dairy Sci.* **2019**, *102*, 961–975. [CrossRef]

35. Forghani, B.; Zarei, M.; Ebrahimpour, A.; Philip, R.; Bakar, J.; Hamid, AA.; Saari, N. Purification and characterization of angiotensin converting enzyme-inhibitory peptides derived from *Stichopus horrens*: Stability study against the ACE and inhibition kinetics. *J. Funct. Foods* **2016**, *20*, 276–290. [CrossRef]

36. Ko, S.-C.; Jang, J.; Ye, B.-R.; Kim, M.-S.; Choi, I.-W.; Park, W.-S.; Heo, S.-J.; Jung, W.-K. Purification and molecular docking study of angiotensin I-converting enzyme (ACE) inhibitory peptides from hydrolysates of marine sponge *Stylotella aurantium*. *Process Biochem.* **2017**, *54*, 180–187. [CrossRef]

37. Salampessy, J.; Reddy, N.; Kailasapathy, K.; Phillips, M. Functional and potential therapeutic ACE-inhibitory peptides derived from bromelain hydrolysis of trevally proteins. *J. Funct. Foods.* **2015**, *14*, 716–725. [CrossRef]

38. Xie, C.L.; Choung, S.-Y.; Cao, G.-P.; Lee, K.W.; Choi, Y.J. In silico investigation of action mechanism of four novel angiotensin-I converting enzyme inhibitory peptides modified with Trp. *J. Funct. Foods* **2015**, *17*, 632–639. [CrossRef]

39. Wang, X.; Chen, H.; Fu, X.; Li, S.; Wei, J. A novel antioxidant and ACE inhibitory peptide from rice bran protein: Biochemical characterization and molecular docking study. *LWT-Food Sci. Technol.* **2017**, *75*, 93–99. [CrossRef]

40. Mirzaei, M.; Mirdamadi, S.; Safavi, M.; Hadizadeh, M. In vitro and in silico studies of novel synthetic ACE-inhibitory peptides derived from *Saccharomyces cerevisiae* protein hydrolysate. *Bioorg. Chem.* **2019**, *87*, 647–654. [CrossRef]

41. Abdelhedi, O.; Nasri, R.; Jridi, M.; Mora, L.; Oseguera-Toledo, M.E.; Aristoy, M.-C.; Amara, I.B.; Toldrá, F.; Nasri, M. In silico analysis and antihypertensive effect of ACE-inhibitory peptides from smooth-hound viscera protein hydrolysate: Enzyme-peptide interaction study using molecular docking simulation. *Process Biochem.* **2017**, *58*, 145–159. [CrossRef]

42. Wang, C.; Tu, M.; Wu, D.; Chen, H.; Chen, C.; Wang, Z.; Jiang, L. Identification of an ACE-Inhibitory peptide from walnut protein and its evaluation of the inhibitory mechanism. *Int. J. Mol. Sci.* **2018**, *19*, 1156. [CrossRef]

43. Priyanto, A.D.; Doerksen, R.J.; Chang, C.I.; Sung, W.C.; Widjanarko, S.B.; Kusnadi, J.; Lin, Y.C.; Wang, T.C.; Hsu, J.L. Screening, discovery, and characterization of angiotensin-I converting enzyme inhibitory peptides derived from proteolytic hydrolysate of bitter melon seed proteins. *J. Proteomics.* **2015**, *128*, 424–435. [CrossRef] [PubMed]

44. Garcia-Mora, P.; Peñas, E.; Frias, J.; Gomez, R.; Martinez-Villaluenga, C. High-pressure improves enzymatic proteolysis and the release of peptides with angiotensin I converting enzyme inhibitory and antioxidant activities from lentil proteins. *Food Chem.* **2015**, *171*, 224–232. [CrossRef]

45. Fan, H.; Wang, J.; Liao, W.; Jiang, X.; Wu, J. Identification and characterization of gastrointestinal-resistant angiotensin-converting enzyme inhibitory peptides from egg white proteins. *J. Agric. Food Chem.* **2019**, *67*, 7147–7156. [CrossRef]

46. García-Mora, P.; Martín-Martínez, M.; Bonache, M.A.; González-Múniz, R.; Peñas, E.; Frias, J.; Martinez-Villaluenga, C. Identification, functional gastrointestinal stability and molecular docking studies of lentil peptides with dual antioxidant and Angiotensin I converting enzyme inhibitory activities. *Food chem.* **2017**, *221*, 464–472. [CrossRef]

47. Saatchi, A.; Kiani, H.; Labbafi, M. A new functional protein-polysaccharide conjugate based on protein concentrate from sesame processing by-products: Functional and physico-chemical properties. *Int. J. Biol Macromol.* **2019**, *122*, 659–666. [CrossRef]

48. Sangsawad, P.; Roytrakul, S.; Yongsawatdigul, J. Angiotensin converting enzyme (ACE) inhibitory peptides derived from the simulated in vitro gastrointestinal digestion of cooked chicken breast. *J. Funct. Foods* **2017**, *29*, 77–83. [CrossRef]

49. Chen, C.; Li, Z.; Huang, H.; Suzek, B.E.; Wu, C.H.; UniProt Consortium. A fast Peptide Match service for UniProt Knowledgebase. *Bioinformatics* **2013**, *29*, 2808–2809. [CrossRef]

50. Adler-Nissen, J. Determination of the degree of hydrolysis of food protein hydrolysates by trinitrobenzene sulfonic acid. *J. Agric. Food Chem.* **1979**, *27*, 1256–1262. [CrossRef]

51. Chatterjee, R.; Dey, T.K.; Ghosh, M.; Dhar, P. Enzymatic modification of sesame seed protein, sourced from waste resource for nutraceutical application. *Food Bioprod. Process.* **2015**, *94*, 70–81. [CrossRef]

52. Marcó, A.; Rubio, R.; Compañó, R.; Casals, I. Comparison of the Kjeldahl method and a combustion method for total nitrogen determination in animal feed. *Talanta* **2002**, *57*, 1019–1026. [CrossRef]

53. Agrawal, H.; Joshi, R.; Gupta, M. Isolation; purification and characterization of antioxidative peptide of pearl millet (*Pennisetum glaucum*) protein hydrolysate. *Food Chem.* **2016**, *204*, 365–372. [CrossRef] [PubMed]

54. Li, F.; Yin, L.J.; Lu, X.; Li, L.T. Changes in angiotensin I-converting enzyme inhibitory activities during the ripening of douchi (a Chinese traditional soybean product) fermented by various starter cultures. *Int. J. Food Prop.* **2010**, *13*, 512–524. [CrossRef]

55. Jain, A.N. Surflex: Fully automatic flexible molecular docking using a molecular similarity-based search engine. *J. Med. Chem.* **2003**, *46*, 499–511. [CrossRef]

56. Jain, A.N. Scoring noncovalent protein-ligand interactions: A continuous differentiable function tuned to compute binding affinities. *J. Comput. Aided Mol. Design* **1996**, *10*, 427–440. [CrossRef]

57. Vaijanathappa, J.; Puttaswamygowda, J.; Bevanhalli, R.; Dixit, S.; Prabhakaran, P. Molecular docking, antiproliferative and anticonvulsant activities of swertiamarin isolated from *Enicostemma axillare*. *Bioorg. Chem.* **2020**, *94*, 103428. [CrossRef]

International Journal of
Molecular Sciences

Article

Structural Analysis and Design of Chionodracine-Derived Peptides Using Circular Dichroism and Molecular Dynamics Simulations

Stefano Borocci [1,2], Giulia Della Pelle [1], Francesca Ceccacci [3], Cristina Olivieri [4], Francesco Buonocore [1] and Fernando Porcelli [1,*]

[1] Department for Innovation in Biological, Agrofood and Forest Systems, University of Tuscia, 01100 Viterbo, Italy; borocci@unitus.it (S.B.); giulia.dellapelle@studenti.unitus.it (G.D.P.); fbuono@unitus.it (F.B.)
[2] CNR—Institute for Biological Systems, Via Salaria, Km 29.500, 00015 Monterotondo, 00015 Rome, Italy
[3] CNR—Institute for Biological Systems, Sede Secondaria di Roma-Meccanismi di Reazione, 00185 Rome, Italy; francesca.ceccacci@cnr.it
[4] Department of Biochemistry, Molecular Biology and Biophysics, University of Minnesota, Minneapolis, MN 55455, USA; colivier@umn.edu
* Correspondence: porcelli@unitus.it; Tel.: +39-0761-357041

Received: 31 December 2019; Accepted: 17 February 2020; Published: 19 February 2020

Abstract: Antimicrobial peptides have been identified as one of the alternatives to the extensive use of common antibiotics as they show a broad spectrum of activity against human pathogens. Among these is Chionodracine (*Cnd*), a host-defense peptide isolated from the Antarctic icefish *Chionodraco hamatus*, which belongs to the family of Piscidins. Previously, we demonstrated that *Cnd* and its analogs display high antimicrobial activity against ESKAPE pathogens (*Enterococcus faecium*, *Staphylococcus aureus*, *Klebsiella pneumoniae*, *Acinetobacter baumannii*, *Pseudomonas aeruginosa* and *Enterobacter* species). Herein, we investigate the interactions with lipid membranes of *Cnd* and two analogs, *Cnd-m3* and *Cnd-m3a*, showing enhanced potency. Using a combination of Circular Dichroism, fluorescence spectroscopy, and all-atom Molecular Dynamics (MD) simulations, we determined the structural basis for the different activity among these peptides. We show that all peptides are predominantly unstructured in water and fold, preferentially as α-helices, in the presence of lipid vesicles of various compositions. Through a series of MD simulations of 400 ns time scale, we show the effect of mutations on the structure and lipid interactions of *Cnd* and its analogs. By explaining the structural basis for the activity of these analogs, our findings provide structural templates to design minimalistic peptides for therapeutics.

Keywords: chionodracines; antimicrobial peptides; circular dichroism; molecular dynamics; peptide-membrane interaction

---

1. Introduction

Antimicrobial resistance towards widely used antibiotics has become a global health crisis [1]. The diffusion of multidrug resistant (MDR) bacteria is widespread and infections caused by human pathogens often require multiple treatment, which not always effective. The United States Food and Drug Administration listed 21 qualifying pathogens that pose serious threats to public health (part 317.2 of CFR Annual Print Title 21 Food and Drugs) [2]. The search for new antimicrobial drugs to replace or integrate classical antibiotics is imperative. Many different alternatives have been proposed as potential drugs for the treatment of resistant bacterial strains such as antibodies, vaccines [3], bacteriophages [4,5], and antimicrobial peptides [6,7]. Antimicrobial peptides (AMP) represent the first line of innate immune system and are present in plants, microorganisms, and animals [8,9]. The Antimicrobial Peptide

Database (APD: aps.unmc.edu/AP) comprises about 3100 entries from six kingdoms (bacteria, archaea, protists, fungi, plants, and animals) with antibacterial, antiviral, antifungal, and antiparasitic activity. Despite their different primary sequences and structures, AMPs share common properties such as cationicity, charge distribution, hydrophobicity, and amphipathicity [10,11]. Amphipathic cationic α-helical peptides have been proposed as potential antimicrobial agents due to their broad-spectrum and instantaneous antimicrobial activity [12–16]. However, a clear relationship between amino acid composition and antimicrobial activity still remains to be elucidated.

Antimicrobial peptides perform their function through several mechanisms such as inhibition of protein and/or DNA synthesis or inhibition of specific enzymatic pathways [17]. However, the majority of AMPs interacts with the anionic bacterial membranes and kills bacteria via cell membrane disruption [18,19]. The amphipathic character of AMP drives the interaction with the bacterial cell membrane and the subsequent disruption that kills the target cells rapidly. Two main mechanisms have been proposed: (a) pore formation and (b) carpet-like model of aggregation. The latter involves essentially four major mechanistic steps: attraction, attachment, insertion, and permeation [20–23]. The first two steps, attraction of peptide toward the cell membrane and attachment, are driven by electrostatic interactions between the positively charged AMP and the negatively charged membrane of bacteria. The two succeeding steps, insertion and permeation, are driven by the amphipathic distribution of hydrophobic and hydrophilic residues on peptide [24,25].

Chionodracine (*Cnd*) is a 22-residue antimicrobial peptide (AMP) isolated from the gills of the Antarctic fish *Chionodraco hamatus* belonging to Piscidins [26], a family of α-helical antimicrobial peptides identified in teleost fish [27]. In previous studies, we isolated *Cnd* [28] and showed that this short peptide binds lipid vesicles, disrupting the outer membrane of *E.coli* and *Psycrobater* sp. Upon membrane interactions, *Cnd* folds into a dynamic α-helix [29]. Based on our findings, we exploited *Cnd* as starting scaffold to design a series of mutants with higher positive charge to increase their antimicrobial activity and selectivity towards prokaryotic organisms. For drug design purposes, an ideal antimicrobial peptide should have maximal antimicrobial activity and minimal cytotoxicity towards eukaryotic cells. We then characterized the interactions of these peptides with model lipid vesicles and showed that they are active against MDR nosocomial bacteria strains [30,31].

In this paper, we studied more in details the structural basis for the differences in the bactericidal activity between *Cnd* and the two promising *Cnd* analogs, *Cnd-m3* and *Cnd-m3a* reported in our previous studies [30,31]. Specifically, we first determined the secondary structure of *Cnd-m3* and *Cnd-m3a* in presence of different LUVs and, afterwards, we performed all-atom MD simulations to elucidate the complex mechanism of peptide-lipid interactions and the associated conformational changes. These data could be of interest to identify the mutations useful to design new analogs with enhanced biologic activity against human bacterial pathogens.

2. Results

2.1. Peptide Design and Physico-Chemical Properties

Cnd-m3 and *Cnd-m3a* are synthetic antimicrobial peptides designed by modifying the parent peptide *Cnd* (Table 1) [30,31] found in the gills of the Antarctic teleost fish *Chionodraco hamatus*.

Table 1. Sequence and physicochemical characteristics of chionodracines.

Peptide	Sequence	Net Charge	pI	(μH)
Cnd	WFGHLYRGITSVVKHVHGLLSG	+2	9.99	0.574
Cnd-m3	WFGKLYRGITKVVKKVKGLLKG	+7	10.58	0.684
Cnd-m3a	WFGKLYRGKTKVVKKVKGLLKG	+8	10.82	0.564

The number of positive charges is critical for the activity of AMP due to the interaction with the negatively charged membranes of bacteria [32]. We modified the parent peptide by introducing charged residues (i.e., Lysines) to increase its net positive charge. For *Cnd-m3*, Ser-11, Ser-22, and three Histidines in position 4, 15 and 17, respectively, were replaced by Lysines. The net positive charge increased from +2 to +7 compared to *Cnd*. For the *Cnd-m3a* mutant, Ile-9, positioned in the middle of the hydrophobic face of peptide, was replaced by Lysine, resulting in a net positive charge of +8. The helical representation of *Cnd* and its analogs is reported in Figure 1. The mutations increased the electrostatic repulsions between charged residues ($i -> i + 3$ and $i -> i + 4$) on the polar face of the putative helical structure but had no effect on the nonpolar face. *Cnd* and *Cnd-m3* displayed 11 hydrophobic interactions. The thermodynamic characterization of the interaction between peptides and vesicles made 100% of POPC and of POPC/POPG (molar ratio 70:30) was carried out and the data show that both peptides display a marked preference for lipid mixtures mimicking the prokaryotic cell membranes (POPC/POPG) [30,31]. Moreover, we showed that the antimicrobial activity of *Cnd-m3* with respect to *Cnd* increased against MDR nosocomial bacteria strains. For instance, we showed that the MIC decreased by two times for MRSA (Methicillin resistant *Staphylococcus aureus*) to 30 times for KPC (*Klebsiella pneumoniae* carbapenemase) [30]. The antimicrobial activity of *Cnd-m3a* decreased as compared to *Cnd-m3*, however, its selectivity ratio increased from 2.4 to 4.9 [31]. Moreover, in silico analysis, carried out using the Peptide Cutter tool available on EXPASY Server (https://web.expasy.org/peptide_cutter/), showed that both mutants were more resistant than *Cnd* to proteolytic degradation, especially towards the chymotrypsin-low specificity.

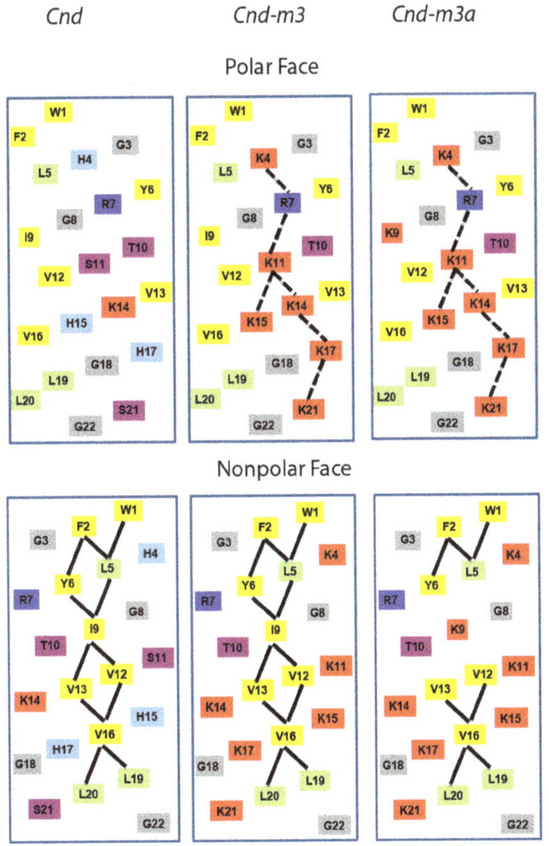

Figure 1. Helical net representation of *Cnd* and *Cnd*-mutants.

2.2. Secondary Structure Analysis by Circular Dichroism

The secondary structure of Chionodracine and *Cnd*-analogues was investigated by CD spectroscopy. The experiments were carried out at 25 °C in buffered LUVs, consisting of either 100% POPC or a 70:30 mixture of POPC/POPG. CD is a rapid and convenient spectroscopic technique for determining the conformational transition of proteins and peptides. We studied the conformational transitions of the three peptides upon interaction with anionic and zwitterionic membrane mimicking prokaryotic and eukaryotic environments, respectively. Typical titration curves in the presence of increasing amounts of lipid vesicles are shown in Figure 2. Increasing peptide:lipid ratios (P:L) were tested in a range between 1:0 and 1:60.

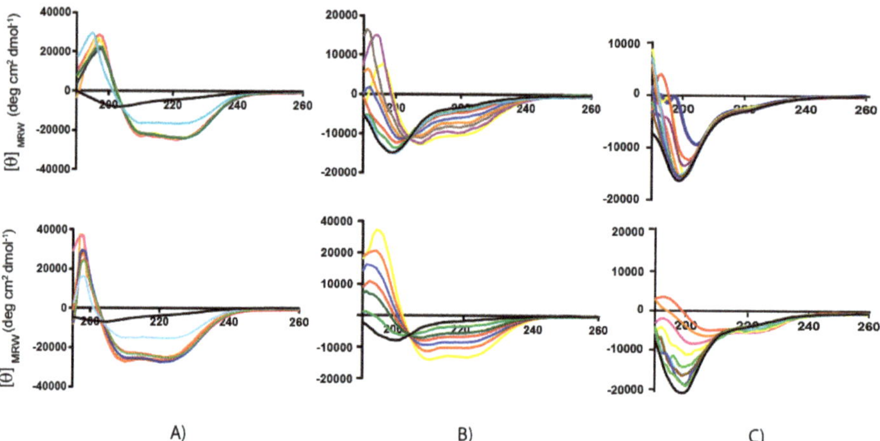

Figure 2. CD spectra in buffer (black line) and in the presence of increasing amount of POPC (Top line) and POPC/POPG (Bottom line) vesicles. (**A**) *Cnd*, (**B**) *Cnd-m3* and (**C**) *Cnd-m3a*. The P:L ratio ranges from 1:0 to 1:60.

In the absence of lipid vesicles (P:L = 1:0), all the peptides were essentially unstructured, with CD spectra showing a minimum at 200 nm characteristic of a random coil structure (Black curve). In the presence of POPC vesicles, mimicking the eukaryotic cell membrane, the CD spectra of both *Cnd* and *Cnd-m3* revealed two minima at ~208 and ~222 nm, which increase with the lipid concentration, and a strong maximum at ~190 nm, typical of α-helix conformation. The rotational strength of the bands at 208 and 222 nm are often used as an index of the presence of helical structures [33,34]: for α-helical polypeptides the intensities of the two bands are almost the same. The analysis of shape of the CD spectrum as well as the ratio $\theta_{222nm}/\theta_{208nm} \sim 1$ confirms that *Cnd* and *Cnd-m3* fold in a canonical α-helix in presence of lipid vesicles. For the *Cnd-m3a* mutant, the CD spectra do not change significantly upon addition of lipids. In the absence of lipids, the spectra show a minimum at ~200 nm, which is typical of a random coil conformation. In presence of increasing amounts of POPC LUVs, this minimum shifts to 206 nm and a very small signal at 222 nm appears only upon addition of a large excess of LUVs (P:L ~ 1:60). The behavior of *Cnd* upon interaction with POPC/POPG (70:30) LUVs is similar to that observed in the presence of POPC vesicles, i.e., the peptide folds in a α-helix conformation at low P:L ratios. In the case of *Cnd-m3*, the value of θ_{222nm} are higher as compared to those observed in POPC, indicating that *Cnd-m3* is more prone to adopt a helical conformation in presence of negatively charged membranes. The helicity becomes even more apparent at a P:L ratio of 1:4 and reaches the maximum at 1:30. From the analysis of CD spectra, it is apparent that in presence of POPC/POPG LUVs the value of ratio $\theta_{222}/\theta_{208}$ can be either less than 1 or greater than 1 depending on the L/P ratio. However, the use of this ratio to distinguish between different helical conformations, such as 3_{10} or α, is not always possible and should be used with extreme caution, since the ratio depends also by the helical chain length, and aggregation to form helical bundle [33,35].

For *Cnd-m3a*, the CD spectra in presence of POPC/POPG (70:30) LUVs are similar to those in POPC, but the θ_{222nm} values are higher. The values of θ_{222nm} and the percentage of helicity are reported in Table 2. The molar ellipticities in POPC correspond to a helical fraction of 0.63, 0.31, and 0.12 for *Cnd*, *Cnd-m3*, and *Cnd-m3a*, respectively. The helical fractions (f_α) in the presence of POPC/POPG were calculated according to equation 1(see Material and Methods) and resulted to be 0.64, 0.79, and 0.14, respectively. For each peptide, the structural differences detected in POPC and POPG are reported as the molar ellipticity ratio at 222 nm in POPC and POPC/POPG lipid systems: $\theta^{PC}_{222nm}/\theta^{PCPG}_{222nm}$ [36]. While for *Cnd* the helicities in POPC and POPC/POPG vesicles were similar, for *Cnd-m3*, the increased amphipathicity ($\mu H = 0.68$) favored the increase of helicity in the presence of negatively charged vesicles (i.e., POPC/POPG). The large decrease of helicity for *Cnd-m3a* with respect to *Cnd* and *Cnd-m3* was probably due to the disruption of the hydrophobic surface caused by the substitution of Ile with the positively charged Lysine in position 9.

These results agree with the partition data previously published, showing partition constants for *Cnd* and *Cnd* mutants higher in POPC/POPG than in pure POPC [30,31].

Table 2. Helicity in presence of different LUVs environment calculated according to the formula (1) derived by Luo and Baldwin.

	POPC		POPC/POPG (70:30)		POPC vs. POPC/POPG
	θ_{222nm}	*% helicity*	θ_{222nm}	*% helicity*	$\theta^{PC}_{222nm}/\theta^{PCPG}_{222nm}$
Cnd	−24,900	62.8	−25,500	64.4	0.98
Cnd-m3	−12,300	30.6	−31,000	78.5	0.40
Cnd-m3a	−5000	11.9	−5100	14.0	0.98

2.3. Steady-State Anisotropy

We also evaluated the interactions of *Cnd* and its derived peptides with lipid bilayers formed by POPC and POPC/POPG LUVs using steady-state anisotropy (r) of DPH (1,6-Diphenyl-1,3,5-hexatriene). DPH is a compound whose fluorescence anisotropy depends on the membrane fluidity [37]. DPH acts as a rod-like hydrophobic probe rotating in the bilayer structure and is located between the acyl chains of the lipids. Due to the presence of the lipid chains, its molecular rotation is restricted, i.e., in the presence of a high organized bilayer the rotational freedom is limited. Changes in the organization of the lipid bilayer, e.g., increasing fluidity, affect the rotational freedom of DPH and the anisotropy will change. In Figure 3, the normalized anisotropy of DPH measured at 25 °C at different concentrations of peptides is reported.

Fluorescence experiments showed that peptides interact with LUVs of different composition perturbing the hydrocarbon chain region in different ways. In the presence of POPC, only *Cnd* is able to perturb the anisotropy of DPH; while *Cnd*-mutants do not. In presence of negatively charged lipid POPC/POPG, both *Cnd* and *Cnd-m3* perturb the anisotropy of DPH while *Cnd-m3a* does not. The detected increase of the anisotropy of DPH suggests that the interaction with *Cnd* and *Cnd-m3* significantly decreased the fluidity of lipid bilayer. This latter effect could be correlated to a moderate structural organization upon interaction with the peptide. For both LUVs, *Cnd-m3a* does not induce changes in membrane fluidity. Thus, the variation of membrane fluidity is not directly correlated with the antibacterial activity. In fact, all the peptides showed antimicrobial activity but different values of DPH anisotropy. However, these differences could be related with the different mode of action of the three peptides. In fact, *Cnd* and *Cnd-m3* are membranolytic peptides [29,30] while *Cnd-m3a* can permeate through the bacteria cell membrane and directly interact with intracellular nucleic acid [31]. These results in combination with CD investigations suggest that the interaction of *Cnd* and *Cnd-m3* with LUVs is different from that of *Cnd-m3a* and such diversity is correlated with a different propensity to adopt a helical conformation.

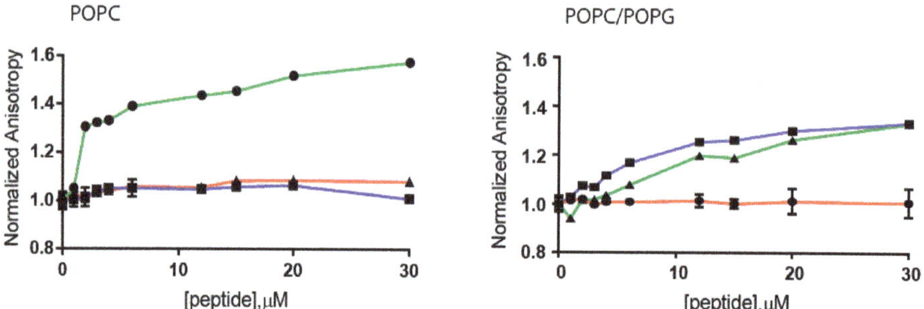

Figure 3. Fluorescence Anisotropy of DPH. Normalized fluorescence anisotropy of DPH in POPC and POPC/POPG lipid vesicles treated with *Cnd* (green), *Cnd-m3* (blue) and *Cnd-m3a* (red). The data are shown as mean ± s.d. from experiments performed in quadruplicate.

2.4. Molecular Dynamics Simulation

To characterize the effects of mutations on *Cnd*-lipid interactions, we carried out atomistic MD simulations in different explicit environments such as water, TFE/water mixture, and lipid membranes composed of 100% POPC and 70:30 mixture of POPC:POPG. In the following sections, we report the results observed in of the different environments.

2.4.1. Water and TFE/Water Solution

Figure 4 shows the time evolution of the secondary structures for *Cnd*, *Cnd-m3*, and *Cnd-m3a* simulated in water and TFE/water (30% *v/v*) solutions. According to the NMR structure of *Cnd* [29] and the model of *Cnd-m3* and *Cnd-m3a* obtained by I-TASSER bioinformatics tool, *Cnd* and *Cnd-m3* were initially in α-helix conformation, while for *Cnd-m3a* the initial structure showed only ~50% of α-helix. The simulation of *Cnd* in water showed a loss of the α-helix conformation (Figure 4A) and the adoption of an unstructured conformation with the formation of short and transient α-helix involving four amino acids. Conversely, in TFE/water mixture, *Cnd* conserved the α-helix throughout the entire simulation (Figure 4B) and the structure is similar to the NMR structure obtained in DPC micelles, with a backbone RMSD of 0.12 nm. In water, *Cnd-m3* is more flexible than *Cnd* and remains essentially unstructured during simulation with the formation of two short antiparallel β-sheets in the C-terminal regions of peptide in the last 10 ns (Figure 4C). Similar to *Cnd*, the simulation of *Cnd-m3* in TFE/water shows the presence of a stable α-helix that is substantially preserved throughout the entire simulation (Figure 4D). The initial structure of *Cnd-m3a* contains 10 residues of the C-terminus in an α-helical conformation; whereas the N-terminus is unstructured. This conformation is persistent both in water and TFE/water (Figure 4E,F).

Overall, all peptides showed that in TFE/water (30% *v/v*) their initial α-helix structures are stabilized and preserved during the MD simulation.

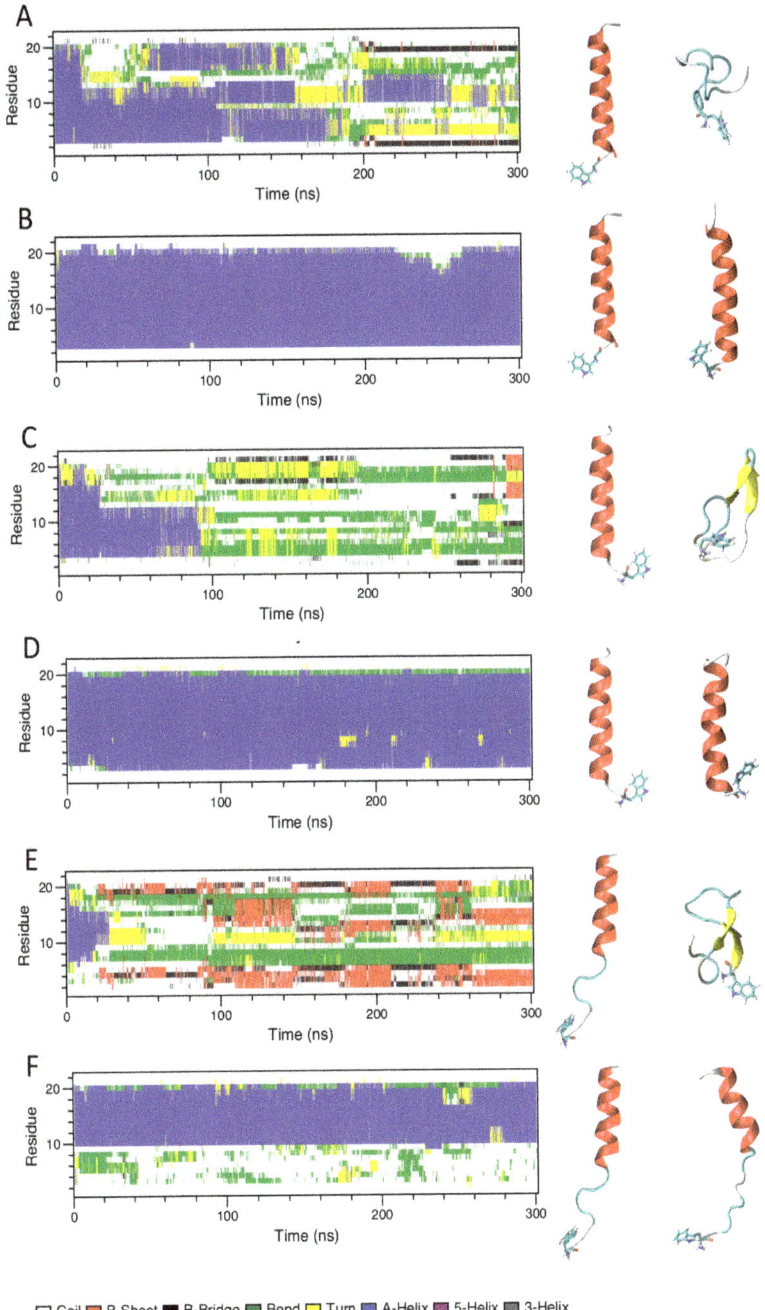

Figure 4. Secondary structure of *Cnd* and *Cnd* mutants in water and in TFE/water as a function of simulating time. (**A**) *Cnd* in water, (**B**) *Cnd* in TFE/water (30% *v/v*), (**C**) *Cnd-m3* in water, (**D**) *Cnd-m3* in. TFE/water (30% *v/v*), (**E**) *Cnd-m3a* in water, (**F**) *Cnd-m3a* in TFE/water (30% *v/v*). The structures on the right are the initial (left) and the final conformations of peptides after 300 ns of MD simulation in the corresponding environment.

2.4.2. Peptides with Lipid Bilayers

The folding of peptide at water/membrane surface is a process with a time scale in the range of milliseconds to seconds. Therefore, to reduce computational cost, we studied the interactions of *Cnd* and its mutants with lipid bilayer starting from their equilibrated conformations in TFE/water. The α-helical of peptides were taken parallel to lipid surface with a distance of 0.5 nm between the center of mass of peptide and the phosphate atoms of lipids. *Cnd* placed close to the surface of bilayer of POPC and POPC/POPG conserves a helicity of 55% ± 3% and 50% ± 5%, respectively (Figure S1). Upon interaction with the POPC head groups *Cnd* is adsorbed on the membrane/water interface, pointing its hydrophobic residues toward the hydrophobic region of bilayer; whereas the charged Lys-14 and Arg-7 and the Histidine residues (His-4, His-15, His-17) are oriented towards the bulk water phase (Figure 5A). In POPC/POPG bilayer, *Cnd* shows a similar orientation of the amphipathic helix as observed for the simulation in POPC. However, the peptide snorkels toward the water/membrane interface due to the presence of arginine and histidine residues that interact with the phosphate groups (Figure 5B).

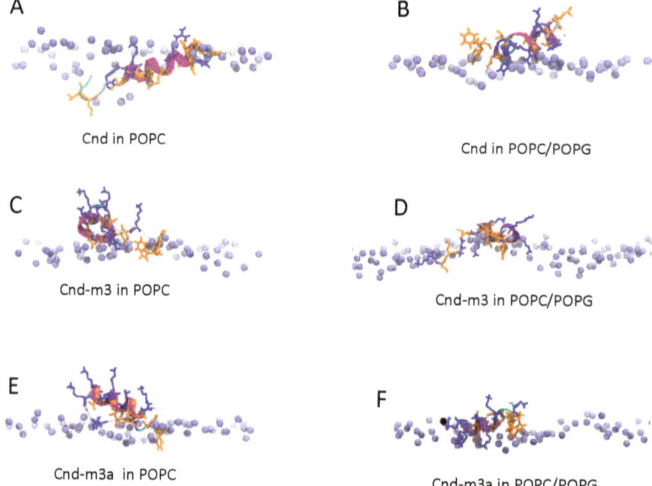

Figure 5. Snapshot MD simulation at 400 ns of (**A**) *Cnd* in POPC, (**B**) *Cnd* in POPC/POPG, (**C**) *Cnd-m3* in POPC, (**D**) *Cnd-m3* in POPC/POPG, (**E**) *Cnd-m3a* in POPC and (**F**) *Cnd-m3a* in POPC/POPG. The phosphate of the upper leaflet of bilayer is represented as violet dots. The residue of lysine, arginine and histidine are represented in blue stick whereas the hydrophobic amino acids are represented as orange stick. The water and counterions are omitted for clarity.

The presence of a high number of lysines, seven in *Cnd-m3* and eight in *Cnd-m3a*, plays a crucial role on the structure of peptides and on their interactions with lipids. *Cnd-m3* and *Cnd-m3a*, located on the surface of POPC bilayer, conserved only a fraction of the initial helicity (25% ± 4% for *Cnd-m3* and 28% ± 3% for *Cnd-m3a*) in the middle region of the peptide. (Figure S1) Lysines and arginines are oriented toward the water phase and the interaction of *Cnd-m3* and *Cnd-m3a* with lipid hydrocarbon chains is due mainly to their hydrophobic residues. In presence of POPC/POPG, the helicity of *Cnd-m3* and *Cnd-m3a* is similar to the simulations in POPC. However, the interactions between lysines and arginines with the phosphate groups of lipid are more persistent (Figure 5C–F). Figure 6 shows the mass density distribution of phosphate and nitrogen atoms of lipids (POPC and POPC/POPG) and the lysine residues of *Cnd-m3* and *Cnd-m3a* calculated, in the final 100 ns of trajectory with respect to the center of the lipid bilayer.

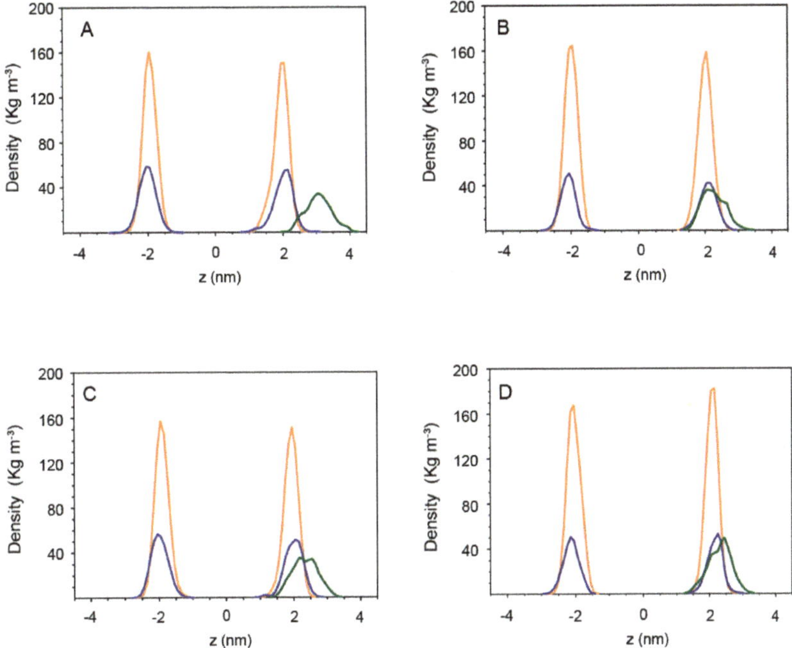

Figure 6. Mass density profiles across lipid bilayer of lipid phosphate (orange), nitrogen atoms of choline (blue) and lysine residues (green) of (**A**) *Cnd-m3* in POPC, (**B**) *Cnd-m3* in POPC/POPG, (**C**) *Cnd-m3a* in POPC, (**D**) *Cnd-m3a* in POPC/POPG. The mass density is calculated with respect to the lipid bilayer center ($z = 0$).

The analysis of the mass density shows that in the interaction with POPC bilayer the l lysine residues of *Cnd-m3* and *Cnd-m3a* are oriented toward the water phase far from the phosphate groups (Figure 6A,C). On the contrary, in the presence of negative charged lipid bilayers (POPC/POPG, Figure 6B,D), the density of lysine residues overlaps with that of the phosphate groups, indicating significant interactions between the charged groups of peptides and lipids. When inserted perpendicularly to the lipid bilayer, the peptides show a more defined conformation with respect to the peptides parallel to the surface of bilayer. *Cnd* embedded into POPC and POPC/POPG bilayer preserves a stable α-helical conformation during the entire simulation (400 ns) with a helicity of 82% ±2% and 81% ± 2% (18 residues), respectively. In POPC and POPC/POPG bilayers, *Cnd* adopts a more helical structure with respect to both *Cnd-m3* and *Cnd-m3a* mutants. The helical content of *Cnd-m3* in POPC is 45% ±2% (10 residues) and increases to 65% ± 3% (14 residues) in the presence of POPC/POPG.

Cnd-m3a, embedded in POPC and POPC/POPG bilayer, shows a stable α-helix in the C-terminus with a helical content of 47% ± 3% in POPC and 44% ± 3% in POPC/POPG. Figure 7 shows the position of the peptides with respect to the lipid bilayers (POPC and POPC/POPG) at the end of each MD simulation. For *Cnd-m3* and *Cnd-m3a*, the insertion of lysine residues into the hydrophobic core of lipid bilayer induces a perturbation of the lipid aggregate. In fact, the hydrophobic side chains of lysines, positively charged, orient toward the negative charged phosphate group of lipids (snorkel effect [38]). As shown in Figure 7, the lipid molecules close to the peptide change partially the orientation to have favorable interactions. The presence of lysine residues in the hydrophobic regions of bilayer determines also the penetration of water molecules (5–7 molecules) that interact with the charged amino acid by hydrogen bonds. The presence of water molecules in the hydrophobic core, together to the presence of charged lysines, contributes to destabilize the lipid bilayer, which shows a reduced thickness in the region near the peptides.

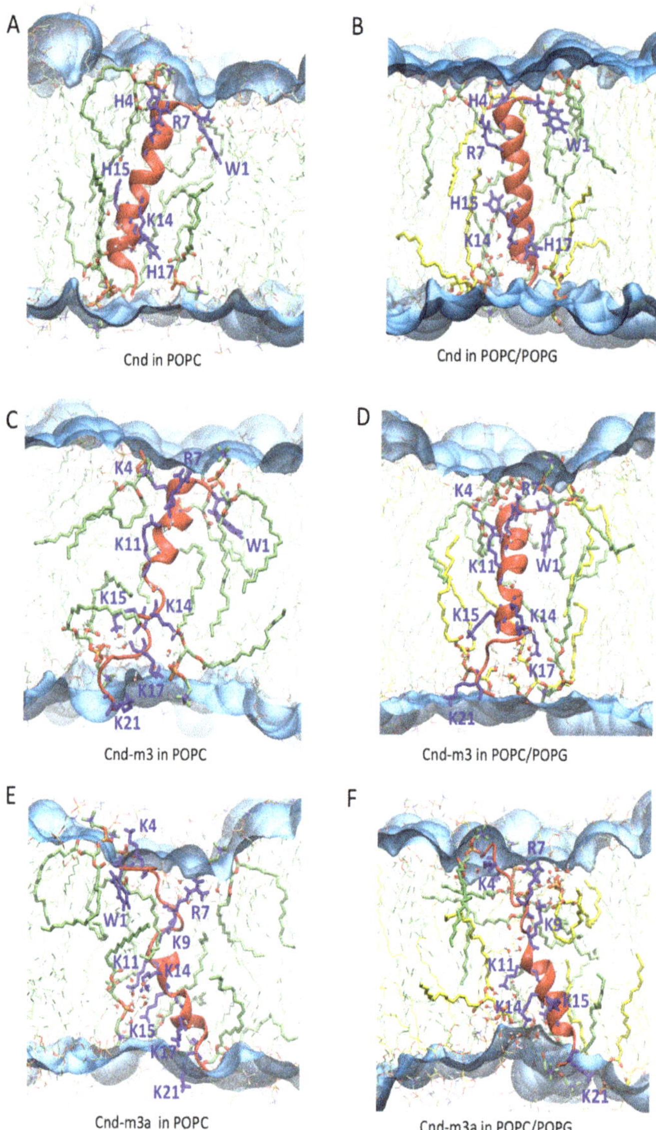

Figure 7. Snapshot at the end (400 ns) of MD simulations of *Cnd*, *Cnd-m3* and *Cnd-m3a* embedded in lipid bilayer: (**A**) *Cnd* in POPC, (**B**) *Cnd* in POPC/POPG, (**C**) *Cnd-m3* in POPC, (**D**) *Cnd-m3* in POPC/POPG, (**E**) *Cnd-m3a* in POPC, (**F**) *Cnd-m3a* in POPC/POPG. The secondary structure of peptides is represented in cartoon and colored in red, lysine arginine, histidine and tryptophan are represented in blue sticks. The lipid molecules are represented in thin sticks and colored according to their atom types (red for oxygen, blue for nitrogen, orange for phosphate atoms, green for carbon atoms of POPC, yellow for carbon atoms of POPG). The lipid molecules within 3.5 Å from the peptide atoms are represented in sticks. Water molecules in the hydrophobic region of lipid are represented in ball-sticks. The water surface is generated using the VMD Surf representation.

The destabilizing effects of *Cnd*, *Cnd-m3* and *Cnd-m3a* on the structure of lipid bilayer can be evidenced by analyzing the change of thickness of lipid bilayer during the last 100 ns of each MD simulations. Figure 8 shows the 3D density maps of time averaged thickness of POPC and POPC/POPG in presence of *Cnd*, *Cnd-m3* and *Cnd-m3a* embedded in lipid bilayer. The blue color corresponds to the region of bilayer with low values of thickness and the red color to the region with high thickness. The peptides embedded in bilayer are localized in the region with the lowest thickness.

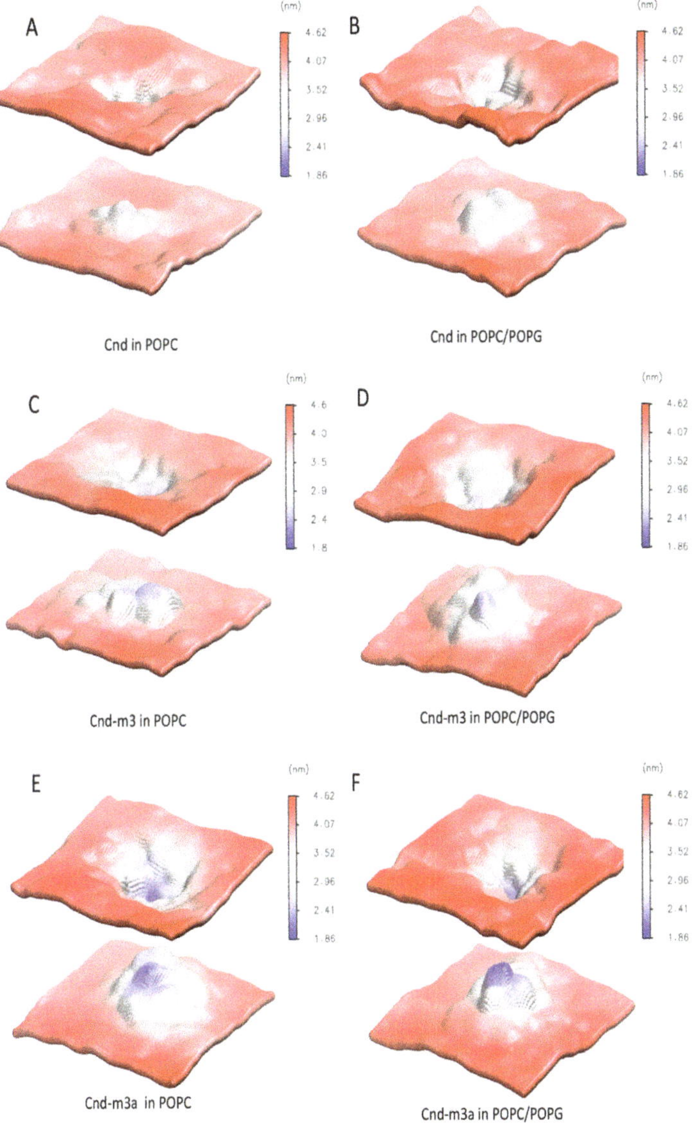

Figure 8. Time average of the last 100 ns of simulations of POPC and POPC/POPG local lipid bilayer thickness in presence of antimicrobial peptides embedded in to lipid bilayer: (**A**) *Cnd* in POPC, (**B**) *Cnd* in POPC/POPG, (**C**) *Cnd-m3* in POPC, (**D**) *Cnd-m3* in POPC/POPG, (**E**) *Cnd-m3a* in POPC, (**F**) *Cnd-m3a* in POPC/POPG. The thickness increases as a blue-white-red color gradient.

As shown in Figure 8 the thickness of lipid bilayer was perturbed around the antimicrobial peptides. The largest structural deformation of lipid bilayer, both in POPC and POPC/POPG, was observed in the case of *Cnd-m3a* with the average bilayer thickness in vicinity of the peptide decreased approximately up to 1.9 nm. In the case of *Cnd-m3* the averaged thickness of lipid bilayer close to peptide was approximately of 2.5 nm. *Cnd* induces a perturbation of thickness smaller with respect to *Cnd-m3a* and *Cnd-m3* with the average bilayer thickness in vicinity of the peptide approximately of 2.9 nm.

3. Discussion

Alternative approaches are needed to neutralize antibiotic resistant diseases. Host defense peptides such as AMPs offer a good starting model to obtain more active antimicrobial agents. More than 3000 AMPs are reported in public database [39] giving a good source of information to delineate strategies for the AMP design. Here, we selected the two *Cnd* analogues, which showed a potent antimicrobial activity against multidrug resistant bacteria and investigated their interaction with model membranes in order to gain insight on such interactions that could be useful for the rational design of novel antimicrobial agents [30,31]. These analogues were designed considering the number of positive charges, the distribution of hydrophobic residues and the location of specificity determinants. According to Hodges and co-workers specificity determinants are positively charged residues in the middle of the non-polar face of the hypothetical amphipathic helix (Figure 1) [40]. The role of these residues is to increase the selectivity between prokaryotic and eukaryotic membranes. Moreover, it has been established that high hydrophobicity on the non-polar face correlates with peptide self-aggregation to form dimers or oligomers and with a decrease in the antimicrobial activity [41]. Selective toxicity, killing bacterial pathogens without damaging eukaryotic cells, is crucial for the development of AMP. Recently, we demonstrated the increase in selectivity of *Cnd* analogs as measured by the selectivity ratios, defined by the ratio between the partition constants measured in POPC/POPG and POPC. Higher values indicate the preference of peptide to partition into anionic vesicles, mimicking the bacterial cell membrane and we found values of 1.43, 2.39 and 4.90 for *Cnd*, *Cnd-m3* and *Cnd-m3a*, respectively [30,31]. We used CD spectroscopy and all atoms MD simulations to elucidate peptide conformational changes the underlay the complex mechanism of peptide-lipid interactions.

CD spectroscopy experiments revealed a difference in the secondary structure adopted by peptides in presence of various membrane mimicking environments. All peptides adopted a random coil conformation in phosphate buffer and undergo a transition to a more ordered conformation in the presence of anionic and zwitterionic LUVs. However, *Cnd* and *Cnd-m3* fold in an α-helical conformation, both in the presence of POPC and POPC/POPG LUVs, whereas *Cnd-m3a* seems to retain its initial random coil conformation in the presence of POPC LUVs and shows a very modest helical content in POPC/POPG LUVs. The percentage of α-helix is much higher in presence of POPC/POPG compared to POPC for *Cnd-m3*, the analog with higher antimicrobial activity, with a θ_{222nm} molar ellipticity ratio of 0.40 (Table 2). For *Cnd* and *Cnd-m3a*, the ratio was ~1 indicating the same percentage of α-helix in both anionic and zwitterionic membranes. Interestingly, *Cnd-m3a* adopts a α-helix conformation, although to a lesser extent ~14% with respect to *Cnd* and *Cnd-m3*. This indicates that the introduction of the Lysine in position 9 in the middle of the non-polar face interfere with the ability to adopt and α-helical structure. Despite the different conformation adopted by *Cnd-m3a* and its different behavior with respect to the other two peptides in the presence of POPC and POPC/POPG vesicles, its antimicrobial activity is preserved. This is in accord with previous studies, which evidenced a different mode of action of the three peptides. In fact, while *Cnd* and *Cnd-m3* have been described as membranolytic peptides, adopting an α-helical conformation to trigger their activity [30], we demonstrated that *Cnd-m3a* is able to traverse inner and outer membrane of *Psychrobacter sp.* targeting intracellular nuclei acids [31]. To get further insights into how primary structure can drive conformations, we carried out molecular dynamics simulations in different environments. MD simulations in presence of POPC and POPC/POPG bilayer agree with the experimental CD results and revealed the atomistic interactions between peptides and membranes. *Cnd* shows a higher content of helical structure with respect to the analogs in presence of both lipid bilayer of POPC and POPC/POPG. *Cnd-m3* shows an increase of helical content moving from POPC to POPC/POPG, whereas *Cnd-m3a* does not show a significant change in the helical content in the presence of anionic component POPG in bilayer. The MD simulation validates that the substitution of Ile-9 with a lysine in *Cnd-m3* with a lysine destabilizes the α-helix conformation in the N-terminal region of *Cnd-m3a*. MD simulations of the peptides absorbed on the surface of POPC bilayers emphasize that the peptides interact with the lipidic surface mainly through hydrophobic contacts. The presence of lysine residues helps to anchor the *Cnd-m3* and *Cnd-m3a* to the surface of POPC/POPG bilayer through the formation of hydrogen bonds and ion-pairs with the phosphate group of lipids (Figure 4). *Cnd*, *Cnd-m3*, and *Cnd-m3a* embedded into bilayer perturb the organization of the lipids in the aggregates, especially for *Cnd-m3* and *Cnd-m3a* due to the presence of lysine residues. The lysine residues in the hydrophobic core of lipid bilayer orient the positively charged head of side chains towards the surface (snorkeling effect) to obtain an energetically favorable interaction with the head groups of lipids. *Cnd-m3* and *Cnd-m3a* induce a reorganization of lipid components close to peptides with a sensible reduction of the thickness of bilayer.

4. Materials and Methods

4.1. Materials

All lipids were purchased from Avanti polar Lipids (Alabaster, AL, USA). All peptides were purchased from CASLO ApS, c/o Scion Technical University of Denmark, with a purity >98%. Peptide concentrations were measured before each sample preparation by UV light absorption at 280 nm. Phosphate buffer pH 7.4 was used to diminish the UV absorbance in CD experiments.

4.2. Preparation of Lipid Vesicles

LUVs (Large Unilamellar Vesicles) were prepared as previously reported [29]. Lipids were dissolved in chloroform/methanol 9:1. The solvent was then removed by rotary evaporation and then overnight under high vacuum. The lipid film was then hydrated by adding 1 mL of buffer (20 mM phosphate buffer at pH 7.4 with 150 mM NaCl and 0.8 mM EDTA) and then subjected to 5 freeze-thaw cycles and vortexed. The multilamellar vesicles (MLVS) formed were extruded through a polycarbonate membrane with pore size of 100 nm using a mini extruder (Avanti Polar). The obtained LUVs were composed of 100% POPC (1-palmitoyl-2-oleoyl-*sn*-glycero-3-phosphocholine) and of a 70/30 (*w/w*) combination of POPC/POPG (1-palmitoyl-2-oleoyl-*sn*-glycero-3-phosphoglycerol). The final concentration of the lipid was 10 mM.

4.3. Circular Dichroism Spectroscopy

CD spectra were recorded in the 190–260 nm spectral range at 298 K on a J 715 JASCO spectropolarimeter equipped with a Peltier device for temperature control (0.5-cm path length quartz cuvette). Secondary structures of the three peptides were determined in phosphate buffer (PB) 0.01 M and EDTA 0.08 M at pH 7.4 in the absence and presence of LUVs of different composition. Peptides were prepared as 1.0 mM stock solution. A stock solution of LUVs (10 mM) was used. For LUVs interaction studies, solutions of 30 μM of *Cnd* and Chionodracine derived peptides were titrated with LUVs of different composition (100% POPC, 70:30 POPC:POPG) with a peptide/lipid ratio ranging from 1:0 to 1:60. Contribution from buffer and LUVs were removed by subtracting their spectra in the absence of peptides. The reported CD spectra are the average of 16 scans with a scanning speed of 20 nm/min, a response time of 8 s, a bandwidth of 1.0 nm and a step size of 0.1 nm. The obtained data in millidegrees (θ) were converted to *mean molar ellipticities* (deg cm^2 dmol^{-1}). The fractional helical content f_α has been estimated using the formula [42]:

$$f_\alpha = \frac{\theta - \theta_{RC}}{\theta_H - \theta_{RC}} \tag{1}$$

where θ is the observed ellipticity at 222nm and θ_{RC} and θ_H are the limiting values for a completely random coil and α-helical conformation, respectively. At the temperature of 25 °C, according to the Luo and Baldwin formula [43], the limiting values of ellipticities are $\theta_H = -39,375$ and $\theta_{RC} = -360$.

4.4. Steady-State Anisotropy Measurements

Lipids were dissolved in chloroform/methanol 9:1 and mixed with stock solution of DPH 1,6-Diphenyl-1,3,5-hexatriene. The molar ratio DPH/Lipids was 1:1000 and the LUVs were extruded as previously described. The anisotropy measurements were performed using a Perkin Elmer LS55. The temperature was set at 25 °C using a thermostatic cell holder controlled by an external bath circulator. The samples were excited at 380 nm and the emission was detected at 420 nm with an emission and absorption bandwidth of 5 nm [44]. Fluorescence anisotropy value, r, were obtained according to the equation:

$$r = \frac{I_{VV} - G^*I_{VH}}{I_{VV} + 2G^*I_{VH}} \tag{2}$$

where I_{VV} is the fluorescence measured with the excitation and emission polarizers oriented in the vertical position and I_{VH} is the fluorescence intensity measured with the excitation polarizer oriented vertically and the emission oriented in horizontal position. G^* is the correction grating factor, defined as the ratio between the vertically and horizontally polarized emission components upon excitation with a horizontally polarized light [45]. DPH acts as a rod-like hydrophobic probe rotating in the bilayer structure and located between the acyl chains of the lipids [46]. Changes in anisotropy values of DPH were used to evaluate membrane fluidity and peptide interaction with LUVs.

4.5. Computational Methods

4.5.1. Preparation of MD Simulations

The initial structure of *Cnd* was taken from the PDB structure resolved by NMR spectroscopy as described previously [29]. The three-dimensional structures of *Cnd-m3* and *Cnd-m3a* has not been resolved yet by NMR spectroscopy. Therefore, we used the fold recognition algorithm implemented in I-TASSER (iterative threading assembly refinement algorithm) [47,48] to predict a 3D model of *Cnd-m3* and *Cnd-m3a* starting from the amino acid sequence of target peptides. The best-scored model structures obtained by the I-TASSER prediction of *Cnd-m3* and *Cnd-m3a* were used.

Titratable residues of peptides were modelled according to their protonation state in water at pH = 7.

The initial configuration of POPC and POPC/POPG (70:30) lipid bilayers was built by using PACKMOL program [49]. Each bilayer system was built with 128 lipids (64 per leaflet) and about 35 water molecules per lipid to have a fully hydrated bilayer. The mixed bilayer of POPC/POPG (70:30) was built by using 92 molecules of POPC and 36 molecules of POPG symmetrically distributed between the two bilayer leaflets. The lipid bilayers were equilibrated for 200 ns before the addition of peptides.

For each peptide six different simulation were performed:

(1) *in water:* the peptides were immersed in a cubic box (5.0 nm) and solvated with at list 4000 water molecules;

(2) *in a mixture of TFE/water (30% v/v):* the peptides were immersed in a cubic box (5.0 nm) and solvated with about 270 molecules of TFE and 2650 water molecules;

(3) *with the peptide absorbed on lipid bilayer surface of POPC:* the peptide was positioned above the pre-equilibrated bilayer surface of POPC with the helical axis parallel to the lipid surface with a distance of 0.5 nm between the center of mass of peptide and the phosphate atoms of lipids. A rectangular box of $6.5 \times 6.5 \times 8.0$ nm^3 was used with at list 5140 water molecules;

(4) *with the peptide embedded in POPC lipid bilayer:* the peptide was embedded in a pre-equilibrated lipid bilayer of POPC with the helical axis parallel to the normal of lipid bilayer in a rectangular box of $6.5 \times 6.5 \times 8.0$ nm^3 with at list 5140 water molecules;

(5) *with the peptide absorbed on lipid bilayer surface of POPC:POPG:* the peptide was positioned above the pre-equilibrated bilayer surface of POPC:POPG with the helical axis parallel to the lipid surface and a distance of 0.5 nm between the center of mass of peptide and the phosphate atoms of lipids. A rectangular box of $6.2 \times 6.2 \times 8.0$ nm^3 was used with at list 4300 water molecule;

(6) *with the peptide embedded in POPC/POPG lipid bilayer:* the peptide was embedded in a pre-equilibrated lipid bilayer of POPC/POPG (70:30) with the helical axis parallel to the normal of lipid bilayer in a rectangular box of $6.2 \times 6.2 \times 8.0$ nm^3 with at list 4300 water molecules.

For the simulation in water and in the mixture of TFE/water the NMR resolved structure of *Cnd* and the I-TASSER model of *Cnd-m3* and *Cnd-m3a* was used. The structure of peptide obtained at the end of the MD simulation in TFE/water was used for the simulations in presence of lipid bilayer of POPC and POPC/POPG. The peptides were immersed in the hydrophobic region of lipid bilayer using the embedding protocol of M. Javanainen [50].

In all systems the appropriate number of chloride ions and sodium ions was used to neutralize the charges of the peptides and POPG molecules.

4.5.2. The MD Systems Setup

All MD simulations were carried out with GROMACS v.5.1.5 package [51]. The GROMOS G54a7 force field [52] was used to model the peptides molecules and ions in conjunction with the GROMOS-CKP [53–55] for the lipids. The phosphatidylglycerol component of the mixed lipid bilayer was a racemic mixture of 1-palmitoyl-2-oleoyl-*sn*-glycero-3-phospho-L-(1-glycerol) (L-POPG) and 1-palmitoyl-2-oleoyl-*sn*-glycero-3-phospho-D-(1-glycerol) (D-POPG). Water was modelled with the SPC model [56] and the force field proposed by Fioroni et al. [57] was used to describe TFE.

Lennard-Jones and electrostatic interactions were calculated using a cut-off of 1.2 nm and the long-range electrostatic interactions were accounted by using the particle mesh Ewald method (PME) [58].

All bonds were constrained using the P-LINCS algorithm [59,60] whereas the geometry of water molecules was fixed with the SETTLE algorithm [61]. The simulations were carried out with a time step of 2 fs.

The molecular systems were energy minimized and equilibrated as follows.

The systems were warmed up with five consecutive MD in the NVT ensemble from 100 K to 298 K in 500 ps with weak positional restraints on heavy atoms of peptides and lipids. Subsequently, 300 ns of unrestrained MD simulation in the NPT ensemble were performed for the peptides in water and in TFE/water whereas 400 ns of unrestrained MD (NPT) were performed for the peptides in presence of lipid bilayer. For each system two simulations with different random initial velocities were performed. The temperature of peptide, lipids, water and ions was kept constant separately at 298 K using the velocity rescale method [62] with a time constant of 0.1 ps. The pressure was controlled by using the Berendsen barostat [63] (P = 1 bar, τ_p = 1.0 ps). The pressure coupling was isotropic in the case of the simulations in water and in the mixture of TFE/water (30% *v/v*). In the simulation with lipid bilayer the pressure coupling was applied semi-isotropic: the z and the xy dimensions were allowed to vary independently.

Periodic boundary conditions were applied in all three dimensions.

A list of the simulations performed and details about the composition of the simulated systems are reported in Table S1.

4.5.3. The MD Analyses

The trajectories obtained by MD simulation were analysed with the GROMACS analysis tools, VMD 1.9.3 [64] and in-house scripts. The analysis of trajectories was performed on the last 100 ns of the simulations.

Analysis of the secondary structure was performed with gmx do_dssp tool of GROMACS using the program DSSP [65]. The time averaged local thickness of lipid bilayer was calculated using the g_lomepro tool considering the distance of phosphate atoms of lipids in opposite membrane leaflets [66].

5. Conclusions

In conclusion, our results highlight the interaction between *Cnd* and its analogues with anionic and zwitterionic membranes. Our studies showed that peptides are unstructured in water and fold, preferentially in α-helical structure, in presence of lipid vesicles of different compositions. Our computational studies confirmed the experimental results through a series of MD simulations of 400 ns time scale and showed at atomic resolution the effect of the mutations on the on the structure of peptides and their interaction with lipid membranes. While *Cnd* adopts a helical conformation both in POPC and POPC/POPG vesicles at low L:P ratio, *Cnd-m3* shows a different behavior with the two types of bilayer, being more prone to adopt an α-helical in the presence of negatively charged vesicles. This correlates with previous partition studies evidencing a preference of *Cnd-m3* toward POPC/POPG vesicles. The nutation of Ile-9 with a lysine in *Cnd-m3*, in the middle of the hydrophobic face, destabilizes the α-helix at the N-terminus and brings to a peptide, *Cnd-m3a* with completely different features. Overall this study lays the groundwork for the design of small peptides that can be used as an alternative to common antibiotics to reduce the risk of antibiotic resistance.

Supplementary Materials: Supplementary materials can be found at http://www.mdpi.com/1422-0067/21/4/1401/s1.

Author Contributions: Conceptualization, S.B. and F.P.; Data curation, G.D.P., F.C. and F.B.; Formal analysis, S.B., C.O., F.B. and F.P.; Investigation, G.D.P. and F.C.; Methodology, S.B., G.D.P., F.C. and C.O.; Software, S.B.; Supervision, F.P.; Writing—original draft, S.B., F.C. and F.P.; Writing—review & editing, S.B. and F.P. All authors have read and agreed to the published version of the manuscript.

Funding: This research was supported by the "Departments of Excellence 2018" Program (Dipartimenti di Eccellenza) of the Italian ministry of Education, University and Research, DIBAF-Department of University of Tuscia, Project "Landscape 4.0—food, wellbeing and environment".

Acknowledgments: In this section you can acknowledge any support given which is not covered by the author contribution or funding sections. This may include administrative and technical support, or donations in kind (e.g., materials used for experiments).

Conflicts of Interest: The authors declare no conflict of interest.

Abbreviations

AMP	Antimicrobial peptide
CD	Circular dichroism
Cnd	Chionodracine
DPH	1,6-Diphenyl-1,3,5-hexatriene
EDTA	Ethylenediaminetetraacetic acid
LUV	Large Unilamellar Vesicles
MD	Molecular Dynamics
MDR	Multidrug resistant
MIC	Minimum inhibitory concentration
MLVS	Multilamellar vesicles
OD	Optical Density
PB	Phosphate buffer
POPC	1-palmitoyl-2-oleoyl-*sn*-glycero-3-phosphocholine
POPG	1-palmitoyl-2-oleoyl-*sn*-glycero-3-(1'-rac-glycerol) sodium salt
TFE	2,2,2-Trifluoroethanol

References

1. Cooper, M.A.; Shlaes, D. Fix the antibiotics pipeline. *Nature* **2011**, *472*, 32. [CrossRef]
2. Administration, F.A.D. CFR Annual Print Title 21 food and Drugs. 2018. Available online: https://www.accessdata.fda.gov/scripts/cdrh/cfdocs/cfcfr/cfrsearch.cfm (accessed on 19 July 2019).

3.   Baindara, P.; Mandal, S.M. Antimicrobial Peptides and Vaccine Development to Control Multi-drug Resistant Bacteria. *Protein Pept. Lett.* **2019**, *26*, 324–331. [CrossRef]

4.   Alvarez, A.; Fernandez, L.; Gutierrez, D.; Iglesias, B.; Rodriguez, A.; Garcia, P. Methicillin-resistant Staphylococcus aureus (MRSA) in hospitals: Latest trends and treatments based on bacteriophages. *J. Clin. Microbiol.* **2019**. [CrossRef]

5.   Ghosh, C.; Sarkar, P.; Issa, R.; Haldar, J. Alternatives to Conventional Antibiotics in the Era of Antimicrobial Resistance. *Trends Microbiol.* **2019**, *27*, 323–338. [CrossRef]

6.   Boman, H.G. Antibacterial peptides: Basic facts and emerging concepts. *J. Intern. Med.* **2003**, *254*, 197–215. [CrossRef]

7.   Zasloff, M. Antimicrobial peptides of multicellular organisms. *Nature* **2002**, *415*, 389–395. [CrossRef]

8.   Hancock, R.E.; Sahl, H.G. Antimicrobial and host-defense peptides as new anti-infective therapeutic strategies. *Nature Biotechnol.* **2006**, *24*, 1551–1557. [CrossRef]

9.   Diamond, G.; Beckloff, N.; Weinberg, A.; Kisich, K.O. The roles of antimicrobial peptides in innate host defense. *Curr. Pharm. Des.* **2009**, *15*, 2377–2392. [CrossRef]

10. Zhang, S.K.; Song, J.W.; Gong, F.; Li, S.B.; Chang, H.Y.; Xie, H.M.; Gao, H.W.; Tan, Y.X.; Ji, S.P. Design of an alpha-helical antimicrobial peptide with improved cell-selective and potent anti-biofilm activity. *Sci. Rep.* **2016**, *6*, 27394. [CrossRef]

11. White, S.H.; Wimley, W.C. Hydrophobic interactions of peptides with membrane interfaces. *Biochim. Biophys. Acta* **1998**, *1376*, 339–352. [CrossRef]

12. Chen, Y.; Mant, C.T.; Farmer, S.W.; Hancock, R.E.; Vasil, M.L.; Hodges, R.S. Rational design of alpha-helical antimicrobial peptides with enhanced activities and specificity/therapeutic index. *J. Biol. Chem.* **2005**, *280*, 12316–12329. [CrossRef]

13. Zelezetsky, I.; Tossi, A. Alpha-helical antimicrobial peptides–using a sequence template to guide structure-activity relationship studies. *Biochim. Biophys. Acta* **2006**, *1758*, 1436–1449. [CrossRef]

14. Bartels, E.J.H.; Dekker, D.; Amiche, M. Dermaseptins, Multifunctional Antimicrobial Peptides: A Review of Their Pharmacology, Effectivity, Mechanism of Action, and Possible Future Directions. *Front. Pharmacol.* **2019**, *10*, 1421. [CrossRef]

15. Latendorf, T.; Gerstel, U.; Wu, Z.; Bartels, J.; Becker, A.; Tholey, A.; Schroder, J.M. Cationic Intrinsically Disordered Antimicrobial Peptides (CIDAMPs) Represent a New Paradigm of Innate Defense with a Potential for Novel Anti-Infectives. *Sci. Rep.* **2019**, *9*, 3331. [CrossRef]

16. Romero, S.M.; Cardillo, A.B.; Martinez Ceron, M.C.; Camperi, S.A.; Giudicessi, S.L. Temporins: An Approach of Potential Pharmaceutic Candidates. *Surg. Infect. (Larchmt)* **2019**. [CrossRef]

17. Jenssen, H.; Hamill, P.; Hancock, R.E. Peptide antimicrobial agents. *Clin. Microbiol. Rev.* **2006**, *19*, 491–511. [CrossRef]

18. Silva, P.M.; Goncalves, S.; Santos, N.C. Defensins: Antifungal lessons from eukaryotes. *Front. Microbiol.* **2014**, *5*, 97. [CrossRef]

19. Wimley, W.C.; Hristova, K. Antimicrobial peptides: Successes, challenges and unanswered questions. *J. Membr. Biol.* **2011**, *239*, 27–34. [CrossRef]

20. Brogden, K.A. Antimicrobial peptides: Pore formers or metabolic inhibitors in bacteria? *Nat. Rev. Microbiol.* **2005**, *3*, 238–250. [CrossRef]

21. Porcelli, F.; Ramamoorthy, A.; Barany, G.; Veglia, G. On the role of NMR spectroscopy for characterization of antimicrobial peptides. *Methods Mol. Biol.* **2013**, *1063*, 159–180.

22. Shai, Y. Mode of action of membrane active antimicrobial peptides. *Biopolymers* **2002**, *66*, 236–248. [CrossRef] [PubMed]

23. Aisenbrey, C.; Marquette, A.; Bechinger, B. The Mechanisms of Action of Cationic Antimicrobial Peptides Refined by Novel Concepts from Biophysical Investigations. *Adv. Exp. Med. Biol.* **2019**, *1117*, 33–64. [PubMed]

24. Huang, H.W. Action of antimicrobial peptides: Two-state model. *Biochemistry* **2000**, *39*, 8347–8352. [CrossRef]

25. Wimley, W.C. Describing the mechanism of antimicrobial peptide action with the interfacial activity model. *ACS Chem. Biol.* **2010**, *5*, 905–917. [CrossRef]

26. Milne, D.J.; Fernandez-Montero, A.; Gundappa, M.K.; Wang, T.; Acosta, F.; Torrecillas, S.; Montero, D.; Zou, J.; Sweetman, J.; Secombes, C.J. An insight into piscidins: The discovery, modulation and bioactivity of greater amberjack, Seriola dumerili, piscidin. *Mol. Immunol.* **2019**, *114*, 378–388. [CrossRef]

27. Noga, E.J.; Silphaduang, U. Piscidins: A novel family of peptide antibiotics from fish. *Drug News Perspect.* **2003**, *16*, 87–92. [CrossRef]

28. Buonocore, F.; Randelli, E.; Casani, D.; Picchietti, S.; Belardinelli, M.C.; de Pascale, D.; De Santi, C.; Scapigliati, G. A piscidin-like antimicrobial peptide from the icefish Chionodraco hamatus (Perciformes: Channichthyidae) molecular characterization, localization and bactericidal activity. *Fish Shellfish Immunol.* **2012**, *33*, 1183–1191. [CrossRef]

29. Olivieri, C.; Buonocore, F.; Picchietti, S.; Taddei, A.R.; Bernini, C.; Scapigliati, G.; Dicke, A.A.; Vostrikov, V.V.; Veglia, G.; Porcelli, F. Structure and membrane interactions of chionodracine, a piscidin-like antimicrobial peptide from the icefish Chionodraco hamatus. *Biochim. Biophys. Acta* **2015**, *1848*, 1285–1293. [CrossRef]

30. Olivieri, C.; Bugli, F.; Menchinelli, G.; Veglia, G.; Buonocore, F.; Scapigliati, G.; Stocchi, V.; Ceccacci, F.; Papi, M.; Sanguinetti, M.; et al. Design and characterization of chionodracine-derived antimicrobial peptides with enhanced activity against drug-resistant human pathogens. *RSC Adv.* **2018**, *8*, 41331–41346. [CrossRef]

31. Buonocore, F.; Picchietti, S.; Porcelli, F.; Della Pelle, G.; Olivieri, C.; Poerio, E.; Bugli, F.; Menchinelli, G.; Sanguinetti, M.; Bresciani, A.; et al. Fish-derived antimicrobial peptides: Activity of a chionodracine mutant against bacterial models and human bacterial pathogens. *Dev. Comp. Immunol.* **2019**, *96*, 9–17. [CrossRef]

32. Dathe, M.; Nikolenko, H.; Meyer, J.; Beyermann, M.; Bienert, M. Optimization of the antimicrobial activity of magainin peptides by modification of charge. *FEBS Lett.* **2001**, *501*, 146–150. [CrossRef]

33. Sudha, T.S.; Vijayakumar, E.K.; Balaram, P. Circular dichroism studies of helical oligopeptides. Can 3(10) and alpha-helical conformations be chiroptically distinguished? *Int. J. Pept. Protein Res.* **1983**, *22*, 464–468. [CrossRef]

34. Manning, M.C.; Woody, R.W. Theoretical CD studies of polypeptide helices: Examination of important electronic and geometric factors. *Biopolymers* **1991**, *31*, 569–586. [CrossRef]

35. Andersen, N.H.; Liu, Z.; Prickett, K.S. Efforts toward deriving the CD spectrum of a 3(10) helix in aqueous medium. *FEBS Lett.* **1996**, *399*, 47–52. [CrossRef]

36. Sani, M.A.; Lee, T.H.; Aguilar, M.I.; Separovic, F. Proline-15 creates an amphipathic wedge in maculatin 1.1 peptides that drives lipid membrane disruption. *Biochim. Biophys. Acta* **2015**, *1848*, 2277–2289. [CrossRef]

37. Plasek, J.; Jarolim, P. Interaction of the fluorescent probe 1,6-diphenyl-1,3,5-hexatriene with biomembranes. *Gen. Physiol. Biophys.* **1987**, *6*, 425–437.

38. Strandberg, E.; Killian, J.A. Snorkeling of lysine side chains in transmembrane helices: How easy can it get? *FEBS Lett.* **2003**, *544*, 69–73. [CrossRef]

39. Wang, Z.; Wang, G. APD: The Antimicrobial Peptide Database. *Nucleic Acids Res.* **2004**, *32*, D590–D592. [CrossRef]

40. Jiang, Z.; Vasil, A.I.; Gera, L.; Vasil, M.L.; Hodges, R.S. Rational design of alpha-helical antimicrobial peptides to target Gram-negative pathogens, Acinetobacter baumannii and Pseudomonas aeruginosa: Utilization of charge, 'specificity determinants', total hydrophobicity, hydrophobe type and location as design parameters to improve the therapeutic ratio. *Chem. Biol. Drug Des.* **2011**, *77*, 225–240.

41. Chen, Y.; Guarnieri, M.T.; Vasil, A.I.; Vasil, M.L.; Mant, C.T.; Hodges, R.S. Role of peptide hydrophobicity in the mechanism of action of alpha-helical antimicrobial peptides. *Antimicrob. Agents Chemother.* **2007**, *51*, 1398–1406. [CrossRef]

42. Fernández-Vidal, M.; White, S.H.; Ladokhin, A.S. Membrane partitioning: "classical" and "nonclassical" hydrophobic effects. *J. Membr. Biol.* **2011**, *239*, 5–14. [CrossRef]

43. Luo, P.; Baldwin, R.L. Mechanism of helix induction by trifluoroethanol: A framework for extrapolating the helix-forming properties of peptides from trifluoroethanol/water mixtures back to water. *Biochemistry* **1997**, *36*, 8413–8421. [CrossRef]

44. Suzuki, M.; Miura, T. Effect of amyloid beta-peptide on the fluidity of phosphatidylcholine membranes: Uses and limitations of diphenylhexatriene fluorescence anisotropy. *Biochim. Biophys. Acta* **2015**, *1848*, 753–759. [CrossRef]

45. Lakowicz, J.R. *Principles of Fluorescence Spectroscopy*; Springer: New York, NY, USA, 2009.

46. do Canto, A.; Robalo, J.R.; Santos, P.D.; Carvalho, A.J.P.; Ramalho, J.P.P.; Loura, L.M.S. Diphenylhexatriene membrane probes DPH and TMA-DPH: A comparative molecular dynamics simulation study. *Biochim. Biophys. Acta* **2016**, *1858*, 2647–2661. [CrossRef]

47. Zhang, Y. I-TASSER server for protein 3D structure prediction. *BMC Bioinform.* **2008**, *9*, 40. [CrossRef]

48. Roy, A.; Kucukural, A.; Zhang, Y. I-TASSER: A unified platform for automated protein structure and function prediction. *Nat. Protoc.* **2010**, *5*, 725–738. [CrossRef]

49. Martinez, L.; Andrade, R.; Birgin, E.G.; Martinez, J.M. PACKMOL: A package for building initial configurations for molecular dynamics simulations. *J. Comput. Chem.* **2009**, *30*, 2157–2164. [CrossRef]

50. Javanainen, M. Universal Method for Embedding Proteins into Complex Lipid Bilayers for Molecular Dynamics Simulations. *J. Chem. Theory Comput.* **2014**, *10*, 2577–2582. [CrossRef]

51. Abraham, M.J.; Murtola, T.; Schulz, R.; Páll, S.; Smith, J.C.; Hess, B.; Lindahl, E. Gromacs: High performance molecular simulations through multi-level parallelism from laptops to supercomputers. *SoftwareX* **2015**, *1*, 19–25. [CrossRef]

52. Schmid, N.; Eichenberger, A.P.; Choutko, A.; Riniker, S.; Winger, M.; Mark, A.E.; van Gunsteren, W.F. Definition and testing of the GROMOS force-field versions 54A7 and 54B7. *Eur. Biophys. J.* **2011**, *40*, 843–856. [CrossRef]

53. Kukol, A. Lipid Models for United-Atom Molecular Dynamics Simulations of Proteins. *J. Chem. Theory. Comput.* **2009**, *5*, 615–626. [CrossRef]

54. Poger, D.; Van Gunsteren, W.F.; Mark, A.E. A new force field for simulating phosphatidylcholine bilayers. *J. Comput. Chem.* **2010**, *31*, 1117–1125. [CrossRef]

55. Poger, D.; Mark, A.E. On the Validation of Molecular Dynamics Simulations of Saturated and cis-Monounsaturated Phosphatidylcholine Lipid Bilayers: A Comparison with Experiment. *J. Chem. Theory Comput.* **2010**, *6*, 325–336. [CrossRef]

56. Berendsen, H.J.C.; Postma, J.P.M.; van Gunsteren, W.F.; Hermans, J. Interaction Models for Water in Relation to Protein Hydration. In *Intermolecular Forces*; Reidel Publishing Company: Dordrecht, The Netherlands, 1981; pp. 331–342.

57. Fioroni, M.; Burger, K.; Mark, A.E.; Roccatano, D. A new 2,2,2-Trifluoroethanol model for molecular dynamics simulations. *J. Phys. Chem. B* **2000**, *104*, 12347–12354. [CrossRef]

58. Darden, T.; York, D.; Pedersen, L. Particle mesh Ewald: An Nlog(N) method for Ewald sums in large systems. *J. Chem. Phys.* **1993**, *98*, 10089–10092. [CrossRef]

59. Hess, B.; Bekker, H.; Berendsen, H.J.C.; Fraaije, J.G.E.M. LINCS: A linear constraint solver for molecular simulations. *J. Comp. Chem.* **1997**, *18*, 1463–1472. [CrossRef]

60. Hess, B. P-LINCS: A Parallel Linear Constraint Solver for Molecular Simulation. *J. Chem. Theory Comput.* **2007**, *4*, 116–122. [CrossRef]

61. Miyamoto, S.; Kollman, P.A. SETTLE: An analytical version of the SHAKE and RATTLE algorithm for rigid water models. *J. Comp. Chem.* **1992**, *13*, 952–962. [CrossRef]

62. Bussi, G.; Donadio, D.; Parrinello, M. Canonical sampling through velocity rescaling. *J. Chem. Phys.* **2007**, *126*, 14101–14107. [CrossRef]

63. Berendsen, H.J.C.; Postma, J.P.M.; van Gunsteren, W.F.; Di Nola, A.; Haak, J.R. Molecular dynamics with coupling to an external bath. *J. Chem. Phys.* **1984**, *81*, 3684–3690. [CrossRef]

64. Humphrey, W.; Dalke, A.; Schulten, K. VMD: Visual molecular dynamics. *J. Mol. Graph.* **1996**, *14*, 33–38. [CrossRef]

65. Kabsch, W.; Sander, C. Dictionary of protein secondary structure: Pattern recognition of hydrogen-bonded and geometrical features. *Biopolymers* **1983**, *22*, 2577–2637. [CrossRef] [PubMed]

66. Gapsys, V.; de Groot, B.L.; Briones, R. Computational analysis of local membrane properties. *J. Comput. Aided Mol. Des.* **2013**, *27*, 845–858. [CrossRef]

Article

Accelerated Molecular Dynamics Applied to the Peptaibol Folding Problem

Chetna Tyagi [1,2,*]**, Tamás Marik** [1]**, Csaba Vágvölgyi** [1]**, László Kredics** [1] **and Ferenc Ötvös** [3]

1   Department of Microbiology, Faculty of Science and Informatics, University of Szeged, Közép fasor 52, H-6726 Szeged, Hungary
2   Doctoral School of Biology, Faculty of Science and Informatics, University of Szeged, Közép fasor 52, H-6726 Szeged, Hungary
3   Institute of Biochemistry, Biological Research Centre, Temesvári krt. 62., H-6726 Szeged, Hungary
*   Correspondence: cheta231@gmail.com; Tel.: +36-705251491

Received: 31 July 2019; Accepted: 27 August 2019; Published: 30 August 2019

Abstract: The use of enhanced sampling molecular dynamics simulations to facilitate the folding of proteins is a relatively new approach which has quickly gained momentum in recent years. Accelerated molecular dynamics (aMD) can elucidate the dynamic path from the unfolded state to the near-native state, "flattened" by introducing a non-negative boost to the potential. Alamethicin F30/3 (Alm F30/3), chosen in this study, belongs to the class of peptaibols that are 7–20 residue long, non-ribosomally synthesized, amphipathic molecules that show interesting membrane perturbing activity. The recent studies undertaken on the Alm molecules and their transmembrane channels have been reviewed. Three consecutive simulations of ~900 ns each were carried out where N-terminal folding could be observed within the first 100 ns, while C-terminal folding could only be achieved almost after 800 ns. It took ~1 μs to attain the near-native conformation with stronger potential boost which may take several μs worth of classical MD to produce the same results. The Alm F30/3 hexamer channel was also simulated in an *E. coli* mimicking membrane under an external electric field that correlates with previous experiments. It can be concluded that aMD simulation techniques are suited to elucidate peptaibol structures and to understand their folding dynamics.

Keywords: accelerated molecular dynamics; alamethicin; membrane; peptaibol

---

1. Introduction

With the growing incidence of antibiotic resistance all over the world, the scientific community is, more than ever, desperate to identify novel, fail-safe ways of treatments for a plethora of devastating diseases. This search is not limited to human pathogens but has also been extended to the problem of agricultural pathogens. The class of antimicrobial peptides (AMPs), defined as short, host-defense peptides found in all life forms, is a promising solution. Many AMPs have already crossed over to clinical trials as novel therapeutics, immunity modulators and wound healing promoters [1]. In this study we focus our attention to a special class of fungal AMPs, known as "peptaibols", owing to their constituent non-standard amino acid residues like α-aminobutyric acid (Aib) and C-terminal aminoalcohols etc. Peptaibols are produced as secondary metabolites and show microheterogeneity. The diversity in their sequence length, hydrophobicity, antimicrobial properties and producing fungal species contribute to a plethora of peptaibols that are yet to be discovered and studied. The increasing gap between known peptaibol sequences and their three-dimensional structures can be reduced using computational modeling and molecular dynamics simulations. The knowledge of peptide structural dynamics is a key to unraveling the mechanisms of their antimicrobial action. In this study we focus on the use of accelerated molecular dynamics (aMD) simulations to obtain conformational ensemble and structural dynamics of alamethicin F30/3 (Alm F30/3). aMD is an enhanced sampling technique that "boosts" the system over energy barriers, thereby fastening the peptide folding process [2,3]. We also show that aMD can be applied to peptide-membrane systems to study the mode of action without introducing significant errors. Alm F30/3 is a 20-amino acid residue long peptide which belongs to the popularly known class of peptaibols, a group of fungal secondary metabolites. Owing to its antibacterial and antifungal activities, a result of its ion-channel forming property, it has been touted as a model system to study AMPs. The majority of Alm sequences consist of hydrophobic residues, the non-proteinogenic residue α-aminoisobutyric acid (Aib), an acetylated N-terminus and the C-terminal phenylalaninol (Pheol).

The discovery of Alm is credited to Meyer and Reusser [4], where it was referred as "antibiotic U-22324" obtained from '*Trichoderma viride*' and classified as a cyclic peptide due to its inability to react with ninhydrin. The anti-bacterial activity against gram-positive strains was highlighted. The correct producer was reidentified later to be *T. arundinaceum* from the Brevicompactum clade of genus *Trichoderma* [5].

The first possible primary structure of Alm was reported by Payne et al. [6], who described it as a cyclic molecule by linking the γ-carboxylate group of the glutamic acid residue to the first proline in the sequence. The sequence was Pro-Mea-Ala-Mea-Ala-Gln-Mea-Val-Mea-Gly- Leu-Mea-Pro-Val-Mea-Mea-Glu-Gln, where Mea is α-aminoisobutyric acid or Aib. They hypothesized a stack-like tunnel formation by the cyclic Alm structure with hydrophobic interior. Martin and Williams [7] later corrected it by describing the structure of Alm as a linear polypeptide by NMR spectroscopy. The new sequence described by them included an acetylated Aib residue at the N-terminus and a phenylalaninol as a side-chain of the 18th Gln residue. They stressed upon the importance of linear Alm structure to be long enough to stretch across a lipid bilayer and rejected the idea of stacked-ring pores. The presence of proline in the 14th position introduces a slight bend in the structure as shown by X-ray crystallography, NMR and optical spectroscopy [8–10]. The helix is formed in such a manner that the polar residues are arranged on one side. Alm is classified as amphipathic due to distinct hydrophobic and hydrophilic faces, which renders the ability to either interact with a membrane horizontally or form voltage-gated ion channels with a vertical insertion. The recent studies to understand Alm conformation and pore formation have been reviewed by Leitgeb et al. [11] and Kredics et al. [12]. Here we provide a short update on various research undertaken on Alm since 2014. The following literature review on Alm is a summary of recent work on the common molecule. Most of these studies were carried out to understand and visualize pore formation in various membranes which formed the basis of our work to study the hexameric Alm pore model embedded in *E. coli* membrane mimic.

Pieta et al. [13] described the pore formation of Alm in a monolayer of 1,2-dimyristoyl-sn-glycero-3-phosphocholine/phosphoglycerol (DMPC/PG) lipids using electrochemical scanning tunneling microscopy (EC-STM). PG lipids constitute the membrane architecture of gram-positive bacteria [14] and the negatively charged PG heads provide the electrostatic surface potential promoting insertion of amphipathic Alm into the membrane [15]. Pieta et al. [13] successfully showed the formation of flower-like hexagonal pattern of six Alm molecules, which supports the barrel-stave model of ion channel formation. Rahman et al. [16] reported the thermodynamics of Alm F30/3 pores ranging from monomer to nonamer of 5 to 11 Å radius in an implicit membrane model. The trimer and tetramer formed 6 Å closed pores whereas the hexamer and octamer formed 8 Å and 11 Å open pores, respectively. Interestingly, all pores formed beyond the pentamer with comparable energy supported the multiple conductance level theory.

Castro et al. [17] also studied the various noncanonical disubstituted amino acids in Alm by replacing one or more Aib residue. Few such substitutions proved to be better alternatives than Aib for inducing well-defined α-helical structures and thermodynamic stabilization. Salnikov et al. [18] characterized a pentameric channel assembly formed by a fluorine-labeled Alm derivative when reconstituted into phospholipid bilayers (POPC) with a peptide/lipid ratio of 1:13. When the ratio was lowered to 1:30, Alm was found to be in dimeric form. The pore formation of Alm has also been studied extensively using computational models. For example, 14 μs long all-atom MD simulations performed by Perrin Jr. et al. [19] revealed that the hexamer pore equilibration was achieved only after 7 μs which remained stable onwards. Another simulation with 10 surface-bound Alm showed membrane binding, but no insertion event was observed for 5 μs. However, simulations at higher temperatures and under an electric field resulted in peptide insertion within 1 μs. Madsen et al. [20] showed that incorporation of silicon-containing amino acids in Alm introduced a 20-fold increase in membrane permeability. Similarly, Das et al. [21] discussed an effective strategy to improve antibacterial activity of peptaibols by capping the N-terminal with 1,2,3-triazole, and with hydrophobic substituents on C-4. Afanasyeva et al. [22] showed that Alm on a POPC membrane, in low concentrations below the threshold, can capture and rearrange free fatty acids which may negatively affect membrane physiological functions.

Recently, Su et al. [23] studied Alm in a 1,2-di-O-phytanyl-*sn*-glycero-3-phosphocholine (DPhPC) bilayer using electrochemical impedance spectroscopy (EIS) and photon polarization modulation infrared reflection absorption spectroscopy (PM-IRRAS) to understand membrane stability and Alm conformation in the bilayer. They demonstrated that at potentials where the bilayer is stable, the Alm peptides assume a surface-bound state with their helices at a large angle from the bilayer normal. Alm peptides are inserted when negative potentials are applied. The ESI data showed the exact potentials at which the helices are inserted and form an open or closed pore, which indicates that the open/closed ion states are potential regulated. Forbrig et al. [24] also observed the mechanism of Alm ion channel formation in lipid membranes tethered on electrodes through surface-enhanced infrared absorption (SEIRA) spectroscopy. It revealed that the peptide monomers interact with each other before transmembrane insertion. Upon evaluation of Alm orientation based on mathematical models, it was proposed that ion-channels may be formed either by N-terminal integration as monomers or as parallel oligomers. Abbasi et al. [25] visualized the formation of Alm pores in floating phospholipid membrane on gold electrodes, which confirmed the hexamer ion channel formation with the diameter of a pore calculated to be 2.3 ± 0.3 nm. A distribution of various Alm aggregates also confirmed the multiple conductivity states theory. Syryamina et al. [26] described Alm (Alm F50/5 spin-labeled analogs) channel formation in POPC membranes using pulsed electron-electron double resonance (PELDOR) spectroscopy, which revealed that it assembled into dimers and higher oligomers can be obtained by increasing peptide concentration. They hypothesized that dimer formation and peptide reorientation must take place simultaneously and is probably the first step in peptide assembly.

Zhang et al. [27] proposed encapsulation of Alm within a γ-cyclodextrin in a way that its hydrophobic residues are buried in the hydrophobic cavity of γ-cyclodextrin, thereby, improving its solubility in aqueous medium, thermal and pH stability and significantly increasing its antimicrobial activity against *L. monocytogenes*. A very recent and interesting work by Abbasi et al. [28] reported the dampening effect of Alm ion channel on conductivity by an order of magnitude upon introducing an amiloride molecule, which does not inhibit the ion channel formation itself. Taylor et al. [29] showed that Alm peptides can induce lipid flip-flop even in a surface-bound state by disordering lipids in the membrane.

In this study, alamethicin was chosen as the model peptaibol to establish the use of accelerated MD techniques in correctly elucidating their conformational ensemble and dynamics. We carried out consecutive aMD simulations with increasing boost parameters to observe the folding progress in comparison to the experimental structure. The highest boost parameters were chosen to observe Alm folding from an unfolded starting structure achieved within 1 μs. The same procedure can be applied for other peptaibols of unknown three-dimensional structure to obtain structural information within a short amount of time. We also discuss the structural convergence of these simulations. The reweighted free energy landscapes have been produced to understand peptaibol folding dynamics.

Reweighting of all distributions was carried out using the Maclaurin series expansion, which is equivalent to cumulant expansion of first order. Miao et al. [30] state that the Maclaurin series expansion method suppresses energetic noise but may give incorrect energy minimum positions on the free energy landscapes. Another method known as exponential average reweighting, which is equivalent to cumulant expansion of third order, may lead to higher energetic fluctuations. They emphasized the accuracy of cumulant expansion of second order as a reweighting method, especially if the boost potential follows close to a Gaussian distribution. However, Jing et al. [31] argued that the boost potential should follow the Gaussian distribution exactly for cumulant expansion of second order to accurately reweight free energy profiles. Along these developments, Miao et al. [32] proposed the Gaussian accelerated MD (GaMD) technique, in which the boost potential is made to follow a Gaussian distribution and accurate reweighting can be achieved through cumulant expansion of second order. It is an update to the existing accelerated MD technique in a way that results in recovery of distinct low energy states that may be lost in the former technique. During this work, we also carried out three consecutive 300 ns long GaMD simulations on Alm F30/3 for comparison, but slight folding could only be observed for the N-terminus and it took longer time. We agree that the lack of more advanced processing units and deeper understanding of the GaMD technique at our end may have been the reason for unsatisfactory results. Nevertheless, we are in the process of optimizing peptaibol structure elucidation using GaMD and its advanced variants like Replica Exchange GaMD [33]. Furthermore, to overcome the barrier of loss of distinct low energy conformations, we carried out the first aMD simulation with low boost parameters which were increased for the other three simulations. Therefore, the combined trajectory likely traverses the whole dynamic pathway of Alm F30/3 folding. The two major conformations obtained through these simulations and their functional significance with respect to previous studies is also discussed in detail. Furthermore, we also show the use of aMD to model Alm F30/3 hexameric pore embedded in an *E. coli* membrane mimic under the application of an external electric field. Even though the free energy landscapes have not been reported for membrane-peptide simulations, no significant deformations were observed due to the application of boost potential. The use of aMD has been highlighted to study such systems within a short period of time with efficacy.

2. Results and Discussion

2.1. Clustering of Three Consecutive aMD Simulations and Comparison with Fourth Simulation: Free Energy Landscapes

The unfolded conformation of Alm F30/3 was used as a starting point for three consecutive ~900 ns (refer to Table 1 for exact timescales) long simulations with increasing "boost" parameters. The first simulation (Sim 1) was carried out with a1, a2 = 0.16 and b1, b2 = 4 which revealed successful folding of the N-terminal segment (Aib1-Leu12) but an incomplete folding of the C-terminus as shown through superimposition of the representative structure of the energy minimum and experimentally known structure (PDB ID: 1AMT) with backbone root-mean-square deviation (RMSD) value of 5.02 Å in Figure 1A. The next simulation (Sim 2) was started from this point with slightly higher boost parameters of a1, a2 = 0.20 and b1, b2 = 4.5 to observe the time length of achieving complete peptide folding. This 950 ns long aMD simulation was clustered into three groups, out of which cluster 2 was closest to the experimental structure (Figure 1B) with RMSD of 1.87 Å. At this point, it was deemed a better choice not to increase the boost further as it may interfere with correct reweighting of energy distribution. These boost parameters were deemed appropriate for fast folding. A third ~900 ns long simulation (Sim 3) was carried out at this point which achieved complete C-terminal folding. The clustering resulted in the most populated cluster whose representative structure is the closest to PDB conformation with a RMSD value of 1.51 Å (Figure 1C). The three trajectories were later combined for most of the analysis. The reweighted torsional (phi-psi) angle distribution for each residue calculated for all these 3 simulations have been provided and discussed in Supplementary Figure S1.

After carrying out three consecutive aMD simulations with increasing boost parameters, we were also curious to observe the extent of folding observed using these boost parameters with a completely unfolded Alm configuration as the starting structure. A separate 1 μs long simulation (Sim 4) was carried out that resulted in highly folded structures with both bent and linear configurations. This simulation shows that slightly aggressive boost parameters used for ~1 µs long aMD carried out using GPUs are sufficient for folding simulations of such short peptaibols. Bucci et al. [34] also reasoned that 1 µs long aMD simulations are sufficient for folding simulations of their modified tripeptide. The representative structure of cluster 5 is closer to the experimental structure with an RMSD value of 1.8 Å between them (Figure 1D).

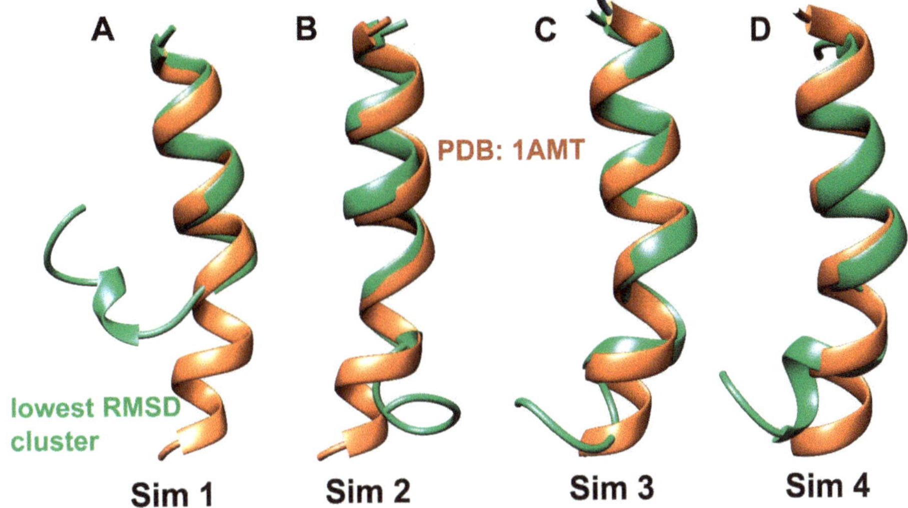

Figure 1. The representative structures of different clusters obtained for individual simulation chosen based on lowest RMSD with X-ray crystallographic Alm F30/3 structure available from PDB: 1AMT.

In conclusion, the combined trajectory of the first three simulations is comparable to the fourth ~1 µs long simulation carried out with aggressive boost parameters. Complete Alm F30/3 folding was achieved within 1 µs, which indicates that this procedure is apt to elucidate short peptaibol structure in a short period of time.

The dihedral principal component analysis (PCA) based free energy landscape (FEL) between PC1 and PC2 has been provided for the combined trajectory of the first three simulations clustered in 10 major representative groups (Figure 2A). The darkest violet regions on the FEL map (Figure 2) show the lowest energy conformation clusters which denote the metastable states of Alm F30/3. As it can be observed on the reweighted maps, at least three distinct clusters can be identified in local energy minimum regions, i.e., the linear helical form (clusters 1, 5 and 9), the bent form (clusters 2, 4 and 6) and the incompletely folded (clusters 3, 7, 8 and 10) conformations. The inter-conversion between the bent and linear forms seems to be energetically allowed within 2 kcal mol^{-1} and the jump between these two is achieved multiple times throughout the trajectory. The C-terminus requires longer time than the N-terminus to completely sample the folded helical configuration. Similarly, the fourth trajectory has been separately projected on to dihedral-based PCs in Figure 2B to avoid missing low energy conformations obtained during previous simulations. The bent conformations of cluster 1 and 2 appear interchangeably for about ~10% and ~8% of 1 µs long simulation trajectory, respectively, roughly from 300 to 800 ns and form the most populated conformational group. The conformation of cluster 5 is closest to the experimental structure. The dihedral angle based FEL plots of the combined trajectory from the first three simulations in comparison with the fourth 1 µs long trajectory shows that a similar conformational space could be covered in less time if slightly aggressive boost parameters are used. However, it comes with the risk of missing distinct low energy conformations. For this reason, it is advisable to carry out few simulations with smaller boost potential.

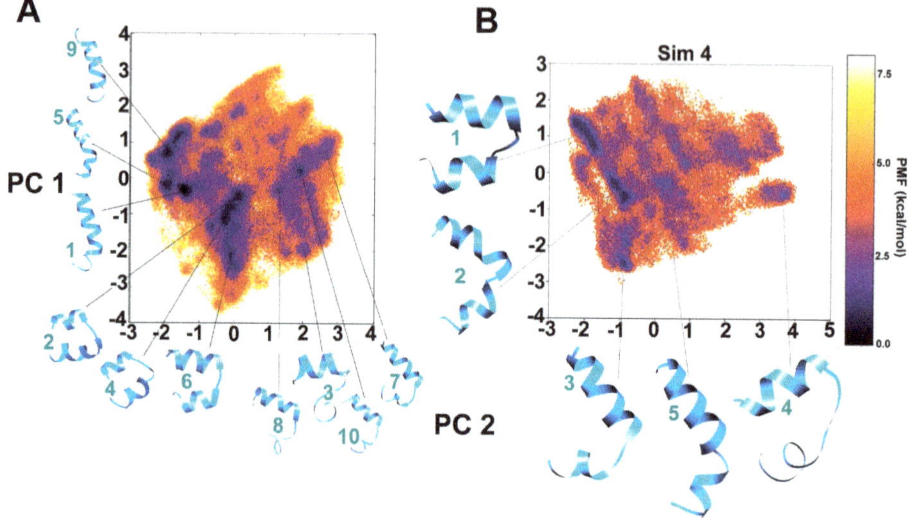

Figure 2. Reweighted potential of mean force (PMF) free energy landscapes of the first two principal components (PCs) calculated from dihedral angles, phi-psi, for better clustering based on internal motions. The deepest blue regions indicate energy minimum. The representative structures of each cluster are provided for (**A**) combined 3 simulations, (**B**) the 4th simulation. The cluster numbers are given in cyan.

In order to observe structural convergence amongst the three independent simulations, the Kullback-Leibler divergence (KLD) was calculated to measure the extent of overlap between probability distribution [35]. This probability distribution is based on PCA of the system Cartesian coordinates instead of dihedral angles. The trajectories are named as "Sim 1" to "Sim 4" and the extent of PC overlap signifies convergence between independent runs. The degree of PC overlap suggests that the independent simulations sampled similar conformational space, hence, convergence. It is based on the idea that any two simulations will eventually sample the same phase space even when started from different starting structures as a measure of true convergence. Figure 3A shows the histograms of projection of coordinates along the first three eigenvectors or modes (accounting for 53% overall motion). The first three PCs obtained from the combined trajectory account for 29%, 16% and 8% (in same order) of overall motions. In case of all PC histograms, the best overlap can be observed between Sim 1, Sim 3 and Sim 4 in comparison to Sim 2 projection which signifies that Sim 2 undergoes a slightly different folding pathway. A better overlap can be observed for all trajectories in case of modes 2 and 3. This divergence can be quantified by calculating KLD vs time (ns).

Figure 3B shows KLD between subsequent histograms as a function of time from four simulations for PCs 1, 2 and 3. KLD:1 is divergence between Sim 1 and Sim 2, KLD:2 between Sim 2 and Sim 3 and KLD:3 between Sim 3 and Sim 4. The rapidly decreasing slope of KLD vs time of any two trajectories indicates the reduction in divergence between sampled conformations. We chose the KLD value of 0.025 as cutoff for convergence based on a previous study [36]. As evident from Figure 3B, convergence for Mode 1 for all four trajectories, except Sim 2, is obtained only after 700 ns. KLD:2 (divergence between first PC of Sim 2 and Sim 3) shows the highest divergence for Mode 1 even though both simulations were started with the same boost parameters but different starting structures.

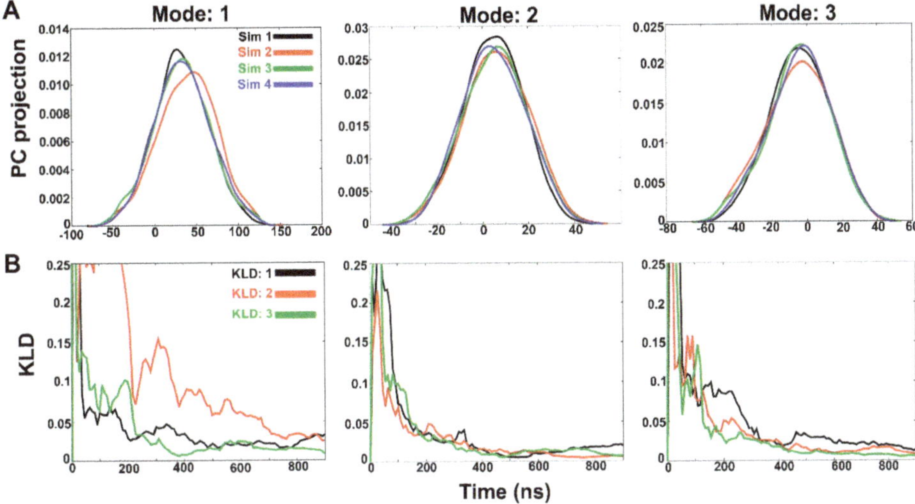

Figure 3. (**A**) Histograms of projection of principal components 1, 2 and 3 for all four simulations in water. Histograms were calculated using a Gaussian kernel density estimator. (**B**) A measure of overlap between histograms from independent simulations calculated using Kullback-Leibler divergence method. The slope values lying below 0.025 indicate convergence between two independent runs.

This also indicates that Sim 2 undergoes a different path of Alm F30/3 folding. KLD:3 (divergence between first PC of Sim 3 and Sim 4) reaches the threshold at ~400 ns. The two trajectories, the former with a semi-folded starting structure and the latter with completely unfolded structure, evolve quickly and sample similar configurations to achieve close to experimental Alm F30/3 structure. In case of Mode 2 and Mode 3, the convergence between all three trajectories is reached at ~400 ns time scale. Using boost parameters a1, a2 = 0.20 and b1, b2 = 4.5 for ~1 μs proved to be appropriate to achieve the near-native Alm F30/3 conformation starting with the unfolded Alm F30/3 structure. A similar approach can be adopted to model other peptaibols of unknown structure. The convergence of all simulations was proven based on the Kullback-Leibler method which showed that Sim 2 follows the most divergent path of conformational folding and takes longer timescales to converge, while the other three show structural convergence within 400 ns.

2.2. Alamethicin Backbone Bending: Functional Importance?

In the previous section we observed at least two very distinct Alm F30/3 conformations: linear and bent helices. Here we discuss whether there is a functional importance of such dynamic structural shift. This knowledge may direct us to understand how these peptides show membrane-perturbing properties. Using paramagnetic enhancements of nuclear relaxation, North et al. [37] demonstrated that the Alm backbone undergoes large structural fluctuations that result in shorter distances between the C-terminus and various positions along the backbone. They linked this observation with the voltage-gating mechanism of the Alm channel. A previous study by Franklin et al. [38] showed that simulated annealing with NMR restraints of Alm peptide bound to micelles yielded both bent and linear conformations, which prompted them to confirm their analysis by appending a spin label to one of the bent conformations and energy minimization using the steepest descent method. The same bent conformation was obtained as the energy minimum each time. Comparing these observations with previous studies, North et al. [39] reasoned that Alm must be in a dynamic equilibrium of linear and bent conformations and that it may provide the "conformational switch" of voltage gating in the Alm channel. In other words, the bent/closed form of Alm bound to membrane may indicate the absence of transmembrane voltage, which—when applied—would allow conversion to the linear and amphipathic Alm conformation. Gibbs et al. [40] reported on the phenomenon of helical bending around residues Aib10-Aib13 of Alm observed during 1 ns long simulation in methanol. The structural states obtained had either Aib10 or Gly11 carbonyl group oriented away from the backbone and did not seem to greatly affect adjacent helix structure. The functional role of helical bending in channel formation was hypothesized.

We were curious to observe the phenomenon of helical bending obtained through aMD. The end-to-end distance (in Å) between the first residue Aib1 (N-terminus) and Pol20 (C-terminus) was calculated for each frame of the combined trajectory. This data was used to calculate the PMF (in kcal mol^{-1}), which is simply the change in free energy as a function of any reaction coordinate. The PMF describes the energy minimum as the most stable state along that function. The end-to-end distance values of 9 and 10 Å designated by highly bent helical conformations (Figure 4) indicate their stability. The distance values of ~25 Å indicated by linear Alm backbone conformations are easily accessible with under ~1 kcal mol^{-1} of energy boost. It supports the idea of a dynamic equilibrium between the two conformations in line with previous studies [39]. Similar results were also obtained in our previous studies with other fungal peptaibols for example, trikoningin KA V [41], the energy minimum conformation of tripleurin XIIc obtained in the hydrophobic solvent chloroform [36], paracelsins B and H, and the newly discovered brevicelsin I and IV molecules [42]. This movement along the backbone from bent to linear conformation is the largest scale of motion as described by the first mode calculated through normal mode wizard (Supplementary Video 1). This indicates that peptaibols like Alm are capable of adjusting their backbone bend in response to bilayer thickness.

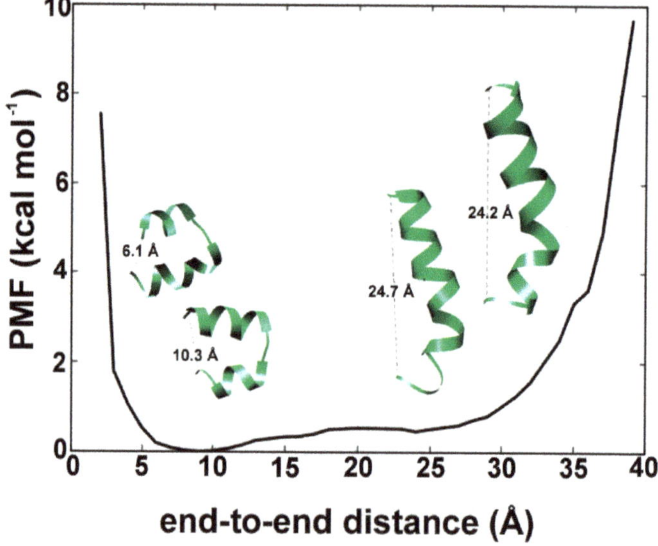

Figure 4. The reweighted potential-of-mean-force values in kcal mol^{-1} as a function of end-to-end distance of Alm calculated for each step in the combined trajectory. The energy minimum lies for a distance value of 10 Å which denotes a highly bent backbone.

The hairpin-like helical conformation, which is representative for clusters 2, 4 and 6 (Figure 2) can be explained due to the presence of the glycine residue at the R11 position. The N-terminal helical continuity in Alm F30/3 always breaks at the Gly11 position. Högel et al. [43] systematically described the local helix bending observed at the glycine position that effectively perturbs the conformational flexibility in transmembrane helices. Glycine does not have a side-chain and can easily conform into many energetically stable Φ-Ψ torsional states as can be seen on the reweighted energy landscape of Gly11 in Supplementary Figure S1. We calculated the PMF for end-to-end distance to find out the lowest energy conformation amongst the bent and linear. The results clearly show that the bent helix lies in the true energy minimum while the linear is easily accessible. It was correlated with the strategic placement of bend-inducing amino acid residues like proline and glycine found almost at the same position in almost all peptaibol sequences.

2.3. Alamethicin Hexamer Pore in a Bacterial Mimicking Bilayer Membrane

It is a well-known fact that Alm shows multiple conductance states, which directly correlates with the number of peptide monomer units. A previous study performed to highlight the difference in channel conductance with quadromer, pentamer, hexamer, octamer and even nonamer Alm F30/3 pores highlighted the occluded state of 3-mer and 4-mer pores, while pores beyond 5-mer were comparable [16]. The hexamer Alm pore is the most widely accepted model of its channel formation. For this reason, we chose to simulate a hexamer Alm model in 1,2-dioleoyl-*sn*-glycero-3-phosphoethanolamine: 1,2-dioleoyl-*sn*-glycero-3-[phospho-rac-(1-glycerol)] (DOPE:DOPG) 3:1 bilayer membrane, a simplistic representation of *E. coli* membrane [44]. DOPE is a cationic or neutral lipid, whereas, DOPG is negatively charged. The electron density profiles for DOPE: DOPG lipids and water calculated across the bilayer normal (Z-direction) has been provided in Figure 5A. A slight increase in water density from 0 to 10 Å in the bilayer signifies water displacement through the Alm pore. The lipid order parameters of the acyl chains were also determined, which can be directly compared with experimental S_{CD} values. S_{CD} is a measure of the relative orientation of the C-D bonds with respect to bilayer normal and can be calculated as $|S_{CD}| = 0.5 <3\cos^2\theta - 1>$, where θ is the angle between bilayer normal and the vector joining C_i to its deuterium atom, where <> means average of all lipid molecules. All contributions from conformational disorder, local tilting known as lipid wobble and collective motions constitute the S_{CD} parameter and thus, can be a measure of membrane fluidity [45,46].

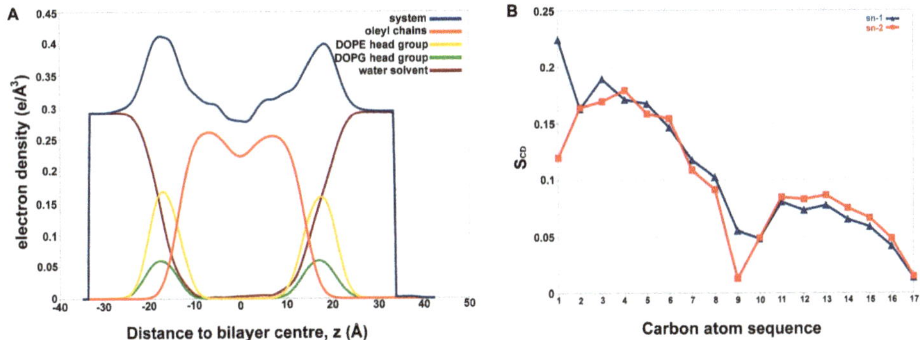

Figure 5. (**A**) The electron density profile for each constituent of membrane system and water calculated across the bilayer normal (Z-direction). A slight rise in the profile of water around 10 Å indicates presence of water in Alm F30/3 pore. Total electron density (in blue), the water density (brick red), the acyl tails of DOPE lipids (orange), the phosphoethanolamine lipid heads of DOPE (yellow) and phosphoglycerol heads of DOPG (green), (**B**) lipid order parameters of the acyl chains, $|S_{CD}|$ can be compared with previous experimental values. The acyl chains show high disorder that may be a result of presence of the Alm F30/3 pore.

Figure 5B shows the average lipid acyl chain order parameters for mixed DOPE and DOPG. It could be compared to the plateau values of the two chains sn1 and sn2 taken from carbon number 4 to 6 for DOPE as 0.211 and 0.215, respectively [47]. The plateau values for sn1 and sn2 in this case (for a DOPE:DOPG mixture) membrane is slightly lower, averaging at ~0.16 for both chains. This could be a result of DOPG mixing or the presence of membrane-perturbing peptide channel. It is evident that the membrane is more disordered than the pure DOPE membrane system. Moreover, another study conducted on *Pseudomonas aeruginosa* mimic membrane system, comprising of DOPE and DOPG with a synthetic lipid, observed an average S_{CD} value for pure DOPE inner membrane as 0.180 and for DOPG as 0.112 [48]. On the other hand, the average value for the membrane system used in this study is 0.10. This clearly shows that the membrane is highly disordered during the course of simulation due to the presence of the Alm F30/3 channel. The primary results of membrane-peptide simulations indicated that the Alm F30/3 hexamer channel increases membrane disorder, which eventually leads to leaking of water molecules and may lead to the disintegration of the bacterial cell.

The diffusion coefficient (DC) of water inside the Alm F30/3 pore was calculated using mean square displacement (MSD $Å^2$ ps^{-1}). It is the average distance that all water molecules travel from their starting position in XYZ direction. The MSD was reported along the z-direction. The speed of water movement can be estimated based on the rise in slope of MSD vs. time plot. The diffusion constant is calculated by fitting a slope to the MSD vs. time plot and multiplying it by $10.0/2 \times N$ (where N is the number of dimensions). The diffusion constant was calculated to be 0.0311 cm^2 s^{-1} and 0.0245 cm^2 s^{-1} for the first and second simulation, respectively. This in comparison to the value of diffusion of water in water (as a liquid) at 0.000023 cm^2 s^{-1}, is much higher.

Conversely, the DC value of water (as gas) in air at 273 K temperature is 0.219 cm^2 s^{-1}. It shows that in membrane simulations, the density of water molecules is more similar to the gaseous phase. Another important fact to note while calculating DC values from simulations is that most force fields, like the TIP3P water model used in this study, overestimate the diffusion coefficient even in bulk solution [49]. It is apparent from DC values for the two trajectories that a higher external electric field value (0.07 V nm^{-1}) causes slightly higher MSD of water molecules across z direction (Figure 6A). The density of water molecules in the z-direction is also shown in Figure 6B by histogramming all water O atoms on a grid with spacing 1 Å. The resultant file can be visualized in Visual Molecular Dynamics (VMD) software. The MSD analysis clearly indicated bulk movement of water molecules through the Alm F30/3 hexamer channel.

Figure 6C is a graphical representation of number of water molecules present in a hypothetical shell. This shell was created around the peptide pore residues at a distance of 3.4 Å. Therefore, the number of water molecules present inside the pore are counted as a function of simulation time. It is clear that from 200 to 230 water molecules are always present at any given time. A 14 μs long all-atom classical MD simulation of the Alm hexamer pore in DOPC membrane studied by Perrin Jr. et al. [19] reported about 40 to 50 water molecules at any given time. The difference in this number can be attributed to multiple factors like the application of an external electric field in our case and the use of accelerated dynamics. They calculated the number of water molecules at a distance of 10 Å from the bilayer center forming a 20 Å region in total (to be in the pore), while our calculation includes the complete length of the Alm F30/3 pore (30 Å) and thus, the surface-bound regions and water molecules are also considered. A diagrammatic representation of the presence of water molecules in the channel is shown in Figure 6D. The peptides present in the front view have been hidden for clear visualization. The displacement of water through the channel can also be visualized in the animation provided as Supplementary Video 2. The high number of water molecules present in the pore as a function of simulation time correlates with the high MSD value discussed above.

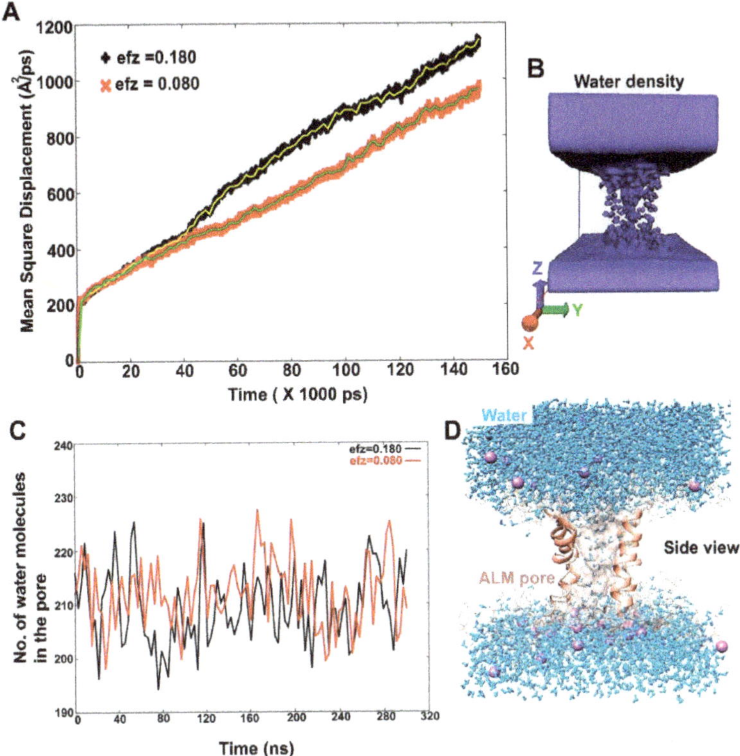

Figure 6. (**A**) The mean square displacement ($\text{Å}^2\ \text{ps}^{-1}$) values of water calculated as a function of simulation time. Using the slope of MSD curve, the average diffusion coefficient was calculated to be $0.0311\ \text{cm}^2\ \text{s}^{-1}$ and $0.0245\ \text{cm}^2\ \text{s}^{-1}$ for Mem-sim1 and Mem-sim2, respectively. Mem-sim1 clearly shows higher displacement of water under the influence of a comparatively stronger external electric field ($0.07\ \text{V nm}^{-1}$), (**B**) A diagrammatic representation of water density around the pore, (**C**) The number of water molecules present in the Alm F30/3 pore at a given time, (**D**) A cartoon representation of water molecules passing through the Alm F30/3pore. The front two monomers have been hidden for visual clarity.

The size of hexamer pore must dynamically change to accommodate the influx of bulk water and thus, pore radius was calculated as function of simulation time. The average pore radius (in Å) as a function of z coordinate was calculated over each frame of the combined 600 ns trajectory (Figure 7A) using HOLE utility from MDAnalysis. It is clear, that the C-terminus (top of Alm F30/3 pore) undergoes stronger deviation than the N-terminus. The funnel-shaped top of the pore undergoes thickening and thinning of pore size continuously. The pore is thinnest at the center of the bilayer. Figure 7B shows the top view of the pore surface surrounded by lipid heads. We were also curious to observe the changes in secondary structure of Alm F30/3 peptides in the hexamer as a function of simulation time. Based on percent α-helicity calculated for each peptide monomer as shown in Figure 7C, it is apparent that helices 1 and 3 undergo major changes. Helices 1, 4 and 6 show an average 50% helicity while helix 2 shows the highest at 65% and helix 3 shows the lowest at 30%. It must be noted that all 6 peptides started with the same conformation and yet undergo vastly varied conformational changes. Perrin Jr. et al. [19] also noted that the average % α-helicity drops to ~47%, which is in line with circular dichroism experiments of Alm channel within DOPC membranes. To correlate this change in percent α-helicity for each monomer, we calculated the angle (in degrees) between a vector passing through the center of mass of N-terminal residues and a vector passing through C-terminal residues (Figure 7D). This angle is mentioned as bend angle from here on. As expected, helix 3 that was observed to have the least percent α-helicity also has the least value of bend angle, followed by helix 5. It means that these monomers undergo extensive backbone bending during simulation, which is not feasible with a strict α-helix conformation. Therefore, these monomers unwind to achieve a more relaxed spiral conformation to accommodate pore transformation. Conversely, helices 1 and 4 show an average α-helicity around 50% and yet their bend angle values are amongst the highest. It is evident that helices 1 and 4 do not undergo drastic backbone bending but lose their helicity indicating towards other factors affecting pore dynamics. One reason could be to accommodate the passage of bulk water molecules. Finally, the last observation regarding helix 2 showing highest α-helicity (65%) but a smaller bend angle of 80∘ indicates that a certain degree of backbone bend is possible without completely losing α-helicity. The percent helicity for each peptide in the Alm F30/3 was calculated and correlated with their backbone bend angle as a function of time. A highly bent conformation is sterically difficult to achieve with the presence of strict α-helix for the whole sequence. Therefore, few monomers show interconversion between α-helix and turns. On the other hand, few other monomers show that a certain degree of backbone bend is possible with strict α-helical conformation.

Alm F30/3 is the most studied molecule in the class of peptaibols, which belong to the antimicrobial peptide (AMP) class. All known antimicrobial peptides are parts of the innate immune responses of all life forms known to us. These may be cationic or anionic in some cases and constitute ~50% of hydrophobic amino acid residues. They can exist in four major structural types, α-helix, β-sheet, hairpin or loop and extended, and may act as membrane-perturbing, pore-forming or immunomodulatory agents. The negatively charged lipid headgroups of bilayer membranes form attractive contacts with cationic AMPs. However, peptaibols lack the presence of positively charged residues and show a surface-bound state before penetrating the membrane in large numbers and form stabilized pores. These pores have peptide backbone carbonyls facing inside that makes it neutral in charge very similar to aquaporin proteins, class of integral membrane proteins, that are also similar in shape [50]. Cationic AMPs, on the other hand, have positively charged side chains lining the pore interior which may make it rigid and narrower and affect water-ion influx.

Another important factor is the presence of both hydrophobic and hydrophilic residues in the chain, folding in such a way that a hydrophobic and a hydrophilic face are formed. It is implicated during pore formation, where the hydrophobic side faces the transmembrane region and the hydrophilic side faces the pore cavity. In such a way, water molecules can easily leak through the stabilized pore. The amphipathic nature is therefore a crucial structural requirement for such peptides to show transmembrane activity. The structure of aquaporins is formed as a bundle of six transmembrane pores in an hourglass shape which is similar to the hexamer Alm F30/3 pore model [50]. However, aquaporins allow only water molecules and sometimes small solutes to pass through the membrane while peptaibol pores are known to cause leakage by also allowing passage of ions across the membrane along with water. This difference has been attributed to the presence of four aromatic/arginine residues that act as selectivity filters in case of aquaporins. Peptaibols generally lack the presence of positively charged amino acid residues like arginine, histidine and lysine and therefore, do not show selectivity. Therefore, only those polypeptide sequences which show such qualities may lead to similar water influx and membrane disintegration. Peptaibols are a special class of AMPs because of the presence of non-standard amino acid residues, which are produced as secondary metabolites and the peptide sequence is arranged through a large assembly of protein subunits known as non-ribosomal peptide synthetases (NRPSs). Although many general AMPs have been clinically approved to be used as therapeutic agents, for example enfuvirtide [51], oritavancin [52] etc., peptaibols present a promising scope as the next line of defense. This study has been carried out as one of the first steps in understanding the potential of peptaibols as therapeutic agents. The knowledge of peptaibol structure and re-construction of their mode of action shall give crucial insight to exploit their antimicrobial properties.

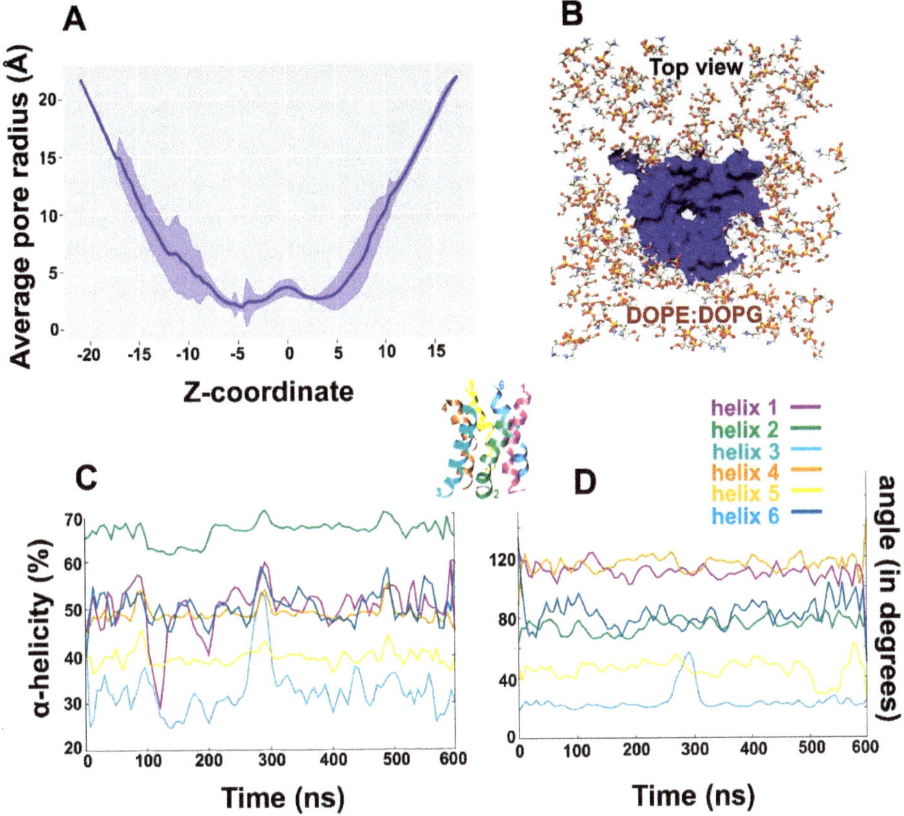

Figure 7. (**A**) The average radius of the Alm F30/3 pore (Å) calculated along transmembrane coordinate. The major fluctuation in the radius is shown by the C-terminus probably to accommodate incoming water flux, (**B**) Cartoon representation of the top view of the Alm F30/3 pore surface in blue, (**C**) percent α-helicity calculated for each Alm F30/3 monomer from the pore with respect to simulation time (in ns), (**D**) the bend angle (calculated as dot product of two vectors passing through the N- and C-terminals of each monomer) as a function of simulation time. Figure 7C,D can be correlated assuming that the higher the bend angle, the lower will be its α-helicity. The helices are color-coded for visual interpretation.

3. Materials and Methods

3.1. Partial Charge Calculation and Force Field Library Generation for Non-standard Residues

As famously known, fungal peptaibols are characterized by their unusual amino acid content. In the Alm F30/3 sequence, Aib and Pheol are non-standard residues. The R.E.D server was used for calculation of their partial charges and creating force field libraries. R.E.D stands for RESP ESP charge derive [53,54]. RESP (restrained electrostatic potential) was used to calculate the charges with a HF/6-31G(d) basis set and Gaussian 09 quantum mechanical program interface [55]. The charges for Aib were calculated along with few other standard amino acids. The charges calculated for standard residues were used to confirm with existing "leap" libraries. For each residue, two conformations, i.e., alpha helix ($\Phi = -63.8$, $\Psi = -38.3$) and beta sheet or C5 ($\Phi = -157.2$, $\Psi = 161.9$) were used. The terminal residue phenylalaninol was also parameterized using two molecules, ethyl alcohol and phenylalanine. The Amber "leap" library is provided in Tyagi et al. [36] for both residues. The sequence of Alm F30/3 is: AcAib^1-Pro^2-Aib^3-Ala^4-Aib^5-Ala^6-Gln^7-Aib^8-Val^9-Aib^{10}-Gly^{11}-Leu^{12}-Aib^{13}-Pro^{14}-Val^{15}-Aib^{16}-Aib^{17}-Glu^{18}-Gln^{19}-Pheol^{20}.

3.2. Accelerated Molecular Dynamics Simulations of Alamethicin

The unfolded Alm F30/3 peptide conformation was prepared by using the 'tleap' module of AmberTools18 and solvated in water (TIP3P water model) as solvent. In total, 4701 water molecules were added with a box size of 48.79 × 71.80 × 54.41 Å and a volume of 190680.596 Å^3. Amberff14SB force field was used to prepare and minimize the system [56]. It was prepared for aMD in six consecutive steps as described by Tyagi et al. 2019 [36]. The unfolded alamethicin peptide was simulated in water using aMD for ~900 ns consecutively with three independent starting structures (~900 ns × 3) making a combined simulation time of 2.7 μs. All simulations were carried out at 300 K temperature, 2 fs time step, and energies and boost information were recorded at every 1000 steps. The electrostatic interactions were calculated using PME (particle mesh Ewald summation) [57] long-range interactions were also calculated with cutoff of 10.0. The temperature scaling was carried out using Langevin thermostat without pressure scaling. SHAKE algorithm was applied on all bonds involving hydrogen. The GPU machines available through the NIIF clusters of Hungary were utilized for all aMD simulations using *pmemd.cuda* implementation of Amber14. aMD can be carried out using three criteria, (i) the whole potential at once (iamd = 1) or (ii) independently boosting the torsional terms of the potential (iamd = 2), and (iii) to boost the whole potential with an extra boost to torsions (iamd = 3). The third criterion seemed to be an appropriate choice, as the dual boost option provides a better reweighting distribution. The extra parameters E$_{dihed}$, α$_{dihed}$, E$_{total}$ and α$_{total}$ were calculated as required in Equation (1):

$$E_{dihed} = V_{avg_{dihed}} + a_1 \times N_{res}, \quad \alpha_{dihed} = a_2 \times \frac{N_{res}}{5}$$
$$E_{total} = V_{avg\_total} + b_1 \times N_{atoms}, \quad \alpha_{total} = b_2 \times N_{atoms} \tag{1}$$

where N_{res} is the number of peptide residues (21, with an addition of acetyl group at the N-terminal), N_{atoms} is the total number of atoms in the system which is 14391. V_{avg_dihed} and V_{avg_total} are average dihedral and total potential energies obtained from the 100 ns long cMD run. The various parameters used for all aMD simulations have been summarized in Table 1 The reweighting procedure and analysis details have been adopted from Tyagi et al. [36].

Table 1. Summary of various accelerated molecular dynamics parameters.

Simulations	Starting Conformation	Time (ns)	a1,a2 Total (kcal mol^{-1})	b1,b2 Dihedral (kcal mol^{-1})	Avg Boost (kcal mol^{-1})
Sim 1	Unfolded	936	0.16	4	11.21
Sim 2	Folded N-terminal	950	0.20	4.5	11.10
Sim 3	Folded with bent backbone	897	0.20	4.5	29.85
Sim 4	unfolded	1000	0.20	4.5	24.09
Membrane-sim 1	Hexamer with efz = 0.180	300	0.16	4	31.71
Membrane-sim 2	Hexamer with efz = 0.080	300	0.16	4	22.60

3.3. Accelerated Molecular Dynamics Simulations of Alamethicin Pore in DOPC Bilayer Membrane

The hexamer pore of Alm F30/3 peptide was obtained through M-ZDOCK server (http://zdock.umassmed.edu/m-zdock/) [58]. This is a Fast Fourier Transform based protein docking program that predicts the structures of cyclically symmetric multimers. The hexameric pore was then embedded into a 3:1 mixture of DOPE and DOPG bilayer membranes which mimics a bacterial (*E. coli*) membrane constitution. This system can be easily prepared in an Amber-ready format by using the '*packmol-memgen*' [59] workflow available with AmberTools18 that uses '*Memembed*' [60] to obtain pre-oriented protein conformation with respect to the membrane. This system was solvated in 4410 water residues. The Alm F30/3 hexamer channel embedded in DOPE:DOPG bilayer system was prepared for aMD simulations starting with minimization followed by two steps of system heating and 10 steps of equilibration. This system was energy minimized for 10,000 steps which switches to conjugate gradient method after 5000 steps of the steepest descent method. The minimization was done at constant volume, no SHAKE algorithm was applied, and the non-bonded cutoff was set to 10.0 Å. The minimized system was set for two rounds of gradual 'heating' to reach the 'production run' temperature. In the first step, the system was heated to 100 K while the second step reaches a temperature of 303 K. The 10-step equilibration was carried out at 303 K temperature for 500 ps each. A short 25 ns long production run at 303 K was carried out with constraints on bond distances to calculate the aMD boost parameters followed by two consecutive aMD simulations of 300 ns each with 2 ps time step. Each aMD simulation was carried out with dual boost (iamd=3) option at 300 K temperature regulated using a Langevin thermostat (Table 1). A weak external static electric field was also applied along the z direction (across membrane) with a value of efz (intensity in kcal $(mol \times A \times e)^{-1}$) = 0.180 and 0.080 for the first and second simulations, respectively [61]. The values of efz were chosen in a way that the magnitude of resulting electric potential is slightly higher than the voltage across plasma membrane. A membrane with 35 Å thickness has a potential of ~70 millivolts (mV) which is 0.07 V per 3.5×10^{-7} cm or 0.02 V nm^{-1} [62]. A 0.180 kcal $(mol \times A \times e)^{-1}$ translates to an electric potential of 0.07 V/nm while 0.080 kcal $(mol \times A \times e)^{-1}$ translates to 0.03 V $\times nm^{-1}$. A distance restraint was applied for all glutamine amino acid residues owing to their importance in Alm F30/3 pore stability. The average pore radius was calculated using the HOLE utility [63] and water density available through MDAnalysis [64,65].

4. Conclusions

This study highlights the use of enhanced sampling molecular dynamics technique known as accelerated MD (aMD) to elucidate fungal peptaibol structures. The gap between the knowledge of these sequences and their structure is ever increasing which requires efficient and cost-effective computational methods to be fulfilled. Through a series of aMD simulations we show that ~1 μs time scale with slightly aggressive boost parameters is appropriate to completely uncover peptide dynamics. The experimental structure of alamethicin F30/3 is compared with representative structures of highly populated clusters in each simulation. aMD was also used to recreate the Alm F30/3 hexamer-membrane system by using *E. coli* membrane mimic under a weak electric field. The various quantities were compared with previous experiments, which suggested that the presence of the Alm F30/3 pore results in membrane disorder and leads to a large influx of water molecules through the channel. The C-terminus seems to fluctuate resulting in changes of the pore size throughout the simulation. The water influx is increased under the application of an external electric field. Overall, this study shows that aMD can be successfully used to elucidate folding dynamics of small, bioactive peptides and to simulate their mode of action over bacteria mimicking membrane systems.

Supplementary Materials: The following are available online at http://www.mdpi.com/1422-0067/20/17/4268/s1.

Author Contributions: Conceptualization, C.T., L.K., and F.Ö.; Methodology, C.T., and F.Ö.; Software, C.T.; Formal Analysis, C.T.; Investigation, C.T., T.M. and F.Ö.; Data Curation, C.T. and T.M.; Writing-Original Draft Preparation, C.T.; Writing-Review & Editing, C.V., L.K., and F.Ö.; Visualization, C.T., and T.M.; Supervision, L.K., and F.Ö.; Funding Acquisition, C.V. and L.K.

Funding: This work was supported by grant NKFI K-116475 (National Research, Development and Innovation Office) as well as by the Hungarian Government and the European Union within the frames of the Széchenyi 2020 Programme (grant GINOP-2.2.1-15-2016-00006). LK is grantee of the János Bolyai Research Scholarship (Hungarian Academy of Sciences) and the Bolyai+ Scholarship (New National Excellence Programme).

Conflicts of Interest: The authors declare no conflict of interest.

References

1. Mahlapuu, M.; Håkansson, J.; Ringstad, L.; Björn, C. Antimicrobial peptides: An emerging category of therapeutic agents. *Front. Cell Infect. Microbiol.* **2016**, *6*, 194. [CrossRef] [PubMed]
2. Hamelberg, D.; Mongan, J.; McCammon, J.A. Accelerated molecular dynamics: A promising and efficient simulation method for biomolecules. *J. Chem. Phys.* **2004**, *120*, 11919–11929. [CrossRef] [PubMed]
3. Hamelberg, D.; de Oliveira, C.A.; McCammon, J.A. Sampling of slow diffusive conformational transitions with accelerated molecular dynamics. *J. Chem. Phys.* **2007**, *127*, 155102. [CrossRef] [PubMed]
4. Meyer, C.E.; Reusser, F. A polypeptide antibacterial agent isolated from *Trichoderma viride*. *Experientia* **1967**, *23*, 85–86. [CrossRef] [PubMed]
5. Degenkolb, T.; Dieckmann, R.; Nielsen, K.F.; Gräfenhan, T.; Theis, C.; Zafari, D.; Chaverri, P.; Ismaiel, A.; Brückner, H.; von Döhren, H.; et al. The *Trichoderma brevicompactum* clade: A separate lineage with new species, new peptaibiotics, and mycotoxins. *Mycol. Prog.* **2008**, *7*, 177–219. [CrossRef]
6. Payne, J.W.; Jakes, R.; Hartley, B.S. The primary structure of alamethicin. *Biochem. J.* **1970**, *117*, 757–766. [CrossRef]
7. Martin, D.R.; Williams, R.J.P. Chemical nature and sequence of alamethicin. *Biochem. J.* **1976**, *153*, 181–190. [CrossRef]
8. Fox, R.O., Jr.; Richards, F.M. A voltage-gated ion channel model inferred from the crystal structure of alamethicin at 1.5-Å resolution. *Nature* **1982**, *300*, 325. [CrossRef]
9. Nagao, T.; Mishima, D.; Javkhlantugs, N.; Wang, J.; Ishioka, D.; Yokota, K.; Norisada, K.; Kawamura, I.; Ueda, K.; Naito, A. Structure and orientation of antibiotic peptide alamethicin in phospholipid bilayers as revealed by chemical shift oscillation analysis of solid state nuclear magnetic resonance and molecular dynamics simulation. *Biochim. Biophys. Acta Biomembr.* **2015**, *1848*, 2789–2798. [CrossRef]
10. Haris, P.I.; Chapman, D. Fourier transform infrared spectra of the polypeptide alamethicin and a possible structural similarity with bacteriorhodopsin. *Biochim. Biophys. Acta Biomembr.* **1988**, *943*, 375–380. [CrossRef]
11. Leitgeb, B.; Szekeres, A.; Manczinger, L.; Vágvölgyi, C.; Kredics, L. The history of alamethicin: A review of the most extensively studied peptaibol. *Chem. Biodivers.* **2007**, *4*, 1027–1051. [CrossRef] [PubMed]
12. Kredics, L.; Szekeres, A.; Czifra, D.; Vágvölgyi, C.; Leitgeb, B. Recent results in alamethicin research. *Chem. Biodivers.* **2013**, *10*, 744–771. [CrossRef] [PubMed]
13. Pieta, P.; Mirza, J.; Lipkowski, J. Direct visualization of the alamethicin pore formed in a planar phospholipid matrix. *Proc. Natl. Acad. Sci. USA.* **2012**, *109*, 21223–21227. [CrossRef]
14. Clejan, S.; Krulwich, T.A.; Mondrus, K.R.; Seto-Young, D. Membrane lipid composition of obligately and facultatively alkalophilic strains of Bacillus spp. *J. Bacteriol.* **1986**, *168*, 334–340. [CrossRef] [PubMed]
15. Salditt, T.; Li, C.; Spaar, A. Structure of antimicrobial peptides and lipid membranes probed by interface-sensitive X-ray scattering. *Biochim. Biophys. Acta Biomembr.* **2006**, *1758*, 1483–1498. [CrossRef] [PubMed]
16. Rahaman, A.; Lazaridis, T. A thermodynamic approach to alamethicin pore formation. *Biochim. Biophys. Acta Biomembr.* **2014**, *1838*, 1440–1447. [CrossRef]
17. Castro, T.G.; Micaélo, N.M. Conformational and thermodynamic properties of non-canonical α, α-dialkyl glycines in the peptaibol alamethicin: Molecular dynamics studies. *J. Phys. Chem. B* **2014**, *118*, 9861–9870. [CrossRef] [PubMed]
18. Salnikov, E.S.; Raya, J.; De Zotti, M.; Zaitseva, E.; Peggion, C.; Ballano, G.; Toniolo, C.; Raap, J.; Bechinger, B. Alamethicin supramolecular organization in lipid membranes from 19F solid-state NMR. *Biophys. J.* **2016**, *111*, 2450–2459. [CrossRef]
19. Perrin Jr, B.S.; Pastor, R.W. Simulations of membrane-disrupting peptides I: Alamethicin pore stability and spontaneous insertion. *Biophys. J.* **2016**, *111*, 1248–1257. [CrossRef]
20. Madsen, J.L.; Hjørringgaard, C.U.; Vad, B.S.; Otzen, D.; Skrydstrup, T. Incorporation of β-silicon-β3-amino acids in the antimicrobial peptide alamethicin provides a 20-fold increase in membrane permeabilization. *Chem. Eur. J.* **2016**, *22*, 8358–8367. [CrossRef]
21. Das, S.; Ben Haj Salah, K.; Wenger, E.; Martinez, J.; Kotarba, J.; Andreu, V.; Ruiz, N.; Savini, F.; Stella, L.; Didierjean, C.; et al. Enhancing the antimicrobial activity of alamethicin f50/5 by incorporating n-terminal hydrophobic triazole substituents. *Chem. Eur. J.* **2017**, *23*, 17964–17972. [CrossRef] [PubMed]

22. Afanasyeva, E.F.; Syryamina, V.N.; Dzuba, S.A. Communication: Alamethicin can capture lipid-like molecules in the membrane. *J. Chem. Phys.* **2017**, *146*, 011103. [CrossRef] [PubMed]

23. Su, Z.; Shodiev, M.; Leitch, J.J.; Abbasi, F.; Lipkowski, J. In situ electrochemical and PM-IRRAS studies of alamethicin ion channel formation in model phospholipid bilayers. *J. Electroanal. Chem.* **2018**, *819*, 251–259. [CrossRef]

24. Forbrig, E.; Staffa, J.K.; Salewski, J.; Mroginski, M.A.; Hildebrandt, P.; Kozuch, J. Monitoring the orientational changes of alamethicin during incorporation into bilayer lipid membranes. *Langmuir* **2018**, *34*, 2373–2385. [CrossRef] [PubMed]

25. Abbasi, F.; Alvarez-Malmagro, J.; Su, Z.; Leitch, J.J.; Lipkowski, J. Pore forming properties of alamethicin in negatively charged floating bilayer lipid membranes supported on gold electrodes. *Langmuir* **2018**, *34*, 13754–13765. [CrossRef]

26. Syryamina, V.N.; De Zotti, M.; Toniolo, C.; Formaggio, F.; Dzuba, S.A. Alamethicin self-assembling in lipid membranes: Concentration dependence from pulsed EPR of spin labels. *Phys. Chem. Chem. Phys.* **2018**, *20*, 3592–3601. [CrossRef] [PubMed]

27. Zhang, M.K.; Lyu, Y.; Zhu, X.; Wang, J.P.; Jin, Z.Y.; Narsimhan, G. Enhanced solubility and antimicrobial activity of alamethicin in aqueous solution by complexation with γ-cyclodextrin. *J. Funct. Foods* **2018**, *40*, 700–706. [CrossRef]

28. Abbasi, F.; Su, Z.; Alvarez-Malmagro, J.; Leitch, J.J.; Lipkowski, J. Effects of amiloride, an ion channel blocker, on alamethicin pore formation in negatively charged, gold-supported, phospholipid bilayers: A molecular view. *Langmuir* **2019**, *35*, 5060–5068. [CrossRef]

29. Taylor, G.; Nguyen, M.A.; Koner, S.; Freeman, E.; Collier, C.P.; Sarles, S.A. Electrophysiological interrogation of asymmetric droplet interface bilayers reveals surface-bound alamethicin induces lipid flip-flop. *Biochim. Biophys. Acta Biomembr.* **2019**, *1861*, 335–343. [CrossRef]

30. Miao, Y.; Sinko, W.; Pierce, L.; Bucher, D.; Walker, R.C.; McCammon, J.A. Improved reweighting of accelerated molecular dynamics simulations for free energy calculation. *J. Chem. Theory Comput.* **2014**, 2677–2689. [CrossRef]

31. Jing, Z.; Sun, H. A comment on the reweighting method for accelerated molecular dynamics simulations. *J. Chem. Theory Comput.* **2015**, *11*, 2395–2397. [CrossRef] [PubMed]

32. Miao, Y.; Feher, V.A.; McCammon, J.A. Gaussian accelerated molecular dynamics: Unconstrained enhanced sampling and free energy calculation. *J. Chem. Theory Comput.* **2015**, *11*, 3584–3595. [CrossRef] [PubMed]

33. Huang, Y.M.M.; McCammon, J.A.; Miao, Y. Replica exchange Gaussian accelerated molecular dynamics: Improved enhanced sampling and free energy calculation. *J. Chem. Theory Comput.* **2018**, *14*, 1853–1864. [CrossRef] [PubMed]

34. Bucci, R.; Contini, A.; Clerici, F.; Pellegrino, S.; Gelmi, M.L. From glucose to enantiopure morpholino β-amino acid: A new tool for stabilizing γ-turns in peptides. *Org. Chem. Front.* **2019**, *6*, 972–982. [CrossRef]

35. Kullback, S.; Leibler, R.A. On information and sufficiency. *Ann. Math. Stat.* **1951**, *22*, 79–86. [CrossRef]

36. Tyagi, C.; Marik, T.; Szekeres, A.; Vágvölgyi, C.; Kredics, L.; Ötvös, F. Tripleurin XIIc: Peptide folding dynamics in aqueous and hydrophobic environment mimic using accelerated molecular dynamics. *Molecules* **2019**, *24*, 358. [CrossRef] [PubMed]

37. North, C.L.; Franklin, J.C.; Bryant, R.G.; Cafiso, D.S. Molecular flexibility demonstrated by paramagnetic enhancements of nuclear relaxation. Application to alamethicin: A voltage-gated peptide channel. *Biophys. J.* **1994**, *67*, 1861–1866. [CrossRef]

38. Franklin, J.C.; Ellena, J.F.; Jayasinghe, S.; Kelsh, L.P.; Cafiso, D.S. Structure of micelle-associated alamethicin from 1H NMR. Evidence for conformational heterogeneity in a voltage-gated peptide. *Biochemistry* **1994**, *33*, 4036–4045. [CrossRef]

39. North, C.L.; Barranger-Mathys, M.; Cafiso, D.S. Membrane orientation of the N-terminal segment of alamethicin determined by solid-state 15N NMR. *Biophys. J.* **1995**, *69*, 2392–2397. [CrossRef]

40. Gibbs, N.; Sessions, R.B.; Williams, P.B.; Dempsey, C.E. Helix bending in alamethicin: Molecular dynamics simulations and amide hydrogen exchange in methanol. *Biophys. J.* **1997**, *72*, 2490–2495. [CrossRef]

41. Marik, T.; Tyagi, C.; Racić, G.; Rakk, D.; Szekeres, A.; Vágvölgyi, C.; Kredics, L. New 19-residue peptaibols from *Trichoderma* clade Viride. *Microorganisms* **2018**, *6*, 85. [CrossRef] [PubMed]

42. Marik, T.; Tyagi, C.; Balázs, D.; Urbán, P.; Szepesi, Á.; Bakacsy, L.; Endre, G.; Rakk, D.; Szekeres, A.; Andersson, M.A.; et al. Structural diversity and bioactivities of peptaibol compounds from the Longibrachiatum Clade of the filamentous fungal genus *Trichoderma*. *Front. Microbiol.* **2019**, *10*. [CrossRef] [PubMed]

43. Högel, P.; Götz, A.; Kuhne, F.; Ebert, M.; Stelzer, W.; Rand, K.D.; Scharnagl, C.; Langosch, D. Glycine perturbs local and global conformational flexibility of a transmembrane helix. *Biochemistry* **2018**, *57*, 1326–1337. [CrossRef] [PubMed]

44. Lombardi, L.; Stellato, M.I.; Oliva, R.; Falanga, A.; Galdiero, M.; Petraccone, L.; D'Errico, G.; De Santis, A.; Galdiero, S.; Del Vecchio, P. Antimicrobial peptides at work: Interaction of myxinidin and its mutant WMR with lipid bilayers mimicking the *P. aeruginosa* and *E. coli* membranes. *Sci. Rep.* **2017**, *7*, 44425. [CrossRef] [PubMed]

45. Pastor, R.W.; Venable, R.M.; Karplus, M. Model for the structure of the lipid bilayer. *Proc. Natl. Acad. Sci. USA* **1991**, *88*, 892–896. [CrossRef] [PubMed]

46. Pastor, R.W.; Venable, R.M.; Karplus, M.; Szabo, A. A simulation based model of NMR T 1 relaxation in lipid bilayer vesicles. *J. Chem. Phys.* **1988**, *89*, 1128–1140. [CrossRef]

47. Venable, R.M.; Brown, F.L.; Pastor, R.W. Mechanical properties of lipid bilayers from molecular dynamics simulation. *Chem. Phys. Lipids* **2015**, *192*, 60–74. [CrossRef]

48. Li, A.; Schertzer, J.W.; Yong, X. Molecular dynamics modeling of *Pseudomonas aeruginosa* outer membranes. *Phys. Chem. Chem. Phys.* **2018**, *20*, 23635–23648. [CrossRef]

49. Shinoda, W. Permeability across lipid membranes. *Biochim. Biophys. Acta Biomembr.* **2016**, *1858*, 2254–2265. [CrossRef]

50. Agre, P.; Kozono, D. Aquaporin water channels: Molecular mechanisms for human diseases. *FEBS Lett.* **2003**, *555*, 72–78. [CrossRef]

51. Tsibris, A.M.; Hirsch, M.S. Antiretroviral therapy for human immunodeficiency virus infection. In *Mandell, Douglas, and Bennett's Principles and Practice of Infectious Diseases*, 1st ed.; Churchill Livingstone: London, UK, 2009; pp. 1833–1854.

52. Rosenthal, S.; Decano, A.G.; Bandali, A.; Lai, D.; Malat, G.E.; Bias, T.E. Oritavancin (Orbactiv): A new-generation lipoglycopeptide for the treatment of acute bacterial skin and skin structure infections. *P & T.* **2018**, *43*, 143–179.

53. Vanquelef, E.; Simon, S.; Marquant, G.; Garcia, E.; Klimerak, G.; Delepine, J.C.; Cieplak, P.; Dupradeau, F.Y.R.E.D. Server: A web service for deriving RESP and ESP charges and building force field libraries for new molecules and molecular fragments. *Nucl. Acids Res.* **2011**, *39*, W511–W517. [CrossRef]

54. Dupradeau, F.Y.; Pigache, A.; Zaffran, T.; Savineau, C.; Lelong, R.; Grivel, N.; Lelong, D.; Rosanski, W.; Cieplak, P. The R.E.D. tools: Advances in RESP and ESP charge derivation and force field library building. *Phys. Chem. Chem. Phys.* **2010**, *12*, 7821–7839. [CrossRef] [PubMed]

55. Frisch, M.J.; Trucks, G.W.; Schlegel, H.B.; Scuseria, G.E.; Robb, M.A.; Cheeseman, J.R.; Scalmani, G.; Barone, V.; Mennucci, B.; Petersson, G.A.; et al. *Gaussian 09*; Gaussian Inc.: Wallingford, UK.

56. Case, D.A.; Ben-Shalom, I.Y.; Brozell, S.R.; Cerutti, D.S.C.; Cheatham, T.E., III; Cruzeiro, V.W.D.; Darden, T.A.; Duke, R.E.; Ghoreishi, D.; Gilson, M.K.; et al. Proceedings of the AMBER 2018, San Francisco, CA, USA, 2018; Available online: ambermd.org/doc12/Amber18.pdf (accessed on 30 August 2019).

57. Essmann, U.; Perera, L.; Berkowitz, M.L.; Darden, T.; Lee, H.; Pedersen, L.G. A smooth particle mesh Ewald method. *J. Chem. Phys.* **1995**, *103*, 8577–8593. [CrossRef]

58. Pierce, B.; Tong, W.; Weng, Z. M-ZDOCK: A grid-based approach for C n symmetric multimer docking. *Bioinformatics* **2004**, *21*, 1472–1478. [CrossRef] [PubMed]

59. Schott-Verdugo, S.; Gohlke, H. PACKMOL-Memgen: A simple-to-use generalized workflow for membrane-protein/lipid-bilayer system building. *J. Chem. Inf. Model.* **2019**, *59*, 2522–2528. [CrossRef] [PubMed]

60. Nugent, T.; Jones, D.T. Membrane protein orientation and refinement using a knowledge-based statistical potential. *BMC Bioinform.* **2013**, *14*, 276. [CrossRef]

61. Escalona, Y.; Garate, J.A.; Araya-Secchi, R.; Huynh, T.; Zhou, R.; Perez-Acle, T. Exploring the membrane potential of simple dual-membrane systems as models for gap-junction channels. *Biophys. J.* **2016**, *110*, 2678–2688. [CrossRef]

62. Lodish, H.; Berk, A.; Zipursky, S.L.; Matsudaira, P.; Baltimore, D.; Darnell, J. Intracellular ion environment and membrane electric potential. In *Molecular Cell Biology*, 4th ed.; WH Freeman & Company: New York, NY, USA, 2000.

63. Smart, O.S.; Neduvelil, J.G.; Wang, X.; Wallace, B.A.; Sansom, M.S. HOLE: A program for the analysis of the pore dimensions of ion channel structural models. *J. Mol. Graph.* **1996**, *14*, 354–360. [CrossRef]

64. Gowers, R.J.; Linke, M.; Barnoud, J.; Reddy, T.J.; Melo, M.N.; Seyler, S.L.; Domański, J.; Dotson, D.L.; Buchoux, S.; Kenney, I.M.; et al. MDAnalysis: A Python package for the rapid analysis of molecular dynamics simulations. In *Proceedings of the 15th Python in Science Conference*; SciPy: Austin, TX, USA; Available online: https://www.researchgate.net/profile/Sean_Seyler2/publication/320267580_MDAnalysis_a_Python_package_for_the_rapid_analysis_of_molecular_dynamics_simulations/links/59d8dcab458515a5bc262281/MDAnalysis-a-Python-package-for-the-rapid-analysis-of-molecular-dynamics-simulations.pdf (accessed on 30 August 2019).

65. Michaud-Agrawal, N.; Denning, E.J.; Woolf, T.B.; Beckstein, O. MDAnalysis: A toolkit for the analysis of molecular dynamics simulations. *J. Comput. Chem.* **2011**, *32*, 2319–2327. [CrossRef]

MDPI
St. Alban-Anlage 66
4052 Basel
Switzerland
Tel. +41 61 683 77 34
Fax +41 61 302 89 18
www.mdpi.com

International Journal of Molecular Sciences Editorial Office
E-mail: ijms@mdpi.com
www.mdpi.com/journal/ijms